5th International Conference on Coastal and Ocean Engineering (ICCOE 2018)

IOP Conference Series: Earth and Environmental Science
Volume 171

Shanghai, China
27 – 29 April 2018

ISBN: 978-1-5108-6781-9
ISSN: 1755-1307

Printed from e-media with permission by:

Curran Associates, Inc.
57 Morehouse Lane
Red Hook, NY 12571

Some format issues inherent in the e-media version may also appear in this print version.

Printed by Curran Associates, Inc. (2018)

For permission requests, please contact the Institute of Physics at the address below.

Institute of Physics
Dirac House, Temple Back
Bristol BS1 6BE UK

Phone: 44 1 17 929 7481
Fax: 44 1 17 920 0979

techtracking@iop.org

Additional copies of this publication are available from:

Curran Associates, Inc.
57 Morehouse Lane
Red Hook, NY 12571 USA
Phone: 845-758-0400
Fax: 845-758-2633
Email: curran@proceedings.com
Web: www.proceedings.com

Table of contents

Volume 171

2018 5th International Conference on Coastal and Ocean Engineering (ICCOE 2018)

27–29 April 2018, Shanghai, China

Accepted papers received: 19 June 2018
Published online: 19 July 2018

Preface

Papers

Marine Geology

Ocean Observation and Information Technology

Wave Mechanics and Fluid Mechanics

Offshore and Deep-sea Engineering

Marine Environment and Climate

IOP Conf. Series: Earth and Environmental Science **171** (2018) 011001 doi:10.1088/1755-1315/171/1/011001

PREFACE

It is my great pleasure to welcome you to the 2018 5th International Conference on Coastal and Ocean Engineering (ICCOE 2018) which was held in Shanghai, China during 27-29 April, 2018. ICCOE 2018 was dedicated to issues related to Coastal and Ocean Engineering.

I, Prof. Hu Yihuai, from Shanghai Maritime University, China, held the keynote speech with title "Research into Sail-Assisting Technology for Ocean-Going Ships". Prof. CHOU Loke Ming from Tropical Marine Science Institute, National University of Singapore, Singapore had a keynote speech with title "Biodiversity Enhancement of the Urbanized Coastal Environment". Prof. Gordon Huang from Faculty of Engineering and Applied Science, University of Regina, Canada had a keynote speech with title "Factorial-Based Ecological Input-Output Model to Support Industrial Policy Formulation". Prof. Koh Hock Lye from Sunway University, Malaysia, held a keynote speech with title "The Life We Want: Achieving United Nation Sustainable Development Goals 2030".

There were four oral presentation sessions and one poster presentation session in which Shaoduo Zheng, Su Yean The, Shudi Dong, Drake Mullens and Zhenlu Wang have received the best presentations awards. Many researchers, engineers, academics as well as industrial professionals from all over the world have presented their research results and development activities to end the whole conference with a great success.

The major goal and feature of the conference was to bring academic scientists, engineers and industry researchers together to exchange and share their experiences and research results, and discuss the practical challenges encountered and the solutions adopted. Professors from China, Canada, Singapore and Malaysia were invited to deliver keynote speeches regarding latest information in their respective expertise. It has been a golden opportunity for students, researchers and engineers to interact with the experts and specialists to obtain their advice on technical matters as well as sales and marketing strategies. And participants have enjoyed the academic visit of studying about Ship Handling Simulator, Marine Engine Room Simulator, Wind Tunnel Laboratory and Hybrid Liquid Cargo Ship Mock-up on April 29 in Shanghai Maritime University, China.

The proceedings present a selection from papers submitted to the conference from universities, research institutes and industries. All papers were subjected to peer-review by conference committee members and international reviewers. The papers selected based on their quality and relevancy to the conference.

This volume presents to the readers the recent advances in the field of Coastal and Ocean Engineering as well as Beach erosion and sediment transport, Climate change and sea level rise, Coastal infrastructure developments, Hydrodynamics of off shore structures, Lowland development and reclamation, etc..

We would like to thank all authors who have contributed to this volume and also the organizing committee, reviewers, speakers, chairpersons, sponsors and all the conference participants for their support to ICCOE 2018.

Prof. Hu Yihuai
Shanghai Maritime University, China
May 16, 2018

IOP Conf. Series: Earth and Environmental Science **171** (2018) 011002 doi:10.1088/1755-1315/171/1/011002

Conference Committee

Conference General Co-Chairs

Prof. YihuaI Hu, Marine Engineering Department, Shanghai Maritime University, Shanghai, China

Prof. CHOU Loke Ming, Tropical Marine Science Institute, National University of Singapore, Singapore

Prof. Hyo Choi, Gangneung-Wonju National University, South Korea

Program Co-Chairs

Prof. Yonghong Wang, Ocean University of China

Assoc. Prof. Teh Su Yean, Universiti Sains Malaysia, Malaysia

Contact Chair

Ms Sophia Du, Hong Kong Chemical, Biological & Environmental Engineering Society, Hong Kong

Technical Committee

Prof. Lo Yat Man Edmond, Nanyang Technological University, Singapore

Prof. Chih-Hua Chang, Ling Tung University, Taichung, Taiwan

Prof. Guanghua He, Harbin Institute of Technology, China

Prof. San-Shyan Lin, National Taiwan Ocean University, Taiwan

Prof. Hsien-Kuo Chang, National Chiao Tung University, Taiwan

Assoc Prof. Dean WU, Hohai university, China

Prof. Ali Moussaoui, University of Tlemcen, Algeria

Assoc. Prof. Baoping Cai, China University of Petroleum, China

Assoc. Prof. Chen Hsien-Wen, Central Police University, Taiwan

Assoc. Prof. Xuan-Kien Dang, Ho Chi Minh City University of Transport, Viet Nam

Dr. Zhaohui Wu, Offshore Oil Engineering Co., Ltd.

Dr. WANG Tianying, Drilling Technology Research Institute, China

Dr. Masoud Dehghani Champiri, University of Houston, USA

Dr. L J Ou, South China University of Technology, China

Dr. Ravindran Balasubramani, National Institute of Animal Science, Korea

Dr. Sreevalsa Kolathayar, Amrita Vishwa Vidyapeetham Deemed University, India

Dr. Ilknur Tuncer Dokuz Eylul University, Turkey

Dr. Cheng Ming Hung, National Taiwan Ocean University, Taiwan

Prof. Snehal Agnihotri, Commerce and Science College, India

Prof. V. Sundar, Indian Institute of Technology Madras, India

Peer review statement

All papers published in this volume of *IOP Conference Series: Earth and Environmental Science* have been peer reviewed through processes administered by the proceedings Editors. Reviews were conducted by expert referees to the professional and scientific standards expected of a proceedings journal published by IOP Publishing.

Experimental Study on Suspended Sediment Concentration and Powder Bed Deformation under Wave-Only and Wave-With-Current Scenarios

Zhaohui Wu[1], Bo Yang[2,4], Yang He[1], Fuxiang Huang[1], Bingchen Liang[2,3] and Dang Zhao[1]

1 Offshore Oil Engineering Co., Ltd., Tianjin 300451, China
2 Engineering College, Ocean University of China, Qingdao, China
3 Shandong Province Key Laboratory of Ocean Engineering, Ocean University of China, Qingdao, China
E-mail: yangbo@stu.ouc.edu.cn

Abstract. In this paper, the suspended sediment concentration and bed geometric characteristics of powder sediment under the wave-only and wave-with-current are described. 9 experimental scenarios with the different wave height H and mean flow velocity V were investigated. It is found that under the wave condition the suspended sediment concentration S increases with the increase of wave height H, however it increases with the increase of flow velocity V under the wave-current condition. The sediment concentration in the near-bottom area is very high for both wave and wave-current condition. The sediment wave length L_{sw} increased slightly with the increasing of the H, but it does not change much in general, while it increased with the increasing of the V. The sediment wave height H_{sw} increased with the increasing H, and it increased firstly and then decreased with increasing V.

1. Introduction

Powder sediment is one of common forms of coastal area, and its motion laws have been studied by many scholars. The geometric and transport characteristics of non-cohesive sand wave generated by waves and currents are conducted by Cataño-Lopera[1], the relationship between the sand wave parameters and the Reynolds number is discussed. The empirical formulas also have been developed by Wu and Lin[2], in order to calculate the nonuniform sediment transport rates and near-bed suspended sediment concentration (SSC) S under non-breaking waves and currents conditions. The suspended sediment transport experiments induced by long waves and wave groups conditions are investigate by Cáceres and Alsina[3], and it was found wave–backwash interactions in the swash zone was a key parameter in controlling the sediment stirring, water velocity magnitudes and the suspended sediment fluxes in the inner surf and outer swash. The motion law of silty sediment under the action of wave and current is different from the above-mentioned, a series of erosion tests conducted by Roberts et al.[4] suggested that the silt-enriched mixtures exhibit cohesive-like behavior. And an expression for sediment incipient motion for both silt and sand under conditions is proposed by using the derivation method for the Shields curve by Zuo et. al.[5], and the formula was verified by a number of experimental datasets as well as field data. A series of flume experiments has been conducted by Yao et.al.[6], in order to investigate sediment transport of sand–silt mixtures in both wave-only and wave-with-current conditions, and a high concentration layer was observed near the bottom together with ripples under wave only conditions. The study of sediment movement not only includes laboratory studies, but also has a lot of field observation. A series of field observations were carried out by Zuo et.al.[7] in the northwest Caofeidian sea area in the Bohai Bay, the water hydrodynamic and sediment concentration process with high spatial and temporal resolution in the bottom boundary layer (BBL) was obtained, and it was found that the suspended load sediment concentration is mainly influenced by wave-induced sediment suspension. A new field data on cohesive sediment erosion was discussed by Debnath et.al.[8], the results show that bed load plays a significant role in cohesive sediment erosion. Except the experimental research, numerical simulation also plays an important role in the study of sediment movement. The numerical simulation of sediment suspension and transport under plunging breaking waves was conducted by Yang et.al.[9], and both instantaneous and statistical results on the sediment concentration with different water depth and wave steepness are analyzed to investigate the processes of sediment suspension and transport. A numerical model to investigate the influences of breakwater layout on coastal waves, wave-induced currents, sediment transport and beach morphological evolution in the vicinity of breakwater are provided by Tang et.al.[10], after verifying the measured data, the model is used to study the influences of breakwater layout on coastal waves, wave-induced currents, sediment transport and beach morphological evolution.

A great deal of contribution has been paid to the sediment motion aforementioned. However, there still lacks adequate powder sediment movement for different waves or currents conditions. The paper studies powder sediment bed deformation and SSC under the wave-only and wave-current conditions experimentally to provide our insight on the powder sediment motion.

2. Experiments

2.1. Experimental Facilities and Setup

A series of experiments were conducted in a wave tank at the Hydraulic Lab of the Ocean University of China, and it is 30 m for length, 0.6 m for width and 1 m for depth. Its bottom was made of steel and its sidewalls were made of the stalinite. A current was produced by recirculating water in the wave tank with a water pump, while regular waves were generated by a piston type wave maker. In the central area of tank bottom, the powder sediment was covered for 0.1 m thickness and 4 m length higher than the wave length of all the experimental scenarios, and two 1:10 slopes made of iron were used to connect the sediment and the bottom smoothly, avoiding the excessive effect of the sediment topography, see figure 1. And the sediment median grain diameter d_{50} is 0.02 mm; however, the sediment contains about 5% organic matter, so that it showed a slight cohesive.

Three wave gauges and one ADV were placed in the wave tank, one wave gauge was placed 6 m away from the wave generator to measure the incident wave height. The ADV was placed in the middle of the sediment bed, and the other two wave gauges were arranged symmetrically in front of and behind it, for measuring the flow velocity and the wave heights, respectively.

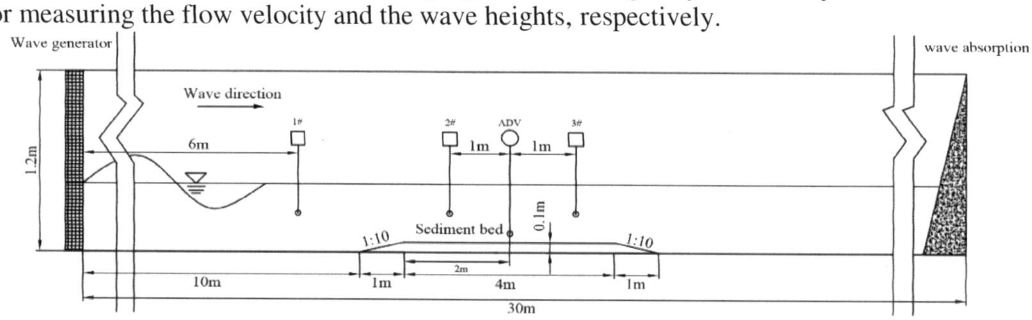

Figure 1. Experiment sketch.

2.2. Experimental Procedure

9 experimental scenarios were investigated with the different wave height H and mean flow velocity V. The still water depth d and wave period T are same 0.4m and 1.6s respectively for all experimental scenarios. The detailed experimental parameters are shown in Table 1.

Table 1. 9 experimental scenarios.

Scenarios	d(m)	T(s)	H(m)	V(m s^{-1})

1			6.538	0
2			9.066	0
3			10.484	0
4			11.492	0
5	0.4	1.6	13.243	0
6			14.046	0
7			8.784	0.03
8			8.569	0.07
9			8.683	0.11

Before the start of an experiment, the topography of the sediment bed must be leveled, and then add water to the tank to a fixed level. After the waves or wave-current have acted on the bed for more than an hour, the SSC at five different water depths is measured by siphon sampling method, and the water in the wave tank is subsequently released. The deformation of bed surface is measured by a laser gauge.

3.Analysis and Result

3.1. SSC

Figure 2 shows the distribution of SSC along the water depth under pure wave conditions with different wave height, and it was found that in the case of small wave height (scenarios 1-3), the SSC does not change much with the increase of water depth, and the SSC at the bottom is slightly larger than that of upper water body, indicating that sediment has not yet reached the start motion conditions. However, as wave heights continue to increase (scenarios 4-6), the SSC in the water body increases sharply, and the sediment concentration in the near-bottom zone is much higher than that in the upper water body, resulting in the formation of high concentrations of powder sediments, which is consistent with the conclusions reported by Yao et. al[6].

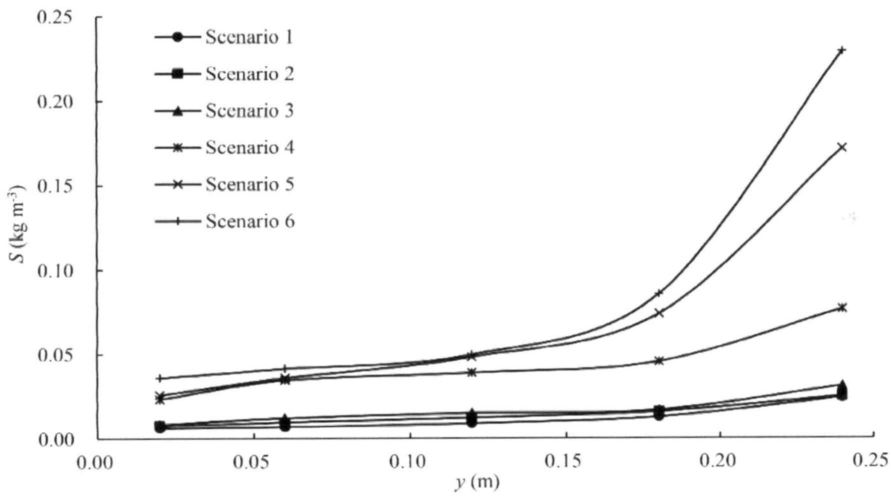

Figure 2. SSC distribution profile under pure wave conditions.

Figure 3 shows the relationship between SSC along the water depth and flow velocity under the condition of wave-current. It was observed that the same trend appears as the pure wave condition, including a sudden increase in sediment concentration with increasing flow velocity and a high sediment concentration at the bottom. Comparing figure 2 and figure 3, it is found that the SSC under the condition of wave-current is slightly larger than that of the pure wave. And even under the condition of smaller wave height, the more obvious SSC distribution can be observed, which indicates that the sediment is easier to start under the conditions of wave-current.

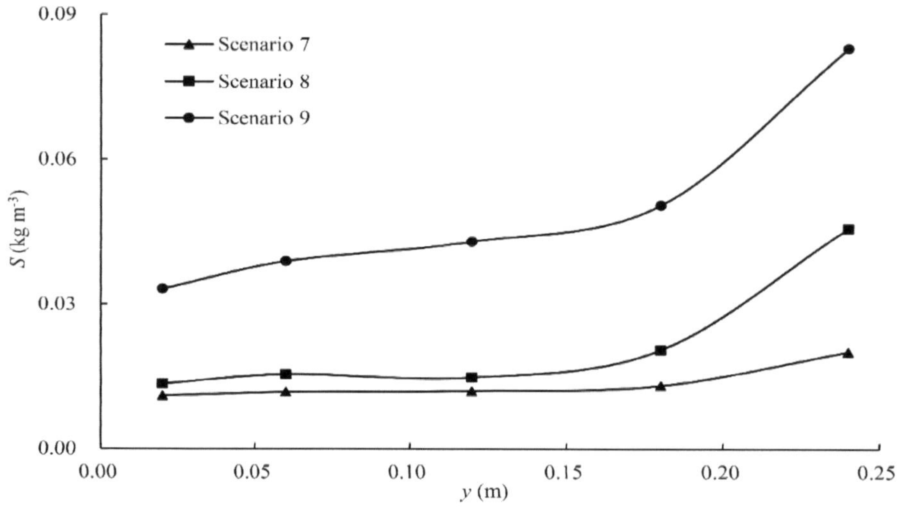

Figure 3. SSC distribution profile under wave-current conditions.

3.2. Sediment Bed Deformation

Figure 4 (a) and (b) show the variation of the sediment wave length L_{sw} with the wave height H and the flow velocity V respectively. It can be seen from figure 4 (a) that the L_{sw} increases slightly with increasing H, except for H=11.5 cm scenario. As can be seen in figure 4 (b), L_{sw} increasing with increasing V. Figure 4 (a)and (b) show that L_{sw} under the wave-current conditions is obviously larger than the pure wave conditions.

Figure 5 (a) and (b) show the variation of the sediment wave height H_{sw} with the wave height H and the flow velocity V respectively. It can be seen from figure 5 (a) that H_{sw} increased with the increasing H. However, H_{sw} increased firstly and then decreased with increasing V as shown in figure 5 (b). By compared with figure 5 (a) and (b), the ranges of variation of H_{sw} differ little for the pure wave condition and wave-current conditions.

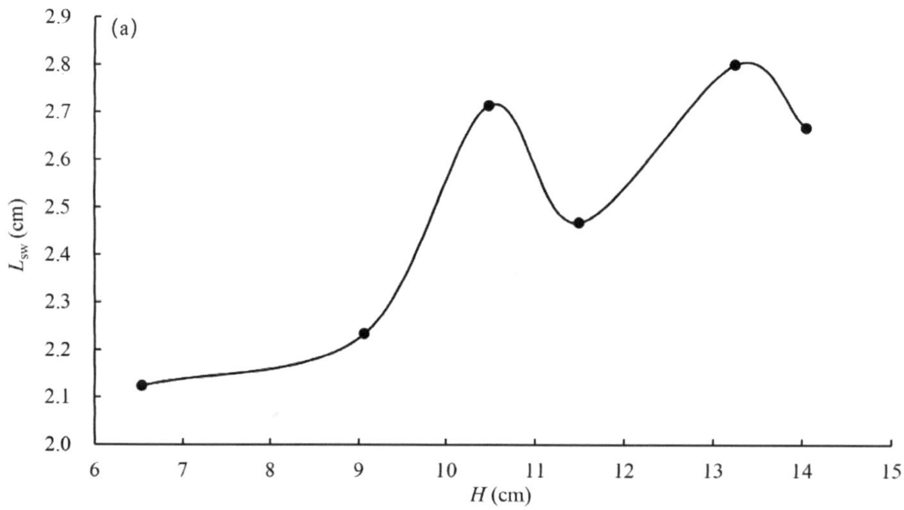

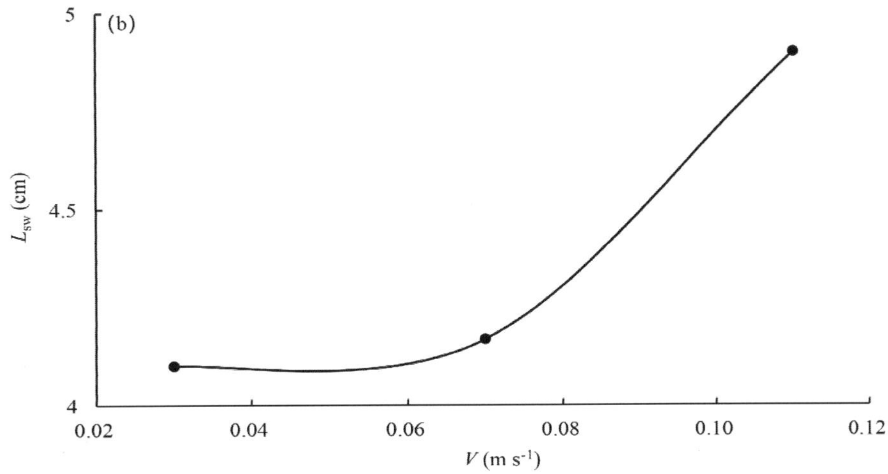

Figure 4. the variation of L_{sw} with H and V.

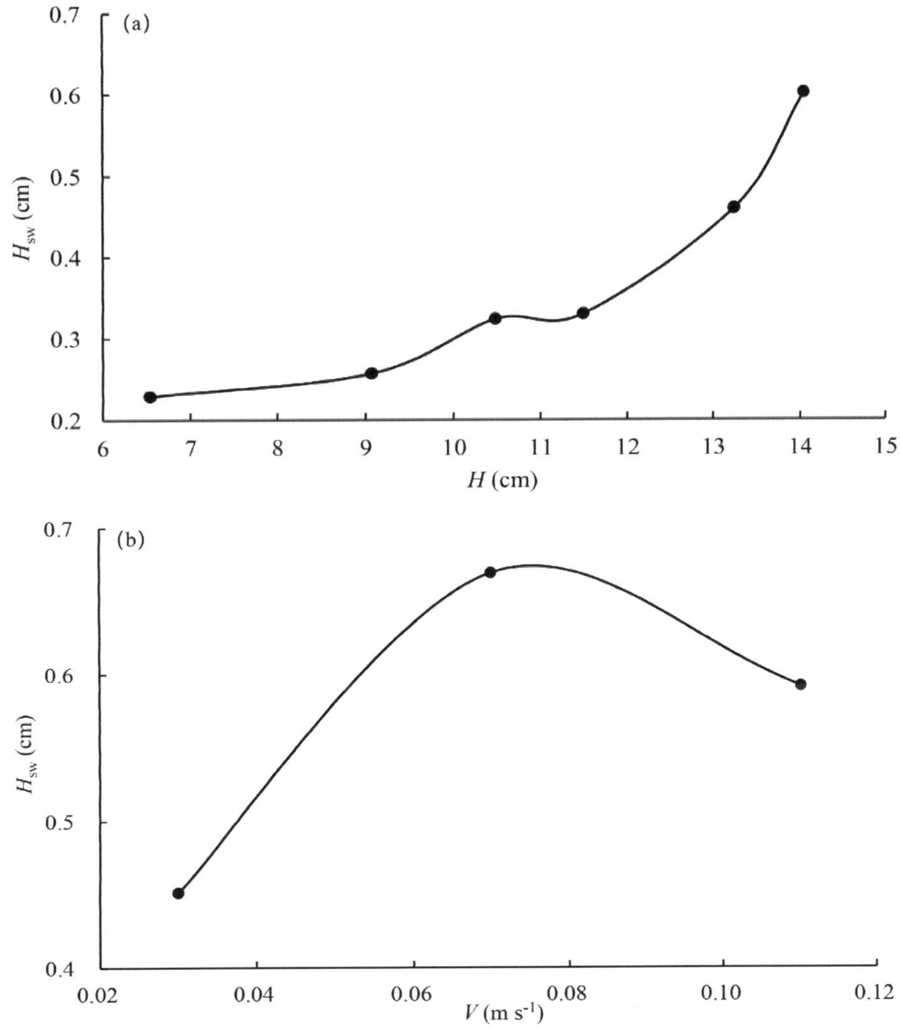

Figure 5. the variation of H_{sw} with H and V.

4. Conclusion

In this paper, the SSC and bed geometric characteristics of powder sediment under the combined effect of waves and currents are described. A series of experimental were conducted with the different pure wave conditions and wave-current conditions. The SSC at 5 different water depths, and the average sediment wave height H_{sw} and wavelength L_{sw} within the range of bed area were measured by siphon sampling method and laser range finder, respectively. It is found that:

(1). The SSC increases with the increase of wave height H under the wave condition. In addition, it increases with the increase of flow velocity V under the wave-current conditions.

(2). SSC in the near-bottom area is very high for both wave and wave-current condition.

(3). The sediment wavelength increases slightly with the increase of the incident wave height, however it increases with the increase of the flow velocity V.

(4). The sediment wave height increased with the increasing H, however it's increased firstly and then decreased with increasing V.

5. References

[1] Cataño-Lopera Y A and García M H 2006 Geometry and migration characteristics of bedforms under waves and currents. Part 1: Sandwave morphodynamics *Coast Eng.* 53(9) 767-780.

[2] Wu W and Lin Q 2014 Nonuniform sediment transport under non-breaking waves and currents. *Coast Eng.* 90 1-11.

[3] Cáceres I and Alsina J M 2016 Suspended sediment transport and beach dynamics induced by monochromatic conditions, long waves and wave groups *Coast Eng.* 108 36-55

[4] Roberts J, Jepsen R, Gotthard D and Lick W 1998 Effects of particle size and bulk density on erosion of quartz particles *J. Hydraul. Eng.* 124(12) 1261-67

[5] Zuo L, Roelvink D, Lu Y and Li S 2017 On incipient motion of silt-sand under combined action of waves and currents *Appl. Ocean Res.* 69 116-125

[6] Yao P, Su M, Wang Z, van Rijn L C, Zhang C, Chen Y and Stive M J 2015 Experiment inspired numerical modeling of sediment concentration over sand–silt mixtures *Coast Eng.* 105 75-89

[7] Zuo L, Lu Y, Wang Y and Liu H 2014 Field observation and analysis of wave-current-sediment movement in Caofeidian Sea area in the Bohai Bay, China *China Ocean Eng.* 28(3) 331-348

[8] Debnath K, Nikora V, Aberle J, Westrich B and Muste M 2007 Erosion of cohesive sediments: Resuspension, bed load, and erosion patterns from field experiments *J. hydraulic eng.* 133(5) 508-520

[9] Yang Z, Lu X H, Guo X, Liu Y and Shen L 2017 Numerical simulation of sediment suspension and transport under plunging breaking waves *Comput. Fluids.* 158 57-71

[10] Tang J, Lyu Y, Shen Y, Zhang M and Su M 2017 Numerical study on influences of breakwater layout on coastal waves, wave-induced currents, sediment transport and beach morphological evolution *Ocean Eng.* 141 375-387

Acknowledgments

The study is financed by the Science and Technology Project of the Ministry of Industry and Information Technology of China "The Simulation Testing Equipment Development and the Key Technology Research on the Safety Operation in the Subsea Engineering" (Grant No. E-0815C003).

Experiment Research on Geometry and Evolution Characteristics of Sand Wave Bedforms Generated by Waves and Currents

Xiaohuan Zhu[1], Zhenlu Wang[2], Zhaohui Wu[1], Feilong Liu[1], Bingchen Liang[2,3], Yang He[1], and Fuxiang Huang[1]

1 Offshore Oil Engineering Co., Ltd., Tianjin 300451, China
2 Engineering College, Ocean University of China, Qingdao, China
3 Shandong Province Key Laboratory of Ocean Engineering, Ocean University of China, Qingdao, China
Email: wangzhenlu_135@163.com

Abstract. Sand waves generated under waves and wave-current have been investigated experimentally in a wave flume. The evolution of bedforms were measured by a Trasonic Terrain automatic Measurement and analysis System (TTMS). With the data measured, the configuration of the bed was analysed. With long time wave action, a large-scale bedforms as sand waves can be formed and many ripples superimposed upon them. Ripples with 2D or 3D patterns can be formed at different locations. The growth rate of sand waves was increased at the beginning and then decreased with the increase of the time. Moreover, it has been found that the sand wave features as sand wave length and height have relationship with hydrodynamics. In general, the sand wave height increases with the increasing of wave height and residual current respectively. The sand wave length increases with the increase of residual current and decrease of wave height. The sand wave growth rate increases with the increase of the hydrodynamic conditions.

1. Introduction

Tidal sand waves are dynamic large-scale rhythmic bed forms observed in many tide-dominated shallow seas that have a sandy seabed [1]. Sand waves develop in response to a dynamic balance between available sand, tidal currents and wave energy [2]. Sand waves can migrate up to tens of meters per year and are formed at a time scale of 1-10 years [3]. The dynamic behavior of sand waves such as migration and growth can cause huge damage because they tend to interfere with cables, pipelines and navigation. It is necessary to require knowledge of sand wave dynamics.

The research methods are mainly theoretical model study, numerical modelling, field survey and physical experiments. The formation of sand waves can be explained as a free instability system of the sandy seabed subject to tidal motion. Hulscher [4] used the shallow water equations to describe the hydrodynamics interaction between tidal bedforms and oscillatory character. And the mechanism giving rise to sand waves formation was the interaction of the tidal current with a bottom perturbation induces the steady recirculating cells. Then Hulscher's work was extended by regarding various physical mechanisms. In order to predict the growth and migration of the fastest growing mode of a random perturbation of the flat sea bed forced by the tidal current and wind waves, a model based on a linear stability analysis was presented [5]. Sand wave migration can be caused by pressure- or wind-driven residual currents [6] and by tidal asymmetry [7]. Numerical simulation models have been performed in recent days, Borsje used a numerical process-based model to investigate the role of suspended load transport mechanisms in formation of sand waves [8]. He found that suspended load transport can cause suppression of long sand waves, resulting in a finite range of wavelengths that experience growth. Later, Gerwen used the model as presented in Borsje [8] to model the effect of suspended load transport and tidal asymmetry on the equilibrium tidal sand wave height [9]. It is shown that both suspended load transport and tidal asymmetry reduce the equilibrium sand wave height. Doré used a 2D RANS numerical model to simulate the morphodynamic evolution of a flat non-cohesive sand bed submitted to a tidal current, reproduces the bed evolution until a field of sand bedforms is obtained that are comparable with observed superimposed ripples in terms of geometrical dimensions and dynamics [10].

To investigate the formation and migration of sand waves with physical experiment is very difficult because the similarity scales are hard to match with the field conditions. Faraci investigated the ripples generated by regular and irregular waves experimentally and analyzed the geometry at the equilibrium [11]. Cataño-Lopera performed a systematic study on the geometric and migration characteristics of sand waves generated under the combined effect of waves and currents in a wave tank [2]. However, the influence of different currents with wave has not been investigated. Jiang develops a numerical model to research the bilateral reverse migration of one-group seabed sand waves around a sand ridge in a small shallow shelf sea of the South China Seas [12]. It is found that long-term sand wave migration trend is mainly controlled by both tidal constituents and residual velocity.

The aim of this paper is to illustrate the formation and evolution of sand waves generated by waves and combined different flows in a wave flume. The paper is organized as follows. Section 2 describes the experiment procedure. The third section and describes the test conditions and some computed parameters. The bedform configurations are also described in section 3. The main findings of this paper are discussed in section 4, and finally the conclusions are presented in section 5.

2. Experiment Setup

A series of experiments was conducted in a wave tank in the Hydraulic Lab at Ocean University of China. Figure 1 shows the schematic diagram of physical experiment layout, the wave tank is 60 m length, 0.8 m width and 1.5 m deep. The waves were generated by a piston type wave maker. The maximum wave height can be generated up to 0.3 m and the wave period up to 3.0 s. The currents with directions of positive and negative were generated by two circulating pumps. Porous medium as wave absorbing was placed at the opposite side of the wave tank. In the central part of the tank, the bottom was covered by a sand bed 15 m in length, 0.8 m in width, and 0.4 m in thickness with uniform grain size sand. Two 1:5 concrete slope was built at both sides of the sand bed in order to make the wave and current more stable. The median grain diameter d_{50} is 0.17mm (figure 2) and the repose angle is 33°.

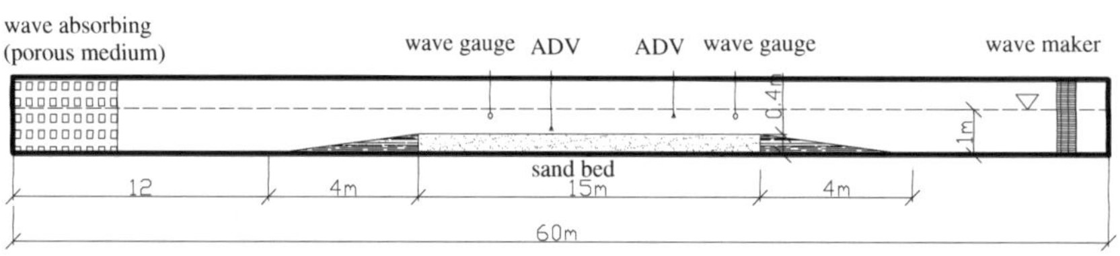

Figure 1. Schematic diagram of model experiment layout.

The depth-averaged velocity was measured by ADV (Acoustic Doppler Velocimetry). The wave height and wave period were determined by two wave gauges with a sampling rate of 50 Hz, which can meet the demands of the measurement with water surface elevation. The evolution of bed features was measured by a Trasonic Terrain automatic Measurement and analysis System (TTMS). The measurement accuracy of TTMS can reach 1mm. The water temperature was measured for each experiment and ranged between 6.3 ℃ and 8.2 ℃ with a mean value of 7.5 ℃. The sand bed must be flat before each experiment. To obtain the stable bedforms, each experiment performs for 5 hours.

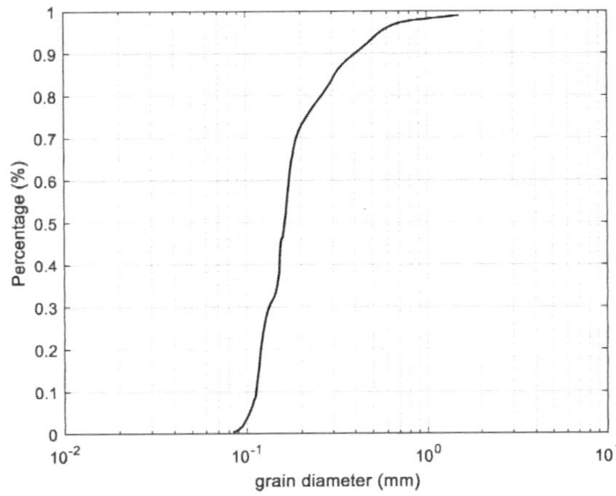

Figure 2. The distribution curve of sediment grain diameter.

3. Results

8 experimental scenarios were designed with different wave heights and currents. The water depth over the sand bed and the wave period are same 0.6m and 2s respectively for all scenarios. More details can be found in table 1. The variables in Table 1 are: H – wave height; T – wave period; u_0 – velocity of flow; u_m – maximum fluid orbital velocity at bed, a – wave orbital amplitude at bed, Re_d – the sediment Reynolds number; Re_w – flow Reynolds number, $Re_w = (u_m a)/\nu$, ν is the kinematic viscosity of water; l_{sw} – measured sand wave length; h_{sw} – measured sand wave height.

Table 1. Hydrodynamic characteristics and bedform characteristics of experiment scenarios.

Run number	H(cm)	T(s)	u_0(ms^{-1})	u_m(cms^{-1})	a(cm)	Re_d	Re_w	l_{sw} (cm)	h_{sw}(cm)
1	10	2	0	23.52	7.49	37.84	17487	228	1.0888
2	12	2	0	28.22	8.98	45.41	25181	220	2.79
3	14	2	0	32.92	10.48	52.98	34274	196	3.1816
4	16	2	0	37.63	11.98	60.55	44767	178	3.7923
5	14	2	-0.2	28.86	9.19	46.44	26335	184	2.6532
6	14	2	0.2	36.93	11.76	59.43	43123	204	3.6675
7	14	2	0.25	37.92	12.07	61.02	45468	241	5.6358
8	14	2	0.35	39.89	12.70	64.19	50310	241	3.8079

From figure 3, which shows the wave height with scenario 3 is no longer linear as the water is shallow. The Cnoidal wave theory should be used to compute all related wave parameters. However, as an approximation the small wave theory was still used here following Cataño-Lopera [2]. So, the orbital velocity and amplitude in Table 1 were still computed assuming that the small amplitude wave theory is applicable.

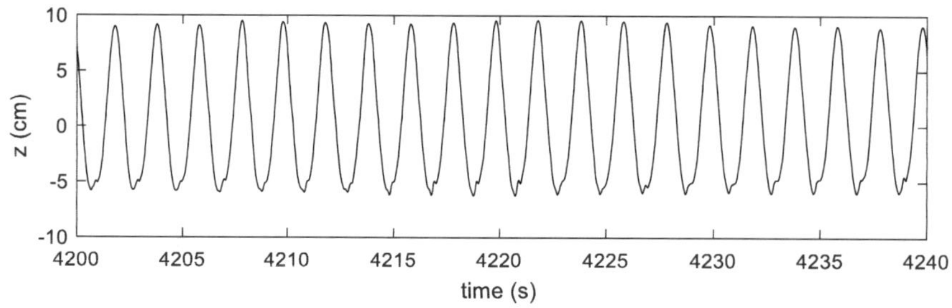

Figure 3. Water elevation measured as the function of times.

Figure 4 shows the bed configuration after 10 hours under the scenario 3 wave condition, and many features can be found from the figure. After a long time wave action, an obvious large-scale bedform as sand wave was formed. The sand wave length is about 200cm, and height is about several centimeters. We can also notice that many ripples superimposed upon sand waves, and ripples present different patterns. In general, 2D ripples often present at the crest and trough location of sand wave while 3D ripples form between the crest and the trough nearer the crest. The ripples at the crest usually larger than that at the trough.

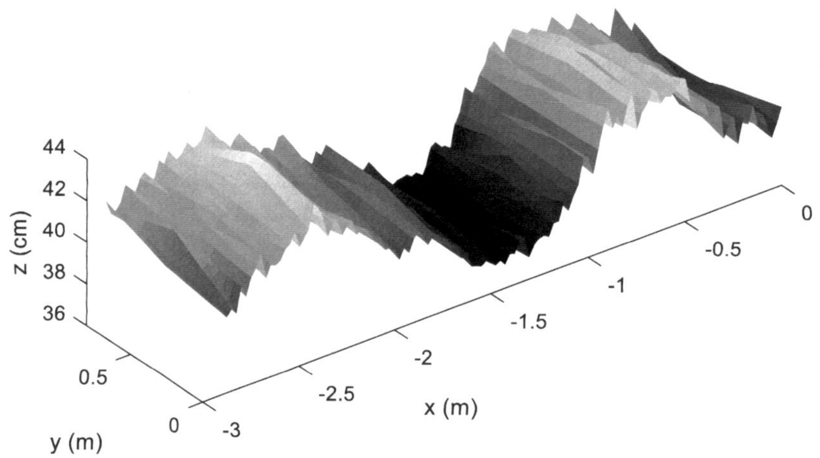

Figure 4. Bed configuration after 10 hours under the wave condition: H =14cm, T = 2s, u0 = 0.

Flat sand bed will form sand wave gradually under some hydrodynamic conditions. From figure 5, sand wave grows obviously in first 5 hours and almost remain the configuration between 5 hours to 10 hours which means that the bedforms has achieved equilibrium state at 5 hours. As the range of Reynolds number is small from about 15000 to 50000, so we performed 5 hours wave action with all other experimental scenarios.

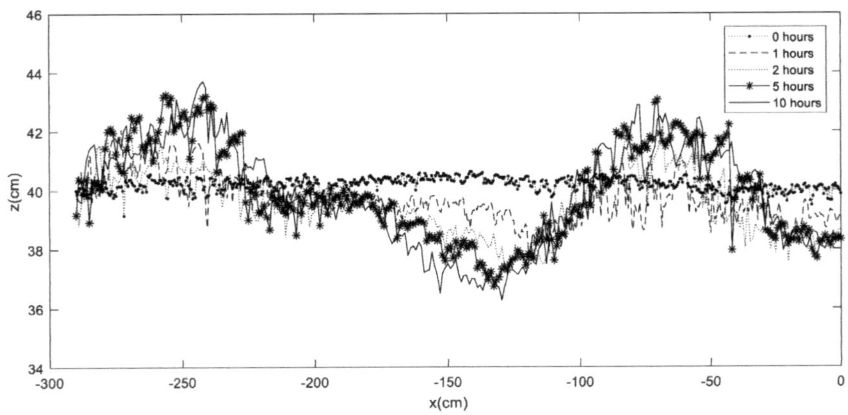

Figure 5. Bottom evolution over time: the generation of sand wave.

Figure 6 shows sand wave growth rate quantitively, in first hours, the sand wave growth rate reaches about 0.5cm/hour, and then the growth rate decreases. After 5 hours, the growth rate reduces to about 0.1cm/hour, and then the sand wave height almost remains the same after 7 hours. So, it is reasonable for 5 hours wave action on sand bed.

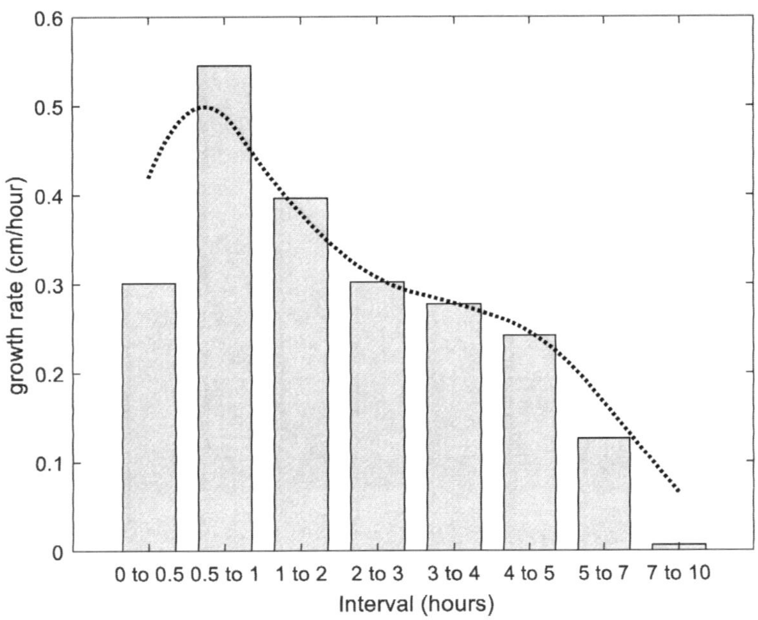

Figure 6. The growth rate of sand wave as a function of time.

4. Discussion

4.1. Configuration of Sand Wave

From figure 7, the relationship between sand wave characteristics and wave height can be discussed. The circle in the figure denotes experiment data while the red line denotes the fitting line. The sand wave height increases with the increase of wave height, and the sand wave length decreases with the increase of wave height.

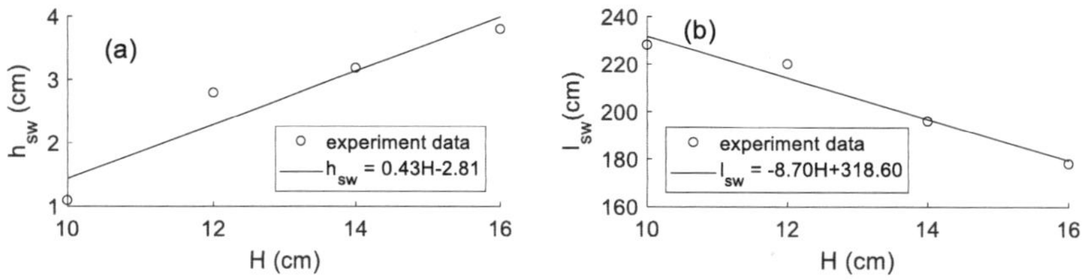

Figure 7. The relationship between sand wave characteristics and wave height.

Figure 8 shows the sand wave characteristics under the same wave condition with different residual current action. With the increase of the velocity, the sand wave height and length both increase respectively though the data is not ideal when velocity is 0.35m/s. While the direction between the current and the wave is opposite, the sand wave height and length both decreased.

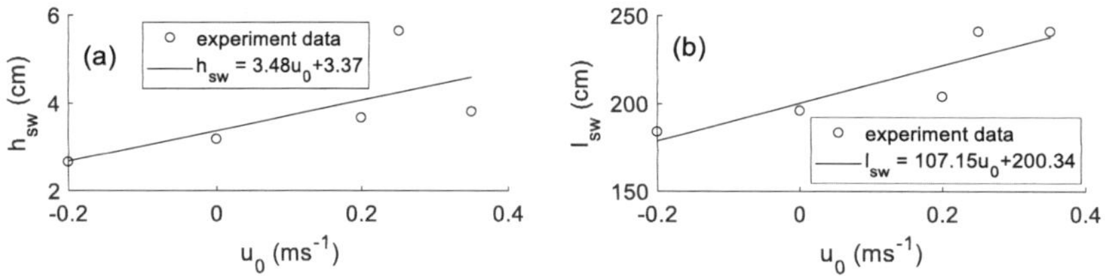

Figure 8. The relationship between sand wave characteristics and current under the same wave conditions.

4.2. *Sand Wave Growth Rate*

Figure 9 shows the sand wave growth rate increases with the increase of the wave height and residual current respectively. Similarly, the growth rate reaches too high when velocity is 0.35m/s which makes the trench not quite ideal. With the current increased, the wave will be deformed more obviously. So, the flow characteristics near the bottom become more complicated which need more detailed measurement data in future.

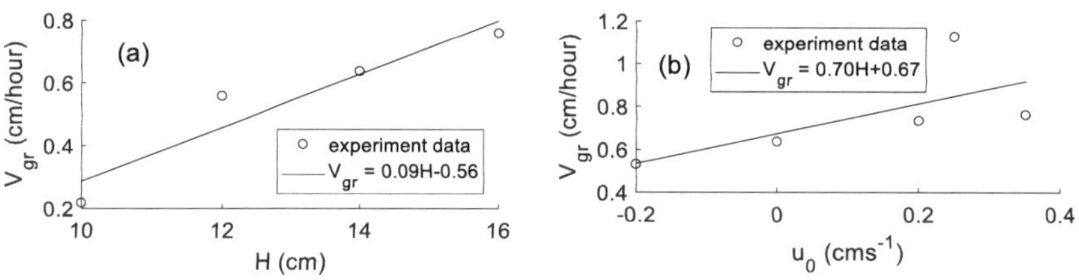

Figure 9. The relationship between sand wave growth rate and hydrodynamic characteristics.

It should be noticed that the fitting equations in figures 7 to figure 9 just show the trend of relationship between sand wave characteristics and hydrodynamics. The equations should be verified by more scenarios.

5. Conclusion

In this paper, the sand wave bed configuration generated by waves alone and combined different flows were experimentally investigated in a wave tank. After several hours wave action, small-scale bedforms as ripples and large-scale bedforms as sand waves coexist on the bed. The main conclusions from the experiments are that:

(1). The growth rate of sand wave increases first and then decreases over time. under the Reynolds number of 50000, it may take about 5 hours to reach the equilibrium state.

(2). In the range of Reynolds number from 15000 to 50000, the sand wave height increases and sand wave length decreases with the increase of wave height respectively. Both the sand wave height and sand wave length increase with the increase of the residual current under the same wave condition.

(3). The sand wave growth rate increases with the increase of wave height and residual current respectively.

6. References

[1] McCave I N 1971 Sand waves in the North Sea off the coast of Holland *Mar. geol.* 10(3) 99-225

[2] Cataño-Lopera Y A & García M H 2006 Geometry and migration characteristics of bedforms under waves and currents. Part 1: Sandwave morphodynamics *Coast. Eng.* 53(9) 767-780

[3] Campmans G H P, Roos P C, de Vriend H J & Hulscher S J M H 2017 Modeling the influence of storms on sand wave formation: A linear stability approach *Cont. Shelf Res.* 137 103-116

[4] Hulscher S J M H 1996 Tidal-induced large-scale regular bed form patterns in a three-dimensional shallow water model *J. Geophys. Res-Oceans* 101(C9) 20727-20744

[5] Blondeaux P & Vittori G 2016 A model to predict the migration of sand waves in shallow tidal seas *Cont. Shelf Res.* 112(5) 31-45

[6] Németh A A, Hulscher S J & de Vriend H J 2002 Modelling sand wave migration in shallow shelf seas *Cont. Shelf Res.* 22(18-19) 2795-2806

[7] Besio G, Blondeaux P, Brocchini M & Vittori G 2004 On the modeling of sand wave migration *J. Geophys. Res-Oceans* 109(C4)

[8] Borsje B W, Kranenburg W M, Roos P C, Matthieu J & Hulscher S J M H 2014 The role of suspended load transport in the occurrence of tidal sand waves *J. Geophys. Res-Earth* 119(4) 701-716

[9] Gerwen W V, Borsje B W, Damveld J H & Hulscher S J M H 2018 Modelling the effect of suspended load transport and tidal asymmetry on the equilibrium tidal sand wave height *Coast. Eng.* 136 56-64

[10] Doré A, Bonneton P, Marieu V, & Garlan T 2018 Observation and numerical modeling of tidal dune dynamics *Ocean Dynam.*

[11] Faraci C & Foti E 2002 Geometry, migration and evolution of small-scale bedforms generated by regular and irregular waves *Coast. Eng.* 47(1) 35-52

[12] Jiang W & Lin M 2016 Research on bilateral reverse migration of one-group seabed sand waves in a small shallow shelf sea *Coast. Eng.* 111 70-82

Acknowledgments
The study is financed by the Science and Technology Project of the Ministry of Industry and Information Technology of China "The Simulation Testing Equipment Development and the Key Technology Research on the Safety Operation in the Subsea Engineering" (Grant No. E-0815C003).

IOP Conf. Series: Earth and Environmental Science **171** (2018) 012003 doi:10.1088/1755-1315/171/1/012003

Analysis of the Structure Suffered Submarine Landslides Using SPH and CEL Methods

Qi Lei[1 2], Liu Zhenwen[1], Xu Hao[1] , Deng Haifeng[2],Li Chun[2]

1 CNPC Engineering Technology Research Company Limited,Tianjin,China
2 Key Laboratory of Offshore Engineering, CNPC,Tianjin , China
E-mail: qilei01@cnpc.com.cn

Abstract. Study shows that the submarine landslides have greatly affect the stability and security of subsea production systems, subsea pipelines and other Marine structures. Landslides can also take place even though the slope Angle is less than 1 °. At present the study of onshore landslide is more, but the study of submarine landslides are very few.The landslides mainly deal with the issues of stability. The analysis of landslide stability mainly use the Galerkin method, discrete element method, SPH and so on.But there is less research on the influence of the subsea structures.When the landslide occurs, Fluid characteristics of the landslide are significant, the soils translate into mudflow or debris flow,which is not solid state but fluid.So both SPH and CEL can be used for simulation, this paper uses the CEL and SPH to study the submarine landslide, the comparative analysis of the advantages and disadvantages of both methods is taken on. Both motheds are introduced at fisrt, Then through calculation, compares the two kinds of calculation methods. It can be seen that the SPH method and the CEL method are basically consistent with the shape of the landslide mass.It shows that both methods are feasible. The results of CEL method is more accurate.When the number of elements of SPH and CEL is few, the calculation time of SPH is shorter than that of CEL, but with the increase of element number, the calculation of the CEL method is more faster. If stress value of Lagrangian structure is required,it is better to choose CEL method .It is useful to further study on submarine landslides on Subsea structures.

1. Introduction

Oil and gas developments often require placing subsea equipment, e.g., subsea wells, pipelines and flowlines, foundation systems for floating structures, in areas with sloping seafloors. Submarine slope failures can occur in such areas and create soil slides. Thus we must be concerned about submarine landslides.There are two issues that we need attention ,One is the stability of submarine slopes ,the other is the effects of submarine landslides on Subsea structures. Both issues must be considered in selecting the site for installing and designing seafloor equipments.

Slope stability analyses were performed to assess the likelihood of slides being triggered by gravity, rapid sedimentation (underconsolidation) and earthquakes. The analyses revealed that it is unlikely that most of the seafloor slope failures were triggered by gravity loads. Earthquake loading was confirmed as a common trigger, and rapid sedimentation (underconsolidation) was also a likely trigger of many slope failures. It is important to note that the study revealed that a relatively large number of submarine slides occurred on much flatter (less than 10 degree) slopes and traveled much greater distances than slope failures on land[1]. Hydroplaning is one mechanism that may explain such large runout distances[2]. Submarine slides will cause damage to distant subsea equipments. So it should be clear that once the submarine landslide occurs, how much impact of submarine structures because of the instability of rocks and soils.

After the landslides, the soils translate into mudflow or debris flow gradually, and evolve into turbidity finally. It is not solid state but fluid. Both smoothed particle hydrodynamics (SPH)[3] and Coupled Eulerian-Lagrangian (CEL)[4] can be used to describe fluid,so this paper will disscuss the influence of landslides on subsea structure by both methods.

2. SPH Method

The SPH method has received substantial theoretical support since its inception (Gingold and Monaghan, 1977)[3].SPH (smoothed particle hydrodynamics) method, a pure Lagrangian, meshless hydrodynamics method, has been widely applied in numerical simulations in many engineering fields. SPH is a numerical method which is belong to one of meshless (or mesh-free) methods. For these meshless methods, the nodes and elements must not be defined as we normally do in a FEA; instead, only a collection of points are necessary to represent a given body. In smoothed particle hydrodynamics these nodes are commonly referred to as particles or pseudo-particles[5].

SPH is a fully Lagrangian modeling scheme permitting the discretization of a prescribed set of continuum equations by interpolating the properties directly at a discrete set of points distributed over the solution domain without the need to define a spatial mesh. The method's Lagrangian nature, associated with the absence of a fixed mesh, is its main strength. Difficulties associated with fluid flow and structural problems involving large deformations and free surfaces are resolved in a relatively natural way.

At its core, the method is not based on discrete particles (spheres) colliding with each other in compression or exhibiting cohesive-like behavior in tension as the word particle might suggest. Rather, it is simply a clever discretization method of continuum partial differential equations. In that respect, SPH is quite similar to the finite element method. SPH uses an evolving interpolation scheme to approximate a field variable at any point in a domain.

Smoothed particle hydrodynamic analyses are effective for applications involving extreme deformation. Fluid sloshing, wave engineering, ballistics, spraying (as in paint spraying), gas flow, and obliteration and fragmentation followed by secondary impacts are a few examples. There are many applications for which both the coupled Eulerian-Lagrangian and the smoothed particle hydrodynamic methods can be used[6-9]. In many coupled Eulerian-Lagrangian analyses the material to void ratio is small and, consequently, the computational effort may be prohibitively high. In these cases, the smoothed particle hydrodynamic method is preferred. For example, tracking fragments from primary impacts through a large volume until secondary impact occurs can be very expensive in a coupled Eulerian-Lagrangian analysis but comes at no additional cost in a smoothed particle hydrodynamic analysis.

SPH method was a newly emerged method for the applications in the geotechnical engineering fields[8].In abaqus, the finite element model (Figure 1) can be converted into the SPH model (Figure 2) easily.

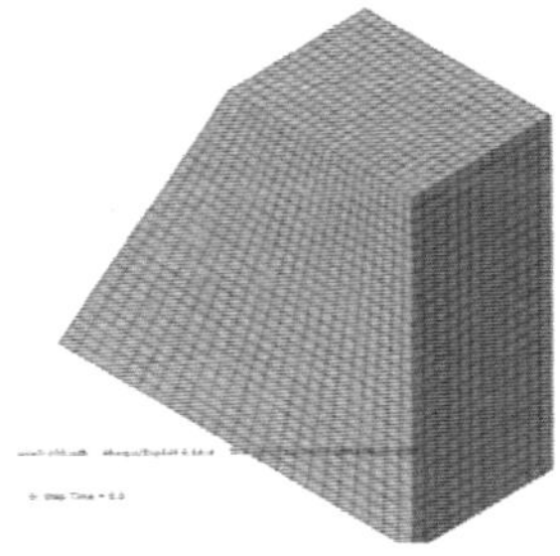

Figure 1. FE model.

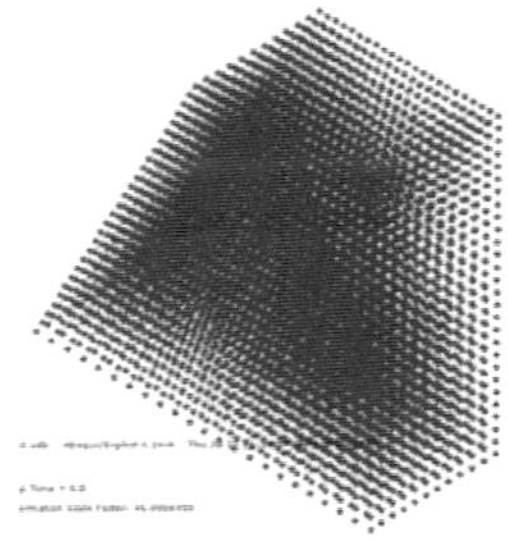

Figure 2. SPH model.

3. CEL Method

In the ninetieths Lagrangian and Eulerian approaches were combined together in the new approach called coupled Eulerian-Lagrangian (CEL)[4]. The CEL method was initially applied to fluid-structure interaction problems. Coupled Eulerian-Lagrangian approach may be also applied to geotechnical problems[10].

If a continuum deforms or flows, the position of the small volumetric elements changes with time. These positions can be described as functions of time in two ways: (a) Lagrangian description: the movement of the continuum is specified as a function of its initial coordinates and time.(b) Eulerian description: the movement of the continuum is specified as a function of its instantaneous position and time.

In simulations with Lagrangian formulation the interface between two parts is precisely defined and tracked. In these simulations large deformation of a part leads to hopeless mesh and element distortion. In Eulerian analysis a Eulerian reference mesh, which remains undistorted and does not move, is needed, to trace the motion of the particles. The advantage of a Eulerian formulation is that no element distortions occur. Disadvantageously, the interface between two parts cannot be described as precise as if a Lagrangian formulation is used. Numerical diffusion can happen during the simulation[11].

In a traditional Lagrangian analysis nodes are fixed within the material, and elements deform as the material deforms. Lagrangian elements are always 100% full of a single material, so the material boundary coincides with an element boundary.

By contrast, in an Eulerian analysis nodes are fixed in space, and material flows through elements that do not deform. Eulerian elements may not always be 100% full of material—many may be partially or completely void. The Eulerian material boundary must, therefore, be computed during each time increment and generally does not correspond to an element boundary. The Eulerian mesh is typically a simple rectangular grid of elements constructed to extend well beyond the Eulerian material boundaries, giving the material space in which to move and deform. If any Eulerian material moves outside the Eulerian mesh, it is lost from the simulation.

Eulerian material can interact with Lagrangian elements through Eulerian-Lagrangian contact; simulations that include this type of contact are often referred to as coupled Eulerian-Lagrangian (CEL) analyses. This powerful, easy-to-use feature of Abaqus/Explicit general contact enables fully coupled multi-physics simulation such as soil-structure interaction.

Eulerian analyses are effective for applications involving extreme deformation, up to and including fluid flow. Eulerian-Lagrangian contact allows the Eulerian materials to be combined with traditional nonlinear Lagrangian analyses.

For general geotechnical problems, a Lagrangian mesh is used to discretize structures, while a Eulerian mesh is used to discretize the soil (Figure 3). The interface between structure and soil can be represented using the boundary of the Lagrangian domain. The Eulerian mesh, which represents the soil that may experience large deformations, has no problems regarding mesh and element distortions.

4. Comparative Analysis of Both Methods

The numerical model is established by ABAQUS software. The slope angle of landslide is 60°, the soil density is assumed to be 2.0g/cm^3, the friction angle is 20°, and the cohesion is 6.19kPa. Assuming that the structure is an elastic steel tube with a density of 7.85g/cm^3, the SPH model and the CEL model are established respectively, as shown in Figure 3 and Figure 4. Both model sizes and the material parameters are the same. And the number of elements divided is the same.

One of the differences between the submarine landslides and landslides on land is that the submarine landslides may have hydroplaning effect, which causes submarine landslides have large runout distances because of low friction in the process [2].so for the convenience of calculation, the seabed is set rigid body, and ignore the friction between the seabed and landslide mass.

The submarine landslides modeled with particles and CEL interact with the subsea structure modeled with finite element via contact. The contact interaction is the same as any contact interaction between a node-based surface and an element-based. Both general contact and contact pairs can be used. And apply gravity to the the whole models.

For CEL , The subsea structure is modeled with conventional Lagrangian continuum elements with reduced integration (S4R), which is a 4-node doubly curved thin shell,whereas the soil body is meshed using Eulerian elements (EC3D8R),which is a 8-node linear eulerian brick ,reduced integration hourglass control.

For SPH , The subsea structure is modeled with conventional Lagrangian continuum elements with reduced integration (S4R), whereas the soil body is meshed using PC3D. Rigid body structures are established to Limit particle overflow.

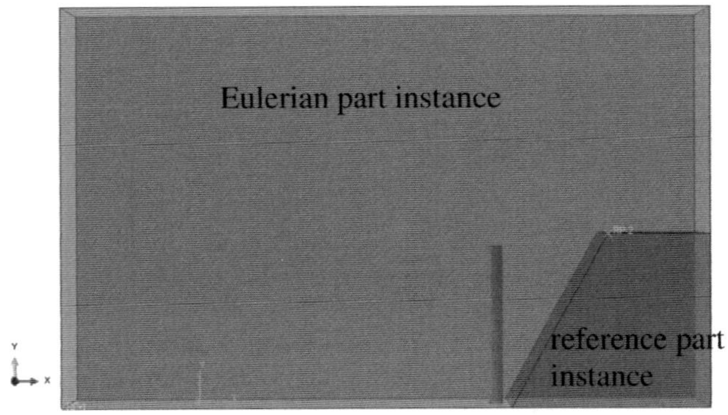

Figure 3. CEL model.

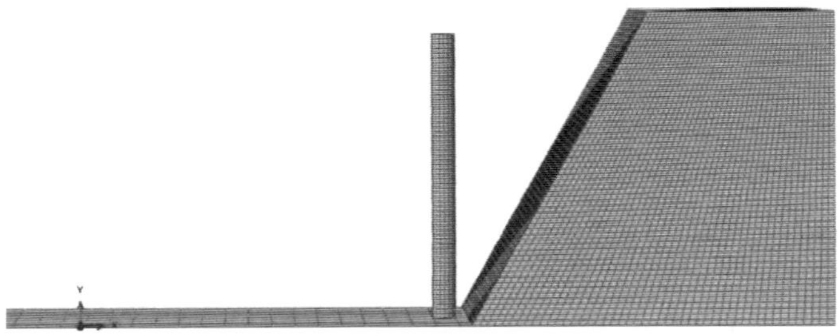

Figure 4. SPH model.

The results obtained by both methods are shown in Figure 5 and Figure 6. The deformation of the landslide mass obtained by both methods at the same time is shown in the following figures.

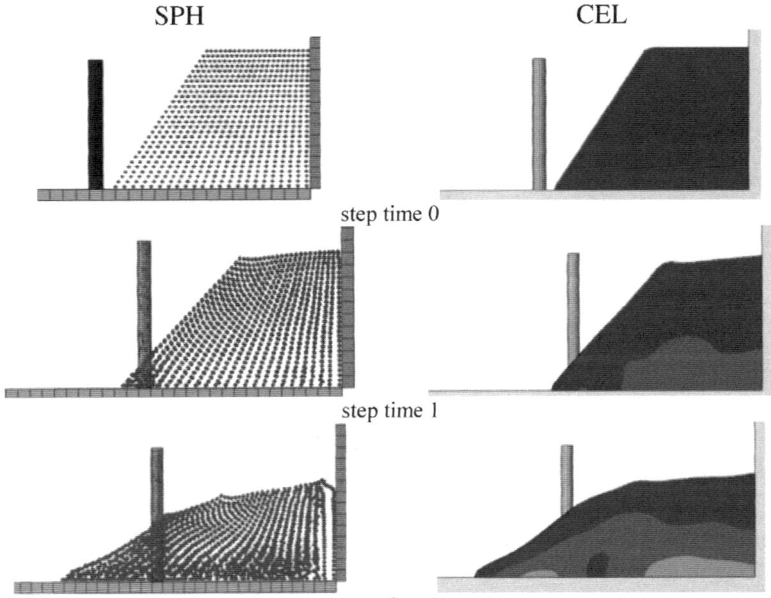

IOP Conf. Series: Earth and Environmental Science **171** (2018) 012003 doi:10.1088/1755-1315/171/1/012003

step time 3

Figure 5. Comparison of calculation results(side view).

SPH CEL

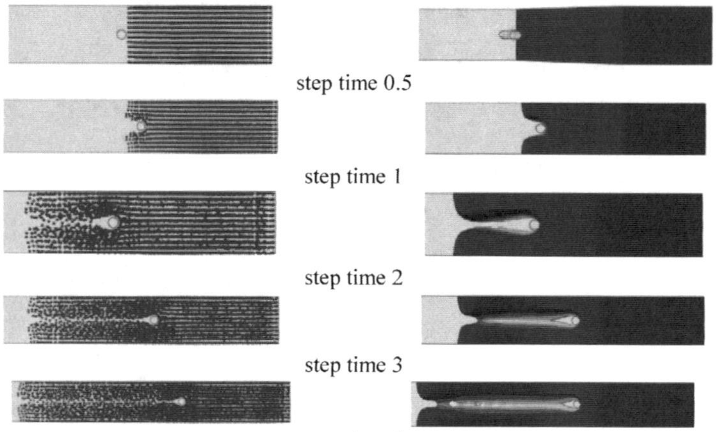

step time 0.5

step time 1

step time 2

step time 3

step time 4

Figure 6. Comparison of calculation results (vertical view).

It can be seen that SPH and CEL are basically consistent with the shape of the landslide mass calculated at the same time. The simulation calculation result is coincidence with the actual state.

The following Figure 7 shows the stress time history curves of the same node on the subsea structure affected by landslide,which is calculated by the two methods.

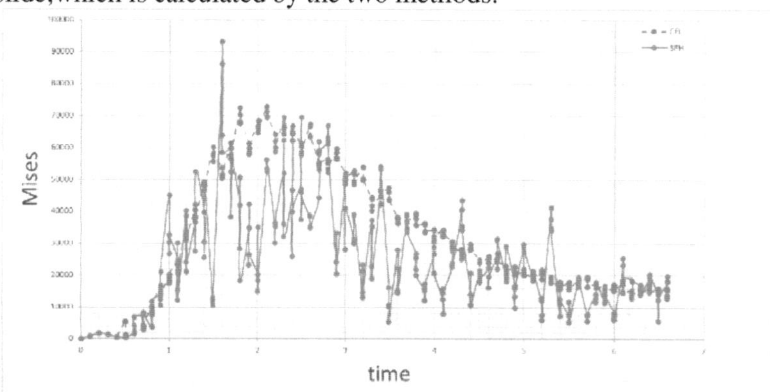

Figure 7. Stress comparison of the same node.

As shown above, the Mises stresses of the same node have the same trend.With the impact of the landslide, the stress increases gradually, reaches the peak and then decrease . But there are also obvious differences between both results. The stress curve calculated by SPH vibrates greatly, There are noticeable noises. But the stress curve calculated by CEL relatively smooth. the maximum stress calculated by SPH is much larger. The stress peaks calculated by SPH appear earlier , and then the stress is lowered and increased again. The stress decreases after the second peak points.But there is only one stress peak for the CEL results. The curve obtained by CEL method is smoother, and there is no maximum change. From the result, it can be concluded that the CEL calculation is reasonable relatively.

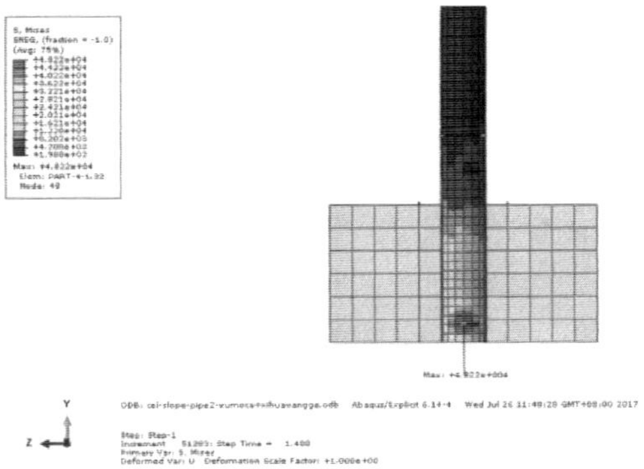

Figure 8. Stress nephogram of structure.

As shown Stress nephogram of structure(Figure 8) ,The maximum stress position calculated by CEL remains in one spot, which is located near the root of the structure faced to the landslide.But the maximum stress position calculated by SPH is in the different postions, and irregular position change.

The table (Table 1) and figure (Figure 9) below shows the calculating time difference between the two methods.

Table 1. Comparison of the CPU time.

Method	Step time	Element	CPU time
CEL	0.1	12	239.6
SPH	0.1	12	164.8
CEL	0.1	24	1143.4
SPH	0.1	24	1429.5

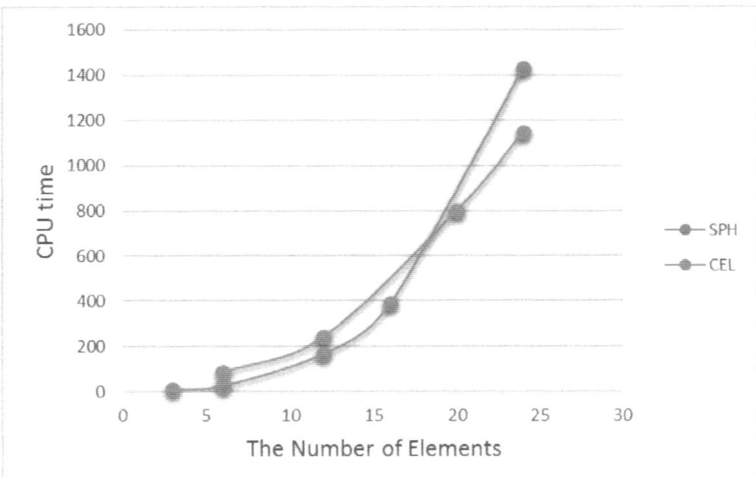

Figure 9. CPU time of different mesh number (SPH and CEL).

When the number of elements of SPH and CEL is few, the CPU time of SPH is shorter than that of CEL, but with the increase of element number, the SPH method will spend more time than the CEL method.

The following figure (Figure 10) shows the contrast of result with different mesh number.

Number of Elements(12) Number of Elements(24)

step time 1(SPH)

step time 1(CEL)

Figure 10. Comparison of calculation results.

After the mesh refinement, the distribution is more regular. At the same time, it also shows that the Mises stress of SPH is lower, and is closer to the result of CEL calculation.

As shown in Figure 11 and Figure 12,by comparing the stress value of the same node under the condition of the different number of elements, it can be seen that the finer the mesh, the smaller the value.

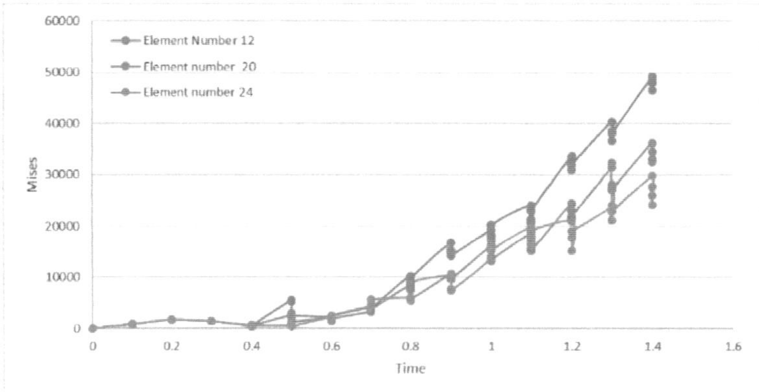

Figure 11. Stress comparison of the different element number(CEL).

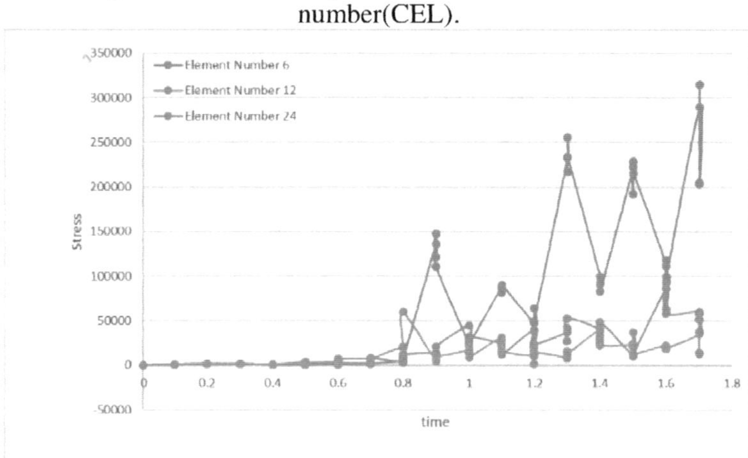

Figure 12. Stress comparison of the different element number(SPH).

As shown in the above,for SPH, although after refined mesh stress is reduced, but the results still have more noise.So if stress value of Lagrangian structure is required,it is better to choose CEL method.

5. Conclusions

It can be seen that the SPH method and the CEL method are basically consistent with the shape of the landslide mass . The computational result is coincidence with the actual state. Both methods can be good computation tools for the submarine landslides simulation.

the Mises stresses of the same node have the same trend,but under the condition of the same elements number,The results of CEL method is more accurate.

When the number of elements of SPH and CEL is few, the calculation time of SPH is shorter than that of CEL, but with the increase of element number, the calculation of the CEL method is more faster.And if stress value of Lagrangian structure is required,it is better to choose CEL method.

This paper try to discuss the calculation method of the influence of the structure due to the landslide,which is useful to further study on submarine landslide on Subsea structures.

6. References

[1] Anders Elverhoi,Hedda Breien,Fabio V Blasio,Carl B Harbitz,Matteo Pagliardi. 2010 Submarine landslides and the importance of the initial sediment composition for run-out length and final deposit *Ocean Dynamics* vol 4 p 1027-1046

[2] FVD Blasio,L Engvik,CB Harbitz,A Elverhøi 2004 Hydroplaning and submarine debris flows *Journal of Geophysical Research Oceans*

[3] Gingold RA and JJ Monaghan 1977 Smoothed Particle Hydrodynamics: Theory and Application to Non-Spherical Stars *Royal Astronomical Society* vol. 181 pp 375–389

[4] BWF Noh 1964 CEL: A time dependent two space-dimensional, coupled Eulerian Lagrangian code Methods of Computational Physics pp 117-179

[5] LD Libersky,AG Petschek,TC Carney,JR Hipp,FA Allahdadi 1993 High Strain Lagrangian Hydrodynamics *Journal of Computational Physics* vol 109 pp. 67–75

[6] GR Johnson,RA Stryk,SR Beissel 1996 SPH for High Velocity Impact Calculations *Computer Methods in Applied Mechanics and Engineering* 139 (1–4) p 347-373

[7] Liang C X,Zhang C Y et al. 2012 Modeling low Reynolds number incompressible flows with curved boundaries using SPH *International Journal for Numerical Methods in Fluids* 68(9) p 1173-1188.

[8] Bui H H, Fukagawa R, Sako K, et al. 2008 Lagrangian meshfree particles method (SPH) for large deformation and failure flows of geomaterial using elastic-plastic soil constitutive model *International Journal for Numerical and Analytical Methods in Geomechanics* 32(12) p1537-1570.

[9] P St-Germain,I Nistor,R Townsend,T Shibayama 2014 Smoothed-Particle Hydrodynamics Numerical Modeling of Structures Impacted by Tsunami Bores *Journal of Waterway Port Coastal & Ocean Engineering* 1(1) p 66-81.

[10] G Qiu,S Henke,J Grabe 2011 Application of a Coupled Eulerian–Lagrangian approach on geomechanical problems involving large deformations *Computers & Geotechnics* 38 (1) p 30-39

[11] F Ducobu,E Rivière-Lorphèvre,E Filippi 2016 Application of the Coupled Eulerian-Lagrangian (CEL) method to the modeling of orthogonal cutting *European Journal of Mechanics - A/Solids* p 58-66

Analysis and Optimization on the Flow Ability of Wave Buoy Based on AQWA

Jinpeng Hu, Ling Zhu and Shihui Liu

Civil and Transportation College, South China University of Technology, Guangzhou 510640, China
E-mail: 1940527624@qq.com

Abstract. Wave buoy is a modern ocean observation facility, and the ability of following wave is an important parameter to reflect the measurement accuracy. This article will focus on the study of hydrodynamics on the ability of three common appearance buoys. By using the hydrodynamic calculation software AQWA, this article will study the heaving response of free buoy in regular wave under frequency domain, it combined with the typical Green function method for solving the wave force and the related hydrodynamic coefficients of floating structure, and it applied the three dimensional potential flow theory and based on the single degree of freedom motion equation. According to the features of short period wave in Chinese sea states, this paper will calculate the heaving response RAOs of 25 buoys on different shapes, weight and shape parameters. Hydrodynamics calculation results show that: under the same conditions in the environment load, the flow-ability of ball float 3-2 has better performance than other buoys in working condition. The research methods of this article has provide good scientific basis for the parameter design on ocean data buoy.

1. Introduction

With the rapid development of the Marine economy in the world, measurement and collection of Marine environmental data is becoming increasingly important for human exploration and exploitation of Marine resources. Ocean data buoy which has the advantages of all-weather, all-day time, fixed point and reliability is widely used in measuring and collecting Marine environmental data. Description of Marine environment need to measure many parameters, and measurement of wave parameter is an important item. In order to ensure accurate measurement of wave parameters in different frequency ranges, oceanic data buoy should have better flow-ability in wave frequency range as large as possible. Flow-ability of buoy is mainly reflected in the heaving response, so the research of heaving motion under different frequency is of great significance for improving the accuracy.

At present, there are two methods to research the buoy motion characteristics at home and abroad: experimentation and numerical calculation. Wang D J et al. [1] calculated the wave exciting loads on a CALM buoy in water of finite depth with linearized potential wave theory. Leonard et al. [2] studied the coupling effect between buoy and mooring system through three-dimensional coupling analysis. Monroy et al. [3] simulated RAOs(response amplitude operator) of the CALM buoy in the regular and irregular wave by using SWENSE method which can simulate the fullly nonlinear effects (including viscous effects) of fluid. Salem et al. [4] studied the linearization method about quadratic damping of the CALM buoy system with the frequency domain analysis method, and estimated the amplitude of pitching by using the linearization method. Zhaozheng Wang et al. [5] designed the disc-shaped data buoy and its mooring system and carried out the related model test. Based on the three dimensional potential flow theory, Quanming Miao et al. [6] calculated the hydrodynamic parameters(additional

IOP Conf. Series: Earth and Environmental Science **171** (2018) 012004 doi:10.1088/1755-1315/171/1/012004

added mass, damping and so on) and motion response of the free-floating buoy, and estimated the motion and tension responses of the three anchors buoy system in different depths under extreme sea conditions. Xiutao Fan, Huajie Wang et al. [7, 8] based on ignoring the influence of mooring, studied the heaving and rolling response of large wave buoys with spectral analysis. Jiming zhang et al. [9, 10] modified the roll damping coefficients of buoy by using model test, and simulated the rolling response of buoy with numerical simulation in frequency domain, which also ignored the effect of mooring.

For Marine floating structures, the research priorities of domestic and foreign scholars focused on six degrees motion responses of the floating body and mooring force, but there are few studies on the heaving response which specially reflects the flow-ability of offshore buoy. In this paper, in order to be able to obtain the accurate prediction result of heaving response, the AQWA hydrodynamic calculation model is used to treat the buoy as a large scale component. Therefore, the hydrodynamic coefficients and wave force functions of the buoy are calculated by the three-dimensional potential flow theory. Based on the heave motion equation, the free floating buoy with different shape parameters under regular wave is analysed in frequency domain, and the heaving response amplitude operator (RAOs) is obtained. Compared with the calculation results, it is concluded that the ball float 3-2 has better work performance, which provides reference for the shape design of wave buoy.

2. Numerical Simulation

2.1. *Frequency Domain Analysis under Linear Regular Waves*

2.1.1. Definition of coordinate system. For convenience of calculations and description, the motion equation of wave buoy with six degree is generally defined as the following two coordinate systems: the reference coordinate system oxyz and the moving coordinate system $o_0x_0y_0z_0$. The origin of reference coordinate system is in the horizontal plane, plane oxy coincide with static water surface, and the z-axis is vertically upward. The moving coordinate system which keeps the same motion as the float and defines a coincidence point between initial static surface and float as the origin of coordinates is fixed on the float. The directions of the two frames that are also consistent satisfy the Descartes right-hand rule. The two frames coincide when the float is stationary.

As shown in Figure 1, the wave force under the regular wave that is decomposed into six directions results in the DOF motion which are rolling, pitching, yawing and surging swaying, heaving. Since the Marine data buoy mainly needs to research its flow-ability, the main considerations in this study are heaving response.

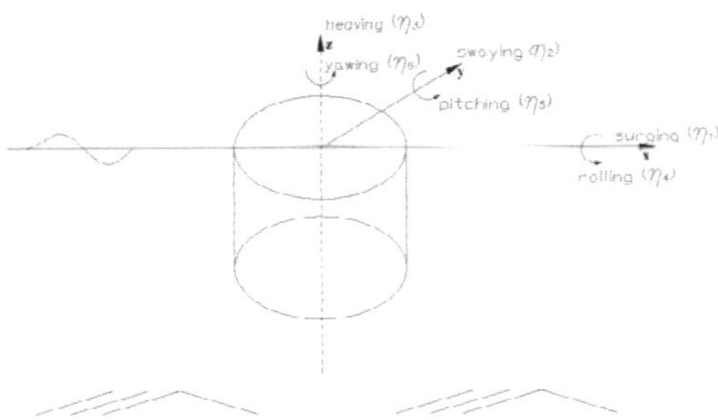

Figure 1. The DOF motion of float

2.1.2. The potential flow theory. Based on the theory of three-dimensional potential flow, the hydrodynamic analysis of float is carried out under regular wave. Need to be satisfied: water is an ideal and incompressible fluid without spinning, the wave is a small amplitude wave. The velocity potential should satisfy the following conditions:

Control equation: $\nabla^2\phi = 0$ (1)

Free surface boundary condition: $-\omega^2\phi + g\dfrac{\partial\phi}{\partial z} = 0, \ z = 0$ (2)

Seabed boundary condition: $\frac{\partial \phi}{\partial n} = 0, \quad z = -h$ (3)

Free surface motion condition after linearization: $\frac{\partial \zeta}{\partial t} - \frac{\partial \phi}{\partial z} = 0$ (4)

Free surface dynamic condition after linearization: $\frac{\partial \phi}{\partial t} + g\zeta = 0$ (5)

In general, the above conditions are solved by using Green function method which establishes the integral equation between the velocity potential and Green function. In solving the integral equation, the boundary element method is used to discrete the object surface for a certain number of units, then to calculate the strength in the center of each unit (called source point) under the boundary conditions, and to obtain the velocity potential at each node with the pulse source distribution.
After the radiation and diffraction velocity potential of the structure are calculated, the first order hydrodynamic pressure can be expressed as:

$$p = -\rho \frac{\partial \phi}{\partial t} \tag{6}$$

The total wave force is obtained by integrating the wet surface of the object:

$$F = \iint_S pn dS \tag{7}$$

Here, S is wet surface of the object; generally speaking, the first wave force consists of three parts:

$$F = F_{HS} + F_R + F_{EX} \tag{8}$$

Here, F is the first wave force; F_{HS} is rigid hydrostatic restoring force; F_R is Radiation force; F_{EX} is the first wave excitation force and moment which are generated in the interaction of wave incidence and diffraction velocity potential.

2.1.3. The frequency domain control equation. In the study of the hydrodynamic response of the floating body, it is generally considered to be a rigid body with six degrees of freedom. According to Newton's second law, the buoy maintains a static equilibrium under multiple forces which are the static water restoring force, the inertial force, the damping force and the wave excitation force. Based on the three dimensional potential flow theory, considering the action of hydrostatic restoring force and wave exciting force (summation of the incident and diffraction forces), the boundary element method is used to calculate the buoy motion response in frequency domain, frequency control equation under linear regular wave is given:

$$(M + \mu(\omega))\ddot{X} + C(\omega)\dot{X} + KX = F(\omega) \tag{9}$$

Here, M is the mass matrix, μ is the added mass matrix, C is the damping coefficient matrix, K is the static water restoring stiffness matrix, F is the wave exciting force matrix, X is the motion response matrix. The additional mass μ, potential flow damping coefficient C and wave excitation force F can be calculated by the potential flow theory.

2.2. Geometric Modeling

2.2.1. 3-D model. Using AutoCAD 2011, three kinds of floating models with different shapes (flat cylindrical, conical bottom cylindrical and ball) were established and the shape parameters were changed. Then, respectively import models into the AQWA module in ANSYS 15.0 for frequency domain analysis.

2.2.2. Mesh generation. Using AQWA to automatic meshing: the maximum size of the grid element is 0.2m, the defeaturing tolerance is 0.1m, the maximum allowed frequency is 1.314HZ, and the meshing type is selected as combined meshing, and the quadrilateral grid is obtained. It is noteworthy that it is not yet possible to employ symmetry mesh in AQWA. The grid divisions of three kinds of float are shown in Figure 2, Figure 3 and Figure 4.

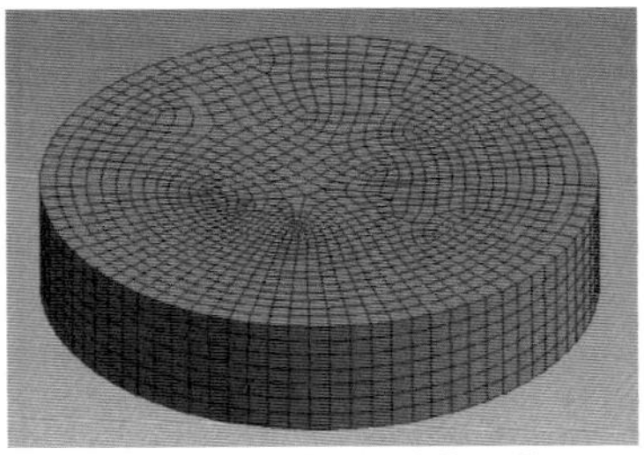

Figure 2. Meshing of flat float

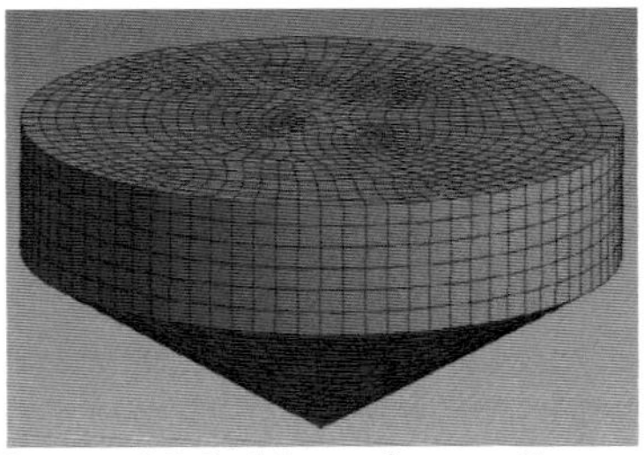

Figure 3. Meshing of cone float

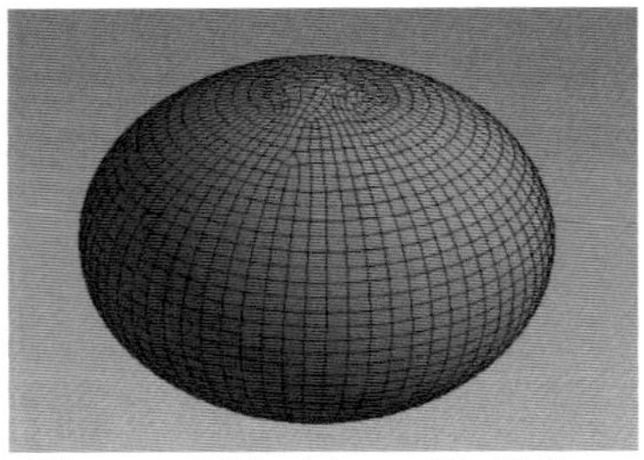

Figure 4. Meshing of ball float

At the same time of external environmental load, the heaving responses of floats with different sizes, different weight and different shapes are analyzed by using the AQWA. Comparing the difference between wave amplitude and heaving response amplitude of various floats, the float model which has the best flow-ability and can be used to determine the optimum shape parameters for oceanic wave buoy is obtained.

2.3. *Calculating Case*

The working performance of wave buoy is influenced by many external factors, such as weight, shape parameter, shape, wave factor and so on. Compared with the European waters, the characteristics of Chinese waters is short period and low-height. In order to achieve more accurate detection in offshore waters of China, we need to design the buoy which is suitable for short period and low height. In this paper, in order to determine the relatively superior floating shape, the floats of 25 working conditions (see Table 1, Table 2 and Table 3) are designed to analyze the floating flow-ability with different parameters under the same conditions of external environmental loads. In the table, the submerged depth refers to the height of the bottom of the float to the static water surface.

Table 1. Working conditions of flat float

Model	Height (m)	Diameter (m)	Weight (kg)	Submerged depth (m)	Grid numbers
1-1	0.8	3.2	1000	0.121	2874
1-2	0.8	3.2	1500	0.182	2904
1-3	0.8	3.2	2500	0.303	2958
1-4	0.8	3.2	3500	0.425	2791
1-5	0.8	3.2	5500	0.667	2791
1-6	0.8	2.4	2500	0.539	1919
1-7	0.8	4.0	2500	0.194	4225
1-8	1.6	3.2	2500	0.303	3295
1-9	2.4	3.2	2500	0.303	4142

Table 2. Working conditions of cone float

Model	Height (m)	Diameter (m)	Taper (°)	Weight (kg)	Submerged depth (m)	Grid numbers
2-1	1.6	3.2	120	1000	0.677	2424
2-2	1.6	3.2	120	2500	0.919	2631
2-3	1.6	3.2	120	4500	1.162	2480
2-4	1.6	3.2	120	6500	1.389	2480
2-5	2.0	3.2	90	2500	1.325	2571
2-6	1.2	3.2	150	2500	0.589	2345

Table 3. Working conditions of ball float

Model	Diameter (m)	Weight (kg)	Submerged depth (m)	Grid numbers
3-1	3.2	1500	0.575	2616
3-2	3.2	2500	0.759	2897
3-3	3.2	4500	1.059	3078
3-4	3.2	7500	1.443	3202
3-5	3.2	11500	1.933	3195
3-6	3.2	16500	2.716	2464
3-7	2.4	2500	0.935	1813
3-8	2.8	2500	0.832	2329
3-9	3.6	2500	0.704	3360
3-10	4.0	2500	0.661	3909

3. Result and Analysis

Set the sea water depth to $h = 15m$, seawater density to $\rho = 1025\,kg/m^3$, the gravity acceleration to $g = 9.8\,N/kg$ and the wave frequency which is calculated to $\omega = 0\sim5rad/s$. And set the wave amplitude to 1m. The RAOs (amplitude response operator) about heaving of each float in different wave frequency is calculated by using AQWA.

3.1. The Influence of Float Weight

Select model 1-1, 1-2, 1-3, 1-4 and 1-5 to analyze the effect of float weight on the heaving motion of the flat-bottomed float, and the RAOs about heaving varies with frequency are shown in Figure 5. When the wave frequency is less than 1 rad/s, the amplitude of the heaving motion of different weight float is consistent with the wave amplitude (1m). But when the wave frequency is between 1 ~3rad/s, only the model 1-3 is still close to the wave amplitude, has better flow-ability and more accurate monitoring of wave data. The more the weight is in the high frequency range, the more likely it is to resonate.

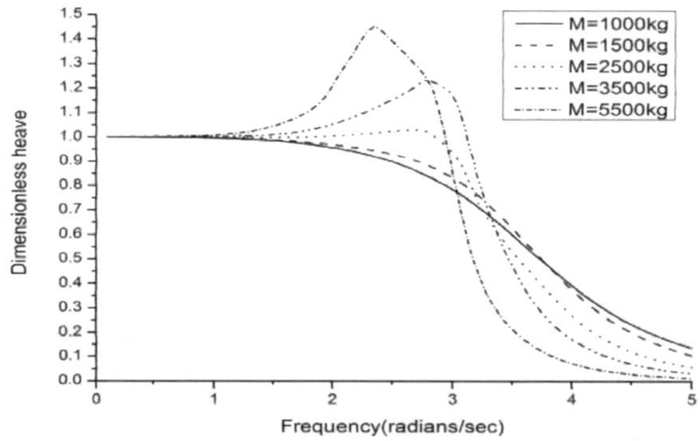

Figure 5. Effect of mass on heaving amplitude of flat float

As shown in Figure 6, model 2-1, 2-2, 2-3 and 2-4 are used to analyze the effect of mass on heaving motion of the cone bottom floater. The tendency of the RAOs varies with the wave frequency is the same as flat bottomed float. When the wave frequency is 1 ~ 3rad/s, only the model 2-2 is still relatively close to the wave amplitude, and the flow-ability is better than others.

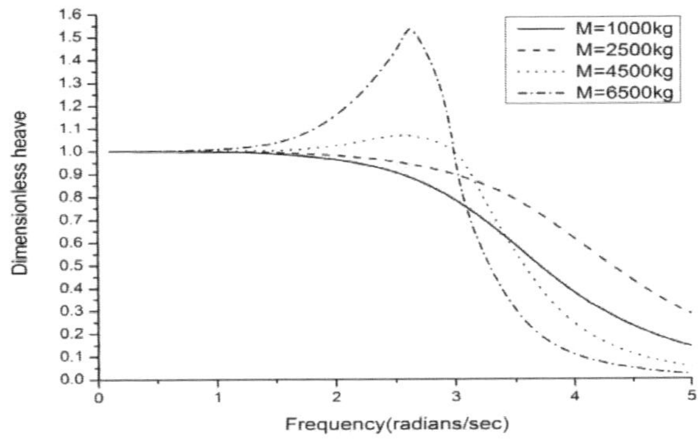

Figure 6. Effect of mass on heaving amplitude of cone float

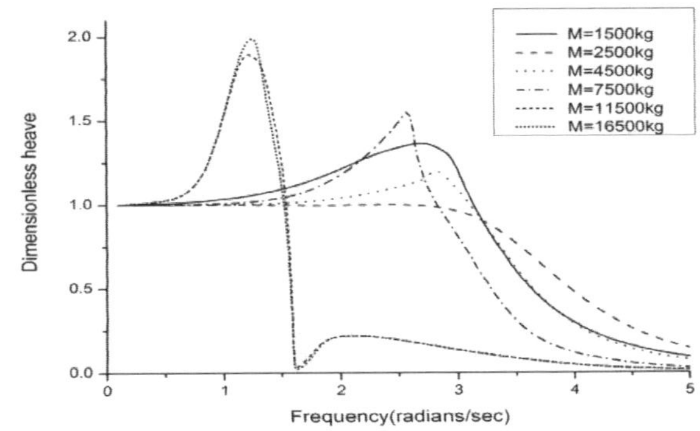

Figure 7. Effect of mass on heaving amplitude of ball float

As shown in Figure 7, model 3-1, 3-2, 3-3, 3-4, 3-5 and 3-6 are chosen to analyze the effect of mass on heaving motion of ball floater. The over-weight ball float resonates with wave in the low frequency region. When the wave frequency is 1 ~ 3rad/s, only the model 3-2 is relatively close to the wave amplitude, and the flow-ability is better than others.

According to the above analysis shows: when the float hydrostatic surface were of round cross-section, whatever shape of the floating bottom are, the trend of float work performance with the weight change rule is roughly same under the same wave frequency, that is to say, the trend of float heaving amplitude with weight is the same. The larger the difference between float amplitude with wave amplitude is, the more inaccurate the buoy monitoring data is. Therefore, the buoy whose heaving amplitude is the closest to the wave amplitude can be chosen, its flow-ability is the best and the motion is relatively stable. Compared with the working performances of 15 selected models, model 1-3, 2-2 and 3-2 are better than others.

3.2. The Influence of Float Shape Parameters

The influence of float diameter on the heaving motion is analysed by selecting model 1-3, 1-6 and 1-7, and the RAOs about heaving varies with frequency are shown in Figure 8. Under the condition of the same quality, the smaller float diameter is, the greater heaving amplitude in high frequency range is, and the more unstable heaving motion is. Float whose diameter is lager is easy to resonate with wave, resulting in inaccurate measurement results. When the wave frequency is less than 2.5 rad/s, the amplitude of a model with a diameter of 3.2m is the closest to wave, and the monitoring result is the most accurate.

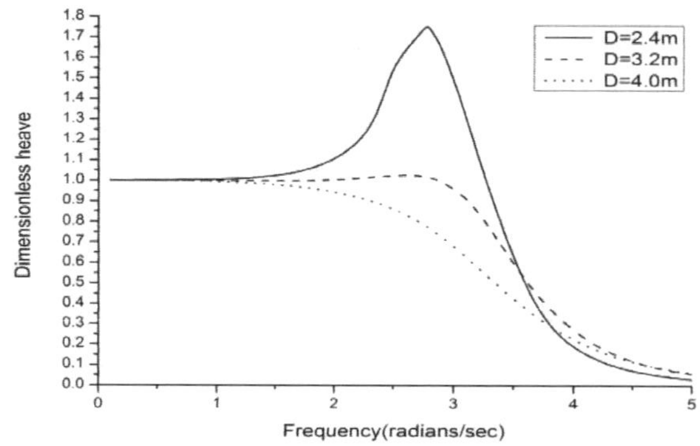

Figure 8. Effect of diameter on heaving amplitude of flat float

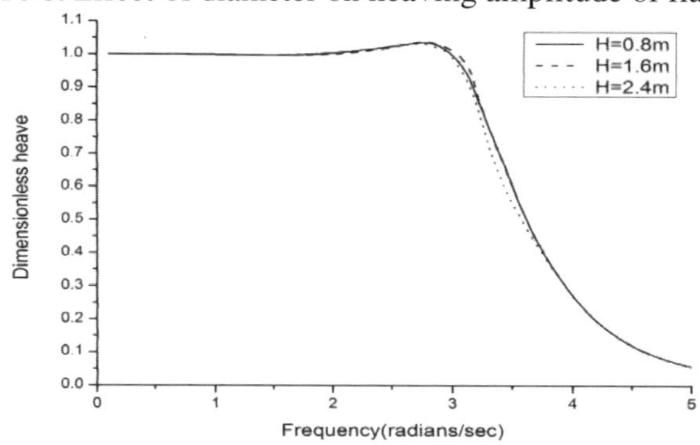

Figure 9. Effect of height on heaving amplitude of flat float

The influence of the float height on the heaving motion is analysed by selected model 1-3, 1-8 and 1-9, and the RAOs about heaving varies with frequency are shown in Figure 9. With the same diameter and quality, the tendency that heaving amplitude of flat float with different heights varies with the wave frequency are roughly coincide. When wave frequency is less than 3 rad/s, calculated results shown that all models are close to wave and monitoring are more accurate. Therefore, the effect of height on float flow-ability can be ignored.

The influence of the float cone angle on the heaving motion is analysed by selecting model 2-2, 2-5 and 2-6, and the results were shown in Figure 10. When wave frequency is 0 ~ 2.8 rad/s and the cone Angle is 120 °, the heaving amplitude is the closest to wave.

The influence of spherical diameter on the heaving motion is analysed by selecting model 1-2, 1-7, 1-8, 1-9 and 1-10, and the results are shown in Figure 11. When the wave frequency is less than 3.0 rad/s, the amplitude of a model with a diameter of 3.2m is the closest to wave, and the monitoring result is the most accurate.

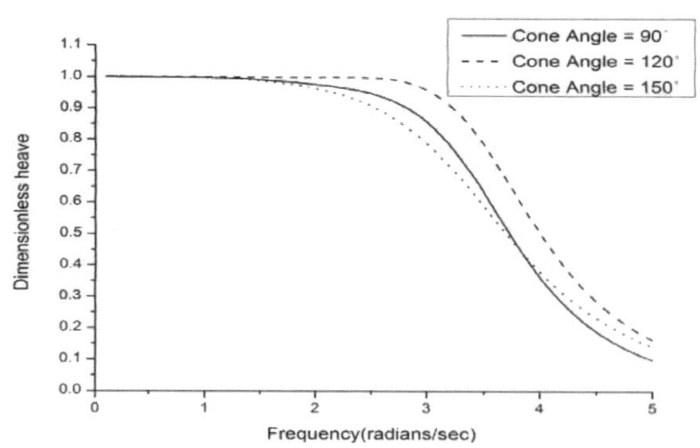

Figure 10. Effect of angle on heaving amplitude of cone float

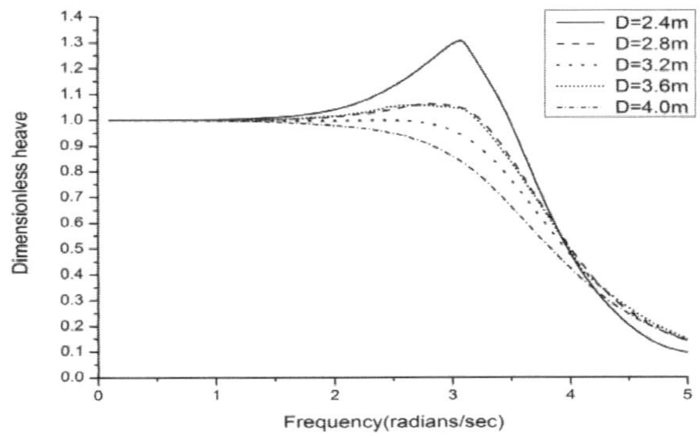

Figure 11. Effect of diameter on heaving amplitude of ball float

3.3. The Influence of Float Shape

Select model 1-3, 2-2 and 3-2 to analyze the effect of float shape on the heaving motion, as shown in figure 12. The distribution curve of floating heaving amplitude with different shapes changing with wave frequency is roughly the same, and the amplitude decreases with the increase of wave frequency. However, the heaving amplitude of 3-2 is more similar to the wave amplitude in the longer frequency range, that is to say, it is more accurate to monitor wave data in short period.

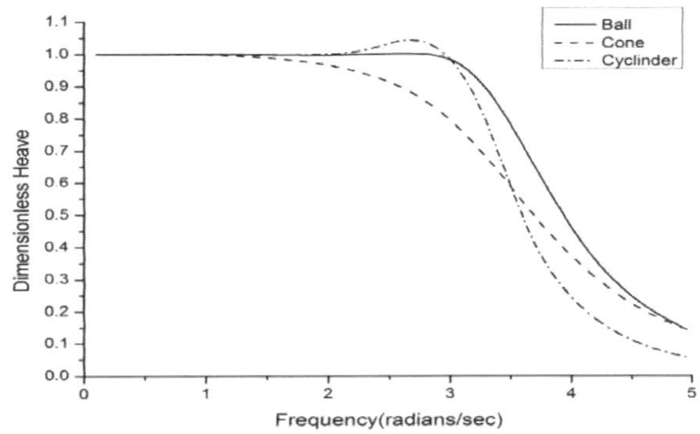

Figure 12. Effect of shape on heaving amplitude of float

3.4. Results Summary

According to the above analysis, the heaving amplitude of each float that is compared with wave amplitude (this paper is set as 1m) when the wave frequency is 2.8rad/s, the results are shown in Table 4, and these results are took as a sign to determine which has better flow-ability (the smaller the difference between heaving motion and wave is, the better flow-ability is) under the high wave frequency.

Table 4. The relative amplitude of each float model

(The relative amplitude refers to the difference between the heaving amplitude and the wave amplitude)

Model	Heaving amplitude	The relative amplitude	Model	Heaving amplitude	The relative amplitude	Model	Heaving amplitude	The relative amplitude
1-1	0.85	0.15	2-1	0.85	0.15	3-1	1.33	0.33
1-2	0.90	0.10	2-2	0.92	0.08	3-2	1.00	0.00
1-3	1.04	0.04	2-3	1.08	0.08	3-3	1.14	0.14
1-4	1.24	0.24	2-4	1.24	0.24	3-4	1.20	0.20
1-5	1.27	0.27	2-5	1.92	0.92	3-5	0.16	0.84

1-6	1.72	0.72	2-6	0.85	0.15	3-6	0.16	0.84
1-7	0.78	0.22				3-7	1.23	0.23
1-8	1.04	0.04				3-8	1.07	0.07
1-9	1.04	0.04				3-9	1.07	0.07
						3-10	0.91	0.09

The statistics of the above table show that: five models (model 1-4, 1-5, 1-6, 2-4 and 2-5) resonate with wave frequency of about 2.5 rad/s and two models (model 3-5 and 3-6) resonate with wave frequency of about 1.2 rad/s, leading to the inaccuracy of measuring the wave data. And only model 3-2 has the smallest relative amplitude in the wave frequency of 2.8 rad/s. Considering the float height has no influence on the heaving amplitude, so the model 3-2 with the best stability is chosen as the initial selection of the oceanic data buoy.

4. Summary and Prospect

This paper selected the float model for biaxial symmetry structure, and the motion of other degrees has no coupling effect on heaving. The single degree of heaving model can accurately describe the heaving response. In this paper, frequency domain analysis of float was carried out under the regular wave with the AQWA hydrodynamics calculation model and the potential flow theory, and the heaving amplitude were calculated by using numerical analysis method. 25 models with three shapes are chosen as examples to carry on the numerical simulation, is used to analysis the effect of float weight, diameter, height, taper angle and shape on the heaving amplitude.

The results show that: the larger the float is, the larger the heaving amplitude is, but the overall trend is roughly the same; the influence of float diameter and bottom taper angle on heaving amplitude is larger, and the variation rule is nonlinear; the performance of float is not influenced by float height; in comparison, model 3-2 has better flow-ability in the longer wave frequency range, which is more suitable for the measurement of China sea with short period wave.

According to the analysis results, this paper completed the preliminary selection of wave data buoy, and laying the foundation for the optimization of the shape parameters.

In this paper, the heaving motion characteristics of float are studied under regular wave, but the real marine condition is extremely complicated. Therefore, it is necessary to further simulate the working performance of float under irregular waves. In addition, this paper only analysed the motion characteristics of free-float in regular wave, without considering the coupling effect of buoy and mooring system.

5. References

[1] Wang D J and Sun S P 2013 *Applied Mechanics & Materials* 477-478 254-258
[2] Leonard J W, Idris K and Yim S C S 2000 *Ocean Eng.* 27(12) 1345-71
[3] Monroy C, Ducrozet G, Bonnefoy F, Babarit L, Gentaz and Ferrantet P 2011 *Int. J. Offshore Polar* 21(4) 264-271
[4] Salem A G, Ryu S, Duggal A S and Datla R V 2009 *J. Offshore Mech. Arct.* 134(1) 907-911
[5] Wang Z Z 1988 *Shandong Sci.* 1988(1) 31-39
[6] Miao Q M, Gu M, Yang Z M and Liu C X 2003 *J. Ship Mech.* 7(5) 21-27
[7] Fan X T, Shao M, Sun J W and Liu S X 2014 *Ocean Technol.* 33(1) 29-32
[8] Wang H J, Sun J W, Shao M and Fan X T 2016 *Shandong Sci.* 6 1-8
[9] Zhang J M, Fan X T, Zhao Q, Li X Q and Zheng S S 2015 *Shandong sci.* 28(4) 8-13
[10] Zhang J M, Fan X T, Zhang S G, Wan X Z and Sun J W 2014 *Shandong sci.* 2 19-24

IOP Conf. Series: Earth and Environmental Science **171** (2018) 012005 doi:10.1088/1755-1315/171/1/012005

A Capable Smart Sensor Interface Module for Interoperable Ocean Observatories

Shijun Lin[1] and Feng Lyu[1, 2]

1 School of Ocean and Earth Science, Tongji University, No. 1239, Siping Road, Shanghai, China
2 E-mail: LF@tongji.edu.cn

Abstract. The rapid development of ocean observatories results in a great challenge to integrate so many sensors into one system. Some solutions have been proposed to address this issue, such as MBARI PUCK, IEEE 1451 and OGC SWE. This paper proposes a novel smart sensor interface module solution which is more capable than PUCK. As the interface standardization is achieved at the instrument-side in our solution, the design of ocean observatory systems is simplified. What's more, the feature of distributed intelligence of the module can benefit many in-situ applications of ocean observatories.

1. Introduction

The rapid development of ocean observatory systems results in dramatic increase in sensor deployments. However, it is a great challenge to integrate so many instruments into one system.
The challenge most comes from the fact that there are no widely recognized interface protocols for undersea sensors. The differences of these interface protocols are summarized as below:

- Electrical interface: Undersea sensors adopt various communication interfaces, such as RS232, RS422, RS485 and Ethernet. So multiple interfaces should be designed and reserved in junction box for compatibility. Their pin definitions and input voltages are also different among sensors.
- Communication protocol: The communication protocols of undersea sensors from distinct manufacturers tend to be totally different. It costs a lot of effort to interact with all sensors in a similar manner.
- Data format: The data retrieved from different undersea sensors, even from different versions of the same sensor, are different in formats. They may be delimiter-separated text, XML-based text or binary data, making it more difficult to integrate the increasing number of data streams into a system.
- Intelligence: Sensors have different levels of intelligence. Some may just output an analog signal; some may support scheduled-sampling, PTP protocol, and even a built-in web server.

So a lot of troubles exist in the design, extension, update, and maintenance of ocean observatory systems. In addition, the manual processes[1] to customize and configure hardware and software for each version of sensors introduce large possibility of human-error. In ocean observatories, the cost of an error could be very high due to the high price of undersea sensors themselves, as well as their recovery and maintenance[2].

The standardization benefits the plug-and-play procedures, and promotes observatories a simpler design and higher reliability.

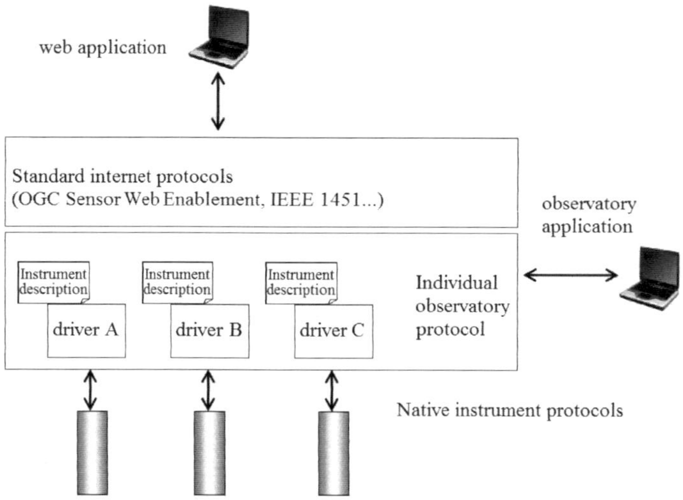

Figure 1. Interoperable instrument protocol layers[3].

Current solution to achieve the standardization of instrument interfaces is to add an interoperable protocol layer to the software architecture of an observatory[3], as shown in figure 1.
Typically, to standardize the interface protocol, we will need "driver" for each type of instruments. Drivers are responsible for translating between native instrument protocols and standard protocols[4]. Customizing driver for native instrument protocols is a major but inevitable cost of this solution. In addition, we need description files to identify each sensor, so that the upper software knows what kind of parameter it measures, its function, what it is, and how to interact with it.
The individual observatory protocol layer is not mandatory; it's just a decision of design. Once everything is standardized in one layer, to implement map between two protocols is easy.
Since it seems unrealistic to persuade all vendors to implement same interface protocol locally in their instruments, we need an alternate to achieve the standardization, which turns out to be the smart sensor interface module.
The smart sensor interface module can be embedded in or attached to the target instruments to allow additional capability.

2. Standard Interface Protocols

2.1. MBARI PUCK

MBARI PUCK (programmable underwater connector with knowledge) is considered to be the minimum implementation of a smart sensor interface module; it does not implement standardization itself, but provides the way to achieve it.
The PUCK defines a simple embedded instrument protocol, so that the host can store and retrieve data from a PUCK-enable instrument. The data consist of a predefined minimal metadata and an optional "payload". The payload can store any data, such as the components of other standards or the executable driver code. So, the description file for implementing interoperable instrument protocol layers is stored in the predefined datasheet and optionally in the payload. The driver code can be stored in the payload or gotten from the database according to the ID of instrument, which is also a solution for description file.

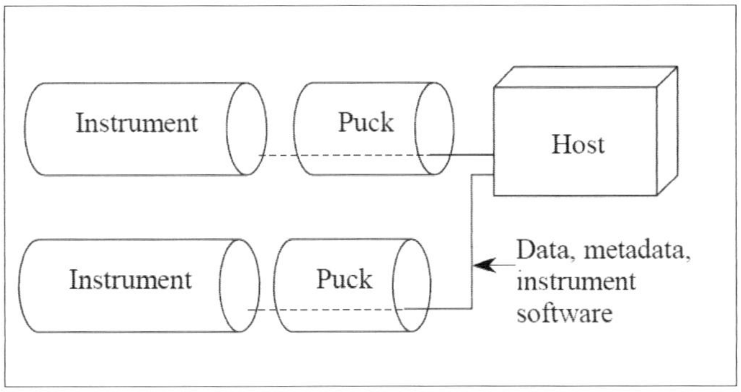

Figure 2. PUCK connecting the host and the instrument[5].

The PUCK can be embedded in an instrument. But for most commercial instruments, it will be a module connecting instruments to a host, as shown in figure 2. It adds additional functions to a traditional instrument. Once a PUCK-enabled instrument is plugged in, the host will detect it by rolling ports, and communicate with it via the PUCK protocol to retrieve data. Once getting all the data needed, the PUCK will be bypassed and the host will communication with the instrument directly[6].

The PUCK essentially doesn't implement standardization itself, but provide a way to identify the instruments; the standardization is completed in the host, where the driver runs. So the reliability and the ability lay on the host and the communication link.

The PUCK has been implemented on RS-232 and Ethernet[7], and is being used on MBARI moored and cable-to-shore observatories [8].

2.2. *IEEE 1451 and OGC-SWE*

The IEEE (Institute of Electrical and Electronics Engineers) 1451 smart transducer interface standards defines a set of open, network-independent communication interfaces for connecting transducers[9]. So the control software can automatically detect a transducer and also issue commands to interact with the transducer. An IEEE 1451 smart transducer has two parts: A Smart Transducer Interface Module (STIM) and a Network Capable Application Processor (NCAP). The predefined description file in STIM, i.e. Transducer Electronic Data Sheet (TEDS), provides the information needed to identify and connect the device.

The OGC-SWE (Open Geospatial Consortium - Sensor Web Enablement) defines an open standard framework to exploit sensors via the Web. Several specifications have adopted in the SWE framework[10], including some XML-based specifications for description, such as SensorML, and some for the standard web service. What's more, the PUCK has been adopted as part of the SWE, and the SWE offers "Hooks" for IEEE 1451 and other standards.

The IEEE 1451 and OGC-SWE are not customized for ocean observatories; they are quite comprehensive but also very complex. Though some work has been done[11], people tend to combine them with the PUCK solution to achieve interoperability [3, 4, 12], rather than locally implement these complex standards completely in the firmware of traditional ocean instruments.

3. Capable smart sensor interface module

Refer to the PUCK protocol, we have designed that a more capable smart sensor interface module, named smart instrument adapter (SIA), which may be more suitable for the complex marine environment.

Suppose the connection to the host is lost or blocked on an ocean observatory, the data retrieved may be lost, but a smart instrument is capable to detect this event and take actions, such as buffering the data or reducing the sampling frequency and waiting for communication restored, and then retransmit the data.

3.1. *The Concept of SIA*

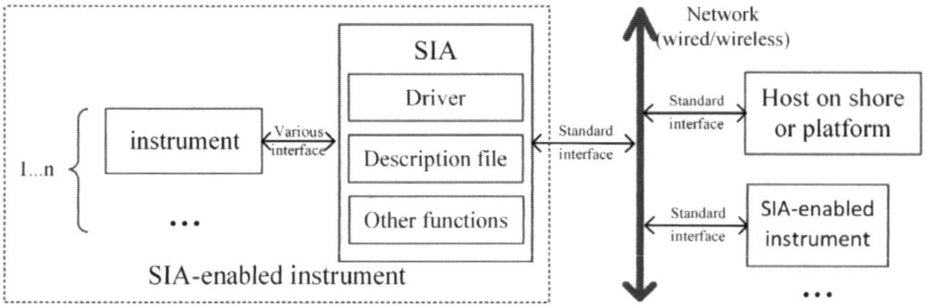

Figure 3. The concept of SIA.

The SIA aims to achieve standardization at the instrument-side, rather than at the host-side as in the PUCK solution, leading to a more capable interface module as shown in figure 3.
Like the PUCK, the SIA can be embedded in instruments or be an external double-end firmware. One end is connected permanently to the corresponding instrument(s) and the other end is used to communicate with the external objects. The SIA will not be passed by as the PUCK do, but be the middleware between the traditional instruments and the observatory network.
To the observatory network, instruments with an SIA are treated as a standard SIA-enabled instrument. The process of adaptation is transparent to the network. So the driver is running on the SIA, the upper software on the host can interact with the SIA-enabled instrument in a standard way, e.g. getting the formatted data. It simplifies the design of the software architecture and enables the "plug-and-play" function.
The SIA is featured by distributed intelligence. The relation between a host and a SIA-enabled instrument is not "master-to-slave" as the PUCK protocol, but "peer-to-peer". From the view of the SIA, there is no difference among the host and other SIAs. So these networked SIAs can interact with each other and that's the reason why SIA-enabled instruments don't connect directly to the host in figure 3.

3.2. The "plug-and-play" process

The "plug-and-play" process of a SIA can be summarized as follow:

1. A SIA-enabled instrument is plugged in and powered.
2. The SIA configures itself and establishes a connection with the host.
3. The host reads the description file from the SIA and knows the capability of the SIA-enabled instrument.
4. The host requests services from the SIA-enabled instrument.
5. The SIA-enabled instrument receives and responses to these requests.

The establishment of connection can be passive by rolling, or be active by connecting to certain IP address or broadcasting.

3.3. Considerations

3.3.1. Connectivity. The "interface" in the figure 3 refers not only to the software protocol, but also the electrical interface. In order to accommodate various electrical interfaces, many interfaces will be reserved and there will be a power adapter module in the firmware of a SIA; but the electrical interface of the other end can be uniform, e.g. 48V and Ethernet ports, leading to the simplified design of the science instrument interface module (SIIM)[13]. The SIIM is only needed to provide standard interfaces to connect SIAs.

The traditional instruments are hard wired to the SIAs, while the connection between the SIAs and the network can be either wired or wireless, e.g. acoustic or optic links.

3.3.2. Standard communication. We have designed a novel observatory protocol for the SIA, and the possibility of implementing part of the IEEE 1451 standard is under evaluation.
Considering the variety of instruments in the ocean observatories, we need a flexible method to describe the interfaces of the SIA-enabled instruments. So in the current, the description file is based on XML, and the XML-based SensorML is also being considered.
The data is formatted to XML and attached with useful meta-data, so that it enable upper software to process the data by a uniform tool set.

3.3.3. Functionality. Benefit from the capable design, the SIA can not only encapsulate the existing functions of instruments, but also add valuable functions. By this way, various instruments can have similar function packages, including but not limited to:

- Positioning and timing: The positioning and timing are the foundation of many important scientific tasks in an ocean observatory system, but so far few instruments support these two functions. The SIA provides a platform to enable precise positioning and timing for each SIA-enabled instrument.
- Reliable transmission: The SIA can ensure a reliable data transmission by buffering data and also the retransmission mechanism.
- Warning mechanism: The SIA can detect the working status of the connected instruments, benefiting the fault detection of the system.
- Hybrid communication: The SIA can use a mixture of communication interfaces, such as Ethernet and acoustic, so that the communication can be more reliable.

3.3.4. Reliability. Thanks to the distributed nature of SIAs, the reliability of the whole system is relative high. One component failure will not crash the whole system; while the failure of the host controller will disable all PUCK-enabled instruments on the platform.
To improve the reliability, the function of remote upgrade is implemented in the software of the SIA. When bugs are discovered, we can fix it remotely.

3.3.5. Power consumption. As the SIA circuit will not be passed by, there will be extra power consumption while operation. But, the SIA can also power down the connected instruments and get into sleep mode to save energy. It is not an issue in a cabled seafloor observatory which has abundant power. But in a moored observatory or in wireless sensor networks, where power is limited, we need to do a trade-off design.

3.4. Compare with PUCK

A comparison between SIA and PUCK is given in table 1.

Table 1. Comparison between SIA and PUCK.

Item	SIA	PUCK
Where to achieve standardization	Instrument-side	Host-side
Numbers of connected instruments	One or more	One
plug-and-play	Support, passive or active	Support, passive through rolling by host
Active time	The whole work time	First time be plugged in, and then be passed by.
Electrical interface	Mixed	RS232 or Ethernet

Intelligence	High, enhance the original instruments. Individual intelligence for each SIA-enabled instrument.	Low, just provide the binding with meta-data. Intelligence can be provided by the software running on the host.
Power consumption	Low, but a little bit higher than PUCK	Low, to be powered down after the information transmitted.
Reliability	High from the prospective of the whole system	High

4. Other applications

Many applications of ocean observation can benefit from SIA.

4.1. Underwater Wireless Sensor Network

Wireless sensor networks have encountered great difficulties when implemented underwater[14]. The SIA meets the need of underwater wireless sensor networks and its capability can be expended to implement the latest acoustic technology of networking, positioning and timing[15]. It can be used as an access point in the underwater wireless sensor network, enabling the traditional instruments to access the wireless network. And, the SIA connected to the seafloor observatory can act as a gateway to provide communications to the nearby access points.

4.2. Cooperative sensing

Because of the communication delay, it is difficult to achieve real-time manipulation of the instruments on the underwater wireless sensor network. If the host loses connection with the underwater instruments, the important data may be lost.

These SIA-enabled instruments can interact with each other, so that they can cooperate to achieve one goal. Based on the concept of swarm intelligence[16], many individuals with small intelligence can interact to find the best answer[17].

In the case of the ocean observatory system, a SIA-enabled AUV can require relevant observation from nearby SIA-enabled instruments. These instruments can response the needed parameters along with their locations, and then the AUV can adjust the moving direction and finally locate the special event, such as algal blooms to be detected.

Another example comes that when an earthquake is detected by a SIA-enabled instrument, it can notify other relative instruments to speed up the sampling, so that scientists can get more valuable data about this event.

5. Conclusion

The paper has proposed a capable smart sensor interface module for ocean observatory system to address the challenge of integration of various traditional ocean instruments. Comparing to the PUCK solution, the capability of the SIA is considered to provide more possibilities for ocean observation. Many applications can benefit from the feature of distributed intelligence, and the SIA promises ocean observatory a simpler and more robust design.

6. References

[1] Toma D M, O'Reilly T, Rio J D and Headley K 2011 Smart sensors for interoperable Smart Ocean Environment *Oceans* p 1-4

[2] Pirenne B 2015 The role of Information Communication Technologies (ICT) for seafloor observatories: Acquisition, archival, analysis, interoperability *SEAFLOOR OBSERVATORIES* (Springer Berlin Heidelberg) p 129-154

[3] O'Reilly T C, Headley K, Edgington D R and Rueda C 2009 Instrument interface standards for interoperable ocean sensor networks *OCEANS 2009-EUROPE* p 1-10

[4] Rı´O J N D, Toma D M, O'Reilly T C, Bröring A, Dana D R, Bache F, Headley K L, Mànuel-Làzaro A and Edgington D R 2014 Standards-Based Plug & Work for Instruments in Ocean Observing Systems *IEEE Journal of Oceanic Engineering* 39(3) p 430-443

[5] Headley K L, Davis D, Edgington D and Mcbride L 2003 Managing sensor network configuration and metadata in ocean observatories using instrument pucks *The International Workshop on Scientific Use of Submarine Cables and Related Technologies* p 67-70
[6] O'Reilly T C, Edgington D, Davis D and Henthorn R 2001 "Smart network" infrastructure for the MBARI ocean observing system *Oceans* p 1276-1282 vol.2
[7] Valls O P 2013 IP PUCK Implementation and Verification *http://www.mbari.org/wp-content/uploads/2015/10/Pallares.pdf*
[8] O'Reilly T C, Headley K, Graybeal J, Gomes K J, Edgington D R, Salamy K A, Davis D and Chase A 2006 MBARI Technology for Self-Configuring Interoperable Ocean Observatories *Oceans* p 1-6
[9] Lee K 2000 IEEE 1451: A standard in support of smart transducer networking *Instrumentation and Measurement Technology Conference, 2000. Imtc 2000. Proceedings of the IEEE* p 525-528 vol.2
[10] Botts M, Percivall G, Reed C and Davidson J 2008 OGC® Sensor Web Enablement: Overview and High Level Architecture 4540 p 372-380
[11] Pearlman J, Jirka S, Rio J D, Delory E, Frommhold L, Martinez S and O'Reilly T 2016 Oceans of Tomorrow sensor interoperability for in-situ ocean monitoring *Oceans* p 1-8
[12] Waldmann C, Delory E, Phillips R, Cater N, Schleisiek K, Río Fernandez J d, O'Reilly T, Headley K, Edgington D and Mihai Toma D 2009 Evaluation of MBARI PUCK protocol for interoperable ocean observatories *Instrumentation viewpoint* (8) p 87-87
[13] Lyu F and Zhou H 2016 Progress of Scientific Cabled Seafloor Observatory Networks *Journal of Engineering Studies* 8(2) p 139-154 (in Chinese)
[14] Heidemann J, Ye W, Wills J, Syed A and Li Y 2006 Research challenges and applications for underwater sensor networking *Wireless Communications and NETWORKING Conference, 2006. WCNC* p 228-235
[15] Heidemann J, Stojanovic M and Zorzi M 2012 Underwater sensor networks: applications, advances and challenges *Philosophical Transactions* 370(1958) p 158
[16] Blum C and Merkle D 2008 *Swarm Intelligence: Introduction and Applications* (Springer Publishing Company, Incorporated) p 43-85
[17] Park H, Han J H and Kim J H 2010 Swarm intelligence-based sensor network deployment strategy *IEEE Congress on Evolutionary Computation* p 1-6

Acknowledgments

This work was supported by Shanghai science and technology innovation action plan fund through project 16DZ1205000 "Research on the power system of large-scale scientific cabled seafloor observatory networks".

Switching High Voltage DC Power in Branching Units of Large-scale Cabled Seafloor Observatories

Shaoduo Zheng [1] and Feng Lyu[1*]

1 School of Ocean and Earth Science, Tongji University, Shanghai, CN
E-mail: LF@tongji.edu.cn

Abstract. Cabled Seafloor Observatories(**CSOs**) are powered by shore stations with DC voltage levels up to -10kV. When a backbone or spur cable fault occurs, the power system must isolate the fault by disconnecting the faulty segment quickly. So the BUs installed in the backbone cable need to carry out the switching operation under high voltages(HV). However, DC circuits have no zero-current points and the fault currents will rise rapidly, bringing a big challenge to the switching on/off operation. In this paper, we have purposed an HVDC circuit breaker based on series-connected IGBTs with voltage balance. Under the backbone voltage of -10kV and a peak current of around 14A, the simulation results show that the proposed design is feasible, with the switching-off time about 1ms.

1. Introduction

Cabled Seafloor Observatories(**CSOs**) can on-line observe the deep ocean with real-time and in-situ experiments. Large-scale CSOs consist of shore stations, primary and secondary junction boxes, instrument platforms and electro-optic cables, as well as branching units(BUs) and repeaters as backbone devices [1]. Shore stations provide abundant power to ensure the normal operation of the system. A BU connects the backbone cable with a spur cable which is linked to a primary junction box. In case of a backbone or spur cable fault occurs, the power switches in the nearest BUs must be opened to isolate the fault so that the system will remain normal operation.

In traditional CSOs, the entire system will shut down due to voltage collapse when a cable fault occurs. Then the system enters a fault location mode, and restarts with the backbone voltage at a low positive voltage of +500V. The shore stations make voltage and current measurements and the system diagnoses and locates the fault cable sections, then enters the fault isolation mode [1]. During the fault locating and isolating process, the undersea observatory will stop working. The system will be restored operation after isolating the fault.

The biggest challenge is to achieve on-line HVDC power switching in each branch node of CSOs, for the system can remain operation during a fault. So the HVDC switching unit in BU is critical for the protection and maintenance of CSOs. In this paper, a switching circuit based on series-connected IGBTs is proposed for fast fault isolation.

2. High Voltage Power Switching in BUs

2.1. Technical Features

During the normal situation, the backbone current of CSOs is about 0.5-10A under -10kV. Designing a switching circuit with the following technical features is quite important for reliability consideration.

- Detecting and isolating faults within a few milliseconds and restoring the energy in the system in a very short time after clearing the fault [2].

IOP Conf. Series: Earth and Environmental Science **171** (2018) 012006 doi:10.1088/1755-1315/171/1/012006

- Cosuming low power during normal operation [3].
- Capable to tolerate short circuit current of dozens amperes.
- Compact and highly reliable.

There are 3 methods to achieve HVDC power switching. The first one is based on HV mechanical switches which have arc when cutting off the current [4]. Due to absence of a natural zero current crossing, it is hard to find a reliable way to extinguish arcs. Each switching operation causes damage to the mechanical switch. Therefore, it is not suitable for subsea equipment which is difficult to maintain. The second one is based on solid-state power electronic switches, which have the advantages of fast breaking speed and no arc operation. But the high power loss limits their use in large current systems [5]. The third one is to combine mechanical and solid-state solutions with their advantages [6]. As the power systems of CSOs have small current and high voltage, the power switching solution based on solid-state power switches may be the best choice. The biggest difficulty is that the maximum blocking Collector-emitter voltage of a single IGBT cannot reach this the voltage level of 10kV.

2.2. High Voltage Power Switching Solution

Figure 1 shows a typical power system feeding loads in parallel through an optic-electric submarine cable. The equivalent inductance of 300km cables is around 120mH, which may generate a high surge voltage across the switch during the current interruption.

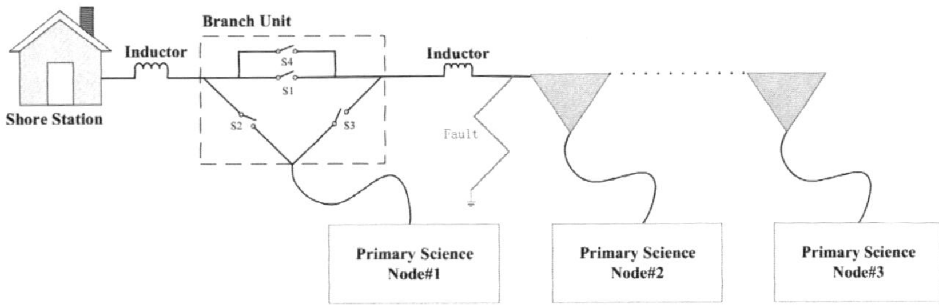

Figure 1. A typical power system of CSOs

The proposed HVDC switching circuit for BUs is shown in Figure 2. It consists of IGBTs, RCD snubber circuits, clamp varistors, a mechanical isolation switch and a control unit. Multiple IGBTs are connected in series to achieve a high blocking voltage. The RCD circuits [7] achieve voltage balance of the series-connected IGBTs. The varistor is a surge arrester used to absorb system energy and clamp the voltage across the switch unit. The mechanical switch enables physical isolation. The control unit sends control signals to each IGBT according to the measurements of current sensors. Each BU consists of four such power switches as shown in Figure 1 [8].

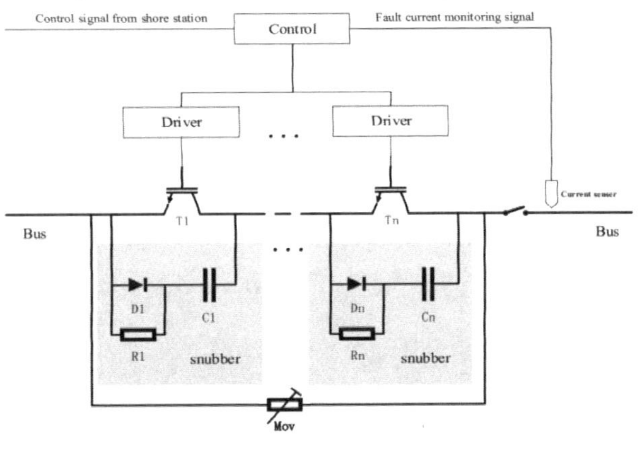

Figure 2. Topology of high voltage DC switching unit

The primary current and voltage waveforms of a breaking operation shown in Figure 3. As soon as a fault occurs, the trunk current begins to rise linearly at the moment of t_0 because of the inductor. After a certain response time, the switch unit detects the fault current and the IGBTs are turned off at $t=t_1$. The current flowing through the IGBTs is transferred to the capacitors in parallel, and the voltages across the capacitors increase during charging. When the voltage across the series-connected IGBTs rises to the preset value of the varistor at $t=t_2$, the varistor's resistance becomes very small, and the falling current transfers to the varistor during t_2-t_3. Then the fault current drops down from t_3-t_4 while the energy in the system is dissipated by the varistor. The current through the switch falls to zero at $t=t_4$ and capacitors discharge until their total voltages are equal to the power supply voltage. At last, the mechanical switch S is turned off without arc to implement physical isolation.

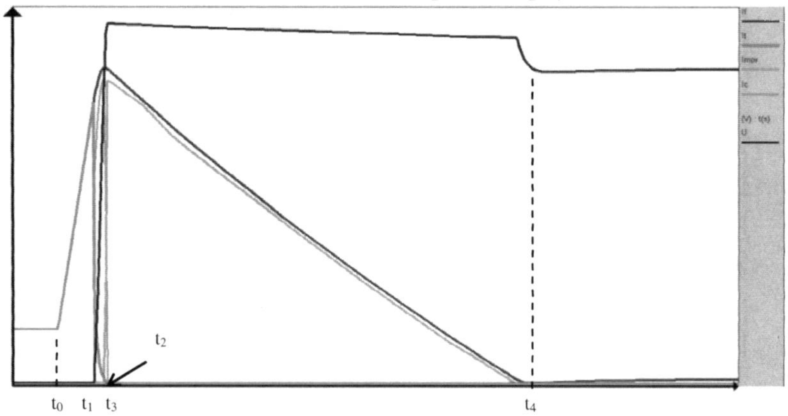

Figure 3. Current and voltage waveforms of a breaking operation

2.3. Principle of Voltage Balancing Circuit

The voltage should be balanced among series-connected IGBTs in both dynamic and static conditions. The static unbalanced voltage is mainly caused by the inconsistency of IGBT leakage currents. Most factors that cause dynamic unbalanced voltage can be attributed to the inconsistencies in the driver parameters of the IGBTs and the dispersivities of the IGBTs.

2.3.1 Analysis of static voltage sharing circuit. When these series-connected IGBTs are off completely, the voltage of each IGBT is different from one another because of the leakage resistances. Each IGBT is parallel to a resistance R_S that is far smaller than the leakage resistance, so that the voltage on each IGBT depends mainly on the voltage sharing resistance in parallel [9]. We should ensure that the unbalanced ratio of the voltage across IGBTs is less than 10% under static condition.

To ensure the blocking voltage of the switch unit, assume the maximum total voltage of the n IGBTs is 2 times of the power supply voltage. So the maximum blocking voltage of a single IGBT is $U_{CES} = 2U_{dc} / n$. Considering the extreme condition, maximum of the parallel voltage sharing resistor is

$$R_s \le \frac{0.2U_{dc}}{n(i_{ces\max} - i_{ces\min})} \tag{1}$$

Where $i_{ces\max}$ and $i_{ces\min}$ are the maximum and minimum leakage current of the n series-connected IGBTs.

Paralleling voltage sharing resistor will make more power loss.

$$P_{Rs} \ge \frac{U_{dc}^{\ 2}}{nR_s} = 5U_{dc}(i_{ces\max} - i_{ces\min}) \tag{2}$$

2.3.2. Analysis of dynamic voltage sharing circuit. The circuit diagram of the IGBTs dynamic voltage sharing based on the RCD circuit (resistance R, capacitance C and diode D) is shown in the snubber part of Figure 2. The current flowing through the IGBT which turns off firstly will transfer to the parallel buffer capacitor, which will charge the capacitor to increase the voltage of capacitor. Because the voltage of the capacitor cannot be changed immediately, it makes the voltage of the IGBT increase relatively slowly so as to balance the voltage of IGBTs [9].

Considering the most severe condition, the first IGBT turns off Δt_{off} earlier than the rest of the IGBTs, where $\Delta t_{off} = t_1 - t_2$. The first one will withstand a higher voltage. In this situation, the maximal voltage difference between the first and final turn-off IGBTs is about equal to [8]

$$\Delta U_c \approx \frac{\Delta t_{off} i_{fault}}{C} \tag{3}$$

Assuming that the maximum allowable voltage difference is 10% of the maximum voltage of a single IGBT, the minimum snubber capacitance should be

$$C \geq \frac{\Delta t_{off} i_{fault}}{10\% \cdot U_{CES}} = \frac{5n\Delta t_{off} i_{fault}}{U_{dc}} \tag{4}$$

The RCD circuit charges and discharges the capacitor during every operation periodically, leading to higher power loss and slower switching speed.

The power loss of a breaking operation is equal to the energy stored in the fully charged capacitors.

$$E_c = \frac{1}{2} nCU_c^{\,2} = \frac{1}{2} n \cdot \frac{5n\Delta t_{off} i_{fault}}{U_{dc}} \cdot \left(\frac{U_{dc}}{n}\right)^2 = 2.5 U_{dc} \Delta t_{off} i_{fault} \tag{5}$$

The turn-off time can be estimated by calculating the time of the capacitor charging.

$$U_c = \frac{i_{fault} t_r}{C} \tag{6}$$

Bringing Equation (4) into Equation (6) yields $t_r = 5\Delta t_{off}$.

The DC power supply voltage U_{dc} is constant and the current i_{fault} takes the maximum value. It can be known from Equation (4), (5) and (6) that selecting a IGBT with a larger blocking voltage can reduce the capacitance with little affecting the energy loss and switching time, because Δt_{off} changes little in practical applications.

2.4. Parameters of High Voltage Switching Circuit

A 10kV solid-state DC switching unit based on series-connected IGBTs should be developed to achieve the current interruption of dozens of amps. Under extreme conditions, assuming a trunk current of 10A and a shutdown response time of 50μs, the maximum fault current at 300km offshore where there is the nearest BU is given by

$$i_{fault} = i_0 + \frac{U_{dc}(t_{br} - t_0)}{L_{dc}} = 10A + \frac{10kV \times 50us}{300km \times 0.4mH / km} = 14.2A \tag{7}$$

Hence, the specification of five 4kV single tube IGBTs in series (IXEL40N400) is used for the design in consideration of the equipment volume, reliability, fault current and spike voltage during the current interruption. The maximum leakage current of this IGBT is 1.5mA in its datasheet. According to the Equation (1) and (2), resistance R_S in parallel with the IGBT and the power loss on it are 260kΩ and 75W.

Assuming the time lag is about 1μs and the fault current is 14.2A calculated by Equation (7), the buffer capacitance and the power loss during each breaking operation can be derived by $\Delta t_{off} = 1\mu s$ and $i_{fault} = 14.2A$ as follows $C = 36\text{nF}$ and $E_c = 3.6J$ according to the Equation (4) and (5).

The voltage across the series-connected IGBTs is clamped by paralleling an arrester. The energy absorbed in the arrester is given in [10]. The lower the clamp voltage is, the more energy the arrester needs to absorb. Considering the blocking voltage and arrester selection, the clamp voltage is determined to be 12kV. A ZnO arrester with a maximum clamping voltage of 12kV is adopted.

3. Simulation of High Voltage Power Switching

3.1. Simulation Models of the High Voltage Power Switching

In order to verify the feasibility of the proposed scheme, the power switching when a short circuit fault occurs in **CSOs** is simulated in the Saber software as shown in Figure 4.

A simulation model of the switching unit is established according to the circuit topology proposed in Section 2.2. The PWL resistance model is used to simulate the load and short circuit fault. The parameters of all components of the simulation model are presented in Table 1.

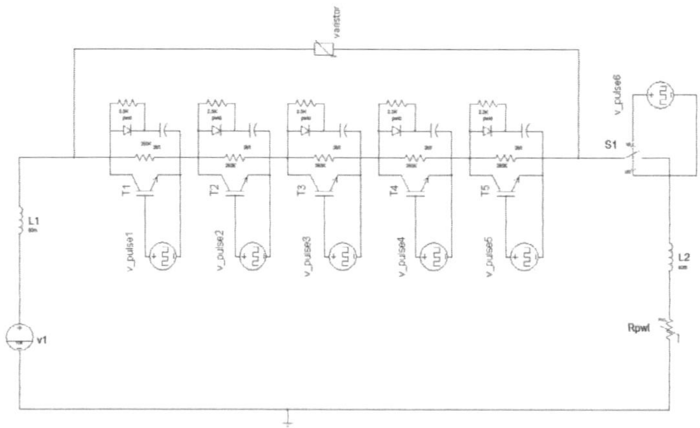

Figure 4. Simulation model of high voltage switching solution

Table 1. Parameters of the simulation model

Part	Component	Parameter
Power system	U_{dc}	10kV
	L_1	80mH
	L_2	40mH
Switching unit	R_s	260kΩ
	C	36nF
	T	IXEL40N400
	R_d	2.5kΩ
	Varistor	1.2kV

3.2. Simulation Results with RCD Buffer Circuit

The simulation results are shown in Figure 5. The switch unit is on at 10μs. Since the effect of the inductor, the current rises to a stable operating point at about 800μs. Then the PWL resistance becomes zero imitating a short circuit at t=960μs. After the 50μs switch unit reaction time, the fault current rises to 14.3A. The series-connect IGBTs are shut off near 1010μs slightly asynchronously. The voltage of the switch unit rises steeply, clamped to the preset voltage of 12kV. From the waveform diagram, the IGBTs' voltage equalizing effect is excellent and the maximum voltage difference is about 320V in case of the RCD circuit, and uneven voltage errors are lower than 10%. It takes about 1200μs to complete a breaking operation.

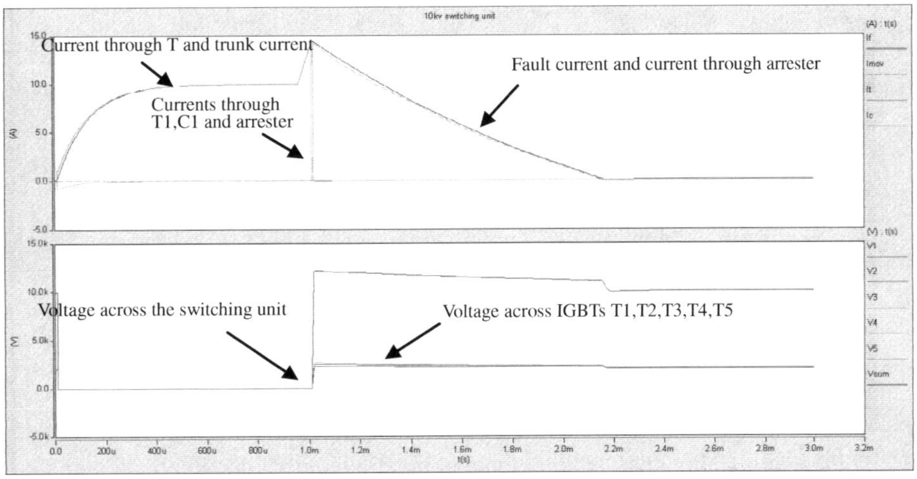

Figure 5. Simulation results of high voltage switching solution

4. Conclusion

This paper has proposed a HVDC power switching circuit for the BUs of CSOs, based on the voltage sharing scheme of series-connected IGBTs. The circuit can cut off the fault current online instead of decreasing the PFE voltage in the traditional offline approach. The feasibility of the switching unit is verified by both theoretical analysis and computer simulation. The result shows that the switching unit can interrupt the fault current within about 1ms and the voltage balancing among series-connected IGBTs performs well.

5. References

[1] El Forjani M and Hashem Z 2013 Detection of fault location, monitoring and control in underwater power system *Power Engineering Conference* (IEEE) pp 1-6

[2] Hajian M, Jovcic D and Wu B 2013 Evaluation of Semiconductor Based Methods for Fault Isolation on High Voltage DC Grid *IEEE Transactions on Smart Grid.* 4(2) 1171-1179

[3] Negari S and Xu D 2014 A new solid-state HVDC circuit breaker topology for offshore wind farms *International Symposium on Power Electronics for Distributed Generation Systems* (IEEE) pp 1-5

[4] Novello L, Gaio E and Piovan R 2009 Feasibility Study of a Hybrid Mechanical-Static DC Circuit Breaker for Superconducting Magnet Protection *IEEE Transactions on Applied Superconductivity.*19(2) 76-83

[5] Kempkes M, Roth I and Gaudreau M 2011 Solid-state circuit breakers for Medium Voltage DC power *Electric Ship Technologies Symposium* (IEEE) pp 254-257

[6] Mokhberdoran A, Carvalho A, Leite H and Silva N 2014 A review on HVDC circuit breakers *Renewable Power Generation Conference* (IET) 2014 1-6

[7] Lu T, Zhao Z, Ji S, Yu H and Yuan L 2012 Parameter design of voltage balancing circuit for series connected HV-IGBTs *Power Electronics and Motion Control Conference* (IEEE) Vol.2, pp.1502-1507

[8] El-Sharkawi, M. A "North east pacific time-integrated undersea networked experiments (NEPTUNE): cable switching and protection *IEEE Journal of Oceanic Engineering* 30.1(2005) 232-240.

[9] Feng L, Gou R, Zhuo F, Yang X and Zhang F 2016 Development of a 10kV solid-state DC circuit breaker based on press-pack IGBT for VSC-HVDC system Power *Electronics and Motion Control Conference* (IEEE) pp.2371-2377

[10] Sano K and Takasaki M 2014 A surgeless solid-state dc circuit breaker for voltage-source-converter-based hvdc systems *IEEE Transactions on Industry Applications.*50(4) 2690- 2699.

Acknowledgments
This work was supported by Shanghai science and technology innovation action plan fund through project 16DZ1205000 *Research on the power system of large-scale scientific cabled seafloor observatory networks*.

IOP Conf. Series: Earth and Environmental Science **171** (2018) 012007 doi:10.1088/1755-1315/171/1/012007

Impacts of Coastal Development on the Shoreline Change of the Eastern Gulf of Thailand

B Bidorn[1,2] and C Rukvichai[1]

1 Department of Water Resources Engineering, Chulalongkorn University, Bangkok, Thailand
2 Research Unit on Technology for Oil Spill and Contamination Management, Chulalongkorn University, Bangkok, Thailand
E-mail: butsawan.p@chula.ac.th

Abstract. The Map Ta Phut (MTP) Port has been developed on the middle portion of the Rayong Bay, Thailand, since 1992. After the expansion of the port completed in 2002, the port has an approximately 3.5 km wide and extending 4.0 km seaward. This paper provides the assessment of the construction of the MTP Port project on the shoreline of the Rayong Bay. Historical shoreline positions were extracted from aerial photography and satellite imagery taken between 1957 and 2016 using ArcGIS version 10.4. The rates of shoreline changes pre- and post-port development were implemented using the Digital Shoreline Analysis System (DSAS). Results indicate that the Rayong Bay shoreline mostly was in dynamic equilibrium with an average rate of less than 0.5 m/y prior the construction of the MTP port. After the completion of the port, a significant shoreline accretion has continuously been found along the shoreline of about 500 m adjacent to both sides of the port. Meanwhile, the Rayong Bay shoreline had been eroded about 28 % of the total shoreline after the port completed in 1992 and increased to 30% after the port expansion in 2002. Coastal protection structures included seawalls, segment breakwater, Y-shaped groins, and jetty were applied along the receded shore. Most of those structures have successfully stabilized the shoreline, and the Rayong Bay has likely reached a new dynamic equilibrium since 2011.

1. Introduction

Coast is one of the most dynamic features on the earth's surface because its shape is controlled by interaction among atmospheric, ocean and land surface systems [1]. Coastline continuously migrates landward or seaward over a various time scale in response to sea level, wave conditions, weather climate, and seasons [2]. In modern society, the coasts are highly valuable in many ways, for instance, as sites for port and harbour facilities, as locations for industrial processes, as sites for tourism, as productive and richest habitats for fish and other marine life, and more than 40% of the world's population nowadays live along the coasts [3]. Such significant coastal developments normally must be protected from any change in shoreline position, and coastal defenses play a vital role in protecting those human assets [1]. However, improper uses of coastal protection structures (seawalls, breakwaters, groins, ripraps, etc.) may produce some adverse consequences in coastline erosion processes and environment [4] [5] [6]. Therefore, post-construction monitoring is an important task to evaluate the impacts of a coastal development project and the effectiveness of erosion-controlling measures. [7] [8] [9].

In the eastern Gulf of Thailand, the Map Ta Phut Industrial Estate project has been developed since 1989 as the greatest base of petrochemical industries in Thailand [10] [11]. The Map Ta Phut (MTP) port was constructed on the middle portion of the Rayong Bay (Figure 1) as a component of the largest industrial estate in the country [12]. After the construction of the port, land reclamation, and infrastructures, severe erosions near the port were reported [11] [13]. To diminish shoreline recession due to existing of the MTP port, many types of coastal protection structures have been applied to the Rayong coast during the period 2000-2004 [14]. The objective of this paper is to document the effects of the construction of the MTP port and its coastal protection projects on the Rayong bay's shoreline using remote sensing technique. Results of this study are intended to facilitate future coastal zone planning and management along the Rayong bay and provide a better understanding of the dynamics of the shoreline in response to the coastal development activities.

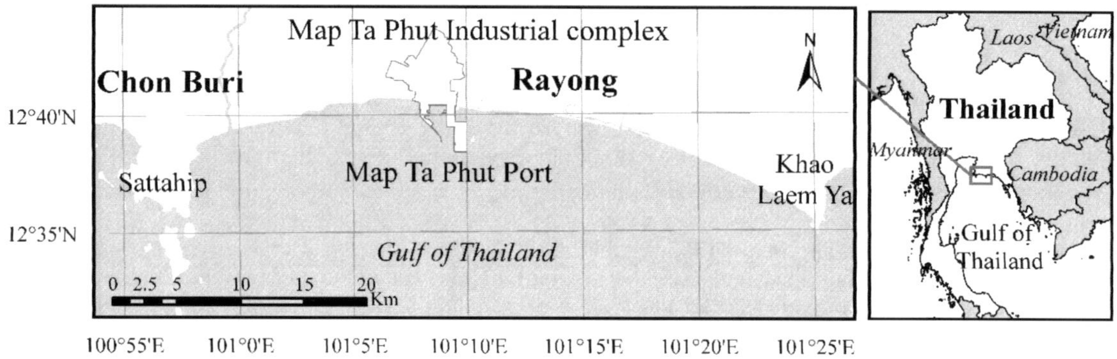

Figure 1. Locations of the Rayong Bay and Map Ta Phut Port, Thailand.

2. Study Area

The study area covers a long stretch of the east coast of Thailand from the Sattahip Peninsula on the west and Khao Laem Ya headland on the east as shown in Figure 1. The Rayong bay shoreline has approximately 60 km long located between latitude 12.58°N to 12.68°N and 100.95°E to 101.42°E. Geological features of the Rayong Bay include granitic rocks, terrace deposits and recent deposit, and the beach material mostly comprises of fine to medium sand with d_{50} of 0.1-0.6 mm [12]. The median grain size of gravels and shell-mixed gravel of about 0.1 mm was found farther of the coast (beyond 10 km from the shoreline)[15]. The sea bottom topography of the bay is generally gentle with the average gradient of about 1/600 [15]. Sea conditions are characterized by the tropical monsoon climate with two major seasons. The Southwest monsoon starting from May through September bring moderate to heavy rain and wind from the South and West, and the Southwest wind generates moderate waves on the Rayong coast with the heights of 1-2 m. The Northeast monsoon blows offshore from December to March producing small wave on along the coast [16]. Wave in this area is characterized by wind-induced waves. The predominant wave height is less than 1 m, and the wave heights more than 2 m is seldom with the prevailing wave period of 5-6 seconds [15]. Tides along the bay are diurnal and semi-diurnal depending on lunar age. The maximum tidal current speed range between 0.3-0.5 m/s and the direction of tidal current mainly is WNW-ESE [15]. The dominant southerly winds in the wet season with an average wind speed of 9.2 knots are much higher than the northerly wind speed in dry season [12]. Prior 1980s', this area was used for mix farming with tree crops and orchard but the productivity was not very high [15].

Regarding the implementation of the Fifth Five-Year National Economic and Social Development Plan in Thailand started in October, 1981, the Eastern Seaboard project was developed as one of its main strategies to accelerate the region's urban-industrial development and to deconcentrate the growth of the Bangkok Metropolitan region [15]. The MTP district was selected as a major base for an industrial port complex of petrochemical, fertilizer, and soda ash industries. To promote the location of heavy chemical industries in MTP, the Industrial Estate Authority of Thailand (IEAT), the MTP project executing agency, has conducted the construction of the MTP Port and the estate development since 1989. The MTP Port was completed in 1992 (considered as Phase I). In response to the rapid increase of petrochemical industries located in Map Ta Phut, IEAT has expanded the Map Ta Phut Port (Phase II), which was completed in 1999, and facilities since 1998. Currently, the expansion of Map Ta Phut Port (Phase III) is under the process of Environmental Health Impact Assessment (EHIA) study.

3. Materials and Methods

Shoreline positions along the Rayong Bay during the past six decades were obtained from aerial photographs (1957, 1967, 1977, 1990, and 1996) and satellite imagery (2002, 2004, 2006, 2011, 2014, and 2016). Geographic Information System (ArcGIS) software version 10.4 developed by the Environmental Systems Research Institute (ESRI) was used to extract shoreline position from these data sources. All imagery data were geo-referenced into Universal Transverse Mercator (UTM) projection with the World Geodetic System 1983 (WGS1983) to eliminate distortion from aerial photographs and satellite imagery [17] [18], and then shoreline positions along the Rayong Bay coast were digitized for each of the 11 time intervals. Typically, shoreline is defined as the boundary between land and water, which represents the mean high water level [17] [18] [19]. Because Rayong Bay currently comprises of several different coast types, different proxies suggested by Zhang et al. [20] are used to define shoreline positions as shown in Table 1.

Table 1. Proxies used for defining coastline positions [20].

Coast types	Features	Description of the position
Rocky coast	Cliff	Base of cliff
	Vegetation	Seaward edge of the vegetation line
Gravel coast	Beach ridge	Seaward edge of the beach ridge
	Cliffs	Line between the cliff and beach
Muddy coast	Vegetation	Seaward edge of the vegetation
Artificial coast	Dike	Seaward edge of the dike top
	Port	Boundary of the ports

The rates of shoreline changes along the Rayong Bay were calculated using the extension of the ArcGIS, Digital Shoreline Analysis System (DSAS) version 4.4 developed by the United States Geological Survey (USGS) [21]. In DSAS, shoreline change can be analysed using the following steps: (1) define a baseline along the coast, (2) generate transects perpendicular to the baseline at user-defined spacing, (3) calculate rates of shoreline change for each time period corresponding to the baseline using either linear regression, endpoint rate, and weighted-linear regression rate. In this study, the baseline was created landward approximately 300 m from the shoreline. Totally 580 transects were generated along the Rayong Bay with 100 m spacing as depicted in Figure 2. The statistic of shoreline change rates along the Rayong Bay during the past 60 years was calculated with 90 percent of the confident interval. Shoreline change rates between pre- and post-project construction were compared to assess the impact of the mega coastal development project on the Rayong Bay.

4. Results and Discussion

The changes in the shoreline along the Rayong Bay during the period 1957- 2017 were determined at each transect as depicted in Figure 3. Short-term rates of shoreline change were analyzed for three different time periods representing the average shoreline changes of the pre-port construction, between MTP Phase I and Phase II port construction, and the post-port (Phase II) construction. The details of shoreline changes as a result of the MTP Port development are discussed in the following subsections.

4.1. Shoreline Changes during the Period 1957-1990 (pre-Map Ta Phut Port development)

Based on the analysis of four shoreline positions of 1957, 1967, 1976, and 1990, it indicates that the Rayong Bay shoreline was mostly stable with the average rate of shoreline change less than 0.5 m/yr. Regarding to the variation of shoreline change in Figure 3 (the red solid line), the moderate rates of shoreline accretion (1-5 m/y) were found locally near the mouths of major rivers and streams, and the maximum shoreline advance of about 4.5 m/y was found at the mouth of the Rayong River (between transects 402 and 417), which was the major river sediment supply to the Rayong Bay. Whereas, shoreline retreat was pronounced at the Khlot Stream (between transects 53 and 74) (Figure 3) with an average recession rate of -2.7 m/y as a result in shoreline adjustment to wave diffraction in the area. Over three decades prior the Map Ta Phut port development, the Rayong bay was relatively stable. The annual land growth and land loss rates between 1957 and 1990 were about 1.6 and 0.67 ha/y, and the pattern of deposition and erosion during this time period was mainly controlled by the riverine system and the coastal processes.

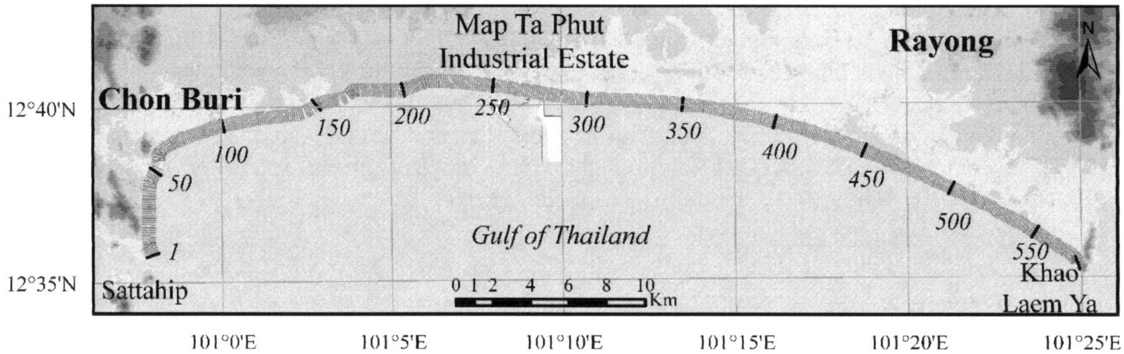

Figure 2. Plot of transects used for the shoreline change evaluation along the Rayong Bay.

4.2. Shoreline Changes during the Period 1990-2002 (during the Map Ta Phut Port development)

After the establishment of the MTP project in 1989, the IEAT has responsibility for developing and expanding the MTP Port and industrial complex. For the first stage of the development (Phase I), the Map Ta Phut Port, which comprised of a multipurpose berth and two liquid cargo berths (Figure 4) was completed in 1992. For the port facilities, a breakwater arranged in an approximately ESE direction was constructed on the west side of the port to protect the berths from SSW waves, and a reclaimed land with the 715 m wide and 1,720 m long was built on the east side (Figure 4). Based on shoreline change simulation studied by Japan International Corporation Agency (JICA) [15], the simulation results suggested that significant shoreline changes in the vicinity of the port would not occur after the Phase I's port construction.

The comparison of the shorelines between 1990 and 1996 indicated that the existence of the MTP Port caused a significant shoreline accretion on the west and east sides adjacent to the port of 4.8 and 2.4 hectares, respectively. On the west side of the port, the beach growth occurred along the coastline between transect 242 and 249 (700 m) with the maximum extent of about 140 m. Nevertheless, the significant shoreline retreat averaging about 6.5 m (-1 m/y) was found along the 3 km of coastline from transects 220 to 241. Regarding the dimension and the orientation of the port, it acted as a large groin or jetty placing in the middle of the Rayong Bay and trapped the eastwardly longshore sediment transport resulting in accretion of the coast during the SW monsoon season. In contrast, the port prevented the longshore sediment moving eastwardly during the NE monsoon resulting in beach erosion with the average rate of -1 m/y along the upcoast of the deposition area. However, the effects of the MTP (Phase I) construction on the western coastline seemed to continue not beyond the Nam Rin Beach (transect 203, about 5 km away westward from the port) at which a bunch of submerged rocks was located. For the rest of the shoreline on the west side of the port, the rates of shoreline change insignificantly varied from place to place with the average rate of shoreline accretion and recession of about 1.7 m/y and -0.4 m/y, respectively.

On the east side of the port, the pattern of shoreline change in response to the MTP Port (Phase I) was similar to the west portion. The significant accretion of coast was found along 450 m of the shoreline adjacent to the port (between transects 277 and 282). Inversely to the west side, the growth of coast in this section caused by the deposition of the littoral drift from the east impeded by the port. About 2 km long of shoreline erosion also was found next to the accretion area further to the mouth of Nam Hu Stream due to the deficit of longshore sediment transport between the SW and NE monsoons. However, the average rate of beach erosion along this segment was about -4 m/y, which was much higher than the erosion rate on the west side of the port (-1 m/y). Meanwhile, the significant accretion of the shoreline was noticed between transects 394 and 409 due to the construction of a jetty (400 m long) at the Rayong River mouth. The shoreline change rates of the remaining coastline was low (less than ±2 m/y on average) except a short section from transects 428 to 452), which was advancing seaward with the moderate rate of approximately 3 m/y as a result of the construction of the TPI port and gas pipeline between transects 436 and 445.

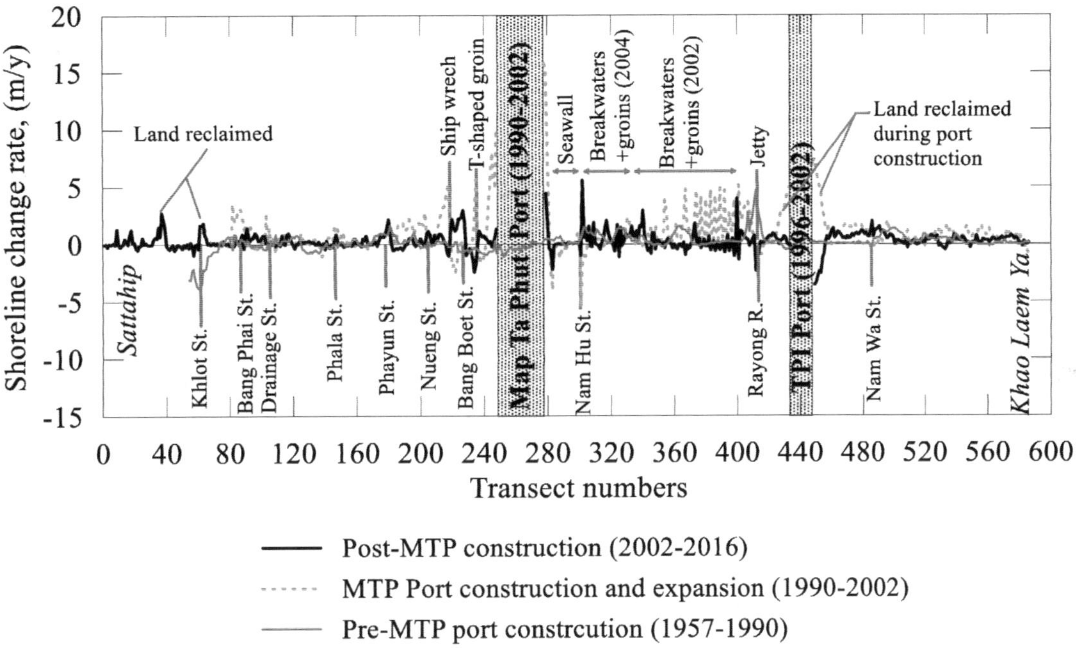

Figure 3. Shoreline variation along the Rayong Bay during the period 1957-2016.

Due to the rapid increase of enterprises located in the MTP Industrial Complex, IEAT expanded the facilities to serve the rapid growth of petrochemical plants, oil refineries, coal-fired power stations, and iron and steel facilities [12]. The expansion of the MTP Complex, considered as "Phase II", included the construction of reclaimed land for 17 berths and its facilities. The reclaimed land of 1 km wide and 2.5 km long was constructed about 1.3 km from the shoreline and next to the east port in the project first phase as depicted in Figure 4. The construction of Phase II port was completed in 2002. In regard to the severe beach erosion after the Phase I port construction plus the expected beach erosion due to the expanded port in Phase II, series of coastal protection structures were used to mitigate the impacts of the construction of the port and its expansion on the eastern shoreline. As shown in Figure 4, the structures on the east side of the port consist of 7 straight groins with 25 m long and 65 m spacing, 22 detached breakwaters with 60 m long and 50 m spacing, and ten Y-shaped groins of 130 m long and 300 m spacing with a 50 m long breakwater placing between each pair of Y-shaped groin. Meanwhile, a straight groin (200 m long) and a T-shaped groin (300 m long) were installed on the west side of the port.

The Analysis of shoreline positions during the Phase I and II development of the Map Ta Phut Port indicated that, on the western shoreline of the port, moderate rates of shoreline accumulation were found near the mouths of natural streams such as Bang Phai, Phala, and Drainage Streams. While significant shoreline changes established near the coastal protection structures and local activities. For instance, the accretion of the coast was noticed between the Nung Stream (transect 209) and transect 219, where a shipwreck was situated (Figure 3). The construction of the T-shaped groin induced a shoreline accretion of 1.2 hectares between transects 232 and 236 with the maximum shoreline growth of about 100 m (Figure 3). Additionally, a groin with 100 m long placed at transect 246 induced shoreline accretion on the west side of the structure with the rate of about 3 m/y at both sides of the structure.

On the east side of the port, the jetty construction at the Rayong River's inlet caused some continuous change of the adjacent shoreline. The typical pattern of accretion was found along 500 m of the western coastline of the jetty (between transects 403 and 408). As the severe erosion occurred between the port and the Rayong River mouth, the coastal protection structures mentioned above were installed as shown in Figure 4. It was found that the combination of a series of Y-shaped groins and breakwaters structures was placed along 2.7 km of shoreline on the west side of the jetty. The structures did not only interrupt longshore sediment but also protect the accreting beach from wave attack resulting in sediment accretion at both sides of each groin including salient developed behind each breakwater as shown in Figure 3. The maximum beach growth varied between 45 and 65 m.

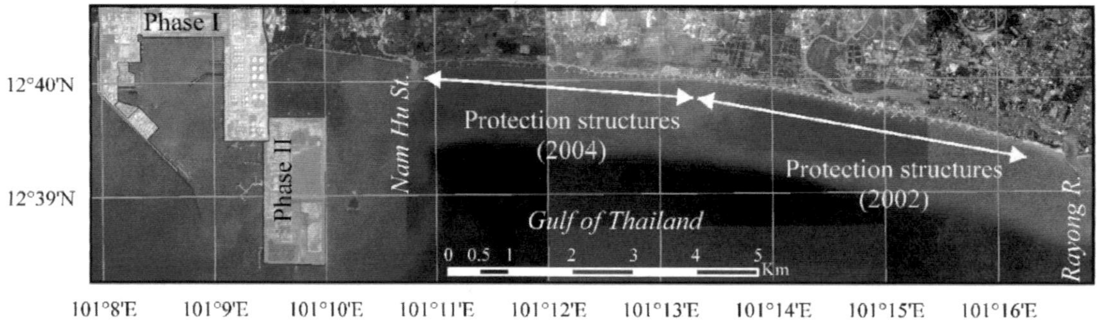

Figure 4. Development of the Map Ta Phut Phases I and II and coastal protection projects.

Even though the amount of sediment accumulation was lower than the prediction in some cases, these structures were successful in building up the beaches. Typically, a groin system caused more downcoast erosion due to upcoast trap of sediment. Therefore, a series of twenty-two segmented breakwaters was installed parallel to the 2.5 km of the shoreline from transects 348 to 373 to stabilize the downcoast of the Y-shaped groins system. These breakwaters reduced wave energy and created a pocket beach or tombolo behind each segment with the area of tombolo ranging between 0.1 and 0.5 hectares. On one hand, this proved that the beach sediment was successfully retained along the protected shore. On the other hand, this meant that there was no sediment moving to the west of the breakwater system, and a significant down-drift starvation ensued. A series of seven straight short groins system with 23-27 m long was used to remedy this impact. However, significant shoreline response, if any could not be noticed downcoast of the straight groin system during this time period due to the ongoing land reclamation and road construction project along the shore between the Nam Hu Stream and Rayong River (transects 302-356).

On the west side of the Nam Hu Stream, as a significant shoreline erosion occurred due to the construction of the MTP Port (Phase I), the 1.7 km of the seawall was constructed to reclaim the land loss between transects 284 and Nam Hu mouth. Results from the shoreline analysis indicated that the construction of the seawall effectively stabilized the protected coastline, and about 3.5 hectares of the beach area was claimed. However, the non-protected coast between the edge of the seawall and the MTP port was reshaping to reach a new equilibrium as a result of coastal development activities in this area. Because the port acted as a very long groin trapping all the westerly longshore sediment transport, the updrift shoreline between transects 278 and 281 (350 m) rapidly grew with the maximum rate of about 16 m/y. Whereas, the downdrift shoreline of the transects 282-284 was experiencing ongoing erosion as a result of less sediment available. The maximum shoreline retreat rate of this cell was approximate -4 m/y. For the17 km coastline from the Rayong River to the eastmost of the Rayong Bay, most of the shoreline statistically remained stable. Only 13% of the shoreline in this section found near the TPI Port (transects 425-459) had significant accretion with the rate greater than 1 m/y due to land reclaimed for TPI port construction.

4.3. Shoreline changes since 2002 (post-Map Ta Phut Port (Phase II) construction),

Subsequent shoreline monitoring from 2002-2016 suggested that almost 30% of the Rayong total coastline experienced changes in shoreline positions in response to the MTP Phase II port construction and other coastal development activities. For instance, the 1.4 m/y of the maximum shoreline accretion rate was observed adjacent to the port on the west. While the shoreline accretion and recession continuously occurred on the non-protected area adjacent to the port on the east until 2014. The maximum rates of shoreline growth was 7.8 m/y, and the shoreline retreat was -2.3 m/y as shown in Figure 3 (black solid line). Due to the construction of coastal protection structures in 2002, the down-drift starvation occurred along the 4.5 km of the unprotected coast (from transects 302 to 347). More segment breakwaters, Y-shaped groins, and seawalls were added to stabilize that eroded portion in 2004. Although the dimension of segment breakwaters and Y-shaped groins of the additional coastal structures were similar to those constructed in 2002, only small salients were formed behind the new structures. The shoreline responded differently to the same kind of structures probably dominated by significant differences in coastal processes such as sediment supply, wave conditions, and beach characteristics.

After 2011, more than 98% of the total shoreline has statistically insignificant changes. As the average rate of shoreline changes for the whole Rayong Bay is currently less than 1 m/y, the Rayong Bay can be considered reaching a new dynamic equilibrium. The construction of the deep sea port with roughly 3.5 km wide, extending 4.0 km seaward directly affected the adjacent shoreline of 4.5 km on the west and of 12 km on the east (about 28% of the total shoreline). By using the appropriate engineering structures, the affected shoreline has recovered from the influence of the MTP Port construction within 2-6 years depending on the local coastal processes. This indicates that coastal engineering approach has successfully diminished the effect of the large deep sea port construction on the Rayong Bay shoreline.

As another expansion of Map Ta Phut Industrial Complex (Phase III) will be started in a very near future, the measures that will be used to mitigate the impact of Phase III port construction should be carefully designed. Moreover, the expected magnitude and time period of changes in coastal processes such as shoreline positions, wave conditions, and nearshore currents, in response to the port expansion should be informed to all stakeholders to minimize negative consequences and conflicts in order to be productive and successful over the long term.

5. Conclusion

The Map Ta Phut deep-sea port was constructed on the middle portion of the Rayong Bay as the major component of the Map Ta Phut Industrial Estate Project which has now been the successful and greatest base of petrochemical industries in Thailand. Historical shoreline change rates before the project development indicated that most of the Rayong Bay shoreline was in dynamic equilibrium with an average rate of less than ±0.5 m/y. After the completion of the Map Ta Phut Port (Phase I) construction in 1992, the significant shoreline accretion was found on the west and east sides adjacent to the port, and the shoreline retreat occurred along 3 km and 5 km of coastline on the west and east sides of the port, respectively. By the end of 2002, several types of coastal protection structures (seawalls, groins, and segment breakwaters, jetty) were built along the eroded areas affected by the Map Ta Phut (Phase I) port, and the 5 km of the eastern shoreline of the port was successfully reclaimed. Upon the expansion of the Map Ta Phut port (Phase II) in 2002, the shoreline recession on the west side of the port extended westward due to down-drift deficit as a result of the coastal protection scheme. More breakwaters and groins were added to mitigate the negative effects of adopted coastal protection structures. At present, the Map Ta Phut Port is 3.5 km wide and extends 4.0 km seaward, about 30% of the total shoreline of the Rayong Bay has been significantly disturbed by the port. However, engineering structures have successfully protected those affected areas, and more than 98% of total shoreline has now reached a new dynamic equilibrium since 2011.

6. References

[1] Davidson-Arnott R 2010 *Introduction to coastal processes and geomorphology* (New York: Cambridge University Press) p 442
[2] Mahapatra M, Ratheesh R and Rajawat A 2014 *J. Indian Society of Remote Sensing* 42 869-876
[3] Cenci L, Disperati L, Persichillo M G, Oliveira E R, Alves F L and Phillips M 2017 *GIScience and Remote Sensing* 1-21
[4] Uda T 2010 *Japan's beach erosion: reality and future measures* (New Jersey: World Scientific) p 418
[5] Balaji R, Sathish Kumar S and Misra A 2017 *Int. J. of Ocean and Climate System* 1-8
[6] ChenthamilSelvan S, Kankara R and Rajan B 2014 *Nat. Harzards* 83 293-308
[7] US Army Corps 1992 *Coastal groins and nearshore breakwaters*. Technical Report EM 1110-2-1617 (Washington, D.C: US Army Corps of Engineer)
[8] Alesheikh A A, Ghorbanali A and Nouri N 2007 *Int. J. Envi. Sci. & Tech.* 4 61-6
[9] Duvat V 2013 *Sustainability science.* 8 363-79
[10] Ariga K and Ejima S 2000 *JBIC Review* 81-115
[11] Siripong A 2010 *Int. Arch. Photo. Rem. Sens. Spat. Info. Sci.* 38 992-6
[12] Industrial Estate Authority of Thailand (IEAT) 1998 Final report: Environmental impact study on the Map Ta Phut port project (Phase II) p 428
[13] Saengsupavanich C, Seenprachawong U, Gallardo W G and Shivakoti G P 2008 *Ecological economics* 67 93-103
[14] IEAT 2004 Environmental monitoring report: Map Ta Phut port project p 390
[15] Japan International Cooperation Agency (JICA) 1989 Final report: The development project of the industrial port on the eastern seaboard in the Kingdom of Thailand. Phase II p173
[16] Harbour Department 1989 Final report: Feasibility study of construction of breakwater, Ban Phe, Rayong Province
[17] Li X, Zhou Y, Zhang L and Kuang R 2014 *J. Hydrology* 511 432-442
[18] Alberico I, Amato V, Aucelli P P C, D'Argenio B, Di Paola G and Pappone G 2012 *J. Coast. Res.* 28 1638-47
[19] Yan X, Hu Y, Chang Y, Zhang D, Liu M, Guo J and Ren B 2017 *Wetlands* 1-13
[20] Zhang X, Pan D, Chen J, Zhao J, Zhu Q and Huang H 2014 *Remote Sensing* 6 9930-50
[21] Thieler E 2011 Digital Shoreline Analysis System (DSAS) version 4.0. An ArcGIS extension for calculating shoreline change: US Geological Survey. Open-File Report

Acknowledgments

This study was financially supported by the Faculty of Engineering Research Grant 2016 (Grant no. 007/2560), Chulalongkorn University, and was partially supported by the Research Assistance Scholarship, Graduate School, Chulalongkorn University.

Study on the Influence of Small Attitude Angle on Lidar Wind Measurement Results

Peilin Dou[1], Xin Zhang[1], Xiaochen Han[2], Yangyang Xue[3]

1 Jiangsu University of Science and Technology, Zhenjiang 212003.China
2 China Sanxia New Energy Co., Ltd. Shandong Branch 250000.China
3 Three Gorges new energy Yangjiang company,Yangjiang 529500.China
Email:hanxiaochenjss@163.com

Abstract. When carrying on wind profile measurement of offshore wind farm by Floating Doppler lidar technique, the platform often produces motion response under the action of ocean environment load. The motion response affects the accuracy of lidar wind measurement.In this paper, the synchronous observer experiment is designed to study the influence of small attitude angle on the accuracy of laser radar's wind measurement,giving an example of comparing the wind measurement data of two lidars, and carrying out the linear regression statistical analysis for all the experimental correlation data.It is found through experiments that the small attitude angle has less influence on horizontal wind speed and wind direction, and has a greater impact on vertical wind speed.

1. Introduction

With the rapid development of global offshore wind power, many countries has also increased the development and utilization of offshore wind energy resources. Offshore wind power industry has a bright future, and it[1] will gradually develop into large-scale and deep sea areas in the future. At the beginning of the development of offshore wind energy, it is essential to collect the short-term and long-term wind profile data of high precision and high reliability over relevant sea areas and in the corresponding height range, which is an important basis for offshore wind energy assessment, wind field location and offshore wind field operation and maintenance.In accordance with the standards of International Electro technical Commission(IEC), the common method of collecting offshore wind profile data is to build a corresponding number of offshore wind towers in the corresponding sea areas. However, the wind towers are only suitable for long-term wind profile data collection, which is obviously inappropriate for short-term measurements, whether from design and construction, costs and difficulty of operation and maintenance, or flexibility of using .

Floating lidar systems gain more and more in importance for the realisation of cost-efficient offshore wind resource assessments. These kinds of systems may not only offer measurement data that are more complete than those from present offshore meteorological masts (further measurement heights due to the used lidar technology) but have also significant advantages in terms of the associated costs for the provision, installation and final removal of the devices.In order to promote the application of Floating lidar system into China's offshore wind power project, reducing the cost of offshore wind resource assessment to bring the greatest benefits to the owners. In this paper, the experiment is designed to study the influence of attitude angle (roll and pitch) on the accuracy of the lidar.Its performance needs to be tested and verified according to a well-defined scheme. G.A Cool[2]used the numerical model programs of lidar to study the influence of ocean parameters on wind measurement results of floating lidar. G. Wolken-Mohlmann[3, 4] used the Floating lidar wind measurement technology to observe the wake wind of the offshore wind field.Songhua Wu [5, 6] was the first one to study the wind profile data of offshore wind fields through Dongfanghong No.2 equipped with Doppler lidar, and accumulated a large number of observation data of horizontal wind speed and vertical wind speed of sea-air boundary layer.Through the design of synchronized observational experiments, Han X[7] studied the influence of dynamic and static attitude angles on the accuracy of lidar.

In order to promote the application of Shipborne lidar system into China's offshore wind power project, reducing the cost of offshore wind resource assessment to bring the greatest benefits to the owners. In this paper, the experiment is designed to study the influence of attitude angle (roll and pitch) on the accuracy of the lidar.

2. Wind Field Inversion Principle

Laser technology is one of the most mainstream laser measurement technologies in the world. With its own technical advantages and characteristics, laser measurement technology not only can simultaneously measure wind speed and direction data at different heights, but also obtains the same wind shear analysis results as the tower.

The working principle of Doppler laser radar: After light incident to an atmospheric molecule or aerosol, the subsequent scattering signal will generate Doppler shift relative to the incident laser signal. When the atmospheric molecules or aerosol particles move along the radial velocity v with the action of wind speed, the frequency f_0 of single frequency laser is scattered with the moving atmospheric molecules or aerosol particles. The frequency of scattered light will produce Doppler frequency shift Δf_D, The size of the Doppler frequency shift is proportional to the radial wind speed v :

$$\Delta f_D = 2 f_0 \frac{v}{c} = \frac{2}{\lambda_0} v$$

3. Synchronous Observation Test

3.1. Test Situation

In CSSC Luzhou Zhenjiang Marine Auxiliary Machinery（E 119° 33′ 63″ ; N 32° 10′ 85″), simultaneously observing the wind speed and wind direction of twelve height layers at the same location through the opening of No.5 and No.6 Doppler lidar. The No.5 Doppler lidar is mounted on a fixed platform, and the No.6 Doppler lidar is mounted on a three degree of freedom platform, and the motion attitude of the three degree of freedom platform is controlled by setting different motion parameters.The three-degree-of-freedom platform always maintains 3 degrees of roll and -3 degrees of pitch.The simultaneous observation test lasted for half a day.

3.2. Test Instrument

Figure 1. Experimental setup

As shown in Fig 1, apparatus for testing mainly includes three degree of freedom platform, wind measurement lidar, and SMC IMU attitude instrument. Three degree of freedom platform system, composed of Stewart mechanism, computer control system, drive system, pump station and other components, is a complex mechanical and electrical integration product. The platform can be used for marine environmental test and multiple motion simulation, custom sea conditions, and custom motion performance;SMC IMU attitude instrument is mainly used to measure the instantaneous attitude of Doppler lidar. The basic working principle of the SMC IMU attitude instrument is based on the Newtonian mechanics law. By measuring the acceleration of the carrier in the inertial reference system, integrating it with time and transforming it into the navigation coordinate system so as to obtain the speed, yaw angle and location information in the navigation coordinate system;Doppler lidar uses the Molas B300 series developed by Nanjing Mulei Laser company, which has the characteristics of high spatial and temporal resolution, high wind measurement accuracy, low power consumption, stability and convenience.The laser radar uses VAD technology to obtain the wind profile, and the measurement of the radial wind speed is carried out simultaneously at 12 height layers with the elevation of 62 ° in 8 directions.Three dimensional freedom platform, inertial navigation system, part of the performance parameters of laser radar as shown in Table 1.

Table 1. Experimental instrument performance parameters

Molas B300 lidar		3-DOF motion platform		SMC IMU attitude indicator	
sampling frequency	1s	Dead weight	3000 kg	Static angle accuracy	0.02 °RMS
range	40-300 m	Roll range	±15 °	Dynamic angular accuracy	0.02 °RMS
Height level	12	Pitch range	±10 °	Roll accuracy	0.03 °RMS
Speed accuracy	0.1 m/s	Heave range	±1 m	sampling frequency	0.01 s
Direction accuracy	1 °	power	341 KW	Roll range	±30°
Speed range	0-60 m/s	Supply voltage	AC380 V	Heave range	±10 m
temperature range	-30-50°	Actuator stroke	2.1 m	Heave accuracy	5 cm
Humidity range	0-100	Cooling method	Water cool	Acceleration accuracy	0.01 m^2/s

3.3. *Test Results*

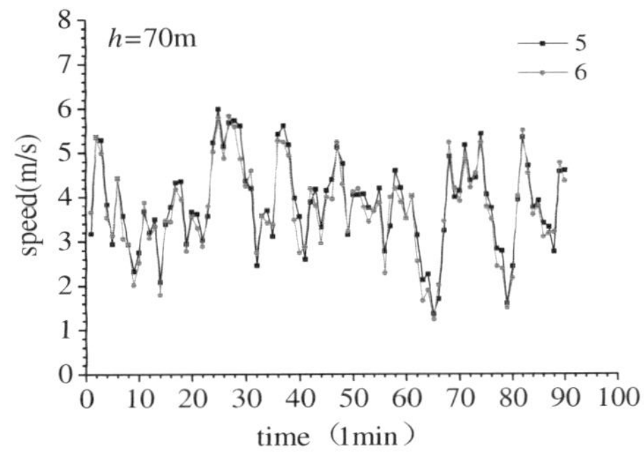

Figure 2. Horizontal wind speed contrast chart

Fig 2 is a random selection of the horizontal wind speed comparison diagram which shows that there is a small difference in the two horizontal wind speed, and the overall trend of the horizontal wind speed and vertical wind speed is the same.

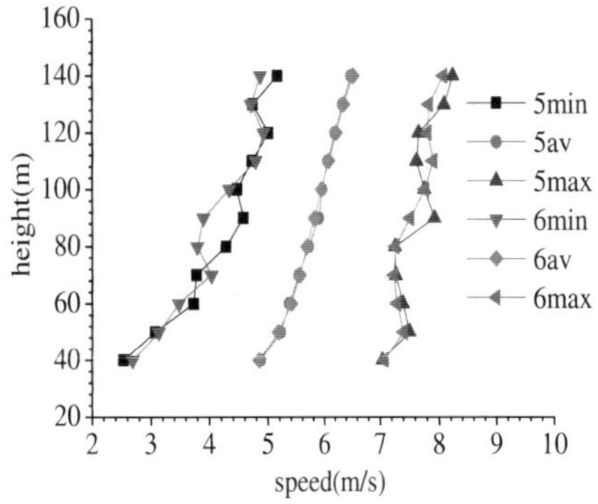

Figure 3. Wind speed contrast chart

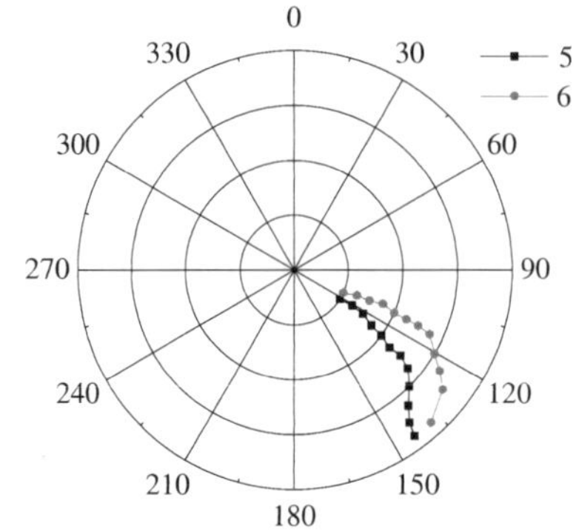

Figure 4. Wind direction rose chart

Fig 3 and fig 4 show the wind profile data observed by two Doppler lidar on June 8, 2017,in which a Doppler lidar is at the condition of rolling 3° and pitching -3° .Seeing from fig 3 and fig 4, maximum wind speed, minimum wind speed and wind direction of ten minutes at the 12 height levels are obviously different, but the average wind speed in ten minutes, two sets of laser radar measurement results have good consistency. It can be concluded that attitude angle has little influence on the average wind speed of lidar wind measurement data, but the influence on wind direction, maximum wind speed and minimum wind speed is relatively large.Using linear regression analysis can clearly see the correlation.

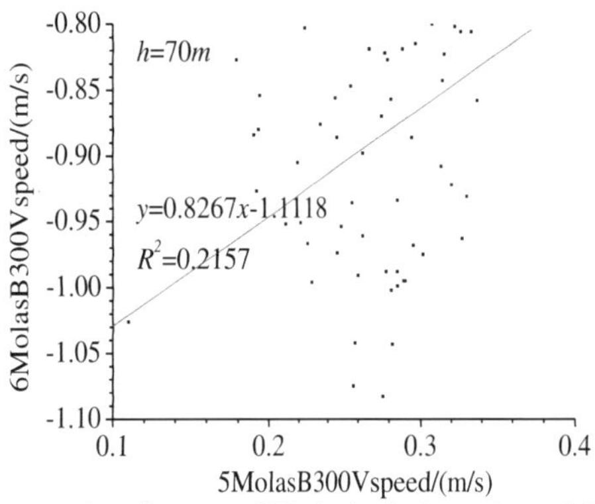

a Linear regression diagram of Vertical wind speed

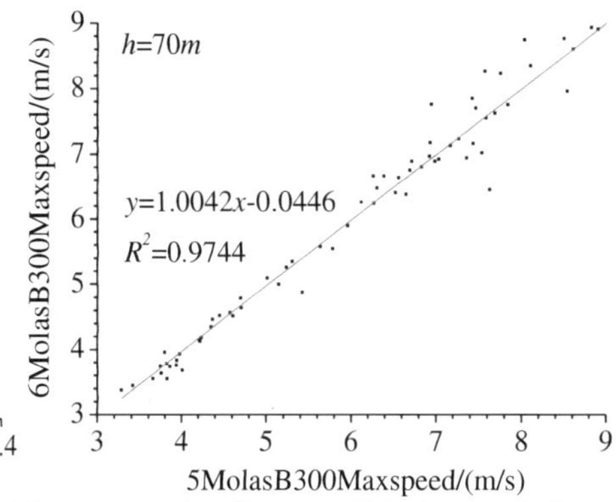

b Linear regression diagram of Maximum wind speed

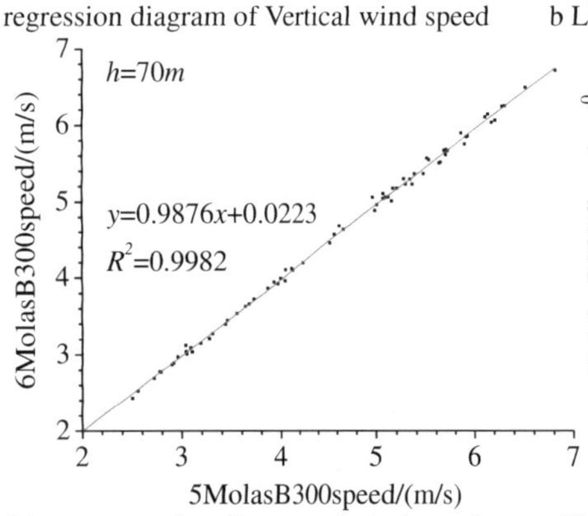

c Linear regression diagram of wind speed

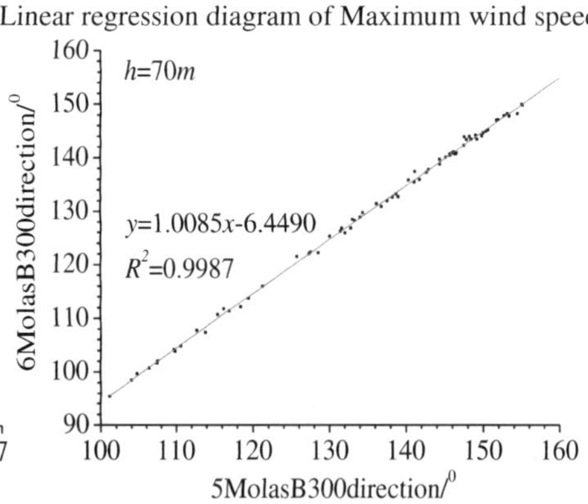

d Linear regression diagram of wind direction

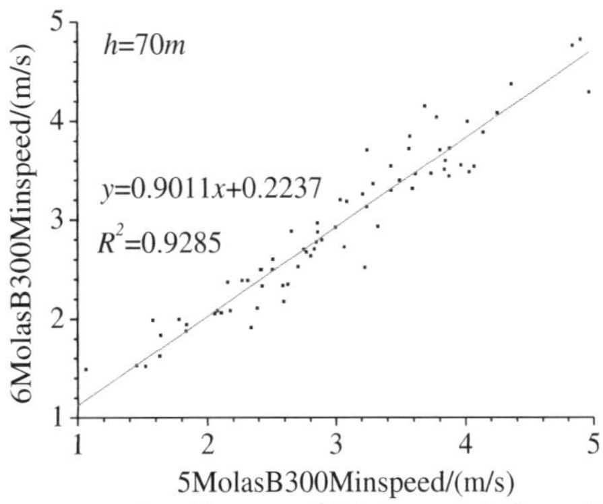

e Linear regression diagram of Minimum wind speed

Figure 5. Linear regression diagram

In this paper, linear regression analysis of experimental data to study the small attitude angle on the lidar wind measuring accuracy.Fig 5 shows the linear correlation between the maximum wind speed, the minimum wind speed, the vertical wind speed, the average wind speed and the wind direction for the two LIDARs at the 70m level from 18:00 on June 8, 2017 to 0:36 on June 9, 2017. It can be seen from Figure 5 that the measured coefficient of vertical wind speed is 0.2157.It Belonging to low correlation.The Correlation coefficient of maximum wind speed, average wind speed, wind direction and minimum wind speed were 0.9744, 0.9982, 0.9987 and 0.9285 respectively, which were highly correlated.The Correlation coefficient of horizontal wind speed and direction of wind is very close to 1, and is similar at other levels.It can be concluded that the correlation between the vertical wind velocities measured by the two LIDARs is very low, which is the primary consideration for the subsequent motion compensation algorithm.Horizontal wind speed and direction have good correlation. There is also a high correlation between the maximum wind speed and the minimum wind speed, but the correlation is slightly lower than the correlation between the horizontal wind speed and the wind direction.

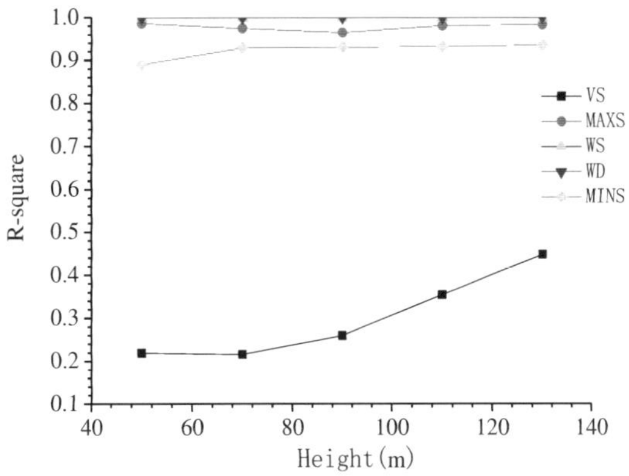

Figure 6. Data **comparison** diagram of R-square

Fig 6 is a statistical comparison of the wind profile data of two laser wind lidars at each high level.As you can see from fig 6:The coefficient of average wind speed and wind direction is above 0.99, which is very close to 1.The coefficient of determination is consistent with the changing trend of height,the wind measurement data of two lidars show a high correlation.The coefficient of maximum wind speed and minimum wind speed is lower than the coefficient of average wind speed and wind direction.The coefficient of the maximum wind speed is obviously higher than the minimum wind velocity measurement coefficient,the coefficients of both of them are above 0.90 and still have good correlation.The coefficient of vertical wind velocity is obviously lower than that of other measured data, and it shows unrelated conditions.However, the coefficient of vertical wind velocity shows an increasing trend with the increase of altitude.It can be seen from the above that in the case of small attitude angles. The small attitudes angle has little effect on the average wind speed and wind direction measured by laser wind radar, but it has less influence on the measurement of maximum wind speed and minimum wind speed. However, the effect on vertical wind speed accuracy is obvious.

4. Conclusions

In this paper, the linear regression analysis of two lidar wind measurement data is carried out by the synchronous observation test. The following conclusions are drawn:

(1)The 3 ° attitude angle has little effect on the average wind speed measured by the lidar, and there is a small difference between the measured maximum wind speed, the minimum wind speed and the wind direction.

(2)The small attitude angle has less influence on horizontal wind speed and wind direction, and has a greater impact on vertical wind speed.

It is worth noting that the conclusions obtained in this paper are based on the current limited experimental data, which will continue to be improved.

5. References

[1] A. Arapogianni, I. Pineda, and J. Moccia, "The european offshore wind industry - key trends and statistics 2012" Tech.Rep[R].The european wind energy association, 2013.

[2] G.A.Cool. Oceanographic parameters influencing accuracy of wind vector reconstruction[D].2016 ,71-83.

[3] G. Wolken-Möhlmann, J. Gottschall, B. Lange. First verification test and wake measurement results using a ship-lidar System [J]. Energy procedia, 2014, (53):146-155.

[4] G. Wloken-Mohlmann,B .Lange. Simulation of motion-induced measurement errors for wind measurements with LIDAR on floating platforms .Proceedings of ISARS conference paris,2010.

[5] Wu S,Yin J,Li R,et al. Observations and Analysis of Turbulent Wake of Wind Turbine by Coherent Doppler Lidar[C] European Physical Journal Web of Conferences, 2016:14007

[6] Feng C,Liu B,Wu S,et al. Characterization of a wind turbine wake evolving over an intertidal zone performed with dual-lidar observations[J].2017:1-19.

[7] Han X, Dou P, Xue Y. Study on the influence of attitude angle on lidar wind measurement results[C] 2017:012175.

A Method for Estimating the Distance of Near-Ocean-Bottom Sources by Combining VLF Underwater Acoustic Field and Scholte Wave Field

Han Zhao[1], Guiqing Sun[1], Hanhao Zhu[2], Junxin Yuan[1]

1 Institute of Marine Information Science and Technology, Ocean College, Zhejiang University, Zhoushan, CN
2 Zhejiang Ocean University, Zhoushan, CN
E-mail: mail_zhaohan@zju.edu.cn

Abstract. In order to solve the problems of passive detection of very-low-frequency (VLF) underwater sound sources, this paper proposes a method of estimating sound source distance based on combining underwater sound field and Scholte wave field. VLF sources near the ocean bottom generate the sound field in the water, then seismic waves are excited by the of near-field effect. The surface wave propagating along the ocean bottom interface is the Scholte wave. By deploying a hydroacoustic&seismic-wave joint detection system on the ocean bottom, the acoustic field and the Scholte wave field excited by the source can be collected. Due to the different physical properties of water acoustics and Scholte waves, the Scholte waves can be distinguished by analysing the polarization characteristics of the captured signals. Due to the different velocity between the underwater acoustics and Scholte waves, the propagation delays to the joint detection system are also different, and the delay difference can be obtained by the correlation method. Combined with the surrounding parameters, the distance of the target source can be determined. Theoretical analysis and finite element simulation experiments show that this method can estimate the position information of near-ocean-bottom VLF underwater sources in the ideal case.

1. Introduction

The stealth technology of modern underwater sound sources has greatly reduced the level of high- and mid-frequency noise compared to the past. At present, the underwater target radiation noise mainly concentrates on the low-frequency and very-low frequency bands. The passive ranging of underwater targets, as the main task of the underwater detection system, has always been a hot and difficult problem in the research of underwater acoustics [1-3]. At present, there are mainly 3 kinds of passive ranging technologies, namely passive array ranging from three-element array, passive ranging from focused beamforming and passive ranging from matching fields. Each of these methods uses a plurality of sub-arrays with very long intervals to measure the azimuth angle or measure the curvature of the wavefront to obtain the distance of the target sound source. For a VLF underwater acoustic field, if a half-wavelength array is used, the array element spacing needs to reach tens of meters. These methods have extremely high requirements for the accuracy of time-measurement and array element arrangement. In engineering, it is not ideal to achieve very low frequency sound source ranging through sonar arrays.

From the 1940s to the present, many researchers have done a series of studies on low-frequency and very-low-frequency sound propagation in the sea. These studies use OBS and hydrophones. In 1947, Scholte [4] discovered a kind of interface waves between solid and liquid that is now called the

Scholte wave. In 1980, Dieter Rauch et al. [5] conducted an underwater experiment for the reception of ocean bottom seismic waves caused by explosion. The time-distance signals, dispersion analysis diagrams, attenuation curves, and particle motion trajectories of seismic waves were obtained, and the characteristics of Scholte waves were summarized. In 2012, Bing Yan et al. [6] of Naval University of Engineering used a single OBS to estimate the target's parameters and obtain the DOA estimation, which is independent with distance.

Over the years, people have made considerable efforts in the interface-wave measurement and inversion for ocean acoustics applications [7-10]. On the other hand, people have made considerable progress in the application of ocean bottom seismometers and acoustic vector sensors for underwater target detection [11-14].

In this paper, a new method for distance estimation of VLF sound sources near the ocean bottom is proposed by obtaining the delay difference between different waveguides which are acoustics and Scholte waves.

2. Signal model and ranging method

2.1. Ocean Bottom Sound Field Analysis

The ocean bottom is the interface between seawater (liquid) and sediment (solid). Scholte waves exist at the solid-liquid interface. Scholte waves' amplitudes decrease exponentially with increasing distance from the interface. Scholte waves have the following properties:

- Its propagation speed is approximately 90% of the shear wave speed.
- Its particle movement trace is rotational.
- It is dispersive when the shear speed varies with depth.

2.1.1. Velocity of Scholte waves. When the solid medium is a homogeneous and isotropic half infinite space, and the liquid medium is homogeneous, isotropic, with finite depth, the formula for the phase velocity of the Scholte wave is as follows [15]:

$$4\left[1-\left(\frac{v_{\mathrm{p}}}{c_{\mathrm{p1}}}\right)^2\right]^{1/2}\left[1-\left(\frac{v_{\mathrm{p}}}{c_{\mathrm{s1}}}\right)^2\right]^{1/2}-\left(2-\frac{v_{\mathrm{p}}^2}{c_{\mathrm{s1}}^2}\right) \tag{1}$$
$$=\frac{\rho_0}{\rho_1}\left(\frac{v_{\mathrm{p}}}{c_{\mathrm{s1}}}\right)^4\left[1-\left(\frac{v_{\mathrm{p}}}{c_{\mathrm{p1}}}\right)^2\right]^{1/2}\left[1-\left(\frac{v_{\mathrm{p}}}{c_{\mathrm{p0}}}\right)^2\right]^{-1/2}\tanh\{\frac{\omega D}{v_{\mathrm{p}}}\left[1-\left(\frac{v_{\mathrm{p}}}{c_{\mathrm{p0}}}\right)^2\right]^{1/2}\}$$

Equation (1) is the dispersion equation for the case with finite water depth. Where the liquid has the sound speed c_{p0} and density ρ_0. And the solid medium has P- and S- wave speeds c_{p1} and c_{s1}, and its density is ρ_1. D is the water depth. When $D \to 0$, this expression becomes Reileigh wave formula. When $D \to \infty$, this expression becomes identical to the expression of equation for the infinite water depth. Equation (1) has always one positive real root, which is the Scholte wave $v_{\mathrm{p}} = v_{\mathrm{sch}}$ and can be found numerically.

The group speed v_{g} of Scholte wave can be found by differentiation - that is, by taking the derivative - and is expressed as

$$\frac{v_{\mathrm{p}}}{c_{\mathrm{p1}}}=\frac{\mathrm{d}\omega}{\mathrm{d}k}=\frac{v_{\mathrm{p}}}{1-\dfrac{\omega}{v_{\mathrm{p}}}\dfrac{\mathrm{d}v_{\mathrm{p}}}{\mathrm{d}\omega}} \tag{2}$$

2.1.2. Polarization properties of Scholte waves. The components of the particle displacement vector of the Scholte wave is given as follows.

The displacement components in the water are expressed as

$$u_{\mathrm{x0}}=u_{\mathrm{x0}}(k,z)\sin(kx-\omega t),(z\le 0) \tag{3}$$

$$u_{z0} = u_{z0}(k,z)\cos(kx-\omega t),(z \le 0) \tag{4}$$

where

$$u_{x0}(k,z) = -kA\exp(\alpha_{p0}z) \tag{5}$$

$$u_{z0}(k,z) = -\alpha_{p0}\exp(\alpha_{p0}z) \tag{6}$$

The displacement components in the bottom are expressed as

$$u_{x1} = u_{x1}(k,z)\sin(kx-\omega t),(z \ge 0) \tag{7}$$

$$u_{z1} = u_{z1}(k,z)\cos(kx-\omega t),(z \ge 0) \tag{8}$$

where

$$u_{x1}(k,z) = \frac{\alpha_{p0}kA}{\kappa_{s1}^2\alpha_{p1}}\{2\alpha_{p1}\alpha_{s1}\exp(-\alpha_{s1}z)-(k^2+\alpha_{s1}^2)\exp(-\alpha_{p1}z)\} \tag{9}$$

$$u_{z1}(k,z) = \frac{\alpha_{p0}A}{\kappa_{s1}^2}\{2k^2\exp(-\alpha_{s1}z)-(k^2+\alpha_{s1}^2)\exp(-\alpha_{p1}z)\} \tag{10}$$

Equations (3) - (10) are parametric representations of ellipses having their main axes parallel to axes of the coordinate system. Where A is Scalar displacement potential amplitude of the water. Since the horizontal wave number, k, is the same for all waves at the interface, the vertical wave numbers describing the vertical decays of the fields have to be

$$\alpha_{p0}=(k^2-\kappa_{p0}^2)^{1/2} \tag{11}$$

$$\alpha_{p1}=(k^2-\kappa_{p1}^2)^{1/2} \tag{12}$$

$$\alpha_{s1}=(k^2-\kappa_{s1}^2)^{1/2} \tag{13}$$

where

$$k=\frac{\omega}{v_p},\kappa_{p0}=\frac{\omega}{c_{p0}},\kappa_{p1}=\frac{\omega}{c_{p1}},\kappa_{s1}=\frac{\omega}{c_{s1}} \tag{14}$$

With increasing distance from the interface, the displacement amplitudes $\hat{u}_{x0}$, $\hat{u}_{z0}$ and $\hat{u}_{z1}$ decrease exponentially without changing sign. The horizontal displacement in the bottom, $\hat{u}_{x1}$, shows the same asymptotic behavior, but with the sign changing at the depth of about one-tenth of the Scholte wavelength. The penetration depth in the water is about one-half of the Scholte wavelength. In liquid, the eccentricity of the displacement elliptical trajectory is close to 0, and the trajectory is close to a circle. In the sediment layer, the ellipse is elongated and the amplitude of the displacement vertical component is greater, i.e. $|\hat{u}_{z1}|>|\hat{u}_{x1}|$. The interface waves are confined to a narrow stratum close to the interface, which means that they have cylindrical propagation loss (i.e., $1/r$) rather than spherical spreading loss (i.e., $1/r^2$). Cylindrical spreading loss indicates that, once an interface wave is excited, it is likely to dominate other waves that experience spherical spreading at long distances.

The interface waves can be excited by a point source close to the interface, that is, as a near-field effect.

2.1.3. Dispersion of Scholte waves. When the shear speed in the bottom varies with depth, the velocity of Scholte waves is not a constant. This is due to the fact that the Scholte wave is related to the shear wave velocity of solid media, while the Scholte wave penetration depth is different for different frequencies. The dispersion properties of the Scholte wave can be expressed by the dynamic stiffness matrix approach [16]. The frequency band most concerned in this study is 30-50Hz and the dispersion of Scholte wave in this band is not distinct. Therefore, this study takes the ideal condition assumption that the Scholte wave velocity does not change with frequency.

2.2. Distance Estimation

IOP Conf. Series: Earth and Environmental Science **171** (2018) 012009 doi:10.1088/1755-1315/171/1/012009

Since underwater acoustics and Scholte waves have different velocities, the propagation delays to the joint detection system are also different. The delay difference can be obtained, and the distance of the target source can be determined by combining with the surrounding parameters.

A VLF hydroacoustic&seismic-wave joint detection system is deployed on the ocean bottom in order to capture underwater acoustics and ocean bottom Scholte waves. The join detection system consists of an acoustic vector sensor (AVS) and an ocean bottom seismometer (OBS). An AVS is able to measure both the acoustic pressure and the three particle velocity components providing an estimate of the full directional acoustic intensity field. An OBS is able to measure the three particle velocity of ocean bottom. By analyzing the signals captured by the joint detection system, the delay difference information for the two types of wavefields can be obtained.

2.2.1. Distinction of acoustics and Scholte waves. According to 2.1.2, the particle displacement track of Scholte wave in the sediment layer is elliptical and the vertical component is greater, i.e. $|\hat{u}_{z1}|>|\hat{u}_{x1}|$. Scholte waves can be detected based on this feature.

2.2.2. Determination of delay difference. In this study, the delay difference of 2 types of waves is the core. The accuracy of delay difference estimation determines the accuracy of ranging. This paper uses classic cross-correlation method to find the delay difference. This method obtains the correlation peak coordinate value as the delay difference by the cross correlation function of the two signals. The advantages of this method are that it can not only solve the delay difference, but also detect the correlation of different signals. This ensures the measured waves are from the same source via different channels.

Let the signal emitted by the sound source in the water be $s(t)$. The received signal of sound pressure channel of the AVS is $r_{\mathrm{p}}(t)$, and the received signal of vertical velocity channel of the OBS is $r_{\mathrm{v}}(t)$. So

$$r_{\mathrm{p}}(t) = a_1(t)s(t-\tau_{\mathrm{w}}) + a_2(t)s(t-\tau_{\mathrm{sch}}) + n_{\mathrm{p}}(t) \tag{15}$$

$$r_{\mathrm{v}}(t) = a_3(t)s(t-\tau_{\mathrm{w}}) + a_4(t)s(t-\tau_{\mathrm{sch}}) + n_{\mathrm{v}}(t) \tag{16}$$

where τ_{w} is the acoustics delay, τ_{sch} is the Scholte wave delay. $n_{\mathrm{p}}(t)$, $n_{\mathrm{v}}(t)$ are noise. $a_1(t)$, $a_2(t)$, $a_3(t)$, $a_4(t)$ are distortions of signals. It is assumed here that they are time invariant, denoted by a_1, a_2, a_3, a_4. In the sound pressure channel of AVS, the direct acoustics signal in the water accounts for a large proportion; in the vertical velocity channel, the Scholte waves signal in the bottom accounts for a large proportion. Which can be recorded as

$$\frac{a_1}{a_2} > \frac{a_3}{a_4} \tag{17}$$

The cross-correlation function of $r_{\mathrm{p}}(t)$ and $r_{\mathrm{v}}(t)$ is denoted as $R_{\mathrm{p,v}}(\tau)$. The delay difference estimate tau can be obtained from coordinates of the correlation peak of the cross-correlation function. For real signals, the cross-correlation function is defined as

$$R_{\mathrm{p,v}}(\tau) = \int_{-\infty}^{+\infty} r_{\mathrm{p}}(t-\tau) r_{\mathrm{v}}(t)\mathrm{d}t = \int_{-\infty}^{+\infty} r_{\mathrm{p}}(t) r_{\mathrm{v}}(t+\tau)\mathrm{d}t \tag{18}$$

Here assumes that $n_{\mathrm{p}}(t)$, $n_{\mathrm{v}}(t)$ is Gaussian white noise, has no correlation with itself, and is uncorrelated with the source signal, so it can be ignored. Substituting equation (15) and equation (16) into equation (18) yields

$$R_{\mathrm{p,v}}(\tau) = (a_1a_3 + a_2a_4)R_{\mathrm{s}}(\tau) + a_1a_4R_{\mathrm{s}}(\tau - \tau_{\mathrm{sch}} + \tau_{\mathrm{w}}) + a_1a_3R_{\mathrm{s}}(\tau + \tau_{\mathrm{sch}} - \tau_{\mathrm{w}}) \tag{19}$$

where $R_{\mathrm{s}}(\tau)$ represents the autocorrelation function of $s(t)$. The peak of the autocorrelation function is taken at the origin. Therefore when $\tau = \tau_{\mathrm{sch}} - \tau_{\mathrm{w}}$, $R_{\mathrm{p,v}}(\tau)$ has a clear peak. Symmetrically along the y-axis, when $\tau = -(\tau_{\mathrm{sch}} - \tau_{\mathrm{w}})$, $R_{\mathrm{p,v}}(\tau)$ has another peak. $a_1a_4 > a_2\, a_3$ is known from inequality (17), so the peak at $\tau = \tau_{\mathrm{sch}} - \tau_{\mathrm{w}}$ is higher. In the ideal case, the delay difference derived from the peak coordinate is the delay difference of waves over the different channels to the receiver. That is, the delay difference between the direct acoustics and ocean bottom Scholte waves.

$$\hat{\tau} = \arg\max_{\tau} R_{\mathrm{p,v}}(\tau) \qquad \text{s.t.}\ \ \tau > \varepsilon \tag{20}$$

Where $\hat{\tau}$ is the delay difference estimate and ε is the time delay for the value of the cross-correlation function to decrease to noise level near zero.

2.2.3. Ranging by delay difference. We consider the situation depicted in figure 1, where the distance from the source to the bottom is D, and the horizontal distance from the source to the receiver is L.

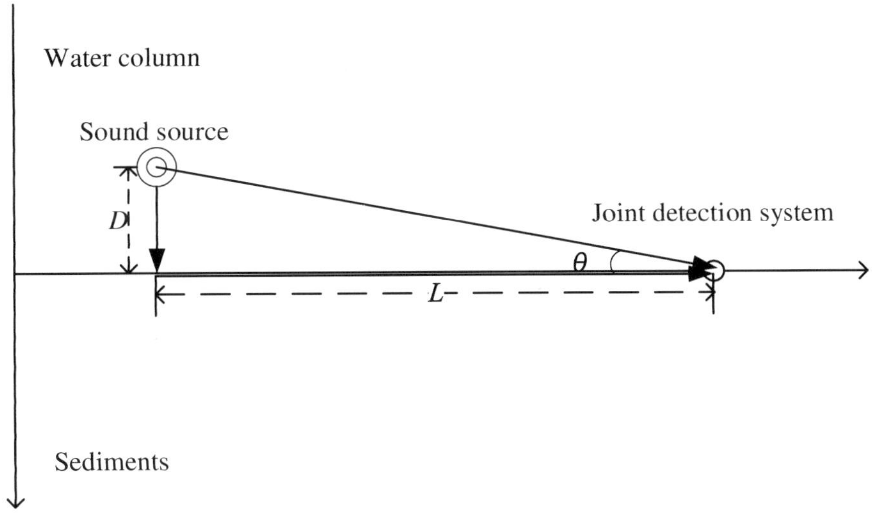

Figure 1. Wave propagation model.

Assuming that the water column and sediments are homogeneous isotropic media, then the delays for the propagation of the signals from the sound source through the direct acoustics in the water and the Scholte wave on the ocean bottom to the joint detection system are

$$\tau_w = \frac{(D^2 + L^2)^{1/2}}{c_{p0}} \tag{21}$$

$$\tau_{sch} = \frac{D}{c_{p0}} + \frac{L}{v_{sch}} \tag{22}$$

Delay difference can be obtained by the difference between equation (21) and equation (22)

$$\tau = \frac{D}{c_{p0}} + \frac{L}{v_{sch}} - \frac{(D^2 + L^2)^{1/2}}{c_{p0}} \tag{23}$$

Because the sound source is very close to the bottom, D is small. The effective range of detection is much greater than the distance from the source to the bottom, i.e. $L \gg D$. The acoustic speed in the water is greater than the shear wave velocity of the sediments and larger than the Scholte wave velocity, i.e. $c_{p0} > v_{sch}$. Therefore, $L/v_{sch} \gg D/c_{p0}$. And because $\theta \to 0$, $\cos\theta \to 1$, $\left(D^2+L^2\right)^{1/2} \to L$. So equation (23) can be simplified as

$$\tau = \frac{L}{v_{sch}} - \frac{L}{c_{p0}} \tag{24}$$

and then

$$\hat{L} = \hat{\tau} \cdot \frac{c_{p0} \cdot v_{sch}}{c_{p0} - v_{sch}} \tag{25}$$

Equation (25) is the distance estimation equation obtained by the delay difference.

3. Simulation Experiment and Model Validation

Finite element method (FEM) is one of the effective methods of sound field modeling. Especially when it involves multiple physical fields, FEM can co-process the coupling conditions of each physical field. This paper uses FEM to model and simulate the ocean bottom environment.

3.1. Model Parameter Settings

3.1.1. Surroundings parameters. The ocean bottom environment model parameters are set as table 1:

Table 1. Properties of the water column and sediment used to model.

Medium	Layer thickness (m)	c_p (ms^{-1})	c_s (ms^{-1})	ρ (kg m^{-2})
Water	20	1500	-	1000
Sediment				
Layer 1	50	2500	400	2000
Layer 2	200	3000	550	2500

3.1.2. Sound source information. A VLF sound source in water is placed 1m from the ocean bottom to excite water acoustics and Scholte waves, i.e. $D = 1$.
The source signal is

$$s(t) = \frac{1000}{\pi} \cdot \text{sinc}(\frac{10}{\pi}t) \cdot e^{j40t} \tag{26}$$

By the Fourier transform, the above equation is converted into the frequency domain form

$$S(f) = \begin{cases} 100, & 30 \le f \le 50 \\ 0, & \text{other} \end{cases} \tag{27}$$

3.1.3. Capture position. The capture position in the water is placed at a distance of 0.5m from the bottom. The capture position under the bottom is placed at a distance of 0.1m from the bottom. This is consistent with existing joint detection system.
The horizontal distance from the sound source to the capture position is from 0m to 5000m, every 10m, i.e. $L = 0, 10, 20, \ldots, 5000$.

3.2. Distance Estimation Results

From time 0, collect 16 seconds of signals. The receiving signals of the water acoustic pressure channel and the vertical component of sediment particle velocity channel are shown in figure 2.

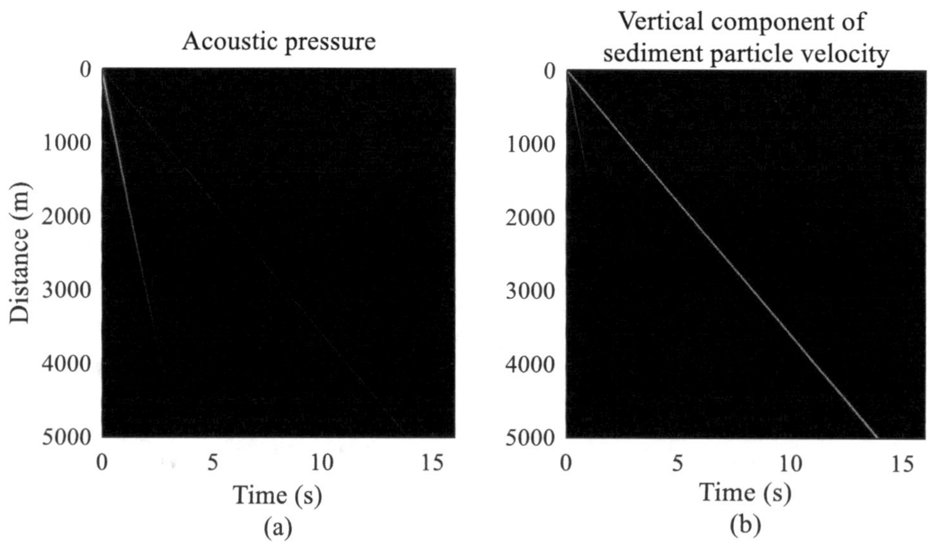

Figure 2. The intensities vary with time and distance.

In figure 2, the lighter the color, the greater the intensity. In both (a) and (b), faster wave (left line) is direct underwater acoustics, and the slow wave (right line) is ocean bottom Scholte wave. In the acoustic pressure channel, the direct underwater acoustics is stronger; in the vertical velocity channel, the Scholte wave is stronger. It can also be seen that the Scholte wave attenuation is smaller than acoustics over long distances.

The signals of the two channels are extracted, and they are substituted into equation (20) to calculate the delay difference of direct underwater acoustics and ocean bottom Scholte wave. And then the distance can be obtained by equation (25). Some of distance estimation results are shown in table 2, figure 3 and figure 4.

Table 2. The results of delay difference and estimated distance.

Actual distance(m)	Delay difference(s)	Estimated distance(m)	Error (m)	Actual distance(m)	Delay difference(s)	Estimated distance(m)	Error (m)
100	0.209	98.32	1.68	2000	4.227	1987.78	12.22
500	1.064	500.15	-0.15	3000	6.345	2983.81	16.19
1000	2.109	991.75	8.25	4000	8.464	3979.84	20.16
1500	3.182	1496.18	3.82	5000	10.6	4984.41	15.59

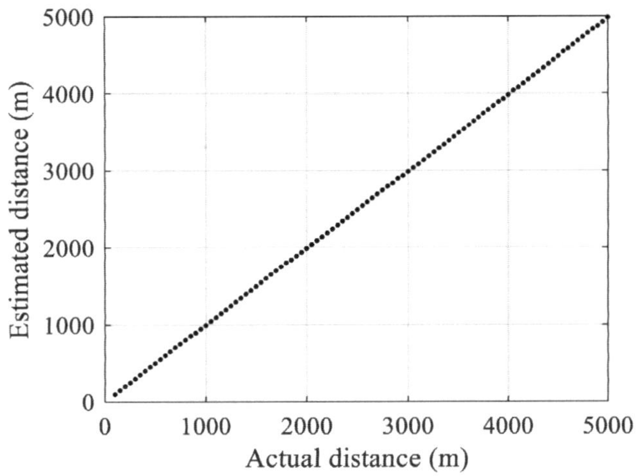

Figure 3. Estimated distance compared to actual distance.

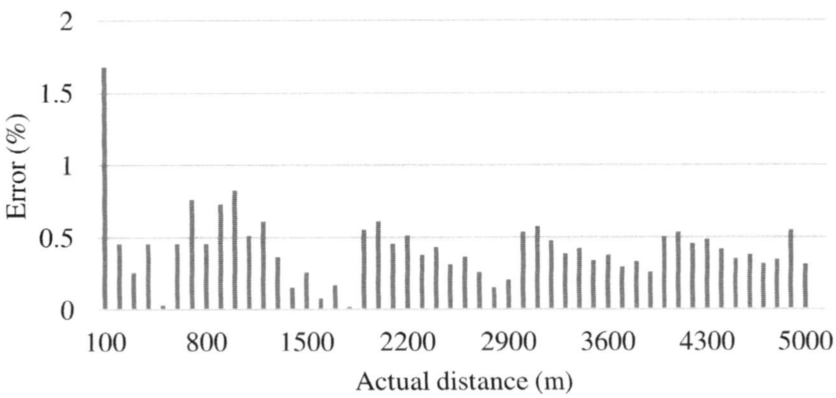

Figure 4. Ranging error ratio.

4. Conclusion

This paper proposes a distance estimation method for VLF sound sources near the ocean bottom. The method simultaneously measures underwater acoustics and ocean bottom Scholte waves, uses the delay difference of the two channels, and combines the parameters of ocean bottom surroundings to estimate the distance.

Under ideal simulation conditions, when the range is from 100m to 5000m, the range error does not exceed 2%.

This method does not require the deployment of sonar arrays as in the traditional method. It is only necessary to deploy an AVS-OBS joint detection system on the ocean bottom. The difficulty in engineering is greatly reduced.

A single AVS has the ability to estimate the DOA of the sound source, but cannot estimate the distance. By combining underwater acoustics and Scholte waves, the distance information can be obtained. Combining the distance and the DOA, the complete position information of the target sound source can be obtained.

5. References

[1] Qihu L I 2015 A new method of passive ranging for underwater target: distance information extraction based on waveguide invariant *Chinese Journal of Acoustics* 34 97-106

[2] Annabattula J, Rao S K, Murthy A S D, Srikanth K S and Das R P 2015 Underwater passive target tracking in constrained environment *Indian Journal of Science and Technology* 8

[3] Song X, Zhao A and Li M 2017 Passive ranging technique using waveguide invariant in shallow water with thermocline *J SYST ENG ELECTRON* 28 244-50

[4] Scholte J G 1947 The range of existence of Rayleigh and Stoneley waves *GEOPHYS J INT* 5 120

[5] Rauch D 1980 Seismic interface waves in coastal waters: A review *Seismic Interface Waves in Coastal Waters A Review*

[6] Yan B, Guo H, Li X and Yu D 2012 Moving target parameter estimation based on single seismometer *Journal of Naval University of Engineering* 24 66-70

[7] Godin O A and Chapman D M F 2001 Dispersion of interface waves in sediments with power-law shear speed profiles. I. Exact and approximate analytical results *J ACOUST SOC AM* 110 1890-1907

[8] Chapman D M F and Godin O A 2001 Dispersion of interface waves in sediments with power-law shear speed profiles. II. Experimental observations and seismo-acoustic inversions *J ACOUST SOC AM* 110 1908

[9] Dong H, Allouche N, Drijkoningen G G and Versteeg W 2010 Estimation of Shear-Wave Velocity in Shallow Marine Environment *Proc. of the 10th European Conference on Underwater Acoustics* (Istanbul) pp 175-180

[10] Dong H, Hovem J M and Frivik S A 2015 Estimation of shear wave velocity in seafloor sediment by seismo-acoustic interface waves: a case study for geotechnical application. *Theoretical and Computational Acoustics* (Hangzhou) pp 33-43

[11] Kalgan A, Bahl R and Kumar A 2015 Source ranging and direction of arrival estimation using accelerometer based implementation of prototype acoustic vector sensor. In: *Underwater Technology* (Chennai: IEEE) pp 1-6

[12] Widmaier M, Fromyr E and Dirks V 2015 Dual-sensor towed streamer: From concept to fleet-wide technology platform *first break* 33 83-9

[13] Felisberto P, Santos P, Maslov D and Jesus S M 2016 Combining pressure and particle velocity sensors for seismic processing. In: *OCEANS 2016 MTS/IEEE* (Monterey: IEEE) pp 1-6

[14] Song X, Zhao A and Li M 2017 Passive ranging technique using waveguide invariant in shallow water with thermocline *J SYST ENG ELECTRON* 28 244-50

[15] Dong H and Hovem J M 2011 *Interface Waves, Waves in Fluids and Solids* (London: InTech) pp 153-176

[16] Potty G R and Miller J H 2012 Measurement and modeling of Scholte wave dispersion in coastal waters *AIP Conference Proceedings* (New York: AIP) pp 500-507

Construction Method of Telecommunication System for Corrective Information Distribution

V V Karetnikov, **S F Shahnov and A A Ageeva**

Department of Navigation on Inland Waterways, Admiral Makarov State University of Maritime and Inland Shipping, 5/7 Dvinskaya Street, St. Petersburg 198035, Russia

Abstract. The paper presents a methodology for optimal construction of telecommunication system for corrective information distribution based on the criterion of minimizing of expendable resources. A calculation method of range of control and correction stations (CCS) action transmitting the corrective information to mobile consumers under various conditions (sea, land areas with various electrical properties) is developed. A construction algorithm of the optimal telecommunication system has been created. The software in the software shell MATHLAB was developed for implementation of the obtained algorithm. The obtained algorithm and the developed program were tested. The results of comparison of calculations with the analytical solution and experimental data for the specific example are given. These results gave a good matching between the computations obtained by the algorithm and the analytical calculations. The developed algorithm and program were applied to optimize the composition of CCS of the local differential subsystem in the rivers basins of Siberia and the Far East. An example of the formation of action zones of CCS in the basin of the Ob River on inland waterways of Russian Federation is given. In conclusion, the inference is made about the effective optimization of the local differential subsystem, as a result of which it became possible to provide all inner waterways in the Ob basin by a continuous field of the differential correction.

1. Introduction

The basis of the technical component of e-Navigation concept implanted in the civil fleet accepted by the International Maritime Organization (IMO) is telecommunication systems for various purposes, one of which is the telecommunication system for providing of corrective information for GNSS GLONASS / GPS [1]. The local differential subsystems (LDSS) based on control and correction stations (CCS) working on medium frequency band (283.5-325.0 kHz), the range of which is about 200 ~ 500 km, are widely used in maritime and especially river conditions as telecommunication systems for providing of corrective information [2]. Such a wide spread is determined by the significant influence of underlying surface on the propagation range of radio waves of this band. In this connection, there is a problem of structure optimization of this telecommunication system (the number and location of the CCS which provide a continuous coverage of navigable sections of river basins of the given region).

Synthesis of river LDSS belongs to the class of tasks at which solution the goal is achieved with a minimum of required resources. In our case, the resource, first of all, means the power and the number of CCS transmitters.

2. Construction methodology of optimal telecommunication system for corrective information distribution

To optimize the structure of the river LDSS, it is necessary to determine the range of CCS including into it. The criterion for the action range calculation is the correspondence of the error probability of piece-by-piece reception of digital message to its permissible value. The estimation methodology is based on the well-known relationship between the error probability of piece-by-piece reception p_{err} and the signal power h^2 under conditions of functional noise impact:

$$p_{err.} = 0.5exp(-0.5h^2). \tag{1}$$

Where the signal power is determined by the relation:

$$h^2 = (P_r T) / \nu^2, \tag{2}$$

where P_r – signal power in a reception point; T – transmission duration; ν^2 – spectral density of white functional noise.

However, it is necessary to consider that the main requirement to the river LDSS is providing of continuous covering of the internal waterways (IW) with the field of the differential correction. In this connection there is a need to overlap the action zones of adjacent CCS, which leads to significant mutual hindrances in the areas of intersection of these zones.

In these areas, the estimation methodology will be based on the coefficient of mutual distinction (CMD) of signal and mutual hindrance [3]:

$$g_{rq}^2 = K_0 \left\{ \left[\int_0^\tau z_r(t) z_{nq}(t) dt \right]^2 + \left[\int_0^\tau z_r(t) \tilde{z}_{nq}(t) dt \right]^2 \right\}, \tag{3}$$

where $z_r(t)$, $z_{nq}(t)$ – structures of the rth version of a signal and the qth concentrated noise; $\check{z}_{nq}(t)$ – function adjoint to $z_{nq}(t)$ by Hilbert; K_0 – normalizing factor.

The normalizing factor K_0 for a digital signal is given by:

$$K_0 = \left\{ \mu_r \mu_{nq} / (TP_r) \right\}^2, \tag{4}$$

where μ_r , μ_{nq} – amplitude coefficients of signal transmission and concentrated noise transmission, respectively.

Then, at non-coherent reception the error probability of piece-by-piece reception of the digital message will be determined by expression:

$$p_{err} = -0.5\exp\left[-\frac{h_{eq}^2}{2} \right] \cdot I_0\left(R_0 h_{eq}^2 \right) + Q\left[\sqrt{\frac{h_{eq}^2}{2}\left(1 - \sqrt{1 - R_0^2}\right)}; \sqrt{\frac{h_{eq}^2}{2}\left(1 + \sqrt{1 - R_0^2}\right)} \right], \tag{5}$$

where Q – Markum function; $I_0(R_0 h^2_{eq})$ – modified Bessel function of the first kind, zero order; h^2_{eq} – Equivalent signal power in the presence of several mutual hindrances; R_0 – radio channel parameter.

For opposite non-fading signals:

$$h_{eq}^2 = h^2 \left[1 - \sum_{q=1}^{N} \frac{g_{0q}^2 h_{nq}^2}{1 + h_{nq}^2} \right] = h^2 (1 - \delta), \tag{6}$$

where g_{0q} – normalized CMD of the signal and the qth mutual hindrance; h_{nq} – power of the qth mutual hindrance; N – number of mutual hindrances.

$$\delta = \sum_{q=1}^{N} \frac{g_{0q}^2 h_{nq}^2}{1 + h_{nq}^2}, \tag{7}$$

$$R_0 = \delta / (1-\delta). \tag{8}$$

3. Method of calculation of structure of the optimal differential subsystem

Depending on whether the signal receiver is under the influence of only functional noise or it is under the influence of mutual hindrances, either (1) or (5) will be used to determine the error probability of piece-by-piece reception.

In both cases, by (2), it is necessary to determine the signal power, and in the second case it is also necessary to determine the power of each mutual hindrance. In order to determine the signal power or hindrance power presented in (2), it is necessary to find the field intensity of the signal at the receiving point $E(r)$, which for emitter in the form of a point dipole is determined by the expression [4]:

$$|E(r)| = -\frac{3\sqrt{10P_{Tr}}}{r}|w(r)|, \tag{9}$$

where P_{Tr} – radiated power of CCS transmitter; r – distance from emitter to receiving point; $w(r)$ – signal attenuation function due to the influence of the underlying surface.

The power at the receiver input P_{res} is related to the field intensity by the known relation:

$$P_{res} = \left(A_{eff} E^2\right) / 240\pi. \tag{10}$$

Where A_{eff} – effective antenna area, defined by expression:

$$A_{eff} = \left(D_{res}\lambda^2\right) / 4\pi, \tag{11}$$

where D_{res} – Directivity factor of receiving antenna; λ – radiation wavelength.

Substituting (9) and (11) into (10) we obtain:

$$P_{res} = (3D_{Tr}D_{res}\lambda^2 P_{Tr} w(r)^2) / 32\pi r^2. \tag{12}$$

Here D_{Tr} –directivity factor of the transmitting antenna used in (9) when the dipole is replaced by a real antenna.

Thence

$$r^2 = \left(3D_{Tr}D_{res}\lambda^2 P_{Tr} w(r)^2\right) / 32\pi P_{res}. \tag{13}$$

Then the action range of CCS will be determined by the expression:

$$r_{max} = B\,|\,w(r)\,|, \tag{14}$$

where B – energy potential of the radio channel [5].

$$B = \frac{3\lambda}{4\sqrt{6}\pi}\sqrt{\frac{P_{Tr}}{P_{res,\min}} D_{Tr}D_{res}}, \tag{15}$$

The value $P_{res,min}$, which determines the minimum permissible sensitivity of the receiver when only the fluctuation noise affects the receiver, is determined from (2) by the specified minimum permissible value of the signal power $h^2_{\min}$, which is calculated from (1) by the maximum permissible error probability of piece-by-piece reception $p_{\text{err,val}}$.

Substituting (2) into (1) taking into account (12), we can express $r_{\max}$ in explicit form:

$$r_{\max} = \frac{3\lambda}{8\pi}\left[\frac{D_{Tr}D_{res}P_{Tr}T}{3|\ln(2p_{err,val})|v^2}\right]^{1/2}|w(r)|, \tag{16}$$

When the mutual hindrances affect the receiver input in general this algorithm for calculation of the CCS action range is not applicable. Here, the maximum permissible value of error probability of the piece-by-piece reception directly appears as the limiting condition. At the same time the action range of CCS $r_{\max}$ corresponds to the condition:

$$p_{err.} = p_{errval.} \tag{17}$$

The error probability in this case is calculated according to (5). The power of signal and mutual hindrances included in (5) are determined by (2), taking into account (9) - (11). The equivalent energy is calculated by (6). The normalized CMD for a signal and a mutual hindrance, having the same structure with it, is given by:

$$g_{0q}^{2} = \left\{ \frac{\sin\left[0.5\Omega_{nq}T\right]}{0.5\Omega_{nq}T} \right\}^{2}, \tag{18}$$

where Ω_{nq} – shift of carrier frequencies of the signal and q^{th} hindrance.
In the explicit form dependence of $r_{\max}$ from p_{errval} at incoherent reception of phase-shift keyed signals can be obtained only under the influence of one concentrated hindrance for the case: a non-fading signal – Rayleigh fading noise. Then the expression for the error probability of the piece-by-piece reception will take the form:

$$p_{err} = 0.5\exp\left[-h^{2}\frac{1}{2+h_{n}^{2}g_{0}^{2}}\right], \tag{19}$$

Taking into account (9) - (11), we obtain:

$$r_{\max} = \frac{3\lambda}{8\pi}\left[-\frac{D_{Tr}D_{res}P_{Tr}T}{\left|\ln(2p_{err,val})\right|v^{2}}\frac{1}{\left(2+h_{n}^{2}g_{0}^{2}\right)}\right]^{1/2}\left|w(r)\right|, \tag{20}$$

The main difficulty at solving of (16) and (20) is the definition of the attenuation function $w(r)$. For distances of 200 ~ 300 km the attenuation function can be described by the simplified Hufford's equation for a spherical surface [6]:

$$w(r) = 1 + i\sqrt{\frac{ir}{\lambda}}\int_{0}^{r} w(x)\left[\frac{1}{\sqrt{\varepsilon(x)}} + \sin\left(\frac{r-x}{2a}\right)\right]\frac{1}{\sqrt{x(r-x)}}dx, \tag{21}$$

where λ – wavelength, a – Earth's radius, $\varepsilon(x)$ – complex dielectric constant of the underlying surface, x – distance from the source to the current integration point, r – distance from the transmitter to the receiving point.
The expression (21) is Volterra integral equation of the second kind:

$$w(r) = f(r) + \rho\int_{0}^{r} w(r)K(r,x)dx, \tag{22}$$

In our case, the free term is identically equal to one, and the core contains a complex variable (dielectric constant $\varepsilon(x)$). Therefore, the solution of (21) will be find in the form:

$$w(r) = Re\left(w(r)\right) + iIm\left(w(r)\right). \tag{23}$$

After simple transformations, we finally obtain:

$$w(r)=1-\int_0^r w(x)\sqrt{\frac{r}{2\lambda x(r-x)|\varepsilon|^2}}\begin{pmatrix}\sqrt{|\varepsilon|-\varepsilon''(x)}+\\ \sin\left(\frac{r-x}{2a}\right)|\varepsilon|\end{pmatrix}dx+i\int_0^r w(x)\sqrt{\frac{r}{2\lambda x(r-x)|\varepsilon|^2}}\begin{pmatrix}\sqrt{|\varepsilon|+\varepsilon''(x)}+\\ \sin\left(\frac{R-x}{2a}\right)|\varepsilon|\end{pmatrix}dx, \quad (24)$$

Herein |ε| – module of complex dielectric constant, which is defined by expression:

$$\varepsilon = \varepsilon' + i2c\sigma/\lambda, \quad (25)$$

where ε′ – relative permittivity; c – light speed; σ – conductivity.

4. Results and discussion

The equation (24) has analytical solution only for a number of special cases. Therefore, its solution was carried out by numerical integration using the quadrature method.

The core of equation (24) has form:

$$K(r,x)=\sqrt{\frac{r}{2\lambda x(r-x)|\varepsilon|^2}}\left[-\left\{\sqrt{|\varepsilon|-\varepsilon''(x)}+\sin\left(\frac{r-x}{2a}\right)|\varepsilon|\right\}+i\left\{\sqrt{|\varepsilon|+\varepsilon''(x)}+\sin\left(\frac{r-x}{2a}\right)|\varepsilon|\right\}\right], \quad (26)$$

Then the numerical solution can be found through the known iterative formula:

$$w_k=\frac{1+\sum_{j=1}^{k-1}A_{kj}K_{kj}w_j}{1-A_{kk}K_{kk}}, \quad (27)$$

where $k = 0,1,2,\dots N$; N – number of knots; $A_{kj}= p_k h$ – set of integration coefficients; h – integration step; $p_k h = Ba_k$ – weight coefficients for different degrees of quadrature formulas; $w = w(r_k)$; $K = K(r_k, x_j)$.

Since the core (26) has singularities at the ends because of the denominator x (r-x), direct use of (27) is impossible. To eliminate singularities, each integral in (24) is divided into three parts. In the first integral, the variable is replaced by the condition $x = y^2$, and in the third integral by $x = r - y^2$. Then we get:

$$\int_0^r K(r,x)w(x)dx=\int_0^{\sqrt{n_1 h}}K_1(r,y^2)w(y^2)dy+\int_{n_1 h}^{r-n_2 h}K(r,x)dx+\int_0^{\sqrt{n_2 h}}K_3(r,r-y^2)w(r-y^2)dy, \quad (28)$$

where n_1 and n_2 – number of points by which the first and the last integrals in (28) are calculated, respectively.

The software for implementation of the obtained algorithm was developed in MATHLAB [7]. The underlying surface is considered as piecewise-homogeneous. The initial data to the program are the Earth's radius, the light speed, the wavelength, the splitting step of track, the number of sections with different permittivity and electrical conductivity and the number of nodal points of each section. In addition, the values arrays of ε' and σ are specified at the nodal points of integration.

Task solution for the track consisting of two sections (well conducting soil and sea), an analytical solution and experimental data on which are given in [4] (p. 310) was obtained for testing of obtained algorithm and developed program. The test results are shown in Figure 1.

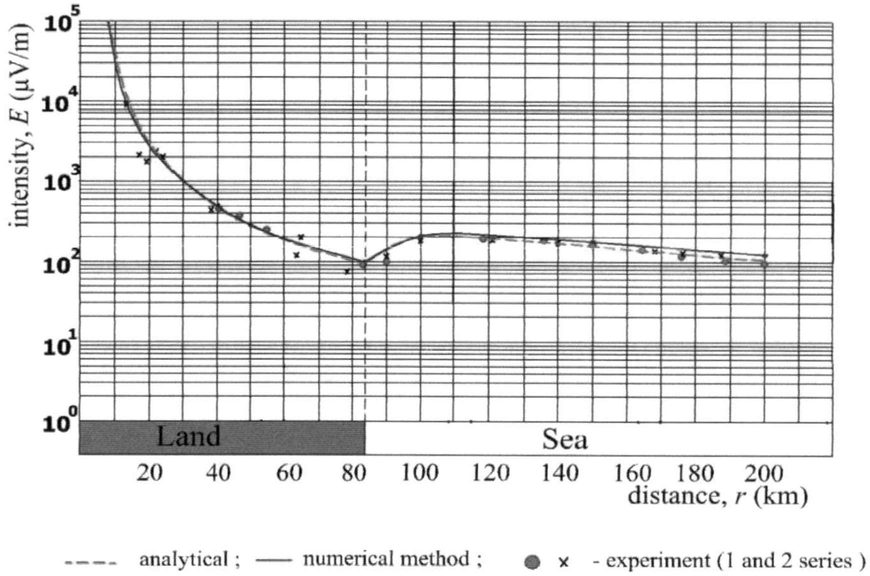

Figure 1. Comparison of theoretical and experimental data with the results of numerical integration.

Initial data for calculation: P_{Tr} =10 kW; λ = 96 m; land area – 84 km; sea section – 116 km; land: ε′ = 10, σ = 0.009 S/m; sea: ε′= 80, σ = 4 S/m.

As can be seen from the plot the developed algorithm of the numerical integration of Hufford's equation gave a good matching with the analytical solution for the spherical surface and the experiment results.

The developed algorithm and program were used to optimize the composition of CCS of LDSS in the rivers basins of Siberia and the Far East. The underlying surface was considered as piecewise-homogeneous. To determine the action zones of the CCS, beams were formed, diverging from the station, along which the dimensions of the sections with different electrical properties were set and in each of them the parameters ε' and σ were set. The size of the action zones for each directions was defined from (16) or (20) after determination of values of attenuation function by the numerical method. Equation (20) was used if in the given direction there were areas of intersection of the action zones of adjacent CCS with the presence of mutual hindrances.

Since (16) and (20) belong to the class of transcendental equations, their solution, and, consequently, the action range of the CCS in each of the given directions are determined by the successive approximation method. In this case, r_{ass} is specified in the first step and $w(r_{ass})$ is determined by (21). Then, r_{max} is determined by (16) or (20). If $r_{max} > r_{ass}$, on the next step r_{ass} increases, if $r_{max} < r_{ass}$, r_{ass} decreases and the calculation process is repeated until the condition $| r_{max} - r_{ass}| < \Delta r$ will be met. Herein the value Δr determines the adjusted accuracy of determination of the CCS action range.

An example of construction of CCS action zones is shown in Fig. 2. The Ob basin of the Russian inland waterways is considered in this figure. The following initial data were used in the calculations: the output power of all transmitters is 400 W; λ = 1000 m; v^2 = 10E-12; T = 0.01 s; $p_{err,val}$ = 10E-3; step of cutting of frequencies of CCS transmitters – 500 Hz; Ω_{nq} = 2÷20 kHz; D_{Tr} , D_{res} = 0.9.

5. Conclusion

As can be seen from Fig. 2, the optimization of the LDSS structure made it possible to provide all inland waterways in the Ob basin by continuous field of differential correction. As can be seen from the results of mathematical simulation, only six CCS (triangles in Figure 2) were required to solve this task.

In addition, it should be noted that, with a rational separation of the carrier frequencies of adjacent CCS, in most cases it was possible to achieve a significant reduction in the influence of mutual hindrances in the areas of intersection of the action zones of adjacent CCS, which allows increasing the range of confident reception of corrective information.
The main problem in determination of the CCS action range was the creation of arrays of electrical parameters of the underlying surface ε' and σ. Existing Russian soil maps are not detailed enough, therefore errors appear. Taking into account these errors at the step-by-step solution of the transcendental equations (16) and (20), it makes sense to specify the accuracy Δr not less than 3 ~ 5 km.

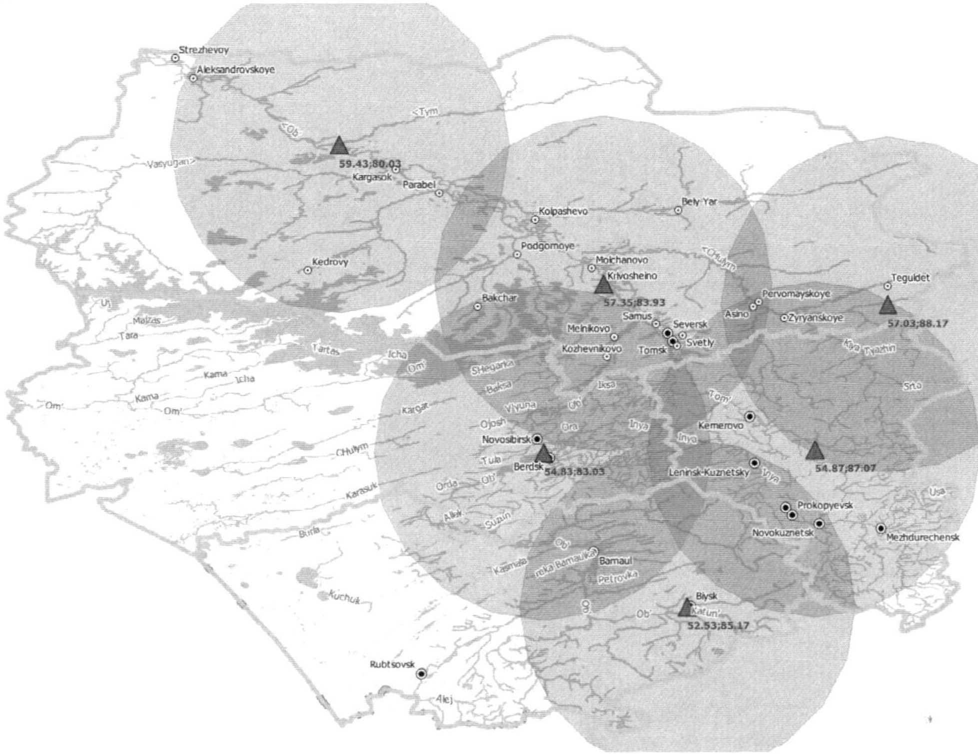

Figure 2. Coverage of the Ob basin the areas of CCS

6. References

[1] Jonas M and Oltmann J-H 2013 IMO e-Navigation Implementation Strategy – Challenge for Data Modelling *TransNav, the International J. on Marine Navigation and Safety of Sea Transportation* **7** no. 2 pp 45–49

[2] Zhang Y and Bartone C 2004 A General Concept and Algorithm of Projected DGPS for High-Accuracy DGPS-Based Systems *Navigation* **51** no. 4 pp 293–309

[3] Sikarev A A and Fal'ko A I 1978 *Optimum reception of discrete messages* (Moscow: Connection) p. 328

[4] Fejnberg E L 1999 *Propagation of radio waves along the earth's surface* 2nd edition (Moscow: Science) p. 496

[5] Karetnikov V V and Sikarev A A 2013 *Topology of differential fields and range of control and correction stations of high-precision positioning on inland waterways* (SPb.: Admiral Makarov State University of Maritime and Inland Shipping) p. 353

[6] Senchenko A A and Salomatov Yu P 2012 A comparison of quadrature methods for a solution of Hufford integral equation *Proc. of Tusur University* **2** no. 2(26) pp 36–41

[7] Shahnov S F 2015 Calculation of function field weakening of the control and correction stations taking into account the influence of the underlying surface *Vestnik Gosudarstvennogo universiteta morskogo i rechnogo flota imeni admirala S.O. Makarova* no.1(29) pp 116–123

IOP Conf. Series: Earth and Environmental Science **171** (2018) 012011 doi:10.1088/1755-1315/171/1/012011

The Experimental Study on the Hollow Wave Dissipation Structure of a Semi-Submersible Embedded Ball Box in the Specification Spectrum Wave

Danian Liu, Zichang Shangguan, Zhiyun Wang

Institute of Ocean and Civil Engineering, Dalian Ocean University, Dalian 116023, China

Abstract. In order to break wave to protect coast without coastal landscape destruction, for better coastal protection and development. Based on the physical experiment mode, a kind of semi-submersible embedded spherical block wave structure is proposed. In this paper, we discussed the effect of this wave dissipation structure. The experimental results show that the wave structure has a good wave dissipation effect. At the same time, the specially designed ball box net box can be used as the incubator to bring economic benefits.

1. Introduction

Before the 1960s, the main protection of the coast was the construction of seawalls, this is a passive form of coastal protection, slthough the coast is reinforced but the development and utilization of the coastal zone is inconvenient.[1]-[3] With the maturity of marine engineering technology, the way to protect the coastline begins from the traditional single way and develops in the form of multi mode combination. In addition, people begin to pay more attention to the environment when they exploit and utilize the resources. The green, environment-friendly and sustainable development projects are becoming more and more popular. The floating breakwater has the advantages of improving the water quality and economy in the harbor area compared with the fixed breakwater.[4]-[6] At present, there are many related researches on the wave performance of floating box structure.[7]-[11] Therefore, a physical model of the hollow block type wave dissipation structure is proposed, which can reduce the erosion of the coastal area and protect the coastal area without destroying the natural environment. Meanwhile, it also brings economic benefits.

The experiments were carried out at the wave flume (length 60m, width 1m, height 1.5m) at Jiangsu experimental base of Chinese Academy of fishery sciences. A wave maker is mounted at one end of the water tank and it can produce stable specification spectrum wave. The other end of water tank has wave screen to reduce the wave reflection.

2. Physical Model and Test Scheme

A physical model is a group wave eliminator consisting of several identical units connected together. The unit body is composed of 8 inner diameter 16mm, an outer diameter 20mm and three through right angle pipe 22, The joint is sealed and made into a six face type block with a side length of 100mm. There is a ball cage with a diameter of 80mm which can be used for breeding inside the permeable block. A small hanging ring is respectively fixed at the lower four corners of the permeable block body for the submergence and fixation of the model. The connection mode between the units is flexibly connected. The model consists of 15 rows and 6 columns, the transverse spacing between the units is fixed to 60mm, longitudinal row spacing has 50mm 100mm and 150mm, Each row of units in

the same dive depth of the regular arrangement and even row, singular row relative to diving 20mm 121 type of dislocation arrangement.
According to the gravity similarity criterion and the test equipment condition, the test scheme is shown in Table 1.

Table 1. Test scheme

Test water depths/m	Wave height/m	Wave cycle/s	Immersion depth/cm	Row spacing/cm	Arrangement mode
0.9	0.05	0.8	0	5	Regulation
	0.06	0.9		10	arrangement,
	0.07	1.0		15	121 arrangement
	0.08	1.1			

The model scale is 1:30.

3. Test Date Analysis

This paper focuses on wave attenuation performance of the hollow wave dissipation structure of a semi-submersible embedded ball box under the action of specification wave by comparing the experimental data with the regular wave and the J spectrum(JONSWAP) wave.
Figure.1-Figure.3 are the experimental result analysis of the influence of wave height, period, longitudinal row spacing and arrangement mode of row on transmission coefficient.

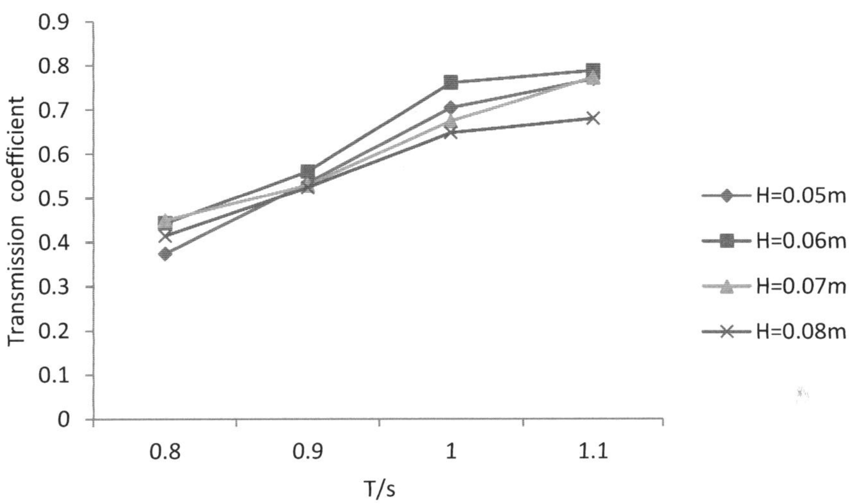

Figure 1. Effect of wave height on transmission coefficient

3.1. Effect of Wave Height on Transmission Coefficient

Figure 1 is the test result of the influence of the transmission coefficient of the semi-submersible embedded ball box with the anti-wave structure of the empty block in the space of 10cm, the descend depth 0cm, and the arrangement of the lines. It can be seen from the figure that, with the increase of wave height, the transmission coefficient is from the larger to the smaller. At wave height 0.06m is the largest transmission coefficient of 0.8. But the range of the numerical changes is not large. It means that wave height has little influence on transmission coefficient.

3.2. Effect of Row Spacing on Transmission Coefficient

Figure 2 is the experimental result of transmission coefficient of the semi-submersible embedded ball box with the anti-wave structure of the empty block under submerged 0cm, wave height 0.7m and row arrangement with regular arrangement under different row spacing. As can be seen from the diagram, the transmission coefficient decreases first and then increases with the increase of row spacing. In the condition of row spacing 10cm, the wave elimination effect is larger than the row spacing, and it is the working condition of 5cm and 15cm. However, the range of numerical change is not large, that is, the longitudinal row spacing has little effect on the transmission coefficient.

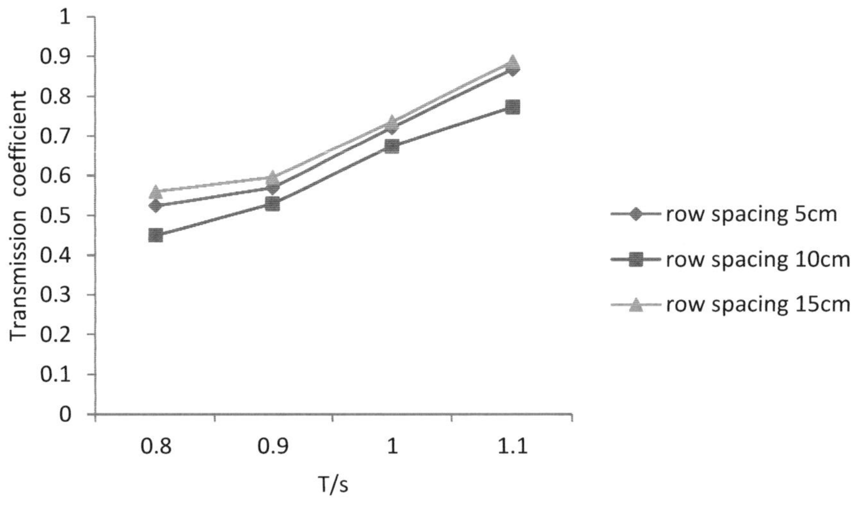

Figure 2. Effect of row spacing on transmission coefficient

3.3. Effect of the Way of Row Array on Transmission Coefficient

Figure 3 is the experimental comparing result between regular arrangement and 121 arrangement under the diving 0cm and row spacing of 10cm. It is shown from the graph that the transmission coefficients of 121 arrangement are smaller than those of regular arrangement. But it is larger in large cycles. It means that the wave dissipation performance of the 121 arrangement is better in small cycle, when the wave become larger the wave dissipation performance of the regular arrangement is better. Comparing the curves of the three wave heights, it can be seen that in the arrangement of the row spacing of wavelets, the range of numerical changes is very small, and the range of variation becomes larger with the increase of wave height. In the wave height 0.08m, the range of variation is about 0.15, that means the longitudinal arrangement has little effect on wave attenuation.

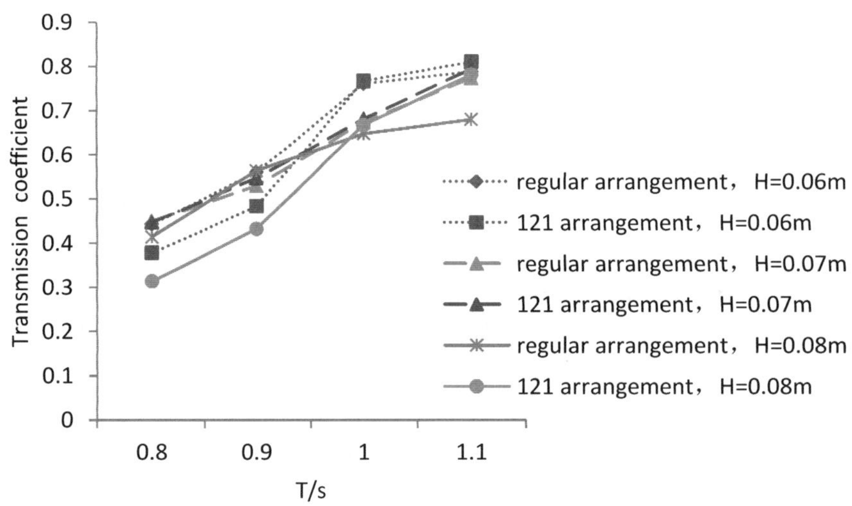

Figure 3. Effect of the way of row array on transmission coefficient

3.4. Effect of Cycle on Transmission Coefficient

According to the results of Figure 1 —Figure 3, the results show that the transmission coefficient increases with the increase of the period under the specification spectral wave. In small cycles, the slope of the change curve of the transmission coefficient is small, and the slope of the curve becomes larger with the increase of the cycle. The wave structure has obvious effect on small cycle. When the cycle reaches 1.1s, the transmission coefficient is 0.8, and there is still a certain wave elimination effect.

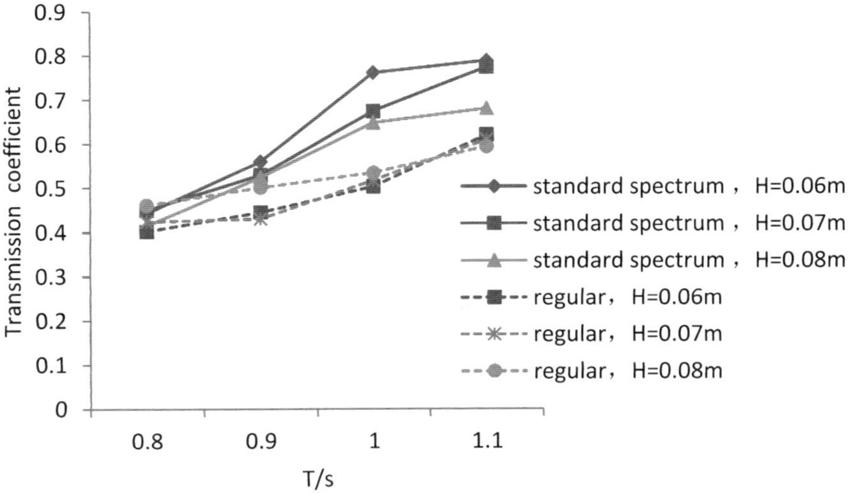

Figure 4. Effect of the cycle between code spectrum wave and regular wave on transmission coefficient

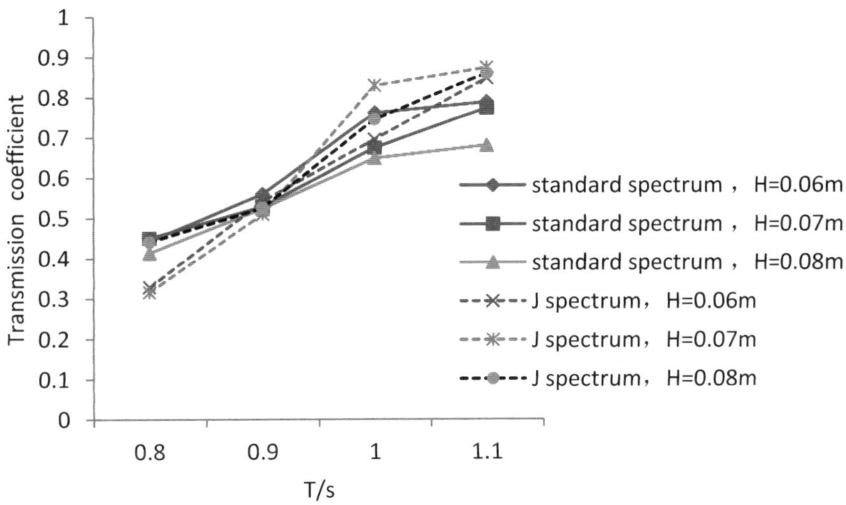

Figure 5. Effect of the cycle between code spectrum wave and J spectrum wave on transmission coefficient

3.5. The Influence of Cycles on Transmission Coefficients under Different Bop

Figure 4 and figure 5 are the contrast of the experimental data of the influence of the period of the gauge spectrum and the regular wave and the J spectral wave on the transmission coefficient,

It can be seen from the figure 4 that the transmission coefficient of J spectrum under small cycle is smaller than that of normal spectral wave. But with the change of the period, the slope of the transmission coefficient change curve of the J spectrum is larger than the slope of the normal spectral wave, At periodic 1.0s, the transmission coefficients of J spectrum waves are larger than those of standard spectral waves.

It can be seen from the figure 5 that the transmission coefficients of the canonical spectrum and regular wave are basically the same at periodic 0.8s, As the cycle increases, the curve of the gauge spectrum increases more than the regular wave. During the cycle 1.0s, the two sets of curves were clearly differentiated.

4. Summary

In this paper, the physical model test of a semi submerged sphere block wave elimination structure under the action of standard spectral wave is studied, analyzed the influence of wave height, period, longitudinal row spacing and arrangement of row arrangement on transmission coefficient. And compare it with the experimental data of periodic influence factors under regular wave and J spectrum wave. The influence of wave height, longitudinal arrangement spacing and arrangement mode on transmission coefficient has a certain regularity, but it has little influence on transmission coefficient. The period is the main factor affecting the transmission coefficient, and there is a similar change rule under the action of specification wave, J wave and regular wave, but there are differences.

5. Reference

[1] Luan Chengjiang, Xie Qingliang 2000 *China J. Coastal erosion and its control countermeasures in China. Journal of soil and water conservation.* 14 pp 44-47

[2] Zuo Shuhua, Li Jiujiang, Chen Shenliang 2006 *China J. Coastal erosion and its causes and protection engineering. Yellow River* 28 pp 23-25

[3] Omer Yuksek, Hizir Onsoy, Alin R B 1995 *J. Coastal erosion in eastern black sea region Turkey. Coastal Engineering*

[4] S.A Sannasiraj, V.Sundar, R.Sundarravadivelu 1998 *J. Mooring forces and motion Responses of pontoon-type floating breakwaters. Ocean Engineering*

[5] Wang Yongxue, Wang Guoyu 2002 *China J. Research progress and application of floating breakwaters,.Shipbuilding of China* 43 pp 69-75

[6] Jiang Muhuai, Zhou Xiaoguo Liu Mingwei 2013 *China J. Research and application progress of wave dissipation effect of permeable breakwater. Chinese water transportation* 13 pp 360-361

[7] K.Murali, J.S.Mani 1997 *J. Performance of cage floating breakwater. Journal of Waterway, Port, Coastal and Ocean Engineering*

[8] Jiang Changbo, Yuan Shengliang, Liu Jing 2012 *China J. Experimental study on wave dissipation characteristics of permeable breakwater with permeable box*. Traffic science and Engineering 28 pp 56-61

[9] Wang Yongxue, Wang Guoyu 2002 *China J. Research progress and application of floating breakwaters,.Shipbuilding of China* 43 pp 69-75

[10] Huang Lu 2013 *D. Ananlysis of wave dissipation performance of T type perforated breakwater.Dalian University of Technology*

[11] He Chaoyong, Wang Dengting, Feng Weibing 2014 Study on wave dissipation performance of rectangular square box floating breakwater J. *Water transport engineering* chapter 1 pp 14-18

Influence Analysis of Blade Fracture of Ducted Propellers of offshore platforms on Flow Field

L J Ou[1], **W Zhang** [2]

1 School of Civil Engineering and Transportation, South China University of Technology, Guangzhou 510640, China;
2 Guangzhou Maritime Safety Administration of the People's Republic of China, Guangzhou, 510700, China;

Abstract. The flow field of ducted propellers in viscous flow is analyzed and calculated with CFD technology, and the flow field and surface pressure distribution of the ducted propeller of offshore platforms with the fracture in different positions of a certain blade are simulated. Firstly, the geometrical models of the ducted propellers are established with UG software and put into the girding caved up by the Gambit. Then the numerical simulations are carried out with CFD, by which the surface pressure distribution of the ducted propellers in open water are calculated using Moving Mesh method. Lastly, the flow field characteristics of ducted propellers with the fracture in different positions of a certain blade are compared, and the result data are analyzed and summarized to gain flow field change laws. The results can be applied in project for reference.

1. Introduction

As special propellers, ducted propellers are widely applied to offshore platform dynamic positioning system. Because ducted propellers are less affected by the sea conditions changes outside and are protected by their ducts and also the ducts can improve the course stability significantly. Now ducted propellers are applied to lots of merchant vessels too. Numerical simulation has advantages of low-cost and short-cycle, and can obtain complete data and simulate the measurement of various data in practical operation process. CFD is widely applied to the numerical simulation of flow field. Because of its adaptability and accuracy00[3][4][5]. CFD gradually becomes an important tool in hydrodynamic research and its application areas have been gradually expanded. Through the analysis of ducted propellers by 3D unsteady CFD, the surface pressure distribution of ducted propellers can be investigated accurately. In the research and application of CFD, Su0 used the potential-based surface panel method numerically forecast unsteady performance of ducted propellers, and the predictions of hydrodynamics are high. But some error existed in surface pressure distribution of the duct surface circulation and pressure distribution of blade surface. In the prediction research of viscous CFD, Lu0 numerically simulated the open water performance of ducted propellers and the performance forecast accuracy with different grid models and turbulence models was compared and analyzed. It was concluded that k-ε turbulence model was more suitable for the numerical calculation of the open water performance of ducted propellers. Yao0 simulated the performance and flow field of ducted propellers with different Froude number using VOF. Li0 numerically researched the effects of tip clearance on duct propeller's hydrodynamic performance, and the results showed that thrust and moment decreased with increasing tip clearance, and tip clearance had obvious effect on the pressure distribution of blade surface near tip. Li0 numerically researched the influence of duct parameters on hydrodynamic performance and analyzed the influence law of length-diameter ratio, contraction coefficient and dilatation coefficients on hydrodynamic performance. Li0 analyzed the PIV experimental data of propeller tail vortex field, including velocity distribution of trailing vortex area, trailing vortex structure, velocity deficit and eddy piece of layered and so on.

At present the numerical simulation methods of hydrodynamic performance of ducted propellers with CFD apply maturely already, so the flow field of the ducted propeller of offshore platforms is simulated using numerical forecast method with CFD in this article.

2. Computational Model and Method

2.1. *Governing Equations and Numerical Methods*

The relative rotation motion between the ducted propeller blade and the duct can be considered as the 3D unsteady flow of incompressible fluid, so it can be represented with the Reynolds-averaged momentum equation:

$$\frac{\partial}{\partial t}(\rho u_i)+\frac{\partial}{\partial x_j}(\rho u_i u_j)=-\frac{\partial p}{\partial x_i}+\frac{\partial p}{\partial x_j}[\mu(\frac{\partial u_i}{\partial x_j}+\frac{\partial u_j}{\partial x_i}-\frac{2}{3}\delta_{ij}\frac{\partial u_l}{\partial x_l})]+\frac{\partial}{\partial x_j}(-\rho\overline{u_i' u_j'}) \tag{1}$$

Where, ρ is fluid density, and $-\rho\overline{u_i' u_j'}$ is Reynolds stress. According to the eddy viscosity assumption by Boussinesq, the relationship between Reynolds stress and average velocity gradient is established, and that is:

$$-\rho\overline{u_i' u_j'}=\mu_t(\frac{\partial u_i}{\partial x_j}+\frac{\partial u_j}{\partial x_i})-\frac{2}{3}(\rho k+\mu_t\frac{\partial u_i}{\partial x_i})\delta_{ij} \tag{2}$$

Where, μ_t represents turbulent viscosity, and it is a function of turbulent kinetic energy k and turbulent dissipation rate ε.

The unsteady turbulent used in this article is calculated with the RNG *k-ε* double equation model and the renormalization group keeps the Reynolds average equation closed, and that is:

$$\rho\frac{Dk}{Dt}=\frac{\partial}{\partial x_j}(\alpha_k\mu_{eff}\frac{\partial k}{\partial x_j})+G_k+G_b-\rho\varepsilon-Y_M \tag{3}$$

Where, G_k is turbulent kinetic energy generated by gradient change, and G_b is turbulent kinetic energy produced by buoyancy change.

$$\rho\frac{D\varepsilon}{Dt}=\frac{\partial}{\partial x_j}(\alpha_\varepsilon\mu_{eff}\frac{\partial\varepsilon}{\partial x_j})+C_{1\varepsilon}\frac{\varepsilon}{k}(G_k+C_{3\varepsilon}G_b)-C_{1\varepsilon}\rho\frac{\varepsilon^2}{k}-R \tag{4}$$

With the amendment of turbulent viscosity in RNG *k-ε* double equation model compared with standard *k-ε* model, rotation and swirling flow in average flux are considered, and the model can hand the flow with high strain rate and large streamline deflection, so it is more suitable to hand the complex flow with the high speed rotating ducted propeller. The wall function is used in the near wall region.

Two-order implicit time marching method is applied in the separated solver to calculate the unsteady turbulent, and the standard law of wall is applied to treat the wall. Governing equations are dispersed in time and space. The finite volume method is used in space discretization transforming the governing equations to calculable algebraic equations. The finite volume method integrates governing equations on every calculating element, dispersed equations show flow filed parameters on every calculating element. Time discretization consists of all the integration of each element in different equations on time in a time step. The pressure term is dispersed with two-order central difference scheme, while others are dispersed with two-order upwind scheme.

2.2. *Computational Modeling*

2.2.1. Geometric model of the ducted propeller. Main dimensions and parameters of the ducted propeller: diameter $D=2.000$ meters; pitch ratio $H/D=0.900$; area ratio $A_Z/A=0.700$; hub diameter ratio $d_0/D=0.167$; blade number $Z=4$; back rake $\varepsilon=0$; stirring paddles $Ka4-70$. The propeller section type is 19A, the chord length is 1.00 meter, the distance Δ between the blade and the duct is 4.0 millimeters, and the inside radius of the draft tube is 1.004 meters. The geometric model of the duct propeller is shown in Figure 1. In order to research the hydrodynamic performance of the ducted propeller with blade fractured, the CFD computational model of the ducted propeller with the fracture at different positions 0.9R, 0.8R, 0.7R, 0.6R, 0.5R of a certain blade, shown in Figure 2,is set to be calculated, compared, analyzed and researched.

The monitoring points are set to simulate the signal extracting of pressure sensors to get the fluctuating pressure of duct wall. By extracting the average pressure in the local region, the pressure variation values of the monitoring points are obtained. And real-time pressure values of the monitoring points are recorded with calculation to simulate the pressure sensors.

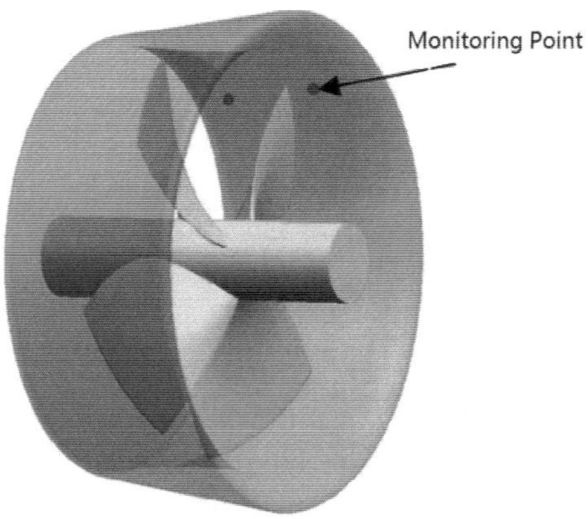

Figure 1. model of ducted propeller (monitoring points).

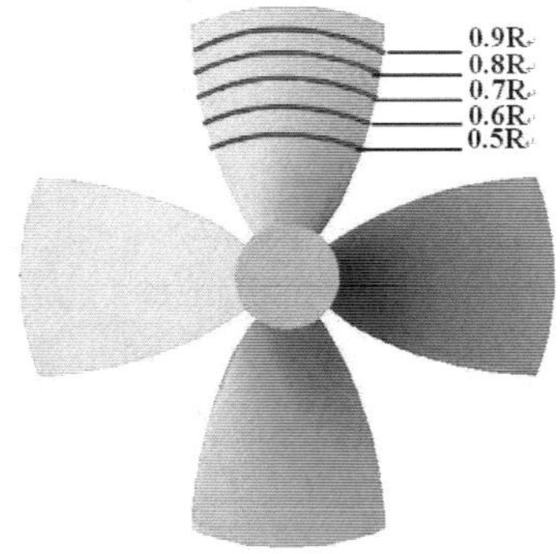

Figure 2. model of ducted propeller with the fracture at different positions.

2.2.2. Computing domain and mesh generation. The computational domain is cylindrical, and the radius R_a is 4 meters. The distance between the front of the computational domain and the propeller disk L_f is 4 meters, while the distance between the end of the computational domain and the propeller disk L_r is 6 meters. Mesh generation is the most important part in CFD simulation and one of key factors that have effect on simulation accuracy and efficiency. Over-sparseness grids result in inaccuracy, while over-density results in the increase of the amount and non-convergence of calculation. When meshing the interior computational domain, "Function{start size 12, growth size 1.4 size limit 20}" is a function defined in order to mesh the propeller tip and the domain with large curvature. The mesh generation of the interior computational domain is showed in Figure 3, and the mesh is 611 000. The mesh generation of the exterior computational domain is showed in Figure 4, and the mesh is 586 000. The data of the interior and exterior computational domain is exchanged through interface, which is meshed by regular grids.

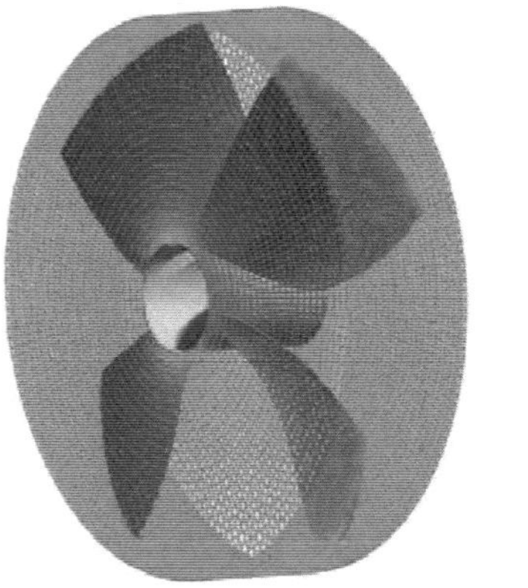

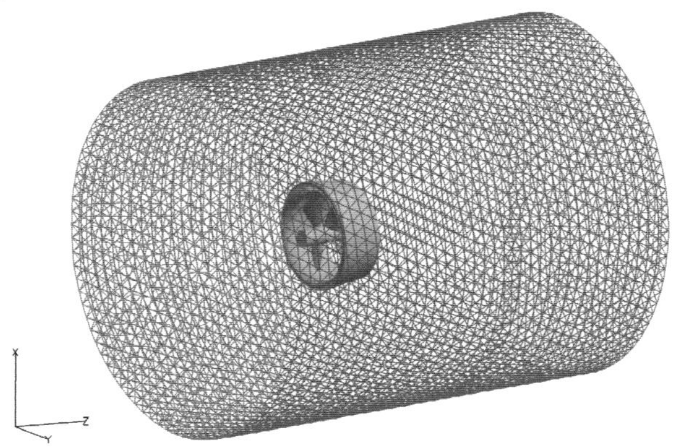

Figure 3. mesh generation of the interior computational domain.

Figure 4. mesh generation of the exterior computational domain

2.2.3. Boundary conditions. In the open water calculation of the ducted propeller, the whole computational domain is in rotation motion relative to some reference frame, and there is no mutual interference around.so MRF model can be used. Velocity inlet boundary conditions are set at the inlet boundary, and velocity components of the uniform inflow are set. The outflow velocity and pressure are unknown before the flow problem is solved, so mass flow inlet boundary conditions are set at the outlet boundary while the cylindrical surface is set as wall and no slip solid wall condition is set at wall. The fluid in the interior computational domain is using MRF model and rotates about axes with the angular velocity 5 revolutions per second (RPS) in the open water calculation of the ducted propeller.

3. Analysis of Duct Inner Wall Fluctuating Pressure and Flow Field Changes of Ducted Propellers with Blade Fractured

3.1. Analysis of Duct inner Wall Fluctuating Pressure of Ducted Propellers with Blade Fractured

Working Propellers are simulated through 3D unsteady calculation. With the duct inner wall fluctuating pressure of normal ducted propellers and the ones with blade fractured, pressure fluctuating graphs are obtained. Due to space limitations, pressure fluctuating graphs of monitoring points of normal ducted propellers and the ones with the fracture at different positions 0.9*R*, 0.8*R*, 0.7*R*, 0.6*R*, 0.5*R* of a certain blade are shown in Figure 5-8, respectively.

In the CFD, the rotation speed of the ducted propeller is 5r/s, which means the period is 0.2s, and the frequency is 5Hz and blade frequency is 20Hz.The results show that the duct inner wall fluctuating pressure of monitoring points varies periodically. The current of fluctuating pressure of normal ducted propellers is sin wave-form, and the change frequency is the same with blade frequency. The current of fluctuating pressure of ducted propellers with the fracture at different positions is deformed sin wave-form, and the change frequency is the same with rotation period of ducted propellers. The more fracture is, the more deformity of the current of fluctuating pressure is.

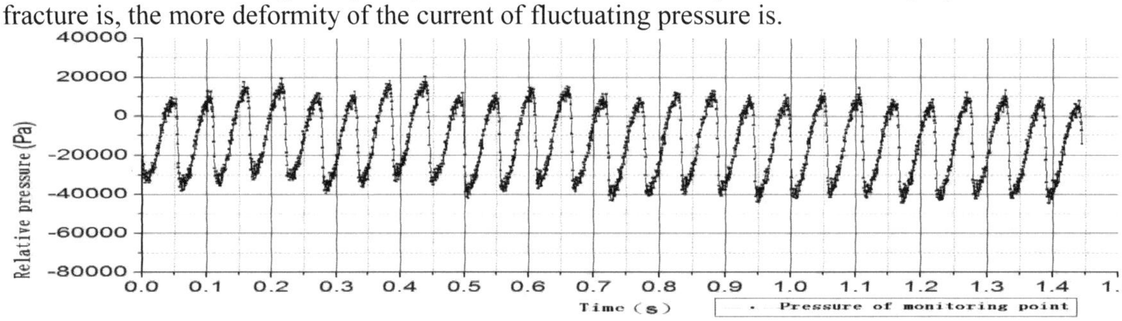

Figure 5. Pressure fluctuating graphs of normal ducted propellers.

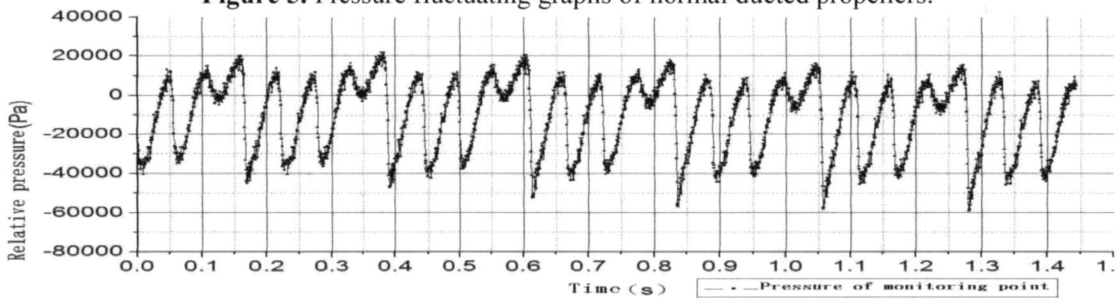

Figure 6. Pressure fluctuating graphs of ducted propellers with the fracture at 0.9R.

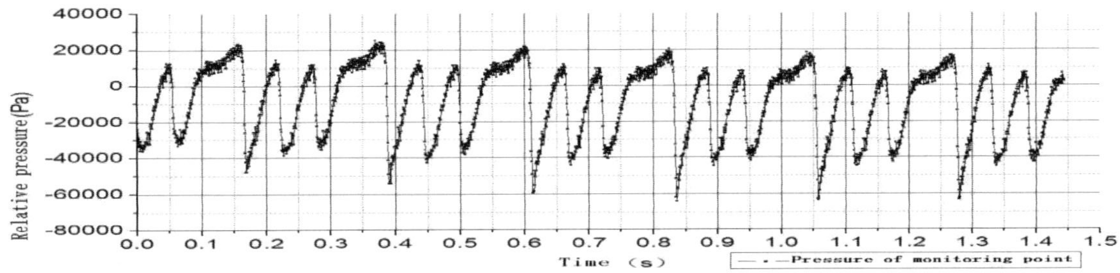

Figure 7. Pressure fluctuating graphs of ducted propellers with the fracture at 0.7R.

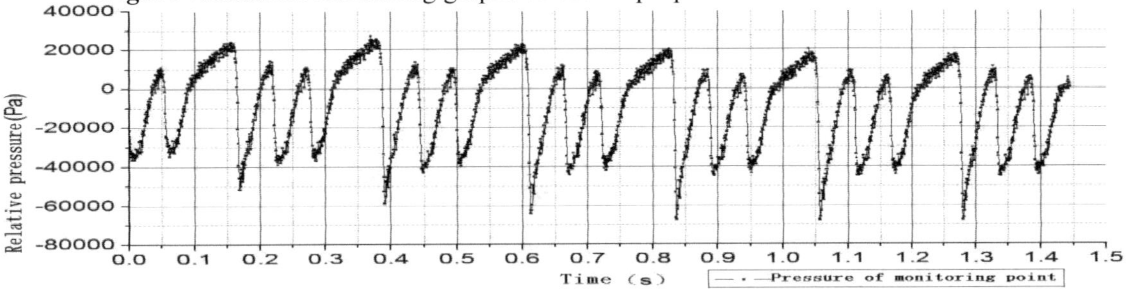

Figure 8. Pressure fluctuating graphs of ducted propellers with the fracture at 0.5R.

3.2. Influence Analysis of Blade Fracture of Ducted Propellers on Flow Field

To study the influence of blade fracture of ducted propellers, the flow field characteristics of normal ducted propellers and the ones with the fracture at different positions 0.7R and 0.5R are analyzed. The pressure of cross section of flow field at computational domain Z=3m, Z=0m, Z=-2.5m, Z=-5.5m is obtained, and the change law is compared and analyzed.

with ducted propellers fractured at different positions 0.7R and 0.5R ,the pressure distribution of cross section of flow field at computational domain Z=3m, Z=0m, Z=-2.5m, Z=-5.5m is shown in Figure 9,10 and 11,while the average relative axially pressure of flow field is shown in Figure 12.

From Figure 9, 10, 11, it is known that the suction of the inflow flow field displays a symmetrical distribution in normal working condition. The eccentric phenomenon of suction of the inflow flow field of ducted propellers appears when blade fractured. Stress nephograms show that the eccentricity lags behind the blade fracture position, which results from viscosity and lag of the flow. In normal working condition, the pressure of the trailing of ducted propellers displays a symmetrical distribution. The pressure of the trailing of ducted propellers with blade fractured displays an unsymmetrical distribution and eccentric phenomenon, which is affected by the fracture blade.

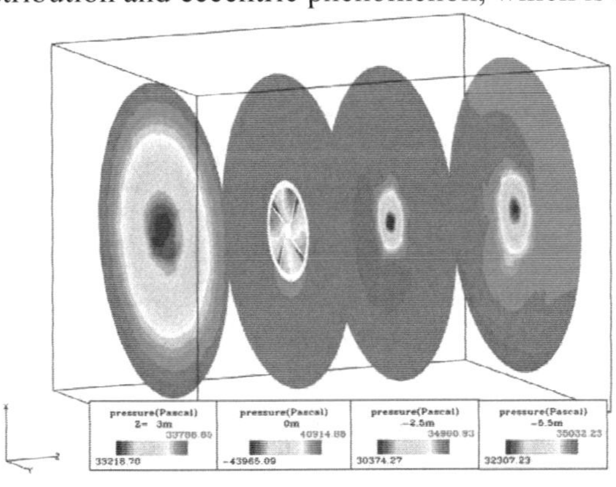

Figure 9. Stress nephograms in normal working condition.

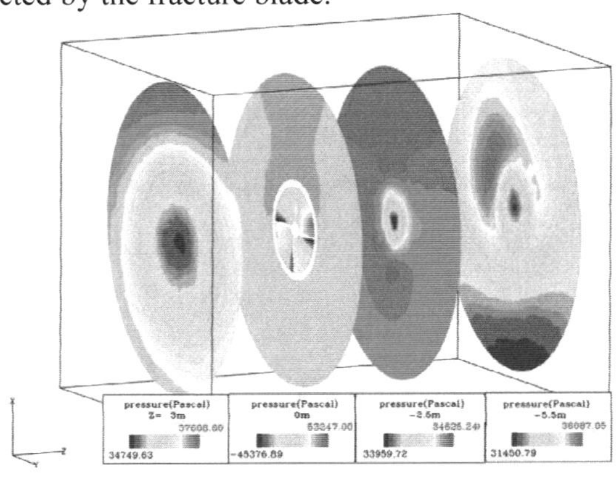

Figure 10. Stress nephograms in normal working condition of ducted propellers with the fracture at 0.7R.

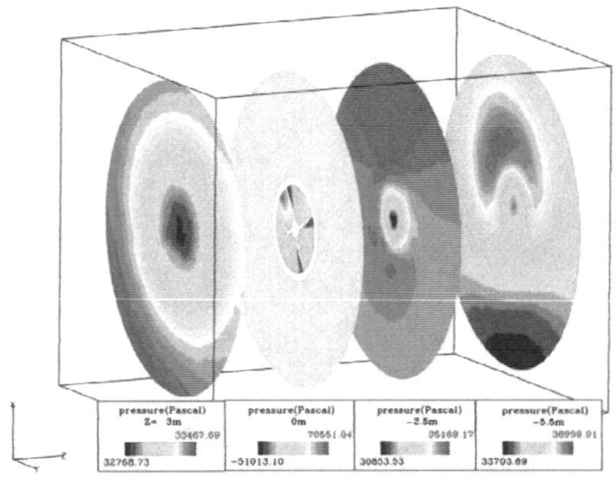

Figure 11. Stress nephograms in normal working condition of ducted propellers with the fracture at 0.5R.

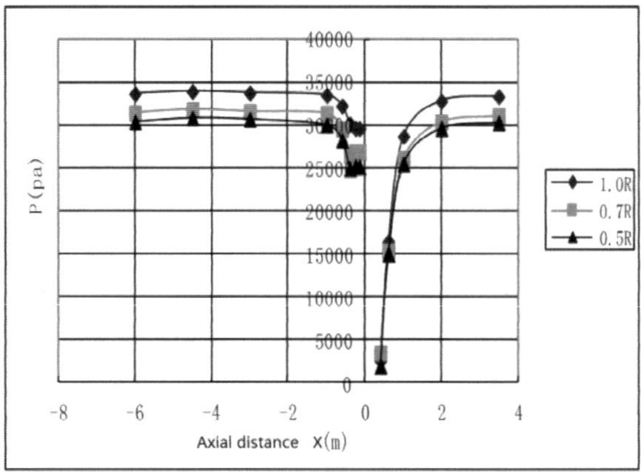

Figure 12. Comparison of pressure of normal ducted propellers and ones with the fracture at 0.7R and 0.5R.

4. Conclusions

(1) The results show that the duct inner wall fluctuating pressure of monitoring points varies periodically. The current of fluctuating pressure of normal ducted propellers is sin wave-form, and the change frequency is the same with blade frequency. The current of fluctuating pressure of ducted propellers with the fracture at different positions is deformed sin wave-form, and the change frequency is the same with rotation period of ducted propellers. The more fracture is, the more deformity of the current of fluctuating pressure is.

(2) The suction of the inflow flow field displays a symmetrical distribution in normal working condition. The eccentric phenomenon of suction of the inflow flow field of ducted propellers appears when blade fractured. Stress nephograms show that the eccentricity lags behind the blade fracture position, which results from viscosity and lag of the flow. In normal working condition, the pressure of the trailing of ducted propellers displays a symmetrical distribution. The pressure of the trailing of ducted propellers with blade fractured displays an unsymmetrical distribution and eccentric phenomenon, which is affected by the fracture blade.

(3) The pressure along the inflow of working flow field of ducted propellers is analyzed and it is found that the pressure of suction surface changes suddenly and the suction reaches maximum and the pressure of pressure surface recovers and increases. So there is pressure difference between the blade of ducted propellers and acting force in the opposite direction of inflow forms, which means the working principle of ducted propellers.

5. References

[1] X M Feng and F M Chen and R Q Cai 2006*Calculation of Propeller Open Water Performance by CFD Soft*ware FLUENT(China: SHIP AND BOAT)

[2] C Wang and S Huang and X S Xie 2008 *Hydrodynamic performance prediction of some propeller based on CFD* (China: JOURNAL OF NAVAL UNIVERSITY OF ENGINEERING)

[3] H Xu and J S Ma 2017 *CFD-based numerical study on the hull-rudder hydrodynamic interaction* (China: CHINESE JOURNAL OF HYDRODYNAMICS 32(3): 273-281)

[4] J Hu and N Wang and Y Hu 2017 *Numerical study on the hydrodynamic performance of ducted propellers* (China: JOURNAL OF HARBIN ENGINEERING UNIVERSIY 38(6): 815-821)

[5] X F Xue and T H Yan and B He 2017 *Modeling and Hydrodynamic Performance Analysis of MAU Propeller* (China: SHIP ENGINEERING 38(1): 38-42)

[6] Y M Su and Y B Liu and H L Shen 2012 *Surface Panel Method-Based Numerical Calculation for Predicting Ducted Propeller Performances* (China: JOURNAL OF HUAZHONG UNIVERSITY OF SCIENCE AND TECHNOLOGY.NATURE SCIENCE40(8): 57-61)

[7] X J Lv and Q D Zhou and G Ji and Y C Pan and B Fang 2010 *Prediction sand Comparison of Open Water Performance of Ducted Propeller* (China: JOURNAL OF NAVAL UNIVERSITY OF ENGINEERING 22(1): 24-30)

[8] Z Q Yao and W Chen 2011 *Numerical Analysis of Viscous Flow Field Around Ducted Propeller Considering Gravity* (China: JOURNAL OF JIANGSU UNIVERSITY OF SCIENCE AND TECHNOLOGY2 5(1): 1-3)

[9] J B Li 2010 *Influences of Tip Clearance on the Performance of Ducted Propeller*(China: SHIP AND OCEAN ENGINEERING 39(3): 36-39)

[10] C L and Y H Xie and L J Wang 2012 *Influence of Duct Parameters on the Hydrodynamic Performance of the Ducted Propeller* (China: SHIP AND BOAT 23(5): 17-22)

[11] G N Li and J Zhang and Z H Chen and Y H Xie 2011 *Propeller Trailing Vortex Analysis Based on PLV Experimental Data*(China: JOURNAL OF SHIP MECHANICS 15(10): 1110-1114)

Acknowledgement

This article is funded by the open project (NO.1605) of the country's key ocean engineering laboratory (Shanghai Jiao Tong University).

Laboratory Study of the Spatial Distribution of Extreme Overtopping Events at Vertical Structures

S Dong and J M Pearson

School of Engineering, University of Warwick, CV4 7AL, Coventry, UK

Abstract. Two-dimensional physical model experiments were conducted on a plain vertical seawall with a 1:20 sloping foreshore. Based on previous research, exponential equations were improved to describe spatial measurements for both impulsive and non-impulsive conditions. Comparisons were made between the spatial distributions from previous research studies, and compared to results of this study. It was observed that overtopping water generally distributes closer to seawall than previous predictions. 90% of overtopping was found generally to land within the distance of 0.04 wave-length under impulsive conditions, and similar results were observed in non-impulsive cases. A comparison between the spatial distribution of extreme events and total overtopping volume are discussed. Exponential equations are improved for fitting the distribution of extreme overtopping waves.

1. Introduction

Previous research primarily focuses on accurate predictions of total overtopping volumes passing the parapet of a seawall [1-3], with little or no guidance on the post-overtopping processes, such as where the thrown wave lands. The landward spatial distribution of overtopping is an important indicator which can help estimate the effects of hazards, such as coastal flooding. However, uncertainties still exist in determining the affected hazard area. This not only results in difficulties in predicting the safe distance behind the seawall crest, but also uncertainties exist in horizontal impact forces.

A number of studies have been performed using empirical prediction tools for estimating the spatial distribution of overtopping water [4-6]. Most of these focus on predicting the distribution of total overtopping water, which can help to understand the influence of the mean overtopping over a typical storm duration. However, the spatial distribution characteristics form an extreme individual overtopping wave is less well understood. For rubble mound breakwaters, Andersen and Burcharth [6] and Andersen et al. [7] presented equations involving the wave steepness, based on both field and laboratory data. The distribution for plain vertical structure was described by an exponential function for both laboratory and field studies [8, 9].

Peng and Zou [10] simulated the spatial distribution of overtopping water over both sloping and vertical structures in a numerical model. The results agreed well with predictions recommended by Andersen et al. [7] and Pullen et al. [9]. Peng and Zou [10] observed that the maximum individual overtopping water was thrown further than the mean overtopping distribution, although the profile had similar characteristics. The spatial distribution of extreme individual overtopping volume still requires further investigations.

It has been demonstrated by previous studies that the scale effects for overtopping discharges for impermeable vertical structure configurations can be neglected [11]. It is also reported by Pullen et al. [9], that the scale effects for spatial distribution is negligible. However, the influence of wind in field on both overtopping discharges and spatial distribution should be considered in predictions, as it may increase both indicators [7, 9, 12].

IOP Conf. Series: Earth and Environmental Science **171** (2018) 012013 doi:10.1088/1755-1315/171/1/012013

This paper presents the spatial distribution of overtopping volume at the plain vertical seawall under impulsive and non-impulsive conditions. In particular, the spatial distribution of extreme overtopping events and total overtopping volume are investigated and discussed.

2. Previous Research

2.1. Overtopping Discharges

Overtopping discharge plays a significant role in assessing the hazardous effects of extreme events at coastal structures. Many research studies have been performed to determine mean overtopping discharges, and results are succinctly summarized in EurOtop II [13]. The empirical equations for the mean overtopping discharge is given by:

For non-impulsive conditions:

$$\frac{q}{\sqrt{gH_{m0}^3}} = 0.05\exp\left(-2.78\frac{R_c}{H_{m0}}\right) \qquad [1]$$

while for impulsive conditions

$$\frac{q}{\sqrt{gH_{m0}^3}} = 0.011\left(\frac{H_{m0}}{\text{hs}_{m-1,0}}\right)^{0.5}\exp\left(-2.2\frac{R_c}{H_{m0}}\right) \quad \text{for } \frac{R_c}{H_{m0}} < 1.35 \qquad [2]$$

$$\frac{q}{\sqrt{gH_{m0}^3}} = 0.0014\left(\frac{H_{mo}}{\text{hs}_{m-1,0}}\right)^{0.5}\left(\frac{R_c}{H_{mo}}\right)^{-3} \quad \text{for } \frac{R_c}{H_{m0}} > 1.35 \qquad [3]$$

The wave conditions are classified as non-impulsive when $h_* = 1.3\frac{h_s}{H_{mo}}\frac{2\pi h_s}{gT_{m-1,0}^2} > 0.23$.

where H_{m0} is the significant wave height and the relative freeboard is described as R_c/H_{m0}. h_s is water depth at the toe of structure, g is gravity acceleration (=9.81 m/s^2), $s_{m-1.0}$, is the wave steepness, and q is the mean overtopping discharge per meter structure width per second.

2.2. Spatial Distribution

Bruce et al. [8] and Pullen et al. [9] described the spatial distribution of mean overtopping water (equation (4)):

$$q^* = e^{(-k(x/L_o))} \qquad [4]$$

where q^* is a percentage of the discharge at the relative distance of x/L_o after seawall crest in total discharge, and L_o is the offshore wave length. The constant k determines the shape of distribution, which also incorporated the wind effects.

3. Experimental Setup

The schematic design of the flume in the School of Engineering at the University of Warwick is shown in Figure 1. The flume is 22.0m long, 0.6m wide and 1.0m high with a 1:20 beach slope. A piston-type wave generator was applied with an active absorption system, and approximately 1000 pseudo-random waves are generated based on the JONSWAP wave spectrum (γ= 3.3), at a scale of 1:50. Due to the limited size of overtopping collecting system and the accuracy of measurements, a small number of tests with large overtopping discharges or long wave period were conducted with approximately 500 - 600 waves. The model vertical seawall was fixed at 12.21m away from wave paddle.

Experiments were undertaken to cover as large relative freeboard range as possible for both impulsive and non-impulsive conditions. **Table 1** presents wave conditions for both impulsive and non-impulsive conditions. The significant wave height ranged between 0.041m and 0.095m, and various wave periods were generated ranging from 1.03s – 1.75s were generated. As a result, generated wave steepness ranged between 0.016 – 0.039, and relative freeboard from 1 to 3.2 were covered in these series of tests. Both total overtopping volume and spatial distributions of overtopping water were measured with a multi-chamber container.

Table 1. Nominal wave conditions used for the physical tests (1:50 scale)

	Vertical seawall condition							
Water depth (m)	0.07	0.09	0.11	0.09	0.16	0.15	0.18	0.21
Input wave period (s)	1.07 - 1.75		1.07 - 1.50		1.03 - 1.20	1.04 - 1.60		
Significant wave height (m)	0.055 - 0.095		0.068 - 0.085		0.041 - 0.052	0.048 - 0.075		
	Impulsive conditions				Non-impulsive conditions			
Relative freeboard	1.05 - 2.54		2.59 - 3.22		0.97 - 1.22	1.35 - 3.24		

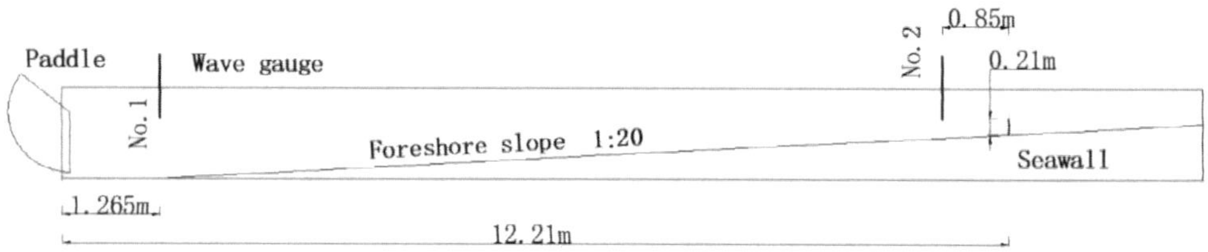

Figure 1. Cross-section of flume

3.1. *Spatial Measurements*

In order to measure both overtopping discharges and spatial distribution of overtopping water, a multi-chamber container was fixed behind the vertical seawall. Figure 2 shows the cross-section of the whole container. Among the 11 chambers, the relatively smaller ones were placed close to seawall while larger ones were located further. This design was selected in order to measure the spatial distribution with a high resolution especially in the area close to the seawall crest. The top of each chamber wall was made into a slope with a sharp edge in order to minimise the overtopping volumes jumping into adjacent chambers. The overtopping volumes were measured with calibrated wave gauges and pressure transducers fixed in chambers.

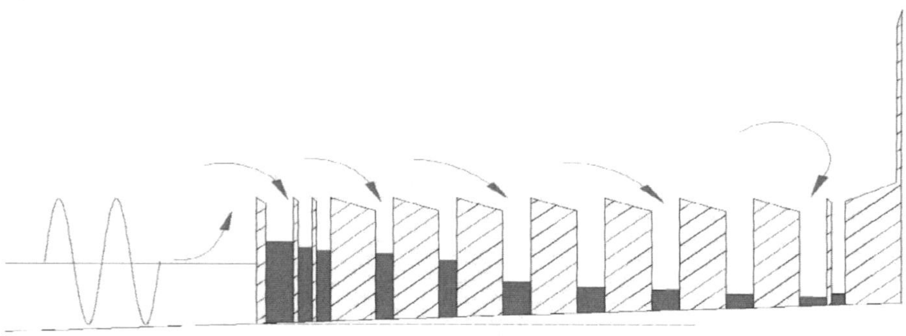

Figure 2. Schematic of the multi-chamber collection container

4. Results

4.1. *Overtopping Observations*

The overtopping water was collected by the multi-chamber container behind the seawall, and overtopping discharges were obtained by volume data converted from water level changes in each container. The measurements of mean overtopping discharges were compared with EurOtop II [13] predictions from equations (1)-(3). The dimensionless overtopping discharges are plotted against relative freeboard in **Figure 3**.

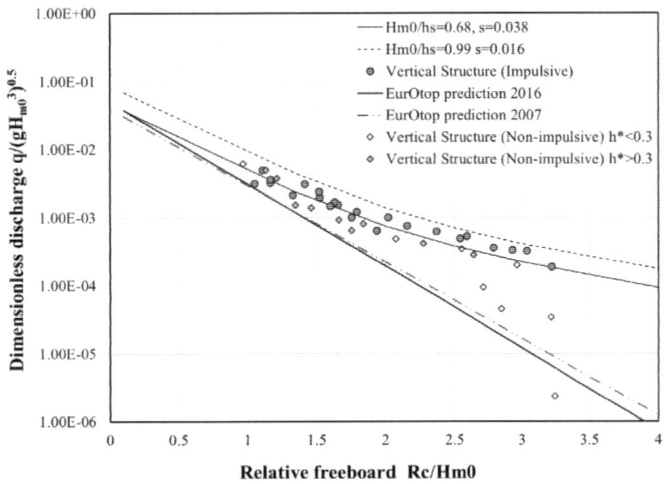

Figure 3. Mean overtopping discharges at plain vertical seawall

Figure 3 shows that the measured mean overtopping discharges from impulsive conditions fit predictions well, however the results from the non-impulsive overtopping events show that the measurements are under predicted. As the relative freeboard increases, the deviation appears to increase. A similar phenomenon was reported by Besley et al. [1], namely that the underestimation in non-impulsive overtopping discharge is noticeable, when the impulsiveness parameter h_* is over 0.3. In addition, the overtopping proportion under non-impulsive conditions was also found be under-predicted, especially in the cases with relative freeboard $R_c/H_{m0} > 1.8$.
A further aim of the study was to elucidate the difference between the spatial distributions of extreme individual overtopping events compared to the mean overtopping distribution. The measured maximum individual overtopping discharges were compared to the equations from EurOtop II [13] (see **Figure 4**). Although there was some scatter for extremely low overtopping events, acceptable agreement was observed between laboratory data and empirical predictions.

4.2. Spatial Distribution of Total Overtopping Volume

The spatial distribution of total overtopping volumes can be predicted by an exponential function of relative distance to seawall (equation (4)). The best fit equation was obtained by the constant k fixed to 29 when there is no wind speed involved in tests [9]. Four examples of the measured spatial distribution are given in **Figure 5**. It is noticeable that the prediction equation from Pullen et al. [9] overestimates the distribution of total overtopping water. According to the predicted distribution curve, 90% of water lands in a distance of 0.08 wave length, however, in terms of measurements from this study, 90% of overtopping water lands in a distance of 0.04 wave length. This over prediction becomes more significant in the cases of extremely low overtopping discharges, while acceptable agreement are observed in tests with larger overtopping volumes. It is therefore assumed that the constant k in the exponent of equation (4) can be related with relative freeboard R_c/H_{m0} when no wind speed is involved. Larger R_c/H_{m0} values are assumed to give a larger k value, and overtopping water lands closer to the crest of seawall.

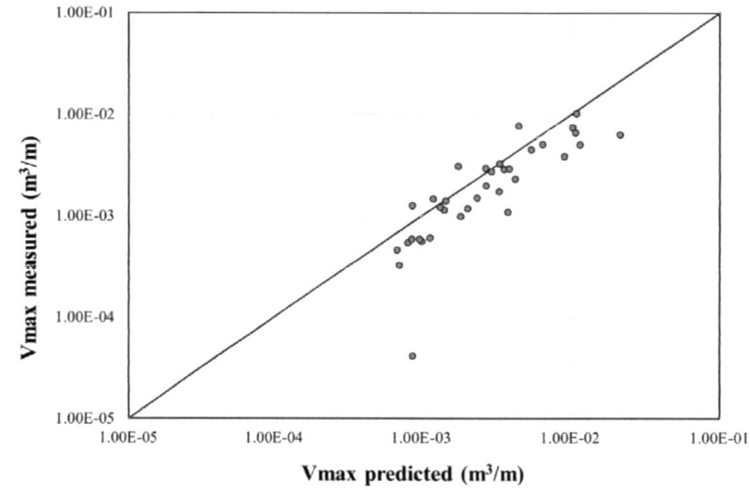

Figure 4. Maximum individual overtopping discharges over plain vertical structure

The spatial distributions given in **Figure 5** (a) and (d) are close to the seawall, which presents difficulties in distinguishing the distribution of total overtopping volume and extreme overtopping events. Thus, the square root of $x/L_{m-1,0}$ was applied to x-axis to improve the results. **Figure 6** gives the new spatial distribution of the four cases shown in **Figure 5**. The new distributions show a clear trend within a distance of 0.14 wave length, while the distributions in further landward of this region from the seawall are less important. Less than 5% of overtopping water landed beyond a distance of 0.14 wave length.

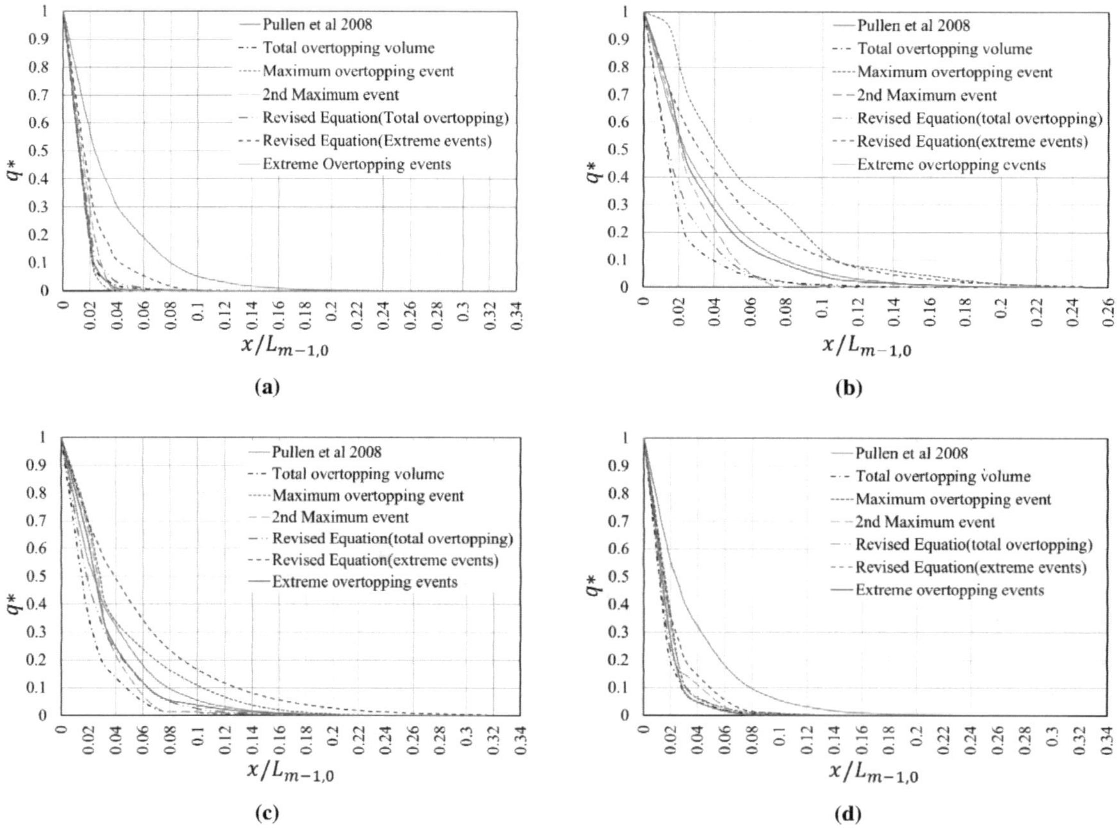

Figure 5. Spatial distribution behind plain vertical seawall. (a) Case R_c/H_{m0}=1.755 (Impulsive) (b) Case R_c/H_{m0}=1.109 (Impulsive) (c) Case R_c/H_{m0}= 1.138 (Non-impulsive) (d) Case R_c/H_{m0}= 2.273 (Non-impulsive).

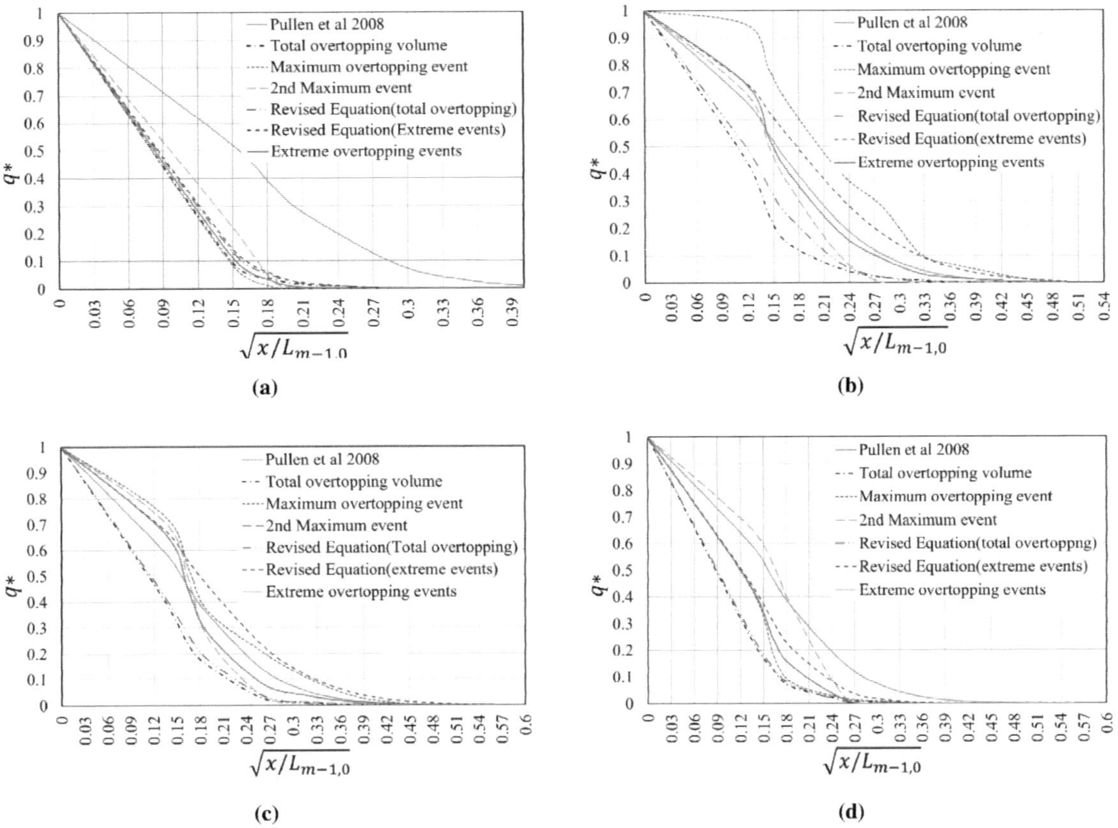

Figure 6. Improved spatial distribution behind plain vertical seawall. (a) Case R_c/H_{m0}=1.755 (Impulsive) (b) Case R_c/H_{m0}=1.109 (Impulsive) (c) Case R_c/H_{m0}= 1.138 (Non-impulsive) (d) Case R_c/H_{m0}= 2.273 (Non-impulsive)

Based on the exponential equation given by Pullen et al. [9], efforts were made to find the best fit equation of spatial distribution in each test case. As no wind speed applied in model tests, constant k was initially fixed to 29, and the distribution of overtopping water close to the seawall crest was given priority [8]. As assumed earlier, the value k is observed to increase with the relative freeboard R_c/H_{m0} . By quantifying the relationship between constant k and relative freeboard (R_c/H_{m0}), it is found that an exponential function is able to predict the value of k with a correlation of approximately R^2=0.93 (see **Figure** 7(a)). When no wind speed is involved, k in exponent for impulsive conditions can be predicted by the expression given as equation (5):

$$k = 22e^{0.85R_c/H_{m0}} \quad [5]$$

For green overtopping events, it is also found that the overtopping volume will influence the landward area behind the crest of structure within a distance of 0.12 of the maximum wave length maximum. The spatial distribution of the overtopping water can also be described by a similar exponential function with a correlation of around R^2=0.91(see **Figure** 7 (b)). The k value can be derived from equation (6).

$$k = 21e^{0.63R_c/H_{m0}} \quad [6]$$

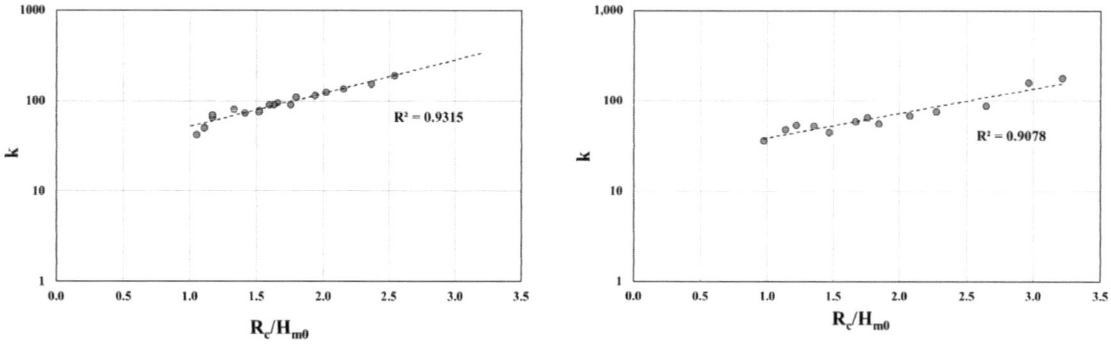

(a) (b)

Figure 7. Relationship between k and R_c/H_{m0}. (a) From Impulsive test conditions; (b) From Non-impulsive test conditions

4.3. Spatial Distribution of Extreme Events

In previous studies, measurements have been conducted on both individual and total spatial distribution on the smooth sloping structures. It is reported that the spatial distribution of maximum and average overtopping volume are similar [7]. Peng and Zou [10] obtained similar results by comparing the distribution of maximum individual overtopping event with the distribution of time average overtopping in numerical simulations. **Figure 5** illustrates that the exponential equation of total distribution is still able to describe the individual overtopping events (purple dotted line). The k value can still be determined by a similar exponential function as total distribution, with correlations of R^2=0.82 and R^2=0.91 for impulsive and non-impulsive conditions respectively (see **Figure 8**). The difference between individual and total spatial distribution is not always significant in all tested configurations. With the increase in relative freeboard, R_c/H_{m0}, the gap between two distributions tends to decrease, as a result of decreasing overtopping events. When the relative freeboard R_c/H_{m0} reaches 2.0, the spatial distribution of maximum individual overtopping events and mean overtopping discharges show similar results, and overtopping water lands immediately after seawall crest.

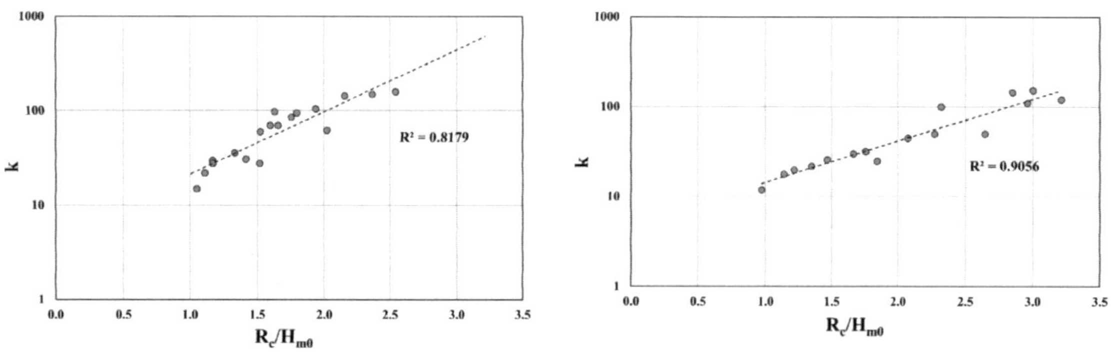

(a) (b)

Figure 8. Relationship between k and R_c/H_{m0} for largest 5 individual overtopping events. (a) From Impulsive test conditions; (b) From Non-impulsive test conditions.

5. Conclusion

This paper presents a series of laboratory experiments on the landwards distribution of overtopping water behind the crest of the plain vertical structure. The spatial distribution of both total and individual overtopping water are analysed, and described by exponential functions. An improved method for the prediction of constant k in exponent is recommended when no wind speed involved, based on research conducted by Pullen et al. [9] and Bruce et al. [8]. This method is also expanded for the predictions of spatial distribution of green overtopping events, as well as the individual overtopping events in the tested cases.

Irrespective of the wind effects, 90% of overtopping was observed generally to land in the distances of 0.04 wave length under impulsive conditions. Similar spatial distribution was found in non-impulsive overtopping events, and overtopping volume is distributed further with increase of relative freeboard. The exponential function of relative distance $x/L_{m-1,0}$ published by Bruce et al. [8] and Pullen et al. [9] can be adopted to predict spatial distribution of overtopping volumes. However, the value of constant k in exponent was underestimated in most cases. Better predictions may be achieved by the application of equation (5) and (6) with reasonable good correlations of R^2=0.93 and R^2=0.91. However, further validation with field data is desirable.

In addition, the individual overtopping distributions are found usually a bit further than the distribution of total overtopping discharge. However, the distribution of extreme and total overtopping volumes do not vary significantly when freeboard increases to over 2.0, while more scatter is observed on the distribution of extreme overtopping volume. No noticeable differences were observed between extreme and total overtopping events for high relative freeboard conditions (R_c/H_{m0} > 2.7).

6. References

[1] Besley P, Stewart T and Allsop N 1998 Overtopping of vertical structures: new prediction methods to account for shallow water conditions *Proc. Coastlines, Structures and Breakwaters, London, UK* 46-57

[2] van der Meer J and Bruce T 2013 New physical insights and design formulas on wave overtopping at sloping and vertical structures *Journal of Waterway, Port, Coastal, and Ocean Engineering* 140 04014025

[3] Allsop N, Bruce T, Pearson J, Alderson J and Pullen T 2003 Violent wave overtopping at the coast, when are we safe. In: *Proc. Conf. on Coastal Management 2003,* pp 54-69

[4] Lioutas A C 2010 Experimental research on spatial distribution of overtopping. In: *Hydraulic Engineering*: TU Delft

[5] Van Kester D 2009 Spatial distribution of wave overtopping. In: *Hydraulic Engineering*: TU Delft

[6] Andersen T L and Burcharth H F 2006 Landward distribution of wave overtopping for rubble mound breakwaters. In: *The International Conferece on the Application of Physical Modelling to Port and Coastal Protection*: Department of Civil Engineering, Aalborg University pp 401-11

[7] Andersen T L, Burcharth H F and Gironella F 2009 Single wave overtopping volumes and their travel distance for rubble mound breakwaters. In: *Coastal Structures,* pp 1241-52

[8] Bruce T, Pullen T, Allsop W and Pearson J 2005 How far back from a seawall is safe? Spatial distributions of wave overtopping. In: *Proc. International Conference on Coastlines, Structures and Breakwaters,* pp 166-76

[9] Pullen T, Allsop W, Bruce T and Pearson J 2009 Field and laboratory measurements of mean overtopping discharges and spatial distributions at vertical seawalls *Coastal Engineering* 56 121-40

[10] Peng Z and Zou Q-P 2011 Spatial distribution of wave overtopping water behind coastal structures *Coastal Engineering* 58 489-98

[11] Pearson J, Bruce T, Allsop W and Gironella X 2003 *Coastal Engineering 2002: Solving Coastal Conundrums*: World Scientific pp 2227-38

[12] Pullen T, Allsop W and Bruce T 2007 *Coastal Engineering 2006: (In 5 Volumes)*: World Scientific pp 4702-13

[13] Van der Meer J W, Allsop, N.W.H., Bruce, T., De Rouck, J., Kortenhaus, A., Pullen, T., Schüttrumpf, H., Troch, P. and Zanuttigh, B. 2016 EurOtop Manual on wave overtopping of sea defences and related structures. An overtopping manual largely based on European research, but for worldwide application. www.overtopping-manual.com.

Acknowledgements
Shudi Dong sincerely acknowledge the financial support for his PhD studies, from the School of Engineering, University of Warwick and Chinese Scholarship Council. Jonathan Pearson sincerely acknowledges the financial support by the Royal Academy of Engineering (RAEng/The Leverhulme Trust Senior Research Fellowships scheme).

A Preliminary Study on Numerical Waves and Its Impact on Horizontal Plate Based on Fluent

Jinpeng Hu , Shihui Liu and Ling Zhu

Civil and Transportation College, South China University of Technology, Guangzhou 510640, China
E-mail: 793495289@qq.com

Abstract. Based on the N-S equations for viscous, incompressible fluid and *RNG* $k-\varepsilon$ turbulence model, a two dimensional Numerical Wave Tank is established, utilizing the secondary development function of Fluent software and the *VOF* method. A two-dimensional linear regular wave and a random wave are generated by defining the motion of moving boundary and adding wave absorbing zone at the end of the wave tank. The resulting numerical wave shape agreed well with the theoretical shape. The simulated wave spectrum of random wave keeps the original structure of the target spectrum. The impact of waves on the horizontal plate is simulated referring to the existing physical model test .Compared with the physical and the numerical pressure data of *11* pressure points under the horizontal plate, it is found that the two datas fit well, indicating the effectiveness of *Fluent* in simulating the interaction between waves and structures, which lays a foundation for further exploring the factors that affect the magnitude of the impact pressure on the horizontal plate.

1. Introduction

With the rapid development of the marine economy, it has become a hot topic to explore the generation and evolution of waves and its impact on marine structures. In order to better study the regularity of wave action on marine structures, low cost and repeatable artificial wave-making methods are often used in the research. The artificial simulation of wave-making methods is divided into two parts: physical test and numerical test. Compared with physical model test, the advantages of computer numerical simulation are mainly economic and small scale effect, and it can effectively avoid the influence of sensor on flow field. Therefore, computer numerical method gets more and more scholars' favor.

At present, the research on numerical simulation mainly focuses on the method of wave making and wave absorbing. Liang[1] imitated physical method of paddle wave maker, combined with the *UDF*(User defined Function) and dynamic mesh function of *Fluent* software, simulating the irregular wave preferably. Based on *Fluent* software, a three-dimensional numerical wave flume was established by Lu[2]. Three dimensional stokes waves in three different working conditions were simulated by source function wave-making method in her test, which certified that *Fluent* has a good effect on wave-making. Based on the *N-S* equation of viscous incompressible fluid, Huang[3] established a two dimensional numerical wave flume by using open source *CFD* code *OpenFOAM*. The results showed that regular waves could be effectively made up by *OpenFOAM*. Li[4] simulated linear regular waves, irregular waves, Stokes waves and solitary waves using both paddle and source wave-making method , and compared the advantages and disadvantages of the two methods. Xin[5] defined the motion law of the wave-making board and successfully absorbed the wave at the wave absorbing area by using the macro function of *UDF*. The simulated numerical wave agrees well with the theoretical solution and the boundary element results.

In the field of wave absorbing, Dong[6] proposed a porous media wave absorbing method which can absorb simulated waves effectively, and gave a best combination of wave making and wave elimination. A conserved absorbing method is proposed Hu[7].By comparing the generated surface histories with the theoretical results, it is found that the conserved absorbing method is efficient for regular waves, random waves and nonlinear waves.

It is a very popular trend to add structures in numerical flume to simulate the effect of waves on marine construction in real ocean environment. Using the secondary development function of *Fluent* software, Zheng[8] put up a submerged seawall in the numerical tank, in order to explore the effectiveness of *Fluent* in simulating the interaction between waves and structures. Nonlinear boundary element method was used by Rolf Baarholm[9] to simulate the process of wave impact on the horizontal plate. Using *Fluent* software, Jin[10] established a two-dimensional mathematical model of interaction between regular wave and flat plate structures. By comparing the numerical simulation and experimental results of different working conditions, the field characteristics influenced by the impact process of waves on the structure was obtained, and the relationship between velocity of water point and impact pressure was got too.

Based on the user defined function (*UDF*) of *Fluent* software, this paper used a push-pedal wave-making method to simulate a two-dimensional linear regular wave and a random wave. A horizontal plate was added to the numerical flume to simulate the impact of regular waves on the horizontal plate. By comparing the numerical test data and the physical test data, the reliability of *Fluent* in simulating the interaction between waves and coastal structures is verified.

2. Mathematical Model

The control equation includes the fluid continuity equation:

$$\frac{\partial \rho}{\partial t}+\frac{\partial \rho u}{\partial x}+\frac{\partial \rho v}{\partial y}=0 \tag{1}$$

N-S equation:

$$\frac{\partial u}{\partial t}+u\frac{\partial u}{\partial x}+v\frac{\partial u}{\partial y}=-\frac{1}{\rho}\frac{\partial p}{\partial x}+\upsilon\left(\frac{\partial^2 u}{\partial x^2}+\frac{\partial^2 u}{\partial y^2}\right)+g_x+S_x \tag{2}$$

$$\frac{\partial v}{\partial t}+u\frac{\partial v}{\partial x}+v\frac{\partial v}{\partial y}=-\frac{1}{\rho}\frac{\partial p}{\partial x}+\upsilon\left(\frac{\partial^2 v}{\partial x^2}+\frac{\partial^2 v}{\partial y^2}\right)+g_y+S_y \tag{3}$$

In the formula, u, v is the speed in the direction of x and y, respectively; g_x, g_y is the acceleration of x and y direction ,and $g_x=0$, $g_y=-g$; p is the fluid pressure; ρ is the density of fluid; υ is the kinematic viscosity of fluid; S_x and S_y are the source term of wave absorbing, which are added only in the wave absorbing region. The expressions are as shown in (4):

$$S_x = \mu(x)\cdot u,\ S_y = \mu(x)\cdot v,\ \mu(x) = \alpha\frac{(x-x_1)}{|x_2-x_1|},\ x_1 \le x \le x_2 \tag{4}$$

In the formula, $\mu(x)$ is the damping coefficient; α is the empirical parameters; x_1 and x_2 are the starting and terminal coordinates of the wave absorbing region, respectively.

VOF method is used to capture the free surface, and the transport equation of fluid volume function is proposed as follows:

$$\begin{cases}\dfrac{\partial \alpha_q}{\partial t}+\dfrac{\partial(u\alpha_q)}{\partial x}+\dfrac{\partial(v\alpha_q)}{\partial y}=0\\ \sum_{q=1}^{2}\alpha_q=1,(q=1,2)\end{cases} \tag{5}$$

In the formula, α_q is the volume fraction of fluid, defined as the ratio of the volume of a fluid in phase q to the volume of the unit. If α_q =1, the unit is filled with q phase fluid, if α_q =0, there is no q

IOP Conf. Series: Earth and Environmental Science **171** (2018) 012014 doi:10.1088/1755-1315/171/1/012014

phase fluid in the unit, and if $0<\alpha_q<1$, the unit is a interface unit. As the numerical simulation, q=1 or 2 represents the air phase and the water phase, respectively.

3. Numerical Simulation

3.1. Geometric Model

Gambit software is used to establish the geometric model, as shown in Figure 1. The size of the numerical flume is *16m×1.2m*. While the depth of water is *0.8m* and the wave absorbing region is *4m* long at the end of the flume. The wave plate is set at the left boundary. Numerical flume mesh generation is shown in Figure 2, and the total grid is about 68900. The size of the grid is *0.05m* in the direction of wave propagation, and the size of the grid at the wave absorbing area is divided by gradual thinning method. Refine the grid at free surface to ensure the accuracy of calculation.

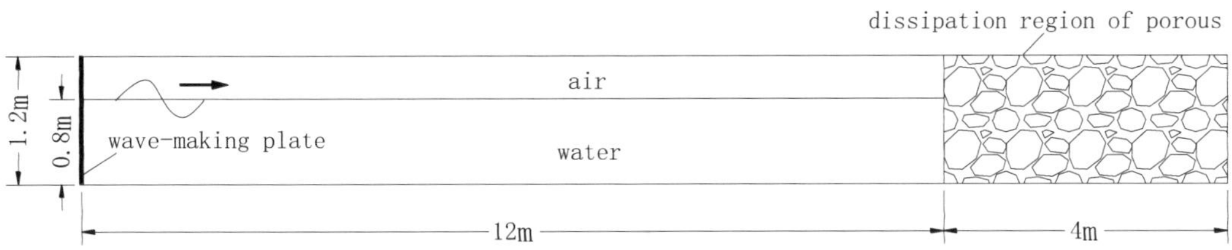

Figure 1. Numerical flume geometric diagram

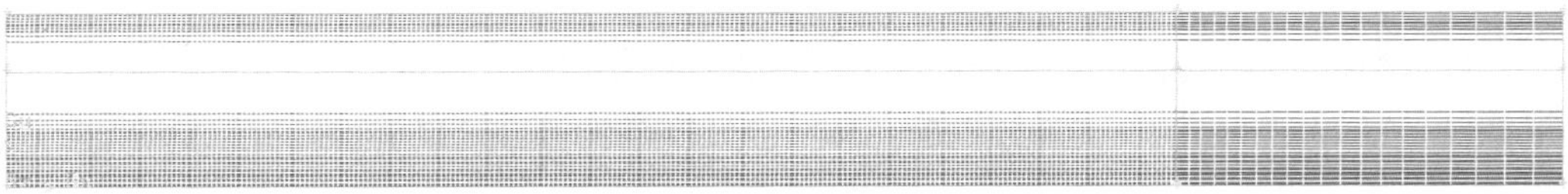

Figure 2. Mesh of numerical flume

3.2. Wave Simulation

The numerical flume is used to simulated regular waves and random waves both. The factors of regular waves are given as follow: wave height *H=0.1m*, periodic *T=1.6s*, depth of water *h=0.8m*, and wave length *L=3.55m*.

The *JONSWAP* spectrum is the target spectrum of simulated random waves. The parameters of *JONSWAP* spectrum are shown as follow: significant wave height *Hs=0.1m,* peak period of spectrum *Tp=1.6s*, depth of water *h=0.8m.*

The motion law of push plate is defined by the *DEFINE_CG_MOTION* macro in *UDF*, and the source item of wave elimination is written through *DEFINE_SOURCE* macro, and finally the macros are imported to *Fluent* through the Define Command Window.

3.3. The Parameter Setting of Fluent

A push plate wave-making method is used to simulate waves, so the left boundary is defined as the dynamic boundary, the top boundary is the pressure inlet boundary, and the rest is the wall boundary. The free surface of water is captured by the *VOF* method, and the *RNG* $k-\varepsilon$ turbulence model is used for turbulence model. The pressure velocity term is calculated by *PISO* algorithm. The layer pattern (*Layering*) is selected to update the grid to realize the motion of the wave-making plate. Time step is *0.005s*, and total calculation time is *50s*.

3.4. Analysis of Calculation Results

The comparison between the theoretical wave duration curve and the simulated wave duration curve of regular wave at one and three times wavelengths are given in Figure 3 and Figure 4 , respectively. As can be seen from Figure 3, the theoretical curve and the simulated curve fit well, the period of simulated wave is basically the same as the theoretical ones, and the phase is almost no deviation. But the wave height is attenuated from Figure 4, and the attenuation degree of wave height increases with the distance of wave propagation. This is mainly due to the fact that the regular waves based on potential flow theory are put forward in the ideal state of irrotational and inviscid, and the numerical wave simulated in this paper takes the effect of viscous and turbulence of water into account.

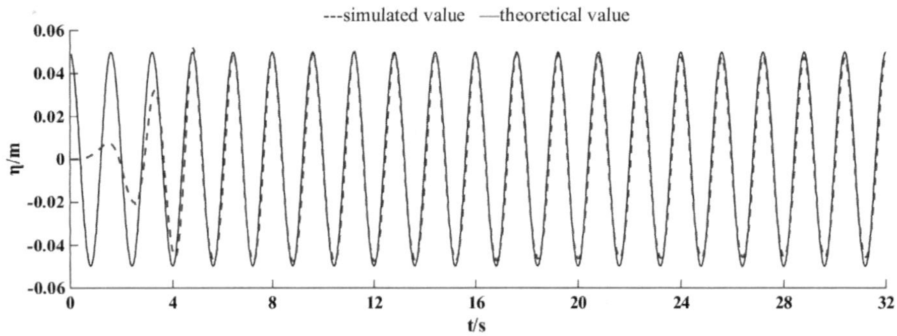

Figure 3. Comparison of theoretical value and simulated value of regular wave at one time wavelength

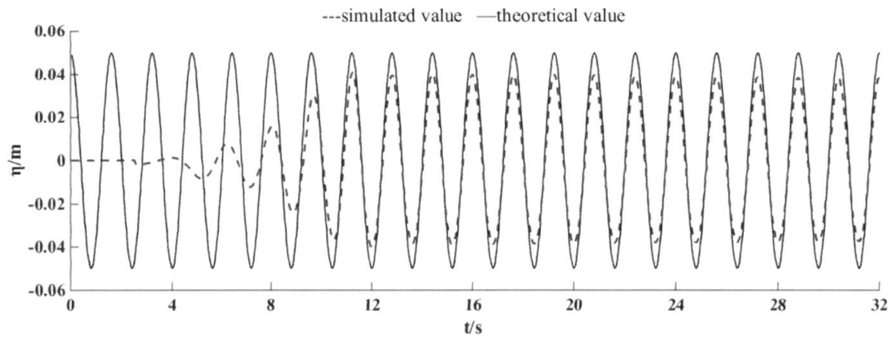

Figure 4. Comparison of theoretical value and simulated value of regular wave at three times wavelengths

The comparison between the theoretical wave duration curve and the simulated wave duration curve of random wave is given in Figure 5. It can be found that the two curves are in good agreement, which shows a good results of *Fluent* in simulating random waves.

The wave duration curve only reflects the external characteristics of the wave, while the wave spectrum can describe its internal structure. The simulated spectrum can be obtained by spectral analysis of waves at the location of 2m.

Comparison between the target value and simulated value of spectral parameters can be seen in Table1. From Table1, we can see the fact that there are little differences between the parameters of simulated and target spectrum.

From Figure 6, we can also get that the simulated wave spectrum keeps the original structure of the target spectrum.

Table 1. Comparison of spectral parameters

	significant wave height	peak period of spectrum

target value	0.1	1.6
simulated value	0.0956	1.587

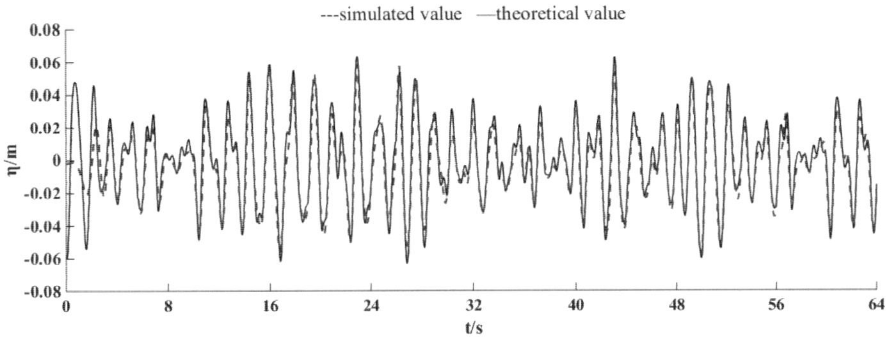

Figure 5. Comparison of theoretical value and simulated value of random wave

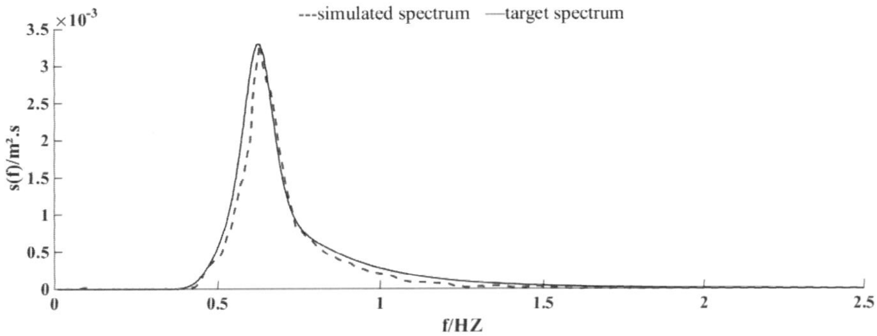

Figure 6. Comparison of target spectrum and simulated spectrum

4. Interaction between Waves and Horizontal Plates

4.1. The Position of the Horizontal Plate and Pressure Measurement Point

Dr. Ren[11] of Dalian University of Technology has done a series of physical model tests on the impact of waves on the horizontal plate. With reference to his physical model test, the geometry of the wave numerical tank and the layout of the horizontal plate are shown in Figure 7. The flume is *16m*×*1.0m*, while the depth of water is *0.6m* and the wave absorbing region is *4m* long at the end of the flume. The center of the horizontal plate is located at *x=8m*.The plate measures *1.0m* long×*0.02m* thickness, and the bottom of the plate is *0.03m* above the free surface.

11 virtual pressure points are arranged at the same distance (*0.09m*) at the bottom of the plate, while the distance between the two end pressure points and the ends of plate both are *0.05m*.

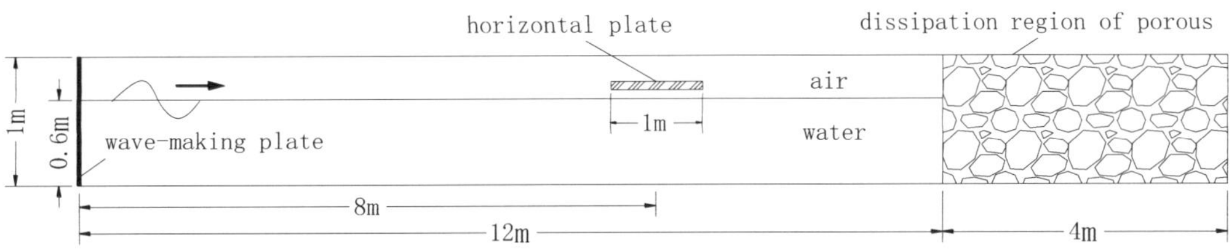

Figure 7. Geometric diagram of numerical flume with horizontal plate

4.2. The Impact of Wave on the Horizontal Plate

Two groups of numerical experiments are carried out according to the difference of the relative plate length. *Case 1:* $L_m/L_1 = 0.33$ $H=0.1m$ $\Delta h/H=0.03$; *Case 2:* $L_m/L_2 = 0.47$ $H=0.1m$ $\Delta h/H=0.03$ .In both case, L_m is the length of the plate, L_1 (L_2) is the wavelength of regular waves in *Case1(Case2)*, H is the wave height and Δh is the distance between the bottom of the plate and the free surface. The simulation time of each case is *30* times of wave period, and each period will have a peak pressure (maximum impact pressure). The average value of the pre *1/3* data of the *30* peak pressure of each measurement point is $P_{C,1/3}$ as the characteristic value, and the results are compared with the pressure data of physical test. The diagram is shown in Figure 8.

It can be seen from the graph that under different working conditions, the distribution law of the characteristic pressure along the bottom of plate in the numerical simulation test is approximately the same trend as that obtained from the physical model test, which indicates that *Fluent* has high accuracy in simulating the impact of waves on the plate structure.

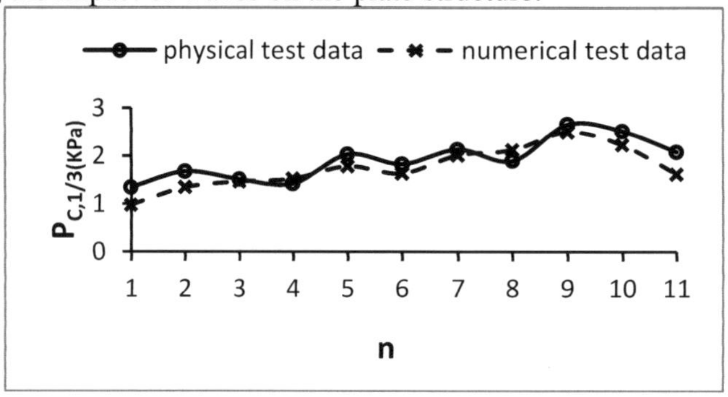

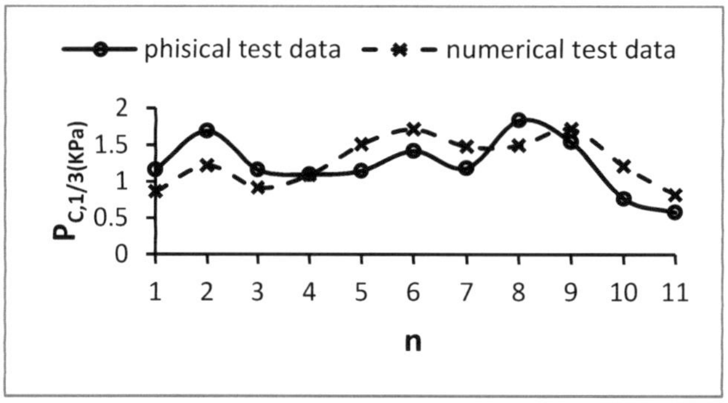

Figure 8. Comparison between characteristic pressure of numerical test data and physical test data at Case1(left) and Case 2(right)

5. Conclusion

Based on the *UDF* function of the *Fluent* software, this paper successfully simulates the regular wave and the random wave. The impact of wave on the horizontal plate is simulated by adding a horizontal plate structure to the numerical tank. Summarize as follow:

(1) In this paper, the *RNG* model and the *VOF* free surface capture method are used to simulate a two-dimensional regular wave and a random wave, and both simulated values are well fitted to the theoretical values. The simulated spectrum keeps the original structure of the target spectrum, and the parameters of simulated spectrum have little difference from target spectrum. The attenuation of wave height reflects the propagation state of real viscous flow in a certain extent.

(2)Compared with the existing physical model test, the impact process of regular waves on a horizontal plate was simulated by *Fluent.* By comparing the characteristic impact pressure values of *11* points under the plate, it is found that the numerical results are of high accuracy. There are some

shortcomings in this article that need further study, for example: the influence of wave height loss and the disturbance of horizontal plate to wave field are not taken into account in the numerical test.
This paper lays a foundation for further study of the influence factors of the impact of waves on the plate structures.

6. References

[1] Liang X F, Yang J M, Li X and Li J 2010 J. Ship Mech. 14(5):481-486.
[2] Lu P 2014 Guangdong shipbuilding. (6):45-47.
[3] Huang P F, Zhang K, Song C W and Fu S Y 2015 Ship Electronic Engineering. 35(6):73-76.
[4] Li H W 2009 *Study on Wave-making Method in Numerical Tank*. Harbin: Harbin Engineering University.
[5] Xin Y 2013 *Application of Fluent UDF Method in Numerical Wave Flume*. Dalian: Dalian University of Technology.
[6] Dong Z and Zhan J M 2009 Journal of Hydrodynamics. 24(1):15-21.
[7] Hu Z, Tang W Y, Xue H X and Zhang X Y 2015 China Ocean Eng. 30(1):137-148.
[8] Zheng Y N, Liu Z, Chen C P and Zhang J X 2015 China Offshore Platform. 30(6):67-71.
[9] Baarholm R and Faltinsen O M 2004 J. Marine .Sci. Technol. 9(1):1-13.
[10] Jin F 2013 Port & Waterway Engineering. (4):52-56.
[11] Ren B 2003 *Study on The Impact of Random Waves on The Superstructure of Different Types of Wharfs*. Dalian: Dalian University of Technology.

Numerical Modelling of Interaction between Solitary Wave and Sloped Seawall

Rui You and Guanghua He*

School of Naval Architecture and Ocean Engineering, Harbin Institute of Technology, Weihai, No.2, Wenhua West Road, Weihai, Shandong, P. R. China
*E-mail: ghhe@hitwh.edu.cn

Abstract. A numerical model for solving interaction between solitary wave and a sloped seawall is established. The CIP (Constrained Interpolation profile) scheme is employed to solve the flow field. THINC (Tangent of Hyperbola Interpolation for Interface Capturing) method is applied for capturing the free surface. Hydrodynamic forces is calculated by integrating the pressure and friction along the seawall surface. First, convergence tests with respect to mesh resolution and time step are examined, and the solitary wave profile is validated by experimental data from the published paper. Then, time series of the free surface elevation at specific location with different heights of incident solitary wave are presented. Finally, hydrodynamic forces acting on the weather side of the seawall are analyzed. The conclusion can be drawn that this numerical model has applicability for strongly nonlinear wave-body interaction problems.

1. Introduction

Seawall has long been concerned about the application in coastal engineering, for its protection mechanism for shore structures. Sloped seawalls are more proper energy dissipaters compared with vertical seawalls, because the considerable wave reflection amplifies the kinematics in front of vertical seawall. This paper employed a type of trapezoid seawall on the sloping beach proposed by Hsiao and Lin [1]. One of the key issue in solving this problem is that the nonlinear phenomenon occurs when solitary waves interacting with ocean structures. Analytical solutions are limited to idealized conditions or simplified geometries. Ma *et al* [2], Wu and Hsiao [3] investigated solitary waves interacted with vertical obstacles through experiments. Recent numerical studies have investigate plate structure as seawall such as [4,5], and vertical seawalls such as [6]. Scholars also carried out experiments to investigate solitary wave interacting with vertical seawalls [7]. More work is need to investigate the solitary wave interacting with sloped seawalls.

The present study establishes a numerical model based on a CIP scheme. The CIP is first developed by Yabe *et al* [8] to solve hyperbolic equation, and then is employed to marine engineering problems by Hu and Kashiwagi [9,10]. To simulate the large surface deformations with high accuracy, THINC (Tangent of Hyperbola Interpolation for Interface Capturing) [11] method is employed. The purpose of the present paper is to examine the effective performance of CIP-based numerical model in solving solitary waves impinging on a trapezoid seawall, and to analyse the difference of free surface due to different incoming wave height. This paper is organized as follows. In section 2, we briefly describe the numerical method employed in this paper. Section 3 provides the numerical experimental setup and the validation of the numerical model. Results and discussion are presented in section 4.

2. Numerical Method

This section briefly describes the numerical method we used to construct the wave-body interaction numerical model. It mainly contains three parts, one part is the CIP based flow solver, one part is the interface capturing method, and the last one is the calculation method for hydrodynamic forces.

2.1. Flow Solver

For incompressible flow, governing equations including continuity equations and Navier-Stokes equations can be expressed as:

$$\frac{\partial u_i}{\partial x_i} = 0 \tag{1}$$

$$\frac{\partial u_i}{\partial t} + u_j \frac{\partial u_i}{\partial x_j} = -\frac{1}{\rho}\frac{\partial p}{\partial x_i} + \frac{1}{\rho}\frac{\partial}{\partial x_j}\left(2\mu S_{ij}\right) + f_i \tag{2}$$

Where u, ρ, p, t and f denote velocity, density, pressure, time and body force, respectively; and $S_{ij} = \left(\partial u_i/\partial x_j + \partial u_j/\partial x_i\right)/2$. The flow solver divides the Navier-Stokes equation into three fractional steps:

$$\frac{u_i^* - u_i^n}{\Delta t} + u_j^n \frac{\partial u_i^n}{\partial x_j} = 0 \tag{3}$$

$$\frac{u_i^{**} - u_i^*}{\Delta t} = \frac{2\mu}{\rho^*}\frac{\partial}{\partial x_j}\left(S_{ij}^* - \frac{1}{3}\delta_{ij}S_{kk}^*\right) + f_i \tag{4}$$

$$\frac{u_i^{n+1} - u_i^{**}}{\Delta t} = -\frac{1}{\rho^*}\frac{\partial p^{n+1}}{\partial x_i} \tag{5}$$

Where, n denotes the present time level, $n+1$ for the new time level. Equation (3) is the advection step, which is solved by the CIP method. Equation (4) and Equation (5) are the Non-advection steps, diffusion term in Equation (4) is treated by central difference method. In Equation (5), the pressure coupled with velocity is solved by SOR (Successive Over Relaxation) iteration.

2.2. Interface Capturing Method

THINC employs a hyperbolic tangent function which makes it a suitable interpolation for the flux computation of a VOF (Volume of Fluid) function [12]. The competitive accuracy compared with other method has been verified by Xiao et al [10]. The hyperbolic tangent function is written as:

$$F_i(x) = \frac{\alpha}{2}\left\{1 + \gamma \tanh\left[\beta\left(\frac{x - x_{i-1/2}}{\Delta x_i} - \delta\right)\right]\right\} \tag{6}$$

Where α, β, γ, δ are parameters can be specified.

2.3. Hydrodynamic Forces

After the calculation of pressures, velocities and density functions, the hydrodynamic forces acting on the solid body can be calculated by integrating the pressure and skin friction along the body surface as follows:

$$F_i = F_i^{(p)} + F_i^{(v)} = \iint_A (-p\delta_{ik}) n_k dA + \iint_A 2\mu S_{ik} n_k dA \tag{7}$$

Where $F_i^{(p)}$ and $F_i^{(v)}$ are the forces due to pressure and friction, respectively. A denotes the surface of solid body, and n_i presents the i-th component of the outward unit normal vector.

3. Numerical Experiment Setup

Figure 1 shows a two-dimensional numerical wave flume constructed in this paper. Left side of the wave flume is a wave maker for generating solitary waves, which is defined as the origin of the Cartesian coordinate system. Reference wave gauge is set at x = 5.900 m, before the starting point of the slope, to measure the incident wave height. The non-dimensional wave height is defined as α =

H/h, where H is the wave height, and h = 0.256 m is the water depth. Four incident wave heights of solitary waves are investigated, including α = 0.1, 0.2, 0.23 and 0.3, in which α = 0.23 is to keep accordance with the experimental data. The parameter of the sloping seawall is listed in table 1. The slope starts at x_{s1} = 7.000 m, and has a trapezoid obstacle composed by x_{s2}, x_{s3}, x_{s4} and x_{s5} on it. Weather side is indicated in figure 1; the hydrodynamic forces acting on it is discussed in section 4.

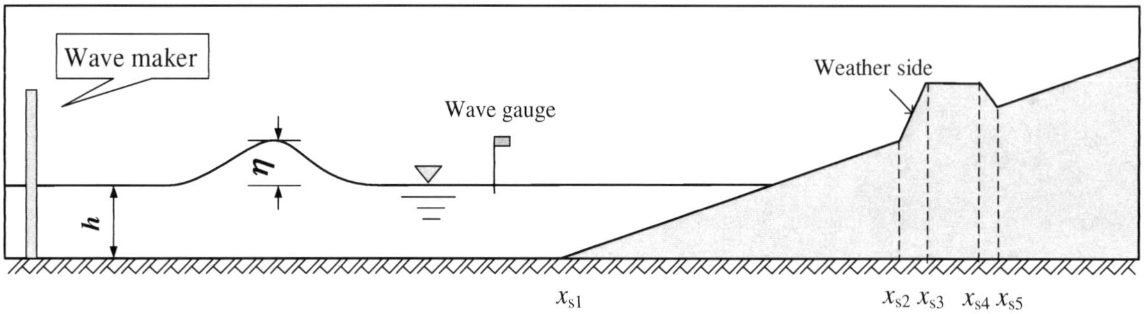

Figure 1. Schematic view of the numerical wave flume.

Table 1. Locations of vertex point.

Vertex point	x_{s1}	x_{s2}	x_{s3}	x_{s4}	x_{s5}
Distance (m)	7.000	10.600	10.900	10.948	11.045

4. Results and Discussion

4.1. Convergence Test

Both mesh convergence and time step convergence tests are examined. Figure 2 shows normalized free surface elevation measured at the wave gauge due to different mesh resolution. Three types of grid are employed; the first number in the legend denotes total grid number in x direction, and the second number denotes total grid number in y direction. For example, "Grid-1: 810 * 220" means Grid-1 has 810 and 220 grids in x and y direction, respectively. The figure indicates that Grid-2 and Grid-3 agree very well with each other, while Grid-1 differs from them. So the conclusion can be obtained that Grid-2 has enough grids for this problem. In the following parts, all data is calculated by utilizing Grid-2. Circle is the experimental data from Hsiao and Lin [1], the main wave crest agrees well with numerical results.

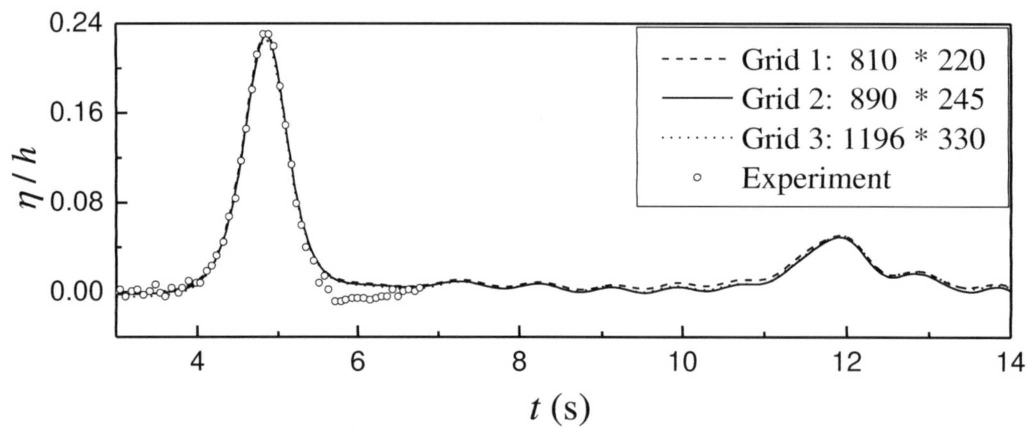

Figure 2. Mesh convergence test. Grid 1 (dash), Grid 2 (solid), Grid 3 (dot); Experiment [1] (circle)

Figure 3 compares the free surface elevation according to different time step. *Dt* denotes time step, *T* denotes truncated solitary wave period. As an example, "*Dt* = *T* / 1000" means calculate 1000 time steps in one wave period. As *Dt* = *T* / 1000, there is a slight discrepancy from another three time step; another three curves reach accordance, we choose the middle one *Dt* = *T* / 2000 as the time step in our computational cases.

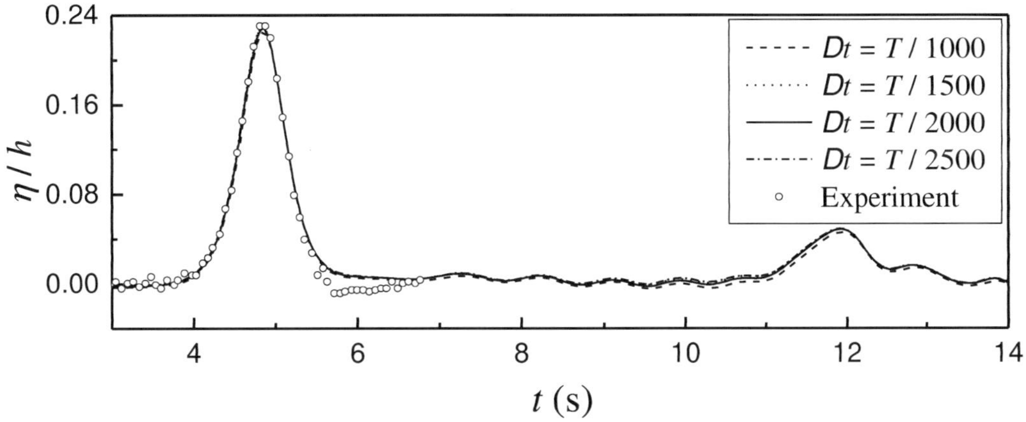

Figure 3. Time step convergence test. *Dt* = *T* / 1000 (dash), *T* / 1500 (dot), *T* / 2000 (solid), *T* / 2500 (dash-dot); Experiment [1] (circle).

4.2. Free Surface Elevation

Figure 4 shows time series of the free-surface elevation, in terms of different incident solitary wave height α = 0.1, 0.2, and 0.3, which are measured at x = 5.900 m. The water depth remains constant which equals to 0.256 m. The abscissa is the time series, while the ordinate is the non-dimensional free surface elevation normalized by dividing the constant water depth. The solitary wave only has a wave crest above the still water level. From the figure it can be seen that with the increase of incident wave height, the main wave crest becomes steeper, and occurs earlier.

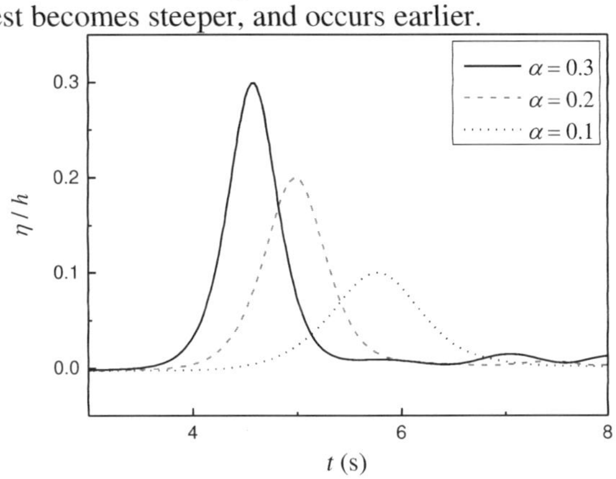

Figure 4. Time series of free surface elevation measured at wave gauge, with normalized incident wave height α = 0.1 (dot), 0.2 (dash), and 0.3 (solid).

4.3. Hydrodynamic Forces

Time series of the horizontal force *Fx* and vertical forces *Fz* acting on the weather side of the seawall is presented in figure 5 and figure 6, respectively. Three different incident solitary wave height $\alpha = 0.1$, 0.2, and 0.3 is compared. Generally, vertical forces are large than horizontal force by almost one order of magnitude. The absolute value of both vertical forces and horizontal forces go up reaching a peak, then drop down. With the increase of incident wave height, hydrodynamic forces increase, and the force curve becomes steeper. The peak force occurred time is illustrated in the figures, for example, "t = 7.413 s" means when non-dimensional incident wave height is 0.3, maximum horizontal force occurs at t = 7.413 s. It can be found that peak force occurs early as incident wave height increase, this is mainly because phase velocity of solitary wave increase as a function of wave height. This phenomenon is accord with the solitary wave propagates faster when wave height increase in figure 4.

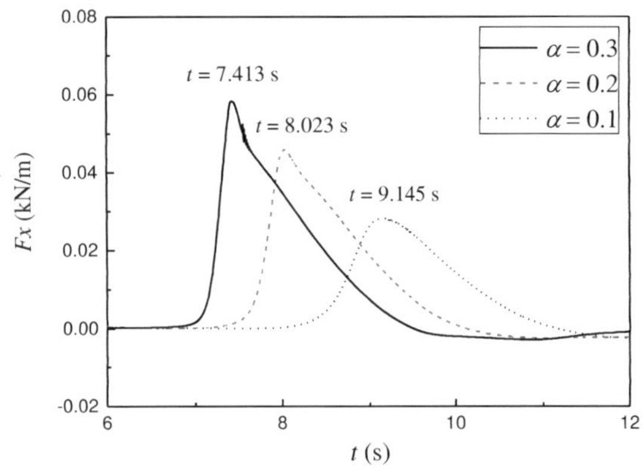

Figure 5. Horizontal forces (*Fx*) acting on weather side, with normalized incident wave height $\alpha = 0.1$ (dot), 0.2 (dash), and 0.3 (solid).

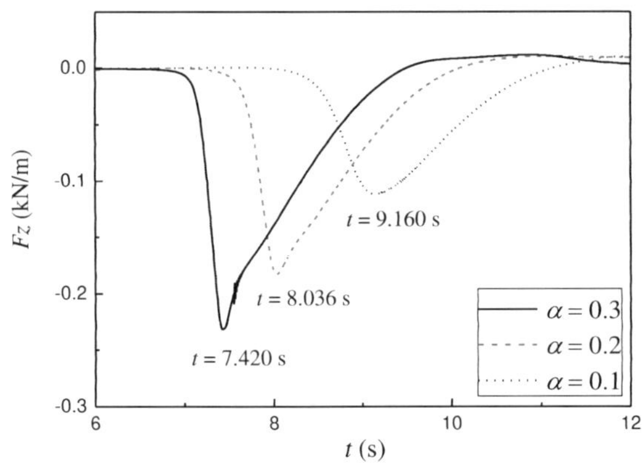

Figure 6. Vertical forces (*Fz*) acting on weather side, with normalized incident wave height α = 0.1 (dot), 0.2 (dash), and 0.3 (solid).

5. Conclusion

This paper implemented a CIP based numerical model to investigate interaction between solitary waves and a trapezoid seawall on a sloping beach. The convergence test and the validation compared with the experimental data confirms that the present numerical model is efficient to calculating wave-structure interaction problems. Different incident wave height was tested to comparing the variation of the free surface elevation and hydrodynamic forces, and the conclusion can obtained that with the increase of incident wave height, the profile of solitary waves becomes steeper; the maximum hydrodynamic force occurs earlier.

6. Reference

[1] Hsiao SC, Lin TC 2010 Tsunami-like solitary waves impinging and overtopping an impermeable seawall: Experiment and RANS modeling *Coast Eng.* 57 1–18.

[2] Ma G, Shi F, Hsiao SC, Wu YT 2014 Non-hydrostatic modeling of wave interactions with porous structures *Coast Eng.* 91 84–98.

[3] Wu YT, Hsiao SC 2013 Turbulence induced by a solitary wave propagating over a submerged object using particle image velocimetry *J Coast Res.* 65 416–21.

[4] Hayatdavoodi M, Ertekin RC 2015 Wave forces on a submerged horizontal plate - Part I: Theory and modelling *J Fluids Struct.* 54 566–79.

[5] Hayatdavoodi M, Ertekin RC 2015 Wave forces on a submerged horizontal plate-Part II: Solitary and cnoidal waves *J Fluids Struct.* 54 580–96.

[6] Rostami Varnousfaaderani M, Ketabdari MJ 2015 Numerical simulation of solitary wave breaking and impact on seawall using a modified turbulence SPH method with Riemann solvers *J Mar Sci Technol.* 20 344–56.

[7] Kihara N, Niida Y, Takabatake, D, Kaida, H, Shibayama, A, and Miyagawa, Y 2015 Large-scale experiments on tsunami-induced pressure on a vertical tide wall *Coast Eng.* 99 46-63.

[8] Yabe T, Aoki T 1991 A universal solver for hyperbolic equations by cubic-polynomial interpolation I. One-dimensional solver *Comput Phys Commun.* 66 233–42.
[9] Hu C, Kashiwagi M 2004 A CIP-based method for numerical simulations of violent free-surface flows *J Mar Sci Technol.* 9 143–57.
[10] Hu C, Kashiwagi M 2009 Two-dimensional numerical simulation and experiment on strongly nonlinear wave-body interactions *J Mar Sci Technol.* 14 200–13.
[11] Xiao F, Honma Y, Kono T 2005 A simple algebraic interface capturing scheme using hyperbolic tangent function *Int J Numer Methods Fluids* 48 1023–40.
[12] Hirt CW, Nichols BD 1981 Volume of fluid (VOF) method for the dynamics of free boundaries *J Comput Phys.* 39 201–25.

Acknowledgments
This work was supported by National Natural Science Foundation of China (51579058), Open Research Fund Program of Key Laboratory of Water & Sediment Science and Water Hazard prevention (Changsha University of Science & Technology), Hunan Province (2015SS02).

IOP Conf. Series: Earth and Environmental Science **171** (2018) 012016 doi:10.1088/1755-1315/171/1/012016

Numerical Study on Influence of Water Exchange for Artificial Island Group

Tao Liu[1], Yuehong Liu[2] and Zhiqiang Hou[1]

1 Coastal & Estuarine Engineering Research Center, Tianjin Research Institute for Water Transport Engineering, Tianjin, China.
2 CCCC First Harbor Consultants Co., Ltd.
E-mail: tksliutao@163.com

Abstract. After the construction of artificial island group, the good water exchange ability is beneficial to improve the water environment quality and improve the landscape effect of the island group. In order to solve the water exchange in the island group, a numerical model based on solving the hydrodynamic and advection-diffusion equations was used to calculate the water exchange of artificial island group in Weifang, China. The water exchange rate was obtained by calculating the concentration variation of dissolved conservative tracer. Based on the analysis of water exchange and influence factors, the suggestion of optimization was proposed. Fist the narrow waterways should be avoided and the most effective way to enhance the hydrodynamics for weak flow region is increasing exchange of channel.

1. Introductions

Nowadays the artificial islands are widely constructed over the world, such as Dubai Lugger Hotel, Japan Osaka Kansai International Airport[1], Dalian artificial island[2], Longkou artificial island and Yangpu artificial island, etc. The influence of artificial island on the local water environment is uncertain, and the factors affecting water exchange are varied[3-6]. This uncertainty could be even more significant for the construction of island group. It may deteriorate the water environment and lead to serious environmental problem. The construction of artificial island group of Weifang in China is an example.

1.1. Background

Weifang artificial island is located at Laizhou Bay and is near the Weifang Sime Darby Port. There are two rivers namely Whitewater River and Yuhe River in this area. Laizhou Bay belongs to the storm surge prone region. Outside the artificial island group, a breakwater was built along the gates of Whitewater River and Yuhe River to resist the storm tide. The water inside the artificial island group has communication with the water of external sea through the gates. The island group consists of more than 20 islands, as shown in Figure 1.

1.2. Hydrodynamic Conditions

Tidal characteristics in the study area are irregular semidiurnal tide[7][8]. The average tidal range is 1.6m. According to the measured data, the average flow velocities of flood and ebb tide range from 0.07 to 0.18 m/s and 0.06 to 0.16 m/s during the spring tide and neap tide, respectively. Since the tidal power is weak, it has adverse effect on the water exchange. The layout of the island group is of importance.

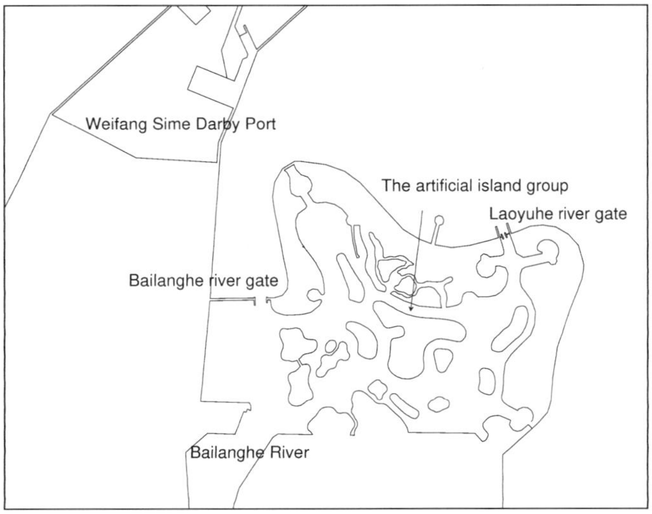

Figure 1. Artificial island group layout

2. Numerical Modelling

2.1. *Mathematics*

The MIKE21 FW module is employed in this study. The governing equations are based on the depth-averaged shallow water equations derived from the original three-dimensional ones and read as follows:

$$\frac{\partial \zeta}{\partial t}+\frac{\partial uh}{\partial x}+\frac{\partial vh}{\partial y}=0 \tag{1}$$

$$\frac{\partial v}{\partial t}+\frac{\partial uv}{\partial x}+\frac{\partial v^2}{\partial y}=-fu-g\frac{\partial \zeta}{\partial y}+\frac{\tau_{sy}-\tau_{by}}{\rho h}+E_y\left(\frac{\partial^2 v}{\partial x^2}+\frac{\partial^2 v}{\partial y^2}\right)-\frac{1}{\rho}\left(\frac{\partial S_{yx}}{\partial x}+\frac{\partial S_{yy}}{\partial y}\right) \tag{2}$$

$$\frac{\partial u}{\partial t}+\frac{\partial u^2}{\partial x}+\frac{\partial uv}{\partial y}=fv-g\frac{\partial \zeta}{\partial x}+\frac{\tau_{sx}-\tau_{bx}}{\rho h}+E_x\left(\frac{\partial^2 u}{\partial x^2}+\frac{\partial^2 u}{\partial y^2}\right)-\frac{1}{\rho}\left(\frac{\partial S_{xx}}{\partial x}+\frac{\partial S_{xy}}{\partial y}\right) \tag{3}$$

where x and y are horizontal coordinates respectively; t is time; h is total water depth; u and v are velocity components; g is gravity acceleration; f is the Coriolis force; τ_{bx} and τ_{by} are bed shear stress components; E_x and E_y are horizontal eddy viscosity components; ρ is seawater density; S_{xx}, S_{xy}, S_{yx} and S_{yy} are wave radiation stress components.

The bed shear stress is calculated using the following equation.

$$\vec{\tau}_b=\rho\frac{g}{C^2}\vec{U}\left|\vec{U}\right| \tag{4}$$

where C is the Strickler coefficient. The eddy viscosity coefficient is calculated by Smagorinsky formula.

Based on the conservative substances dissolved in the water like the tracers, an advection-diffusion model is set up. The water exchange rate (WER) is calculated by observing the variation of tracer concentration. The governing equation is as follows:

$$\frac{\partial hC}{\partial t}+\frac{\partial huC}{\partial x}+\frac{\partial hvC}{\partial y}=\frac{\partial}{\partial x}\left(hD_x\frac{\partial C}{\partial x}\right)+\frac{\partial}{\partial y}\left(hD_y\frac{\partial C}{\partial y}\right)-FhC\cdot \tag{5}$$

where *C* is the concentration of the tracer; *F* is the decaying factor, which can be set to zero for conservative matter; D_x and D_y are dispersion coefficients for *x* and *y* directions, $D_x=D_y=0.01m^2/s$, respectively.

During the numerical simulation, the initial tracer concentration is set as 1.0 in the area of artificial island group and as 0.0 for other water areas (see Fig. 2). The simulation period is 30 days with continuous tide movement. The WER is defined as the percentage of the mass of tracer transporting out of the artificial island group area over the mass of initial tracer in the seam area. The WER is calculated by the following equation.

$$WER(t_j) = 1 - \left(\frac{\sum_{i=1}^{N} C_i(t_j) h_i(t_j)}{\sum_{i=1}^{N} C_i(t_0) h_i(t_0)} \right) \times 100\% \quad (6)$$

where i is the index of a specific grid; N is the total number of grid inside the area of artificial island group; j is the index of a specific time step.

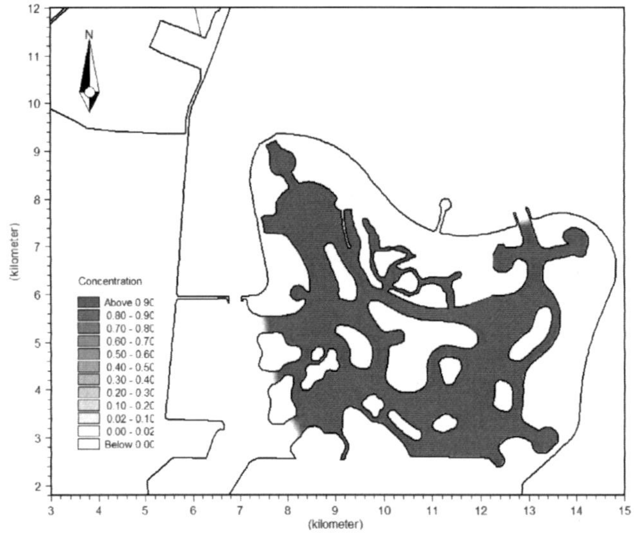

Figure 2. Initial condition of concentration for simulation

2.2. *Model Set Up*

The range of model is the whole Laizhou Bay. Triangle mesh is used to divide the computational domain. The maximum grid size is 3000 m. The grid is refined in the area of island group, where the minimum grid size is 10 m. The computational time step is 0.5 s. The tidal levels at the open boundaries are extracted from the Chinese sea tide model. Figure 3 shows the mesh arrangement for the study area.

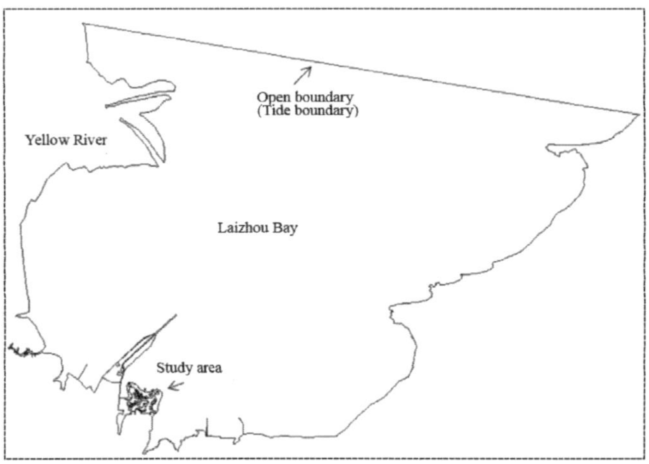

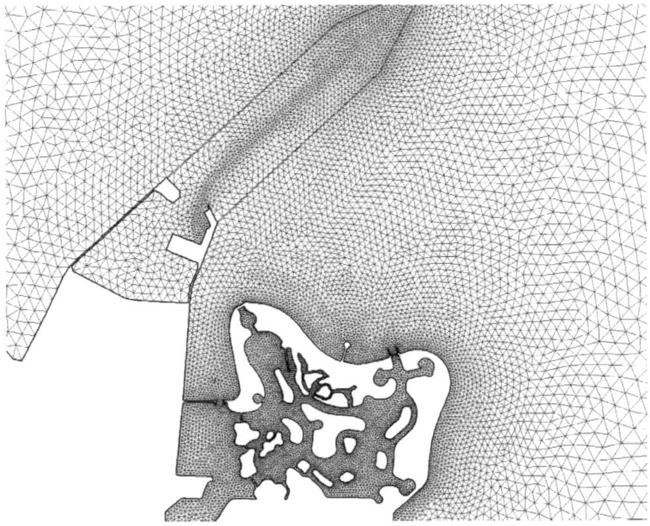

Figure 3. Model area and Mesh arrangment of study area

3. Results and Discussions

The present hydrodynamic model of the Laizhou Bay has been verified in the previous study, where the computed tidal level, current speed and current direction were validated against the field-measured data. Figure 4 shows the time history of the water exchange and velocity field for the original planning. Based on the analysis of numerical results, the main findings are summarized as follow.

The tidal current movement in the study area demonstrates typical reciprocating feature. The main flow direction during ebb tide is northeast. It is west southwest during flood tide. After the artificial island group and the breakwater are built, the tidal current becomes reciprocating along the waterways due to local topography and constructed structure. The average flow velocity changes from 0.01~0.25m/s to 0.01~0.93m/s. Increase of flow velocity can improve the capability of water exchange. The gates of Whitewater River and Yuhe River are the main flow channel, where the flow velocity significantly increases. During the flood tide, water from the external sea flows into the breakwater area and then dilutes the tracer concentration between island groups. During the ebb tide, the water carrying high concentration of tracer flows out. Exchange between the internal and external water continues under the action of tide. Therefore, the WER increases as time gradually increases. The WERs over 4, 15, 30 days are approximately 43%, 71% and 80%, approximately.

The change of tracer concentration shows that the WER is poor in the backwater area. Take the northwest and southeast sides of islands as examples, the flow velocity approaches zero. The WER is less than 10% over 30 days. The backwater will cause the deterioration of the water quality and then affect the whole water quality and landscape. The original planning needs to be modified.

To modify the layout of island group while avoiding large change of original planning, the feasible approach is to increase the hydrodynamic force in the weak flow region and prevent the forming of backwater. The proposed optimization planning is shown in Figure 5.

Figure 6 shows the time history of water exchange and velocity field for optimization planning. Figure 7 compares the time history of WERs over 30 days for the original and optimization planning. After the layout of island group is optimized, the hydrodynamic in the weak flow region is strengthened. The backwater disappears. The WER increases substantially. The WERs of the optimization planning over 4, 15 and 30 days are nearly 56%, 82% and 92%, respectively. Compared with the original planning, the water exchange capability increases 30% in the early stage and 15% after 30 days. Increase of tidal capacity has proved to be more effective measure for improving water exchange of island group.

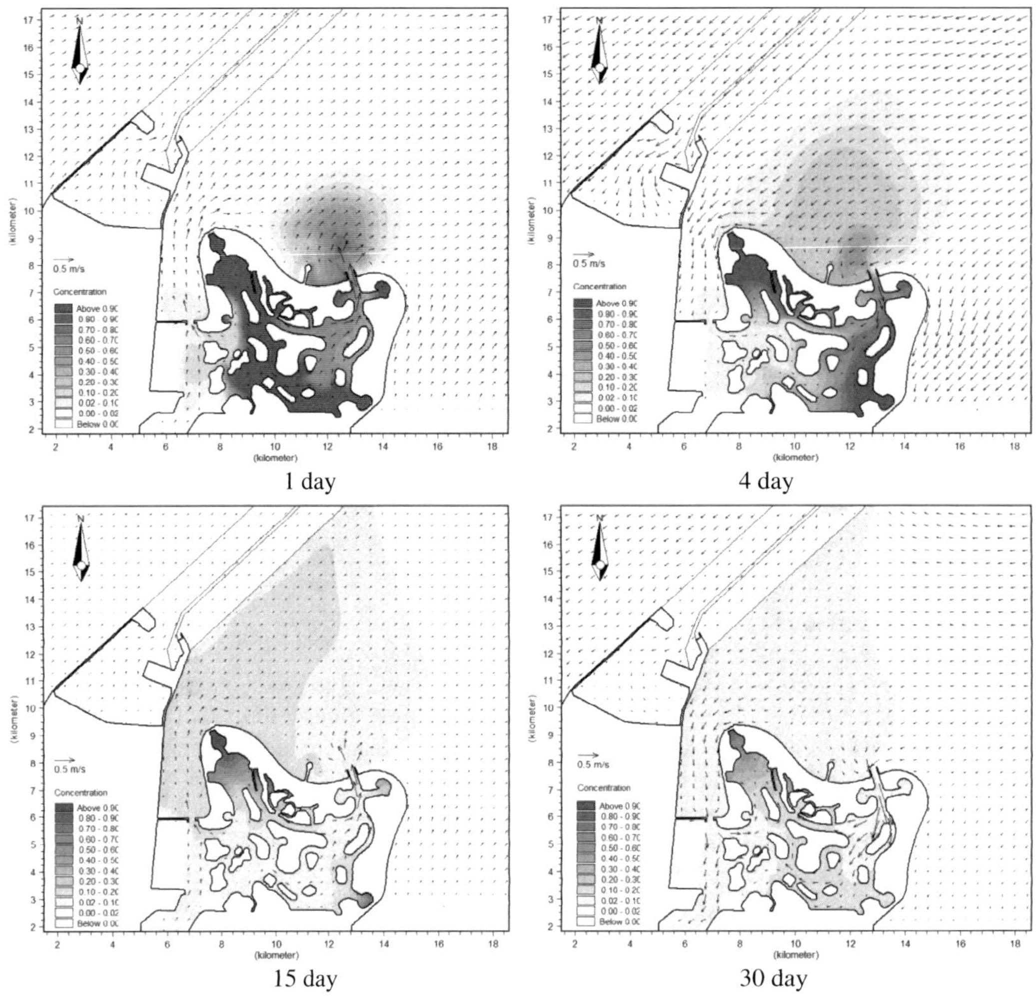

Figure 4. Time history of water exchange and velocity field for original planning

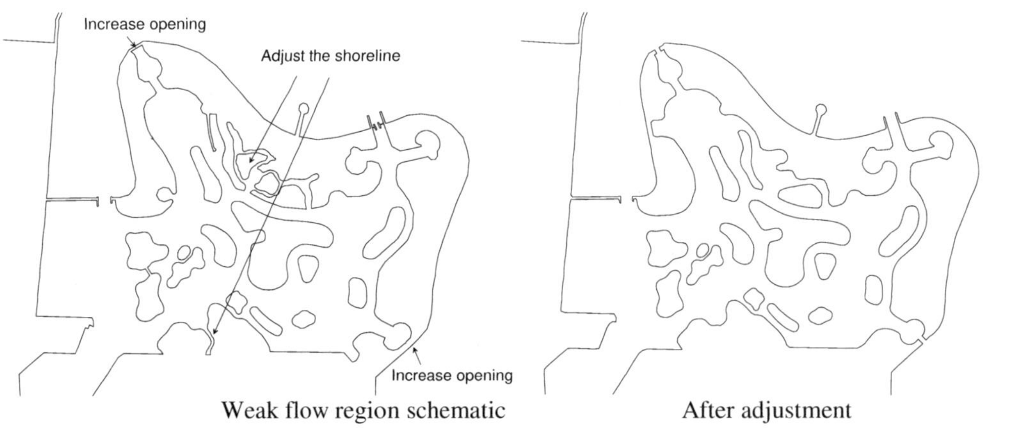

Figure 5. Optimization planning

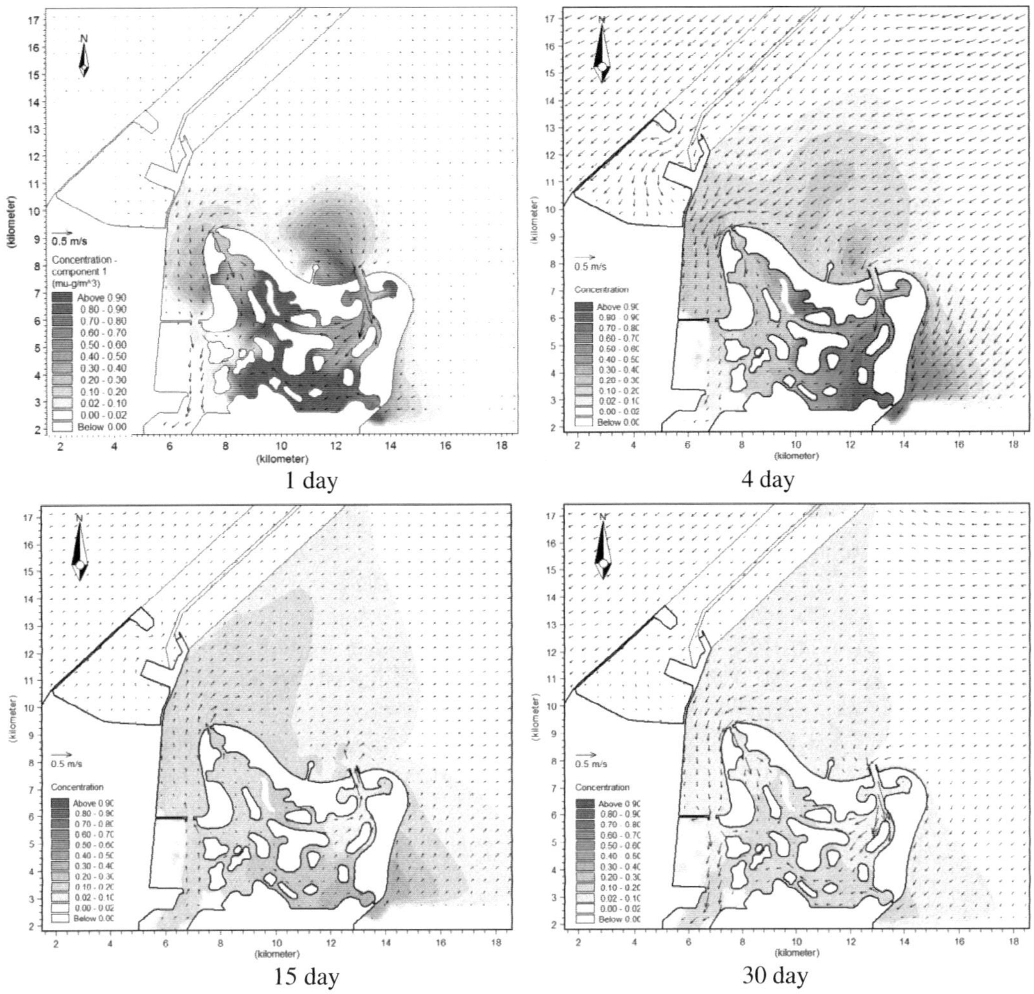

Figure 6. Time history of water exchange for optimization planning

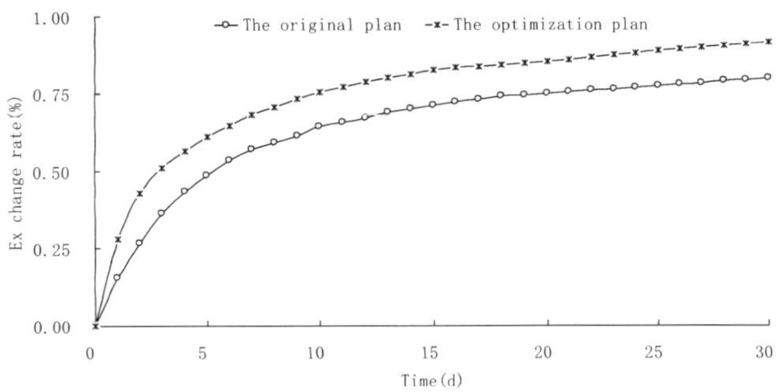

Figure 7. Time history of WER for different cases

4. Conclusion

The artificial island group is a complex system. With the rational arrangement of islands, the flow velocity between island groups can be effectively improved to provide favorable conditions for the water exchange. The present study shows that:

From the aspect of improving water exchange, to avoid the backwaters is the most desirable; increase of tidal capacity has proved to be the effective measure for improving water exchange of island group. Increasing exchange of channel is the easiest and most effective way to enhance the hydrodynamics for weak flow region; narrow waterways can lead to siltation, so it should be avoided.

In addition, based on the study of Xu[9] and Yao[10][11] et al, the water exchange can be benefited from a reasonable design of water depth. Water depth of island group should be minimized under the condition of meeting planning requirements and security conditions.

5. References

[1] ZHONG Lai MING Nan 1994 *The Construction of Kansai International Airport* (Qingdao: Coastal Engineering) 13(2) pp 197-201

[2] YOU Shuang 2013 *Numerical Study on Water Environment in Jinzhou Bay affected by Dalian Offshore Airport* (Dalian: Dalian University of Technology)

[3] ZHANG Jing ZHANG Yuping 2017 *Study on Exchange of Water Mass in Shuangdaowan Bay of Weihai Port Engineering Technology* (Tianjin: Port Engineering Technology) 54(2) pp 6-8

[4] HUANG Shaobin LI Kaiming JIANG Guoqiang 2013 *Research on Water Exchange in Pearl River Estuary based on MIKE3 Model* (Harbin: Environmental Science and Management) 38(8) pp 134-140

[5] JIANG Hengzhi etc 2017 *Numerical study of water exchange in the Pulandian bay* (Dalian: Marine Environmental Science) 36(1) pp 44-47

[6] HAN Weidong ZHANGWei CHEN Zhen 2015 *Research on the Influential Factors of the Water Exchange Effect for an Encircled Basin* (Beijing: Science Technology and Engineering) 15(9) pp 258-265

[7] WU Mingyang FENG Yulin 1996 *Study on sediment siltation for extension project of Weifang harbour* (Beijing: Journal of Sediment Research) 6(4) pp 34-42

[8] YAN Bing LIU Tao HOU Zhiqang 2011 *Water flow and Sediment Mathematical Model Test Research Report of Weifang Binhai city planning* (Tianjin: Technical Report of Tianjin Research Institute for Water Transport Engineering)

[9] XU Ting HAN Zhiyuan WEN Chunpeng 2016 *Numerical simulation of water exchange capability in Qinhuangdao West Port* (Tianjin: China Harbour Engineering) 36(1) pp 10-14

[10] YAO Shanshan ZHANG Peng CHEN Chun 2017 *Numerical study on water exchange ability of planning water system project in Haitang Bay* (Tianjin: Journal of Waterway and Harbor) 38(3) pp 252-257

[11] YAO Shanshan XIE Mingxiao 2013 *Study on the improvement method of water exchange rate for Shuidong bay using numerical model* (Switzerland: Advanced materials research) 765(7) pp 283-8

IOP Conf. Series: Earth and Environmental Science **171** (2018) 012017 doi:10.1088/1755-1315/171/1/012017

Numerical Simulation of Wind Wave in Bohai Sea Induced by Cold Wave

Chen Chun and Huang Yu Xin

Associate Researcher, Tianjin Research Institute for Water Transport Engineering, Key Laboratory of Engineering Sediment, Ministry of Transport, Tianjin, China
E-mail: chenchun9999@163.com

Abstract. This paper using WRF mesoscale atmospheric model, the reanalysis wind field data provided by ncep global forecast field as the initial condition and boundary condition with time, calculated the wind field with sufficient precision, provided the driving wind field for the wave numerical calculation, calculated the corresponding wave field with model SWAN, and simulated the wind wave process caused by a cold wave wind in the Bohai sea area from 5 to 7 November 2015. The calculated model was verified by the measured wind and waves data at the stations of Huanghua and Dongying. The results show that the simulation values of wind and wave elements are in good agreement with the measured values. The model can reproduce the characteristics of wind and wave distribution during the cold tide in Bohai Sea.

1. Introduction

The cold wave usually brings strong winds, is one of the worst disastrous weather that affects our country in the winter half year, Bohai Sea is in the northern part of the China sea, cold wave strong wind is more frequent. The high storm and storm surge caused by cold wave have serious influence on marine engineering, shipping operations, channel silting and coastal economy. So the wind and waves numerical simulation of the Bohai Sea during the cold wave is of practical significance.

The disaster caused by cold wave is more serious, the impact of cold wave cannot be ignored[1]. Many scholars have done a lot of research on the wave characteristics of the Bohai Sea[2-6], and the cold wave is relatively scarce, so the numerical simulation of the wave field under the cold wave has certain significance.

With the rapid development of wave theory and computer technology, it is more effective to study the wave propagation law by numerical model. In this paper, the wind and waves caused by a cold wave wind in the Bohai sea between November 5 and 7, 2015 are simulated and validated by the wind and wind data measured at the observation stations of huanghua and dongying. The results show that the simulation values of wind and wave elements are in good agreement with the measured values. The model can reproduce the characteristics of wind and wave distribution during the cold tide in Bohai Sea.

2. Wind Field Numerical Calculation

2.1. Atmospheric Patterns

WRF is a mesoscale atmospheric prediction model developed by NCAR and NECP. It has been widely studied and applied in the world[7-9].

2.2. Initial and Boundary Conditions

The initial conditions and boundary conditions in the WRF model are provided by the NCEP historical reanalysis wind field data with a time accuracy of 6h and a spatial resolution of 1.0 ° X 1.0 °

2.3. Model Area

The WRF model is calculated using double mesh nesting technique. See Figure 1. A quadrilateral grid is used. The latitude and longitude range of the large model is 114.05 ° ~ 130.95 ° E, 24.05 ° ~ 41.95 ° N, and the grid number is 170 x 1 The longitude and latitude range of the small model A2 is 117.03 ° ~ 128.98 ° E, 29.03 ° ~ 40.98 ° N, with a grid number of 240 x 240.

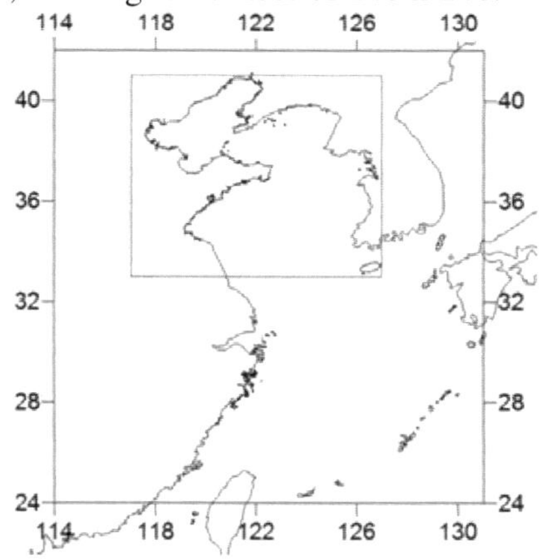

Figure 1. Computational domain

2.4. Parameter Settings

There are many kinds of physical modes to choose from for each kind of meteorological physical process provided by WRF mode. In this calculation, the vertical stratification is taken to 35 layers, the maximum pressure is taken to 1000 Pa at the top, the integral time step is 30 S and the output time interval is 1 H.

2.5. Wind field Calculation and Verification

The cold and windy process from 5 to 7 November 2015, lasted 48 hours and above, with an initial wind direction of E, then ENE and finally NE.

To verify the validity of the wind field model, Figure 2 show the measured wind speed and wind direction data of Huanghua and Dongying stations during the cold wave storm in November 2015 were compared with the model.

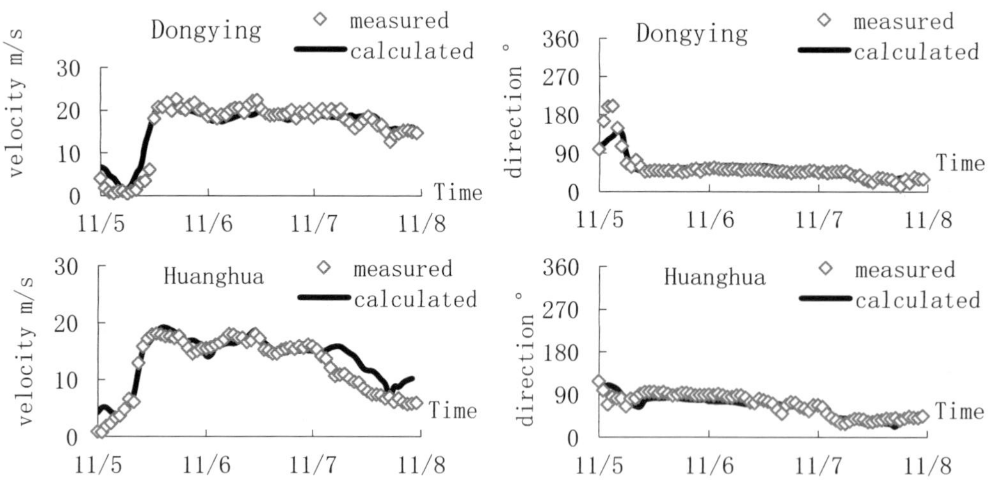

Figure 2. Comparison of calculated wind velocity and measured

The results show that the calculated and measured values are well verified in terms of magnitude and trend. WRF model can reasonably describe the wind speed and wind direction during the cold wave wind process. It can provide high precision wind speed and ensure the rationality of wave field simulation.

Figure 3 and Figure 4 show the calculated domain wind velocity isobaths and wind vector field at 2015-11-5 23:00 UTC. At this time, the sea area of the Bohai Sea is basically SE direction wind, wind speed distribution is small in the Northeast and large in the Southwest, the maximum wind speed is about 20 m/s.

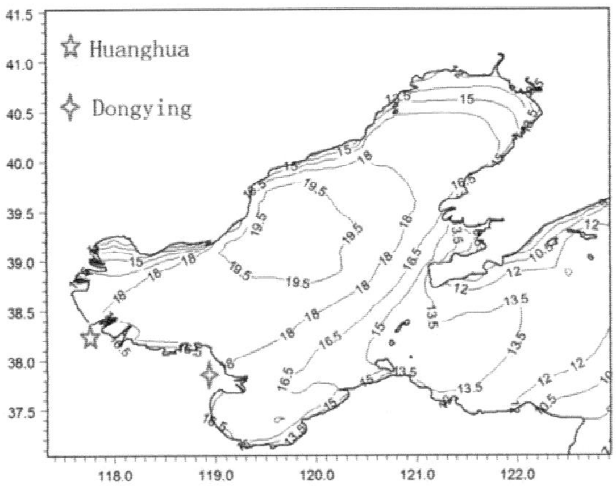

Figure 3. wind field contour (2015-11-5 23:00 UTC)

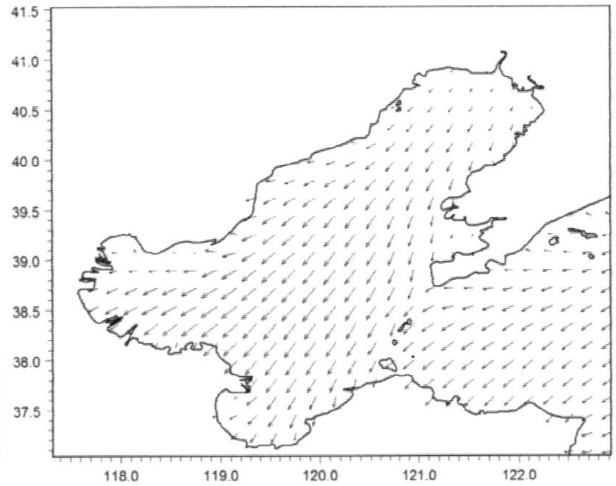

Figure 4. wind field vector (2015-11-5 23:00 UTC)

3. Numerical Calculation of Waves

3.1. Model Area

SWAN is the third generation of coastal shallow water wave numerical computing model developed by Delft University of Technology in the Netherlands. After years of improvement, SWAN has gradually matured and has been widely used in many shallow sea numerical studies[10,11].

Waves are calculated using locally encrypted unstructured grid terrain, as shown in Figure 5, covering the area of the Yellow Bohai Sea with a single node number, with a maximum grid space step length of 0.1 ° and a minimum of 0.001 ° The high resolution wind field calculated by WRF is used to calculate the parameters of wave height, mean period and wave direction in the region of concern.

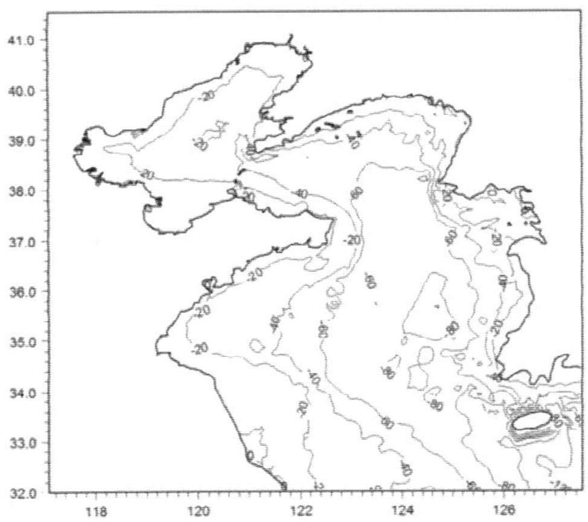

Figure 5. Computational domain of wave

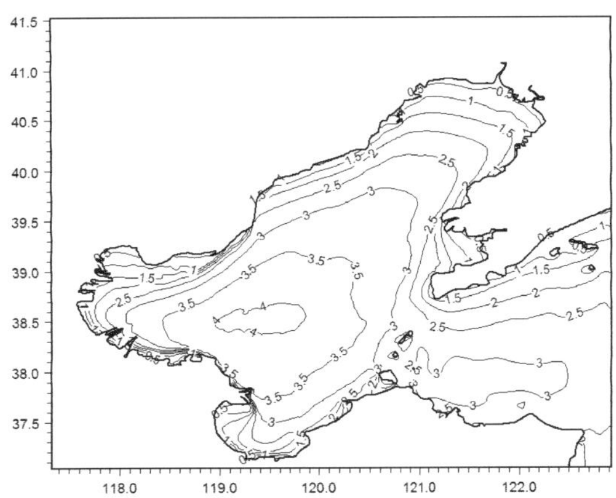

Figure 6. Wave field (2015-11-5 11:00 UTC)

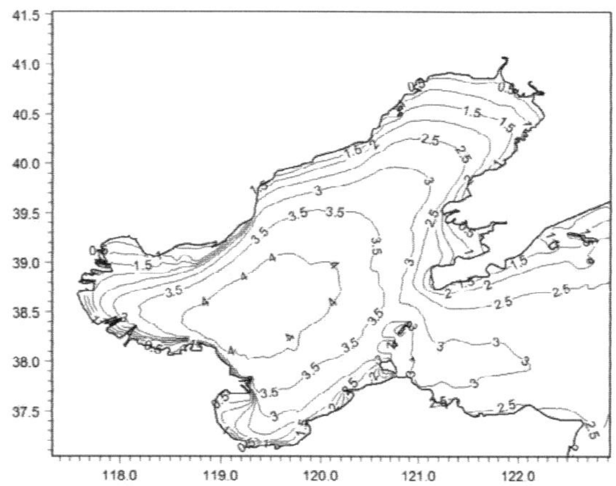

Figure 7. Wave field（2015-11-5 23:00 UTC）

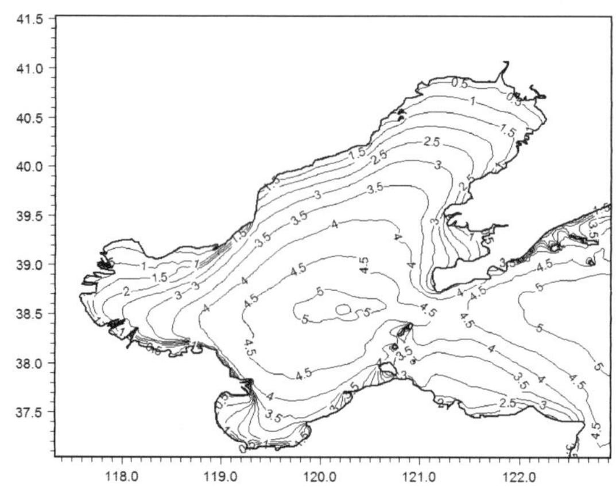

Figure 8. Wave field（2015-11-7 2:00 UTC）

3.2. Wave Calculation and Verification

Figure6~8 gives the calculation domain effective wave high isobath wave field at 2015-11-5 11:00 2015-11-5 23:00 2015-11-7 2:00 UTC time during cold and strong winds. During high winds, the wave field in the Bohai Sea was small in the Northeast and large in the Southwest, and the maximum effective wave was above 5.5m/s.

The calculation value of Dongying station model is compared with the measured wave data(Figure 9). The calculated values of wave height and wave direction and measured values are well proved in terms of magnitude and time variation.

This cold wave wind is mainly ENE ~ NE, the wave distribution in the Bohai sea area is small in the Northeast, increasing gradually to the Northwest sea area, the maximum effective wave height is more than 4 meters, the largest wave reaches 3.9 m in Dongying, From the wind energy, the reconstruction period is about 25 a once. The channel of Huanghua port and Binzhou port caused serious siltation. The study of wave growth process and wave field distribution during the cold-tide wind can provide accurate wave dynamic force field for the model of sudden deposition in the high-wind.

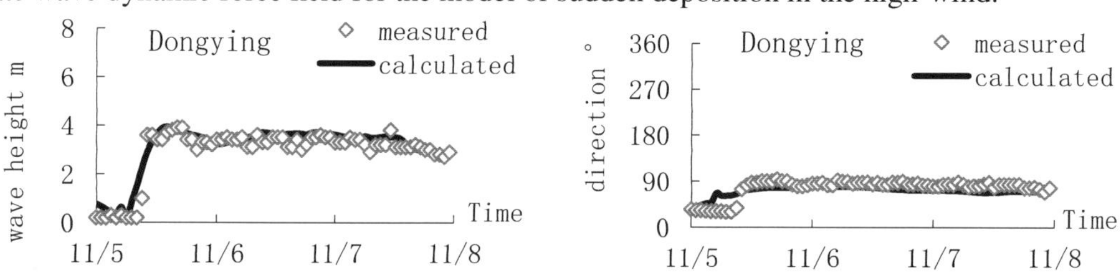

Figure 9. Comparison between calculation and measured effective wave height

4. Conclusions

By using WRF mesoscale atmospheric model, the wind field with sufficient precision in the Bohai Sea area is calculated, which provides the driving wind field for the numerical calculation of the waves.

(1) The calculation model was verified by the measured wind and waves data at the observation stations of Huanghua and Dongying. The results show that the simulation values of wind and wave elements agree well with the measured values.

(2) The cold and strong winds are mainly ENE ~ NE, and the wave distribution in the Bohai Sea is small in the Northeast, increasing gradually to the Northwest, the maximum effective wave height is over 4 meters.

(3) The study of wave growth process and distribution of wave field during the cold wave and high wind can provide accurate wave dynamic force field for the model of sudden deposition in the high wind.

5. References

[1] CAO Haifeng 1998 *Effect of cold storm surge and cold waves can not be ignored*(Beijing: Marine Forecasts) vol15(2) pp 43-46.
[2] ZHENG Guo-dan et al 2010 *Numerical simulation of wind waves in Bohai Sea induced by "98·04" cold wave*.(Beijing: Port & Waterway Engineering) vol2 pp36-39.
[3] MENG yanjing et al 2014 *Computational comparison of wave characteristics in northern Bohai Bay and northern South China Sea*(Xiamen: Sixth Cross-Strait Forum 2014 People's Weather Forum) pp182-192.
[4] LI Daming et al 2014 *A study of the numerical forecast model of wind waves in Bohai Bay* (haerbin: Journal of Harbin Engineering University) vo35(1) pp132-140.
[5] WANG Dianzhi et al 2004 *Modeling of Wind Waves in the Bohai Bay* (China. MARINE SCIENCE BULLETIN) vol 23(5): pp10-17.
[6] Bo wen bo 2014 *Numerical simulation and Predictions of the Bohai Sea Surface Winds* (Qingdao: OCEAN UNIVERSITY OF CHINA).
[7] TAN Feng et al 2012 *Numerical simulation of WIPHA typhoon waves using WRF-SWAN model*(Tianjin: Journal of Waterway and Harbor) vol33(1) pp14-18.
[8] QI Qinghui et al 2015 *Numerical simulation on typhoon waves in Lianyungang sea area by W R F and MIKE-SW* (Beijing: Port & Waterway Engineering) vo505(7) pp36-40.
[9] PENG Wei et al 2017 *Numerical simulation of wave field around artificial island under typhoon* (Wuhan: China Water Transport) vo17(6) pp168-171.
[10] LIANG Shuxiu et al 2015 *Influence Factors of Typhoon Wave Forecast in the South Sea by SWAN Model* (Nanjing: ADVANCES IN WATER SCIENCE) vo33(1) pp19-30.
[11] YIN Hongqiang et al 2014 *Calculation of Typhoon Haiyan Wind and Wind Based on SWAN Mode* (Wuhan: China Water Transport) vo14(6) pp105-111.

Acknowledgements
This work is supported by the basic Funding of the Central Public Research Insititutes (TKS150207)

A New Concept of Jackup Combined Semisubmersible Multifunction Platform

T Y Wang

Drilling Technology Research Institute, Shengli Petroleum Engineering Ltd., SINOPEC, China
E-mail: wangtianying7072@126.com

Abstract. In order to overcome the shortcomings of the existing semisubmersible platforms, based on care investigation into various factors affecting the quality of floating platforms, a new concept of jackup combined semisubmersible multifunction platform named JCSM was developed. The JCSM integrates the advantages of the jackup platform and the semisubmersible platform. Also, the comprehensive performance of the JCSM is improved by optimizing the details of the topological structure. Therefore, the JCSM is well-suited for either wet tree or dry tree developments in harsh environments. Meanwhile, it possesses a variety of attractive functions covering drilling, early production, workover, oil and gas treatment, oil storage and unloading, and etc. Thus, a JCSM can be used throughout the entire life cycle of deepwater oil and gas field development, which can greatly help reduce the capex and opex and obtain higher economic benefit. The JCSM is an innovative alternative to the conventional semisubmersibles and a new choice provided for the development of offshore oil and gas fields in the future. The advantages of the JCSM was elaborated and analysed from mechanics perspective in the paper. This study will provide reference for the development of creative floating platforms.

1. Introduction

With the declining of oil and gas resources on land and offshore regions, deepwater oil and gas exploitation has become one of the focuses of international competition in energy fields. In general, the deepwater floating structures available mainly include SPARs, TLPs and Semis. Compared with other floating systems, Semis have many attractive traits, such as good stability, high wind/wave resistance ability, spacious deck, large payload capacity, adaptable to a wide range of water depths, etc., and thus are the most widely used and promising floating platforms for deepwater and even ultra-deepwater applications worldwide 0. However, the conventional semis have some shortcomings 0, including low stability margin due to relatively high center of gravity; inefficient utilization owing to deficient functions; shortage of oil storage capacity; too large heave motion to serve as the host of the dry-tree well system; and so on.

To mitigate the weaknesses of the conventional semis, many new concepts have emerged in recent years 0. Typical versions include the truss semisubmersible platform (Truss Semi) and the extensible draft semisubmersible platform (ESEMI) presented by US FloaTEC company, the expandable semisubmersible platform (EDP) proposed by France Technip company, the FourStar Semi by the United States SBM Atlantia company, and the semisubmersible platform with free suspended solid ballast tanks (FHS Semi) by Mansour and Huang 0. The main design goal of those new-type semis was to reduce the platform's heave motion. Unfortunately, although they have indeed improved the heave performance, meanwhile a series of new challenges followed with them, such as vortex-induced

vibration caused by using long columns, difficulties in installment and towage also due to long columns, and unsatisfactory pitch/roll motion performance caused by deep draft design, sharply rising costs because of deep draft configuration, and etc. On the other hand, the new concepts and the matching techniques have been patented, forming the monopoly in deep water technology. By contrast, the types of the semis in China are seldom, and the research and application is still mainly concentrated on the drilling ones. Efforts on the other function of the semis are far from sufficient, which has been one of the bottlenecks of China's development of deepwater oil and gas fields. Thus, it is urgent to design the new-type semisubmersible platforms with own intellectual property rights and applicable in the harsh environment for the South China Sea in order to accelerate the development of deepwater oil and gas in China.
The uncertainty of the geological reservoir and the high costs of the conventional facilities determine the high risk of deepwater oil and gas field development. Especially under the current international situation that the oil price has been being long-term sluggish, how to maximize the platform utilization is the key measure to reduce the costs of the deepsea oil and gas exploration. Therefore, the multifunction platform has become the best choice for deepsea oil field development.
In consideration of the current situation, this paper proposed a new concept of the semisubmersible platform, which overcomes the shortcomings of the existing semis, adaptable to dry tree, wet tree and the combination of the both, also possessing multiple functions, such as drilling, early production, workover, oil and gas treatment, oil and gas storage as well. It will provide a new low-cost life-long solution for the future deepsea oil and gas field development.

2. Theoretical Basis

2.1. The Affecting Factors of Heave

Hydrodynamic response is one of the main indexes to evaluate a floating platform's behaviour. It is not only related to operation efficiency and economic benefit, but also directly affects platform safety. Generally, the heave motion of the conventional semi is large due to its natural heave period close to wave period, so it's difficult for the conventional semi to meet the movement requirements for dry tree with the top tension riser. Dry tree system is economical and practical, easy to operate and maintain. The amplitude of heave determines whether or not the dry tree system can be accommodated on a floating platform. Therefore, it's the most concerned topic to reduce the heave through careful studying the parameters affecting the platform's heave.
Ignoring the coupling effect between the heave and the motions in other directions, the heave motion equation of the semisubmersible platform can be written as 0

$$(\mathrm{M}+\mathrm{A}_{33})\ddot{Z}+(\mathrm{C}+\mathrm{B}_{33})\dot{Z}+\rho g\,\mathrm{A}_{\mathrm{w}}\,Z=\mathrm{F}_3(t) \tag{1}$$

where M is the platform mass, A_{33} is the heave additional mass, Z is the heave amplitude, C is viscous damping coefficient, B_{33} the heave damping, A_{w} the waterplane area, and F_3 the vertical excitation force on the platform.
Assuming that the free surface raise at the center of the platform is $\varsigma=\varsigma_\alpha\sin\omega t$. k is a small number and z_{m} is the height of the geometric center of the lower floating body from the waterplane. The heave amplitude is derived as follows:

$$Z=\frac{\varsigma\cos\left(\mathrm{k}B_{\mathrm{f}}/2\right)}{\exp(kz_{\mathrm{m}})}\left[1-\frac{kz_{\mathrm{m}}}{\left(\omega/\omega_n\right)^2-1}\right] \tag{2}$$

where B_{f} is the width of the lower floating body and ω_n is the natural heave frequency.
From equations (2), it can be seen that the heave response of the platform is proportional to the wave height, and the heave amplitude decreases with the increase of z_{m} or B_{f}.
The formula for the natural heave period of the platform is

$$T_{n3}=2\pi\left(\frac{\mathrm{M}+\mathrm{A}_{33}}{\rho g\,\mathrm{A}_{\mathrm{w}}}\right)^{1/2} \tag{3}$$

As shown in equation (3), the main factors affecting the natural heave period of the platform is the waterplane area and the heave additional mass when the mass of the platform is settled. The smaller the waterplane area is, the larger the natural heave period of the platform is; while the greater the heave additional mass is, the larger the natural heave period is. In addition, increasing the mass of the platform also can increase the natural heave period, which can achieve by enlarging the platform scale or adding the clump weight.

In summary, the heave motion of the platform within the range of wave frequencies can effectively reduce by increasing the damping and the natural heave period to keep it away from the wave energy range, and reducing the heave excitation force, and so on. Moreover, under a given marine environmental condition, the main affecting factors of the heave response are the draft of the platform, the width of the lower floating body, waterplane area and the heave additional mass of the platform. The new platform concept will be carefully designed from all the details of the topology to improve the heave performance.

2.2. *The Affecting Factors of Roll and Pitch*

Pitch and roll are important indicators of the operating performance of a floating platform. The swing amplitude will affect the working status of the production facilities, the processing quality of the crude oil and the comfort of the crew's living environment.

Due to the symmetrical structure of the floaing body in the horizontal plane, Just need study the case of the roll motion here. According to the theory of hydrostatic mechanics 0, there are

$$T_{n5} = 2\pi\left(\frac{I_x}{\nabla h}\right)^{1/2} \tag{4}$$

$$r = \frac{I_x}{\nabla} \tag{5}$$

$$h = z_{\mathrm{B}} + r - z_{\mathrm{G}} \tag{6}$$

$$\mu_\theta = \frac{N_\theta}{(I_x \nabla h)^{1/2}} \tag{7}$$

$$\theta = \frac{\alpha_0}{\left[(1-\Lambda_\theta^2) + 4\mu_\theta^2\Lambda_\theta^2\right]^{1/2}} \tag{8}$$

$$I_x = \frac{\nabla}{12}(B^2 + 4z_G^2) \tag{9}$$

where T_{n5} is the natural roll period of the floating body, ∇ is the displacement, h is the initial stability height, $I_x = I_{55} + \mathrm{A}_{55}$ is the total inertia moment, herein I_{55} and A_{55} are the roll inertia and the roll additional mass respectively. According to the engineering experience of the semisubmersible platform based on the potential flow theory, the roll additional mass A_{55} is about 1/3 of the total inertia moment. r is the initial stability radius, z_{B} is the vertical coordinates of the buoyancy center, z_{G} is the vertical coordinates of the gravity center, μ_θ is the dimensionless damping coefficient, N_θ is the damping coefficient, θ is the roll angle, and B is the width of the floating body.

It can be seen from equation (4) that when the displacement is given, the increase of roll natural period can be achieved by reducing the initial stability height or increasing the roll moment of inertia.

As shown in equations (4) ~ (6), when the displacement is given, a smaller full-loaded waterplane area can make I_x reduce rapidly, accordingly make r decrease. Moving the buoyant center down simultaneously, h value will be smaller, so the larger T_{n5} value will achieve. Moreover, as seen from equations (7) and (8), reducing h will cause μ_θ to increase and the roll/pitch amplitude to decrease. On the other hand, the trouble with the decreasing waterplane area leading to the larger heave can be settled by setting some devices to increase the viscosity damping. From the equations (4) ~ (9), it also can be seen that increasing the width B of the floating body can induce I_x and T_{n5} to increase while

μ_θ to decrease, and then the roll/pitch amplitude of the floating body increases. Therefore, it is necessary to consider various factors in an integrated manner when carrying out the optimal design.

In summary, using the smaller waterplane scheme and the larger width is the effective measures to overall improve the roll performance of the floating body. In addition, using the inclined side edges is also helpful to increase the diffraction damping, thereby reducing the heave excitation force and the heave motion of the platform 0.

2.3. The Affecting Factors of Stability

Harsh and complex deepsea environments put forward higher requirement for the safety and weatherliness of the semisubmersible platform, so the reserve stability of the platform needs to be improved. As can be seen from Section 2.2, there are many measures to increase the stability of a floating body. For example, the increase of the width of the floating body can lead to the increase of the transverse waterplane moment of inertia of the floating body and the static moments of the wedges above and under the waterplane, and then the metacentric height and the arm of static stability will be larger, so the stability of the floating body is increased. The stability of the floating body can also be effectively enhanced by reducing the gravity center or modestly increasing the moulded depth of the floating body. In addition, the flaring-type design of the profile above the waterline can increase the righting moment when the inclination is large, so it can also enhance the stability of the floating body.

The initial stability is an important factor determining the swaying frequency of a floating body. The swaying period of the floating body with small initial stability is long and the swaying is gentle; otherwise, the swaying period of the floating body with large initial stability is short and the floating body will violently sway under wind and storm. As known from the analysis above, reducing the initial stability can improve the motion performance; however, reducing the stability height is contradictory with the stability requirement. Therefore, various factors must be comprehensive measured and wholly considered throughout the design process. The rational procedure is using collaborative optimization principle 0 to obtain the ideal stability under the premise of ensuring excellent motion performance of the floating body through adjusting the design parameters effecting performance of the platform, making both the stability and hydrodynamic performance be well satisfied.

3. JCSM Platform Concept

3.1. The Sketching of New Platform

Based on the detailed analysis of the shortcomings of the existing semis and the factors affecting the platform performance, combining the advantages of the jackup platform and the semi platform, the new concept of jackup combined semisubmersible multifunction platform (JCSM for short) has been put forward, which has been submitted to China's State Patent Office for examination as a patent application (Patent application number: 2017102910467). An example for the technical solution of the new concept platform is illustrated by Figure 1. As shown in Figure 1, the main body of the JCSM is composed of a box deck, a lower floating body and at least three jackup legs. Thereinto, the lower floating body is made up of at least three outward-tilting columns from its bottom and an annular pontoon connecting the lower portions of the columns. The jackup legs connect the box deck and the lower floating body and support the topsides. The deck can be raised or lowered along the jackup legs to the required height through lifting device which is driven by electric hydraulic, electric gear rack or electromagnetic lift technology. Figure 2 shows the elevation view of the platform with the oil and gas production module on the deck. The platform is positioned using the mooring system, dynamic positioning system or the combination of the both according to the environmental condition and water depth of the working sea area.

Taking full advantage of the length adjustability of the jackup legs and great ballast capacity of the lower floating body, the gravity center of the JCSM can be changed according to the working conditions and the marine environment in order to improve the stability and hydrodynamic performance, particularly to reduce the heave amplitude so as to meet the requirements for dry tree oil production. Huge internal space of the annular pontoon can be divided into storage tanks, ballast tanks, pump tanks and others for production. The design of the jackup leg makes the sizes of the topside module and the lower floating body not affected by column spacing, and convenient to adjust the distance between the deck and the lower floating body, so quayside integration and overall towage can be conveniently implemented.

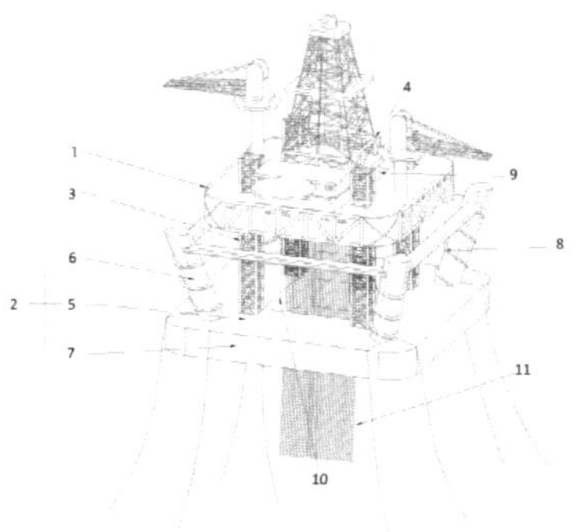

Figure 1. Conceptual diagram of JCSM.

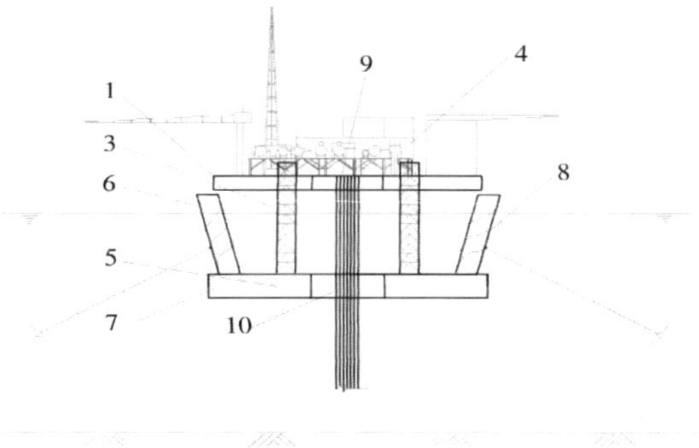

Figure 2. The elevation view of JCSM with production facilities on the deck.

(1 box deck; 2 lower floating body; 3 jackup leg; 4 drilling rig and oil & gas processing module; 5 annular pontoon; 6 outward-tilting column; 7 out-extending part of pontoon; 8 spiral side plate; 9 moon pool; 10 central well; 11 oil and gas production riser / drilling riser)

3.2. Components of the Platform

3.2.1. Box deck. The box deck is equipped with skid-mounted modular units, including oil and gas treatment system, dry and/or wet oil wellhead, drilling equipment, living quarter, power modular, mooring system, moon pool, and etc. The box deck has a 1-2-meter-high double bottom and 2-storey decks with height of 3-4 meters each. The double bottom enhances the global structural strength of the topside. Laying pipelines inside the double bottom contributes to the unified planning and layout for the pipelines. Also, the space inside the double deck can arrange production and drilling equipment, and more topside area will be saved for laying out increasing functional facilities in the future.

3.2.2. Lower floating body. The lower floating body includes the annular pontoon and the columns, two parts providing buoyancy and stability together. At the center of the annular pontoon is a cylindrical or prismatic central well. The interior space of the annular pontoon is divided into ballast tanks, oil storage tanks, production compartments, drilling equipment compartments and pump compartments. Among them, some of the ballast tanks are permanently fixed ballast with weights to ensure that the floating center is higher than the gravity center of the platform. At the top of the annular pontoon are at least three symmetrically arranged columns inclined outwardly along the vertical axis of the geometric center of the annular pontoon. The cross section of the columns is round, fillet square or the other fillet polygon. The outer inclined columns are each arranged near the outer side of the annular pontoon. The lateral side of each column sets up a spiral plate, and the internal space of each column is divided into several watertight compartments. The central well of the annular pontoon is aligned with the moon pool of the box deck in the vertical direction to facilitate the oil & gas rig and the production & drilling risers passing through. The jackup legs together with the central well protect the risers from collisions and provide lateral support for the risers.

3.2.3. Jackup leg. The JCSM uses the truss-framed jackup legs to connect the box deck and the annular pontoon and support the topsides, thus, the lower floating body is not directly connected with the box deck. Therefore, the length of the jackup legs between the box deck and the annular pontoon, the number of the jackup legs, the cross-section dimension and the location can be adjusted according to the configuration of the jackup legs, the weight of the upper module and installation requirements of the platform. Similarly, the deck size is not limited by the leg spacing and can be defined according to the area requirements of the deck. The size of the annular pontoon, the spacing and size of the columns can also be optimized to ensure favorable motion and stability performance.

3.3. Advantages of JCSM

3.3.1. Easy to Optimize. The deck of the JCSM is directly supported by the jackup legs and without direct contact with the lower floating body, so the deck design will not be limited by the column size. Therefore, the designer can independently define the size and shape of the deck to make the facilities on it meet the operational process requirements better, and then improve the operation efficiency of the platform, the safety and reliability of the topside facility as well. The topsides hardly interact at all with the lower floating body, so the column spacing, waterplane area and displacement can also be optimized in a wide range, thereby improving the global performance of the platform.

3.3.2. Strong adaptability to environment. The pontoon is designed as a box-shaped structure with fillet edges to reduce the wave drag force. Some of the ballast tanks in the pontoon are permanently fixed ballast filled with the iron ore or other heavy objects to significantly reduce the gravity center of the platform and then ensure the floating center is above the gravity center at any time. The others are water ballast tanks, which can dynamically adjust the draft to maintain the platform a deep draft state during the service to reduce the heave exciting force. The combination of the two types of ballasts achieves unconditioned stability of the JCSM in the marine environment. Furthermore, the weights with a heavier dense increase the heave natural period and improve the heave motion of the platform. The powerful ballast system allows the platform to have a larger adjusting range of drafts to better suit the marine environmental conditions and different operating modes in different marine areas. Besides, the deck can move up and down along the jackup legs to a reasonable distance between the deck and the pontoon through the lifting device. Combined with adjustment of the ballast volume, the optimization of the gravity center, displacement, inertia moment and waterplane area of the JCSM is achieved, so that it can be better adapted to the operating conditions in different sea areas.

3.3.3. Good self-maintaining capability. Variable load is an important indicator measuring performance of a deepwater semisubmersible drilling platform. Large variable load can reduce the frequency of transport supplies and operating costs to ensure continuous operation and improve economic efficiency. The annular pontoon extends outside from the bottom of the columns and forms the out-extending part of the annular pontoon, which increases the variable load, displacement and oil storage capacity, reduces the gravity center of the platform, and makes the annular pontoon double as a heave plate as well. The out-extending part of the annular pontoon increases the ratio of variable load to platform weight, allowing the platform to meet the requirements for the self-sustaining capability, and then adapting to deep or ultra-deep water, all-weather and long period working.

3.3.4. Excellent structural stiffness and strength. The annular pontoon not only serve as the support structure to provide sufficient buoyancy for the platform, increase loading capacity and then enhance self-sustaining ability of the platform, but also make the substructure become an integral frame structure to transfer the environmental load acting on the columns to the global structure, thereby effectively resisting the environmental impact on the platform and then enhancing the strength of the platform, so the JCSM has a good hull safety and anti-storm capacity. At the same time, the annular pontoon structure can also effectively transmit interaction between the columns caused by the wave and the imbalance load of the platform. Therefore, the JCSM has excellent global structural stiffness and strength and effectively decrease the fatigue stress of the connected joints, reduce the welding and building difficulty, and improve the fatigue life of the platform structure. Obviously, navigation resistance of the annular pontoon is significantly larger than that of the double lower shells, so the platform need be placed on a large barge to tow.

3.3.5. Adapted to dry tree system. The out-extending part of the pontoon not only enlarge the compartment volume, increase the payload, lower the gravity center and improve stability of the platform, but also function as a heave plate to increase the added mass and the added damping, thereby increasing natural roll/pitch period and natural heave period and then improving hydrodynamic performance of the platform. Thus, the JCSM is allowed to use the dry oil production systems. In addition, through optimization design of the internal compartments, the displacement can be separated into several parts to allocate reasonable positions, and then the platform can obtain more added mass than the conventional semis with the same displacement.

3.3.6. Tilt columns with obvious advantages. Large stability margin is key to enhancing anti-storm ability and safety of a floating platform. The outward-tilting columns make the column spacing larger at the waterline, so the inertia moment of the platform at the waterplane is larger and then the initial stability height and righting moment are larger, which effectively reduce pitch/roll motions as well as the mooring line tensions.
Moreover, the columns can be designed as a smaller size as they do not bear the topside load, which makes the waterplane area of the columns smaller. Wave diffraction effect on the columns is significantly reduced because the column diameter is very small compared with the wavelength, and then effectively reducing the surface wave load acting on the platform. At the same time, coupling hydrodynamic response is also reduced between the columns as well between the column and the jackup leg. And the natural heave period further increases the motion performance of the platform is improved. The tilt sides of the columns also help to increase the radiation damping, thereby increasing the natural heave period and reducing the vertical motion of the platform. The tilt column has a better structural strength than the conventional vertical column, and can better resist the bending moment caused by the waves and the gravity of the platform, so that the global structure strength increases. In addition, the tilt design of the column together with the spiral side plate can effectively reduce vortex-induced response of the platform caused by the current.
Furthermore, in the process of optimization design, according to the principle of external forces suppressed each other 0, the tilt angle of the column can be adjusted to make the column spacing more reasonable so as to offset some of the external forces such as the gravity, buoyancy, wind, wave force, inertial forces and others acting on the platform, thereby further reducing the motion of the platform. If the spacing of the columns is designed as half of a wavelength, the forces acting on the columns and the lower floating body are equal and opposite each other and then the motion of the platform is reduced. When the crest is located in the canter line of the platform, the left and right column together with the lower floating body will suffer from the force outwards which are equal and opposite; otherwise, when the trough is located in the center line of the platform, the left and right column together with the lower floating body will suffer from the force inwards which also balance each other. If the spacing of the columns equals to a wavelength, when the crest is located on the left and right columns, the upward buoyancy of the columns can be offset a portion so as to reduce the motion of the platform; and vice versa.

3.3.7. Truss-framed jackup legs with obvious advantages. Relative to conventional columns, the truss-framed jackup legs are transparent to waves and wind, save steel as well, facilitate the current through in serious sea flow and reduce drag force; thereby reduce the mooring line tension and the platform motion. Moreover, they can also reduce the wave climbing up the leg and effectively increase airgap of the platform.

3.3.8. Easy to install and transport. The JCSM has smaller draft but larger displacement than other semis with the same amount of steel, so it is fully capable of quayside integration. This effectively reduces the difficulty of installation and minimizes installation risk. The JCSM can also achieve significantly simple offshore transportation without large floating crane. As shown in Figure 3, while towing or shifting operations, the ballast water in the pontoon is discharged and the deck is lowered along the jackup legs to near the pontoon in order to minimize the gravity center of the platform, reduce the windward area, improve towing stability, reduce towing drag, and then avoid the risk of installation at sea. After arriving at the specified working site, the deck is raised to the designed height along the jackup legs, and then the risers are installed for operation. At this point, the stability is plentiful because the floating center is above the gravity center. In sum, quayside integration and overall towing of the JCSM greatly reduce offshore working hour, cost and risk.

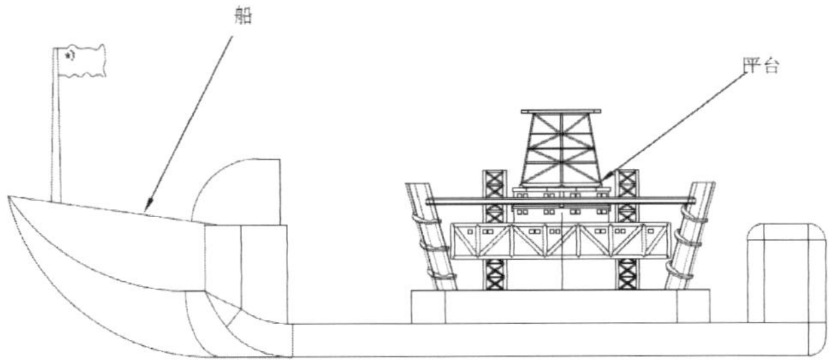

Figure 3. JCSM in the course of towage.

3.3.9. Modular design and construction. The JCSM has symmetrical configuration and simple structure, so it is suitable for modular design and construction to reduce cost and shorten construction time. In addition, the structure of each module is highly similar, which further reduces the difficulty of design and construction, and thus the serialization of design and construction of the JCSM can be easily achieved.

4. Conclusion

In this paper, a new concept of jackup combined semisubmersible multifunction platform named JCSM was developed. The JCSM overcomes the disadvantages of the available semis and possesses a lot of advantages: large stability margin, superior hydrodynamic characteristics, adaptable to the dry tree well system, large variable load, heavy oil storage capacity, simple structure, excellent global structure strength and stiffness, easy to construct and install, and high operation efficiency as well. Meanwhile, the JCSM has multiple functions with drilling, workover, oil production, and can be used throughout the life process of deepwater oil and gas field development. The JCSM with satisfactory comprehensive performance and high utilization can greatly reduce the overall development cost of the oil field, thereby obtaining the higher value of the investment.

The JCSM fully expands the application of the semis. It is invented as a low-cost solution for the oil production in deep water, which can meet the requirements for complicated deepwater environment. With the continuous exploitation of deepwater oil and gas fields in China, the JCSM will have a good application prospect. In the follow-up study, in order to verify the applicability of this novel platform in the South China Sea, it is necessary to carry out a series of studies on its key technologies such as stability, hydrodynamic, mooring system design by numerical calculation and model testing, and etc.

5. References

[1] Xie B, Wang S S, Feng W and Fu Y J 2008 Summarization of key technologies of semisubmersible drilling platform in 3000 m-depth water. High-technology and industrialization, 12: 34-36. (in Chinese)
[2] Cao H and Shao W D 2014 Development of Novel Dry Tree Semisubmersible Platform. Ship Standardization Engineer, 4: 27-32. (in Chinese)
[3] Jiang Z, Xie B and Xie W H 2011 A review of novel semisubmersible production platforms. The ocean engineering, 29(3): 132-38. (in Chinese)
[4] Mansour A M, Huang E W. 2007 H-shaped pontoon deep-water floating production semisubmersible. Proceedings of the 26th International Conference on Offshore Mechanics and Arctic Engineering. California, USA, OMAE-29385.
[5] Faltinsen O M. 1998 Sea Loads on Ships and Offshore Structures. London: Cambridge University Press, 10-70.
[6] Wu X H, Zhang L W and Wang R K. 1988 Ship maneuverability and seakeeping. Beijing: China Communications Press, 165-200. (in Chinese)

[7] Sheng Z B and LIU Y Z. 2003 Principle of ship (volume One). Shanghai: Shanghai Jiao Tong University Press, 46-78. (in Chinese)
[8] Lin R, Song Y Q, Zhang B and Li D W 2013 Research on large ship concept design platform based on multidisciplinary design optimization. Ship standardization engineer, 46(6): 33-35. (in Chinese)
[9] Meng Z Y and Ren G Y 1995 Working principle and structural characteristic analysis of Semisubmersible platforms. China offshore platform, 10(1): 35-37. (in Chinese)

Acknowledgments
The Author acknowledges the financial support from Sinopec Research Project and Shandong Province Innovation Project. (NO JP16008)

IOP Conf. Series: Earth and Environmental Science **171** (2018) 012019 doi:10.1088/1755-1315/171/1/012019

Bayesian-Networks-Based Risk Evaluation Approach of Offshore Platform

Xiaobing Yuan[1], Baoping Cai*[1], Xiaoyan Shao[1], Xiangdi Kong[1], Yonghong Liu[1] and Guoming Chen[1]

1 College of Mechanical and Electronic Engineering, China University of Petroleum, Qingdao, Shandong, 266580, China
*Corresponding author E-mail: caibaoping@upc.edu.cn

Abstract. This paper presents a Bayesian-networks-based risk evaluation approach of offshore oil platform, through the investigation of offshore oil platform is prone to accidents, to understand the structure of the relevant subsystems and collect the prior probability of related events. Then the events are divided into a three-tier structure, and the typical Bayesian network model of fire and collision accident in offshore oil platform is established. The corresponding parameter analysis is made to complete the verification and analysis of the model. The results show that the method based on Bayesian network can carry out effective risk analysis, and the risk assessment is feasible and advantageous by using this method.

1. Introduction

The offshore platform is the basic facility for the development of offshore oil and natural gas resources, which integrates drilling, oil testing, oil production, underground operation, oil and gas gathering, preliminary processing, storage and transportation. Flammable and explosive materials raise the operational risk. In the process of offshore oil extraction, many countries have encountered all kinds of problems, some of them are major accidents. For example, in 2010, the Gulf of Mexico deep-sea horizon drilling platform blowout and explosion, 11 people were killed, 17 people were injured, resulting in huge property losses, lots of oil spills and so on. With the occurrence of accidents, people pay more attention to engineering safety work.

In the early stages of the safety assessment of marine structures, the method used was mainly QRA (Quantitative Risk Assessment). In 1981, the safety assessment work was maturing, and the Norwegian Petroleum Board of Directors issued a guidance document for the safety assessment of the platform concept design. These regulations point out that the platform design should consider the marine platform accident that occur every 4 to 10 years. The accident met the above provisions were selected to prepare for the implementation of preventive measures and the remaining risk assessment, and then compared with the probability of a priori [1]. The Bayesian network is a graphical model used to represent the probability of connection between variables, provides a method of expressing causal information and directed acyclic network composed of nodes, directed arcs and conditional probability distributions [2]. Bayesian theory gives a mathematical function of the trust function in mathematics, has a solid mathematical foundation. It also describes the consistency of trust and evidence, trust with the evidence of incremental learning characteristics and uses the relationship between the data that is described by the weight of the probability measure. The semantic is clear and the comprehension is strong, which is helpful to use the causal relationship between the data to predict and analyse [3]. Cai used Bayesian network on underwater BOP for risk analysis and reliability assessment [4,5]. Wang used logic tree and Bayesian network carried out the offshore platform fire

Probability analysis. Meng used Bayesian network for offshore oil cranes operating risk analysis. In general, the Bayesian network technology developed in recent years has a strong ability to describe, both for reasoning and for diagnosis, and is well suited for risk assessment.

2. BN-Based Risk Analysis Methodology

Figure 1 shows a model that can be divided into three layers. The three-tier structure divides the whole network node into cause, incident and accident, which is the central idea of constructing this model. Cause refers to the behaviour that may trigger an abnormal event, which may or may not be wrong. Incident refers to those who can cause accidents of abnormal events, sometimes it does not intuitively show serious consequences. Accident refers to those with serious consequences, threatening personal safety, resulting in economic losses. When there is a cause, the likelihood of an incident will increase. The incident is something people do not want to see, and once the event has not been ignored by people, and cannot be self-regulation, then an accident may occur.

The graphs of the network nodes represent the variables, and the arrows with directions indicate the dependencies and causal relationships between the nodes [6]. This quantitative relationship is established by the conditional probability table. In general, the conditional probability table is determined by historical data and expert judgment. At the same time, once the corresponding nodes change, the nodes associated with them also change accordingly, the real-time status of each node can be changed by the influence of each other.

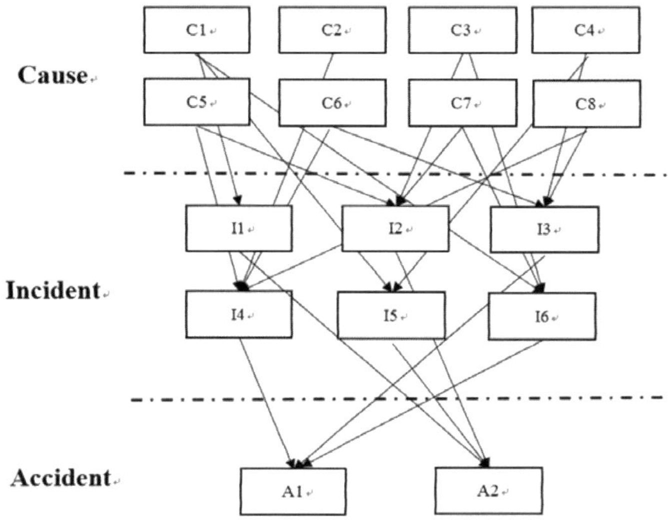

Figure 1. Typical three-tier Bayesian network.

3. Risk Analysis for Offshore Platform

3.1. Modelling Methods

After determining the events that may occur, the next step is to divide them into the three-tier structure of the cause, the incident and the accident. At the same time, based on the interconnection of nodes between each layer, for them to establish the corresponding Bayesian network fill [7]. Figure 2 shows the complete Bayesian network model.

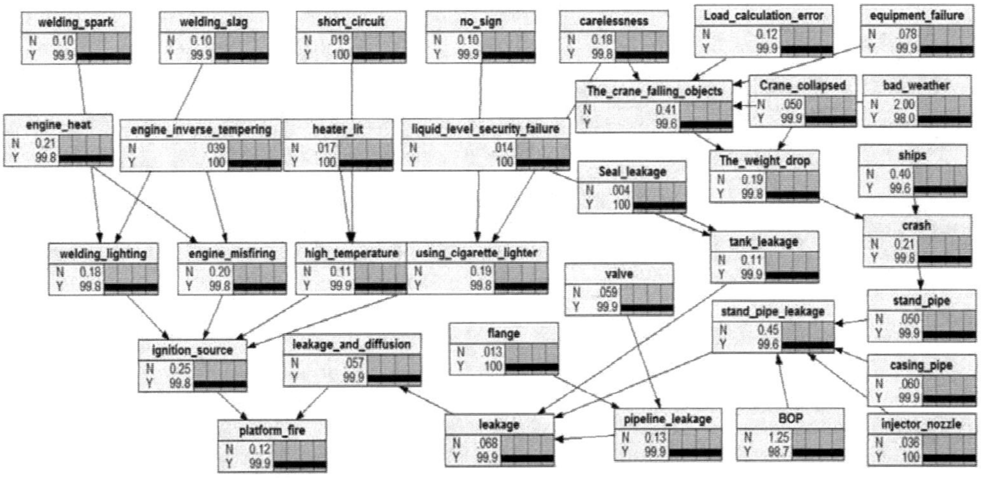

Figure 2. Complete Bayesian network model.

(1) Cause: in the offshore platform, the cause of the accident can be divided into three categories. Three different causes are summarized as follows: The first cause is human error. This is the most influential factor because people are the subjective operator of the whole unit. More specifically, human error can be further divided into organizational errors, group errors and individual errors. Team transfer, lack of knowledge and skills, mismanagement, high work pressure, misunderstanding and poor attitudes can lead to unusual events. The second cause is caused by the failure of the equipment. Such as prolonged exposure, rainwater corrosion, and prolonged use may cause corrosion and aging of the equipment. During the work of the equipment, the device may also be faulty. When these events occur, the likelihood of an accident will increase significantly. Therefore, when these events are found, an effective means should be taken to remedy the incident to prevent a more serious accident. The third cause is the cause of nature, for example, too high temperatures can cause spontaneous combustion of flammable substances. In addition, there are other harsh conditions such as earthquakes, rain and snow, wind and lightning and so on. In this paper, some typical causes for the offshore platform are analysed and transformed into a Bayesian node as shown in Figure 3.

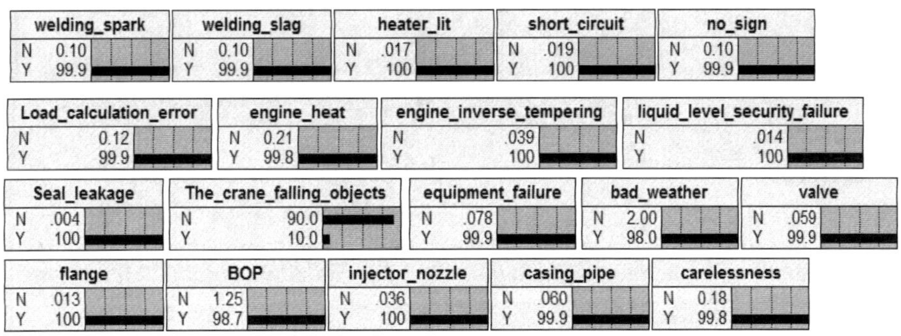

Figure 3. cause layer.

(2) Incident: Some Incident that occur on the offshore platform can lead to some more serious consequences. Such as equipment failure and abnormalities can be considered an incident. The operation of the offshore platform should ensure that the various subsystems are stable, and if the equipment is abnormal, it may cause some danger, resulting in equipment damage and system failure, or even a fire or explosion. The Bayesian network nodes of some typical incidents in the offshore platform are shown in Figure 4.

IOP Conf. Series: Earth and Environmental Science **171** (2018) 012019 doi:10.1088/1755-1315/171/1/012019

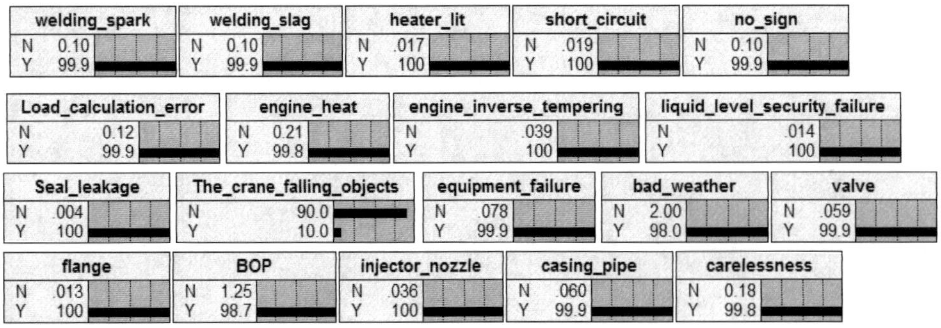

Figure 4. incident layer.

(3) Accident: The consequences of the accident are worse than other things, especially in the offshore platform, as mentioned above, although the accident does not occur often, but it brings the loss is huge. Some of the typical accidents in the offshore platform are fires, explosions and collisions. After these accidents, not only bring immeasurable economic losses, resulting in a huge waste of energy, but also hurt the staff on the offshore platform and even the people around the surrounding areas. This paper mainly studies two kinds of accidents such as fire and collision. The Bayesian network node is shown in Figure 5.

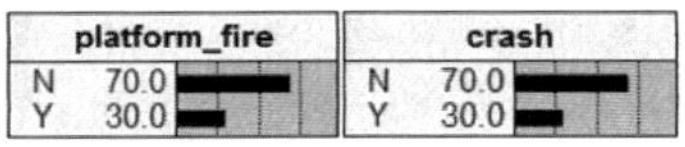

Figure 5. accident layer.

Table 1. "cause" probability table in Bayesian network.

	Node	Probability	State
C1	welding_spark	0.1	N
C2	welding_slag	0.1	N
C3	heater_lit	0.017	N
C4	short_circuit	0.019	N
C5	no_sign	0.1	N
C6	carelessness	0.18	N
C7	Load_calculation_error	0.12	N
C8	equipment_failure	0.078	N
C9	bad_weather	2	N
C10	casing_pipe	0.06	N
C11	injector_nozzle	0.036	N
C12	BOP	1.25	N
C13	valve	0.059	N
C14	flange	0.013	N
C15	liquid_level_security_failure	0.014	N
C16	Seal_leakage	0.004	N
C17	engine_heat	0.21	N
C18	engine_inverse_tempering	0.039	N

Table 2. “incident” probability table in Bayesian network.

	Node	Probability	State
I1	engine_misfiring	0.2	N
I2	welding_lighting	0.18	N
I3	high_temperature	0.11	N
I4	using_cigarette_lighter	0.19	N
I5	ignition_source	0.25	N
I6	The_crane_falling_objects	0.41	N
I7	Crane_collapsed	0.05	N
I8	ships	0.4	N
I9	The_weight_drop	0.19	N
I10	stand_pipe	0.05	N
I11	tank_leakage	0.11	N
I12	stand_pipe_leakage	0.45	N
I13	pipeline_leakage	0.13	N
I14	leakage	0.068	N
I15	leakage_and_diffusion	0.057	N

Table 3. “accident” probability table in Bayesian network.

	Node	Probability	State
A1	crash	0.21	N
A2	platform_fire	0.12	N

Based on the research data of several scholars and the expert’s judgment [8], this paper determines the prior probabilities of a series of causal events and the conditional probability between events. Table 1 represents the prior probabilities of “cause”, Table 2 represents the prior probability of “incident”, and Table 3 represents the prior probability of “accident”. By using the auxiliary software Netica to enter the prior probabilities into the Bayesian initial event [9]. For the initial event, for example, the probability of an “N” state of the welding spark is 0.1%, which indicates that the spark is abnormal, that is, the probability of a dangerous spark is 0.1% and the probability of normal welding is 99.9%.

For the conditional probability table, the quantitative relationship between the nodes is determined by querying the relevant literature. For example,

Pipeline leakage: If the valve and flange are abnormal, these two causes may lead to pipe leakage. They are the parent nodes of the pipeline leak, and Table 4 is the conditional probability table for the pipeline leakage. When there is only one reason for the occurrence of an exception, then the state of the node is “N”, the remaining node status “Y”.

Table 4. Conditional Probability of pipeline leakage.

C13	C14	N	Y
N	N	70	30
N	Y	40	60
Y	N	30	70
Y	Y	0.1	99.9

Another example: ignition source, for example, engine misfire, welding lighting, high temperature, using cigarette lighter are likely to lead to the occurrence of ignition source. Table 5 shows the effect of the probability that the source is abnormal when the factors are in different states. "N" in the table represents the appearance of abnormal state, "Y" represents the normal state.

Table 5. Conditional Probability of ignition source.

I1	I2	I3	I4	N	Y
N	N	N	N	90	10
N	N	N	Y	70	30
N	N	Y	N	69	31
N	N	Y	Y	50	50
N	Y	N	N	66	34
N	Y	N	Y	45	55
N	Y	Y	N	50	50
N	Y	Y	Y	20	80
Y	N	N	N	67	33
Y	N	N	Y	50	50
Y	N	Y	N	44	56
Y	N	Y	Y	25	75
Y	Y	N	N	44	56
Y	Y	N	Y	20	80
Y	Y	Y	N	20	80
Y	Y	Y	Y	0.1	99.9

3.2. Results and Discussion

After deducing the interrelationships between accidents, the most sensitive events can be identified by calculating mutual information and variety of beliefs. Mutual information and variance of beliefs show the correlation between nodes. By calculating mutual information and variance of beliefs can find the most likely event associated with the incident. In addition, it can help people to speculate on the cause of the accident [10].

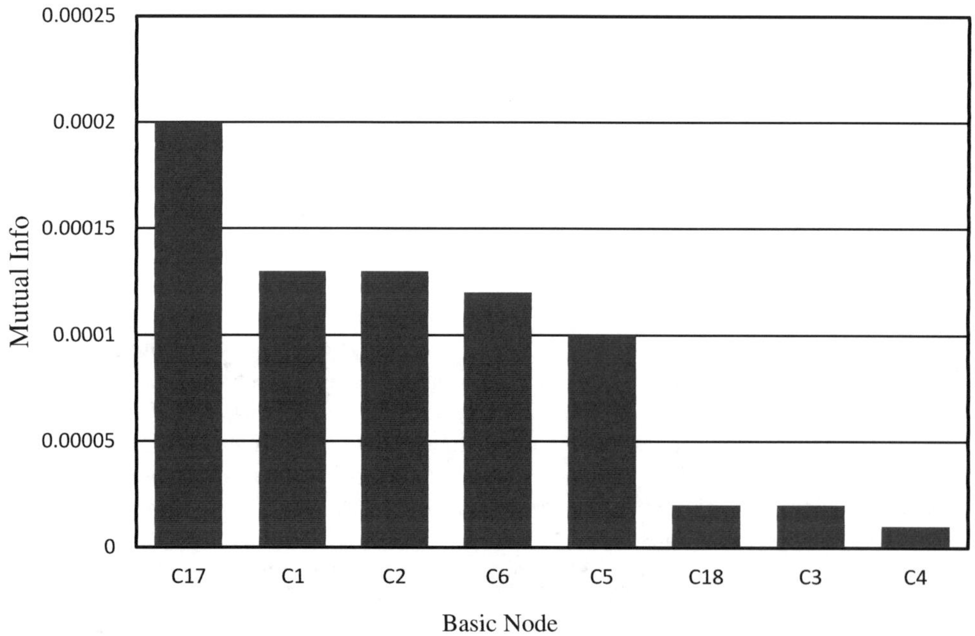

Figure 6. Mutual Info of "platform fire" at other nodes.

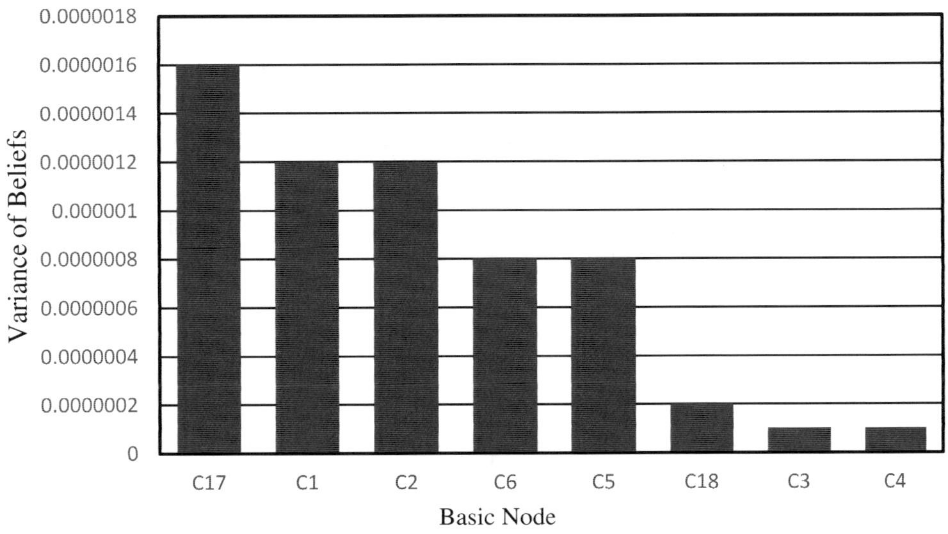

Figure 7. Variance of Beliefs of "platform fire" at other nodes.

Using Netica to compute platform fire mutual information and variance of beliefs, and the data for each base node is shown in the table. Figure 6 shows the mutual information of each node. Figure 7 shows the results of the variance of beliefs. From the comparison we can easily see that engine heat is the most relevant event with the accident which means that engine heat is the most likely cause of the fire accident. Therefore, during the work of the offshore platform, the operator should pay special attention to control the engine exhaust and monitor timely to maintain the surrounding environment temperature control to keep the engine temperature maintained at a relatively healthy level to avoid the occurrence of platform fire.

In addition, other events may also lead to the occurrence of fire accidents. It can be seen welding spark and welding slag two parameters are relatively high, indicating that welding spark and welding slag are factors that cannot be ignored in the fire accident. In the work of the platform, this aspect of monitoring should be strengthened to reduce the probability of accidents due to the abnormalities of these nodes.

Offshore platforms often increase the reliability to improve the accuracy of risk forecasts to avoid the risk of the occurrence of the purpose. However, in the daily work of some abnormalities is inevitable. Therefore, after knowing the occurrence of the basic event, through the Bayesian network model can predict the probability of risk. At this point, the operator needs to give timely preventive measures to reduce the risk of occurrence of the probability. Figure 8 shows the probability of occurrence of the cause node and the probability of a collision event after the exception of the crane falling objects.

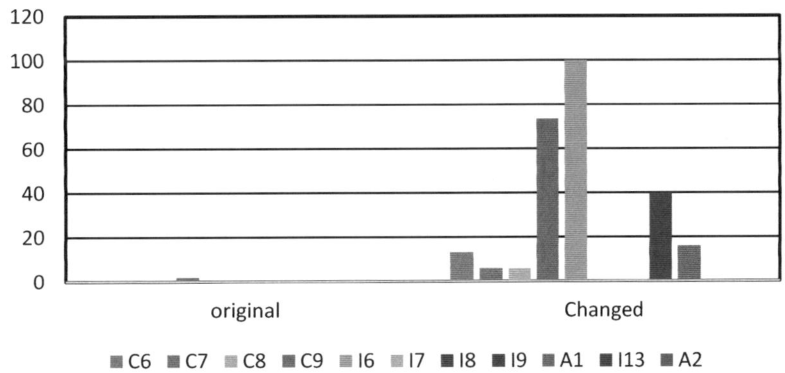

Figure 8. Possible accident when "the crane falling objects" error occurs.

It is easy to see that when the crane falling objects is abnormal (the probability of the occurrence of the crane falling objects becomes 100%), the change in bad weather is most pronounced. It is most likely that the cause of this anomaly may be bad weather. In addition, the probability of the crash has also significantly improved (from 0.21% to 16.1%), indicating that after the occurrence of the crane falling objects there will be a great chance of collision accidents. Therefore, after detecting this anomaly, the staff should take effective safety measures to prevent the occurrence of the collision accident and, if necessary, stop the crane operation. It is worth mentioning that, after the exception of the crane falling objects, the probability of stand pipe leakage changed from 0.45% to 0.84%, the probability of fire from 0.12% to 0.39%, which shows that in this unusual case, the probability of fire also increased accordingly, increasing the probability of occurrence of platform accidents, so the appropriate prevention and remedial measures should be made promptly.

4. Conclusion

This paper presents a case study and completes the validation and analysis of the model to prove the feasibility of this method. For the offshore platform, the study shows that engine heat, welding spark and welding slag become the main causes of the fire accident, so in the process of offshore platform working people should strengthen the engine and welding process monitoring and management, and found such problems timely handling will be able to greatly reduce the occurrence of the fire accident. By the probability change of the crane falling objects after the bad weather event, it can infer that the main cause of the collision accident is bad weather, so it is necessary to keep abreast of weather information and respond to bad weather.

5. References

[1] Fuentes-Bargues J L, González-Gaya C, González-Cruz Ma C, and Cabrelles-Ramírez V 2016 *Journal of Loss Prevention in the Process Industries* 44 44-52.
[2] Khakzad N, Khan F, and Amyotte P 2011 *Reliability Engineering & System Safety* 96(8) 925-932.
[3] Yeo C T, Bhandari J, Abbassi R, Garaniya V, Chai S, and Shomali B 2016 *Journal of Loss Prevention in the Process Industries* 41 259-269.
[4] Cai B, Liu Y, Liu Z, Tian X, Hang L, and Ren C 2012 *Journal of Loss Prevention in the Process Industries* 25(6) 1044-1054.
[5] Cai B, Liu Y, Liu Z, Tian X, Zhang Y, and Ji R 2013 *Risk Analysis* 33(7) 1293-311.
[6] Darwiche A 2009 *Modeling and reasoning with Bayesian networks.*
[7] Montani S, Portinale L, Bobbio A, and Codetta-Raiteri D 2008 *Reliability Engineering & System Safety* 93(7) 922-932.
[8] Keeney R L, and Winterfeldt D V 2002 *IEEE Transactions on Engineering Management* 38(3) 191-201.
[9] Weber P, and Jouffe L 2006 *Reliability Engineering & System Safety* 91(2) 149-162.
[10] Dimattia D G, Khan F I, and Amyotte P R 2005 *Journal of Loss Prevention in the Process Industries* 18(4) 488-501.

Acknowledgements

This work was supported by National Natural Science Foundation of China (No. 51779267), Fundamental Research Funds for the Central Universities (No. 17CX05022 and No.14CX02197A), and Program for Changjiang Scholars and Innovative Research Team in University (IRT_14R58).

IOP Conf. Series: Earth and Environmental Science **171** (2018) 012020 doi:10.1088/1755-1315/171/1/012020

Effect of Touchdown Trench Modelling on SCR Fatigue Life

G. Hu[1], C. Shi[2, 3], and Y. Zhao[2]

1 Dalian Maritime University, Dalian, China
2 School of Petroleum Engineering, China University of Petroleum, Qingdao, China
E-mail: shichen2004@hotmail.com

Abstract. Touchdown zone (TDZ) is one of the most critical fatigue hot spot for a Steel Catenary Riser (SCR). In current industry practice, a flat profile is always used when modelling the seabed surface. However, survey data shows that trenches are formed at SCR touch down zone after riser installation. It is important to understand the impact of trench modelling for fatigue life at TDZ. This paper presents a SCR TDZ fatigue life comparison between a flat and trenched seabed profile using Finite Element Analysis (FEA) models based on linear soil stiffness. The trenched seabed is modelled in its assumed equilibrium condition using an analytical equation derived from non-linear hysteretic analysis in the literature. Further analyses are conducted for more parameters. The objective of this study is to determine if a linear seabed model with trenched profile can improve TDZ fatigue life of an SCR when compared to an SCR on a flat seabed surface.

1. Introduction

Steel Catenary Riser (SCR) is one of the most widely used riser concepts in deepwater for its simplicity. A general SCR configuration is illustrated in Figure 1. One of the most critical fatigue hot spot for a SCR is touchdown zone (TDZ)[1]. Understanding SCRs touchdown zone fatigue performance under environmental loading and host vessel motion can be very challenging. One of the challenges is the modelling of riser interaction with seabed. Current industry practice typically uses a flat seabed surface assumption when conducting SCR analyses[2]. This approach is considered to be conservative and it is believed that incorporating a trenched profile into an FEA model can help improve TDZ fatigue life[3]. However, very limited studies have shown the actual impact of seabed trench modelling on SCR fatigue life at TDZ. One of the reasons is that implementing trench profile in the FEA model can be very challenging.

In this study, a SCR model on a seabed with trench profile is developed. And the fatigue results based on this model is compared to the one with flat seabed. The trench theory, field data, design parameters, model development, and fatigue results are presented in the following sections, followed by conclusions.

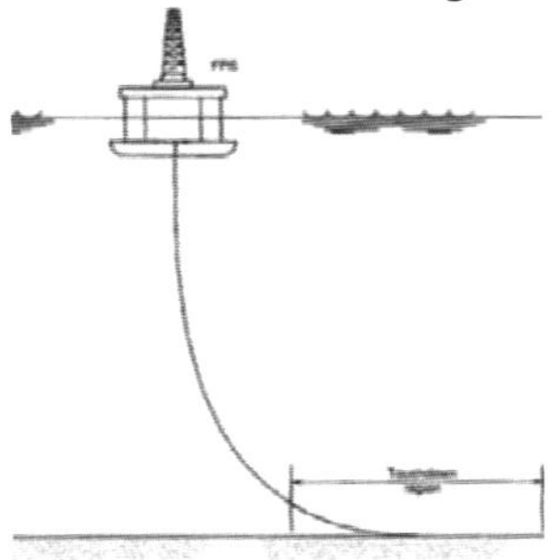

Figure 1. Typical SCR Layout Showing the Touchdown Zone

[3] Any correspondence should be addressed to the author (C. Shi, shichen2004@hotmail.com)

2. Trench Theory in Literature

A theoretical trench shape includes a few key points [4]:

- Touchdown Point (TDP), where the SCR reaches the nominal level of the seabed;
- Trench Bottom Point (TBP), where the trench reaches its maximum depth;
- Trench Surface Point (TSP), where the SCR reaches essentially the zero gradient towards the anchored end of the riser.

A theoretical trench shape schematic is shown in Figure 2.

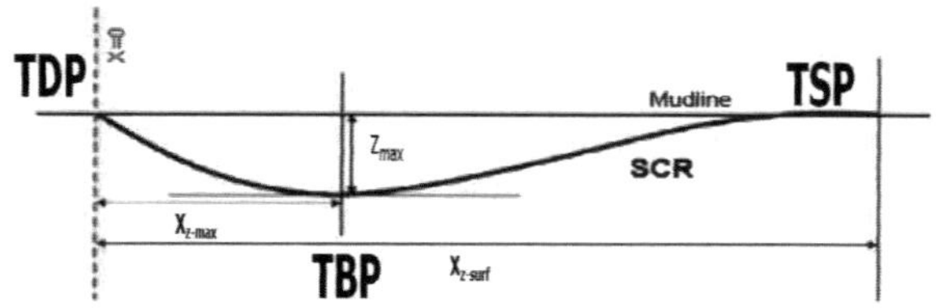

Figure 2. Theoretical Trench Shape [4]

Shiri [4] concluded a mathematical equation to approximate the SCR trench profile based on SCR trench development simulation on hysteretic non-linear seabed for a range of cycles, as shown in Figure 3. A quadratic exponential equation is used to mirror the Abaqus model at 1,000 cycles, as shown in Figure 4.

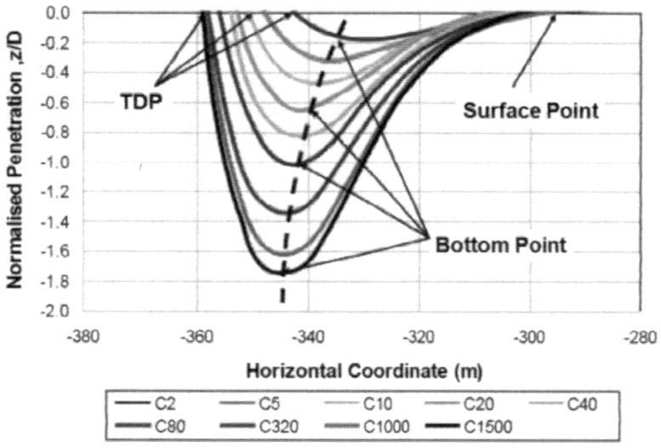

Figure 3. Gradual Displacement of Trench Key Points with Trench Deepening [4]

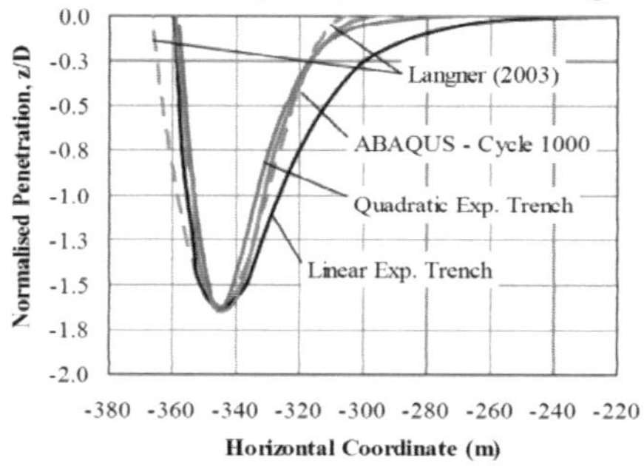

Figure 4. SCR Trench Mathematical Approximation [4]

The quadratic exponential approximation is given as follows:

$$z = -z_{max}\left[\left(\frac{x}{X_{z-max}}\right)e^{\left(1-\frac{x}{X_{z-max}}\right)}\right]^2 \qquad (2)$$

Where:

- x – Horizontal distance from the TDP;
- z – Vertical distance below the mudline;
- z_{max} – Maximum trench depth (i.e. TBP);
- $X_{z\text{-}max}$ – Horizontal distance from the TDP corresponding to the maximum trench depth.

Shiri [4] assumes the TSP is located where z falls to 1% of z_{max}, therefore, the relative distance from the TDP to the TSP may be approximated as follows:

$$X_{z-max} = \frac{1}{5} * X_{z-surf} \qquad (3)$$

One consequence of using the analytically derived trench seabed profile in a global FEA model is that inconsistencies between the riser's natural catenary and the seabed may be present, leading to varying stress ranges in the TDZ, as shown in Figure 5 [4].

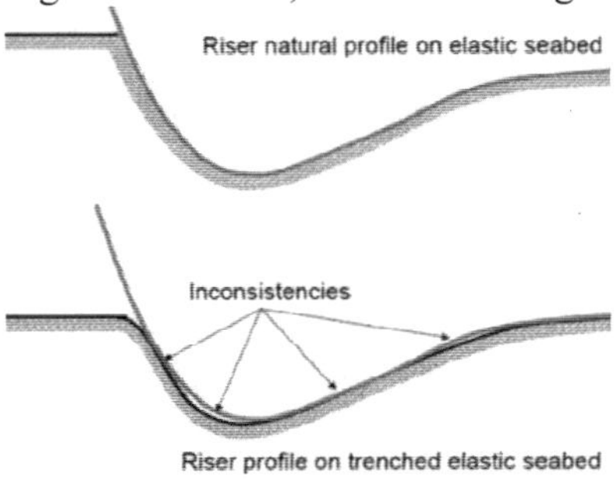

Figure 5. Riser-trench Contact Geometry with Various Trench Locations [4]

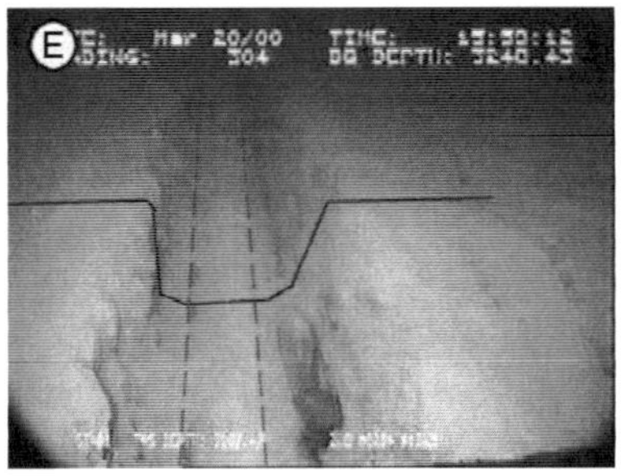

Figure 6. Gas Export SCR TDZ Trench in the GoM [5]

3. Field Data

Gulf of Mexico (GoM) has the most SCR application in the world, relevant survey data in GoM is reviewed to determine appropriate trench dimensions. Bridge's study [5] presents field data in the GoM that is utilized in this study to determine approximate trench sizes for modelling purposes. A trench from a 12.75 inch gas export SCR in the Allegheny field in the GoM seven months after installation is shown in Figure 6 [5]. The survey showed that the trench is approximately four diameters in depth and 60m long [5]. Dimensions in this range are considered for this study.

4. Design Parameters

This study is conducted for a South China Sea field in approximately 1,000 meter water depth using a semi-submersible vessel. Regular wave analysis is performed using a high probability low amplitude seastate of the type that dominates the fatigue damage at the TDZ. The vessel's response to waves is extracted by its response amplitude operator (RAO) and applied to the riser models in OrcaFlex.

The hang-off angles considered is 11 degrees from the vertical for both flat and trenched seabed models. SCR properties are listed in Table 1 and relevant environmental data is detailed in Table 2. The API X' fatigue curve with an SCF of 1.1 is chosen for this study[6].

Table 1. SCR Model Design Data

Parameter	Value
OD, in. (m)	10.75 (0.273)
ID, in. (m)	7.75 (0.197)
Wall Thickness, in. (m)	1.50 (0.038)
Internal Fluid Density, lb/ft^3 (kg/m^3)	10 (160)
Internal Pressure at Water Surface, psi (MPa)	500 (3.45)
Hang-off Angles (°)	11
Material	X65
Yield Strength, ksi (MPa)	65.3 (450)

Table 2. Environmental Data

Parameter	Value
Water Depth (m)	1,015
Hs (m)	1.8
Tp (sec)	7.5
Wave Type	Regular

5. Model Development

Analysis is carried out using the non-linear time domain finite element analysis program OrcaFlex [7]. The models are constructed with equivalent single string elements. The properties of the strings are captured using general pipe element.

Details of the seabed parameters are given in Table 3. Trench lengths considered are based on survey data and also by the hang-off angle which determines how the SCR's natural catenary would lay in the trench. The SCR model in the trenched TDZ for both soil stiffness values considered is shown in Figure 7. The green line is the SCR and the orange line is the seabed. The vertical lines represent the SCR nodes, where white indicates contact and green indicates a gap between the SCR and seabed. Contact discontinuities are present between the SCR and seabed when the larger seabed stiffness is used, but no discontinuities are present for the smaller seabed stiffness. Another representation of the SCR in the trenched profile is shown in Figure 8. The SCR model resting on the flat seabed surface for both seabed stiffness values is shown in Figure 9, where no contact discontinuities are present.

Table 3. Seabed Properties

Seabed Profile Type	Seabed Soil Stiffness ($kN/m/m^2$)	Trench Length from TDP to TSP (m)	Trench Depth
Flat	60 120	N/A	N/A
Trenched	60 120	100 100	3*OD

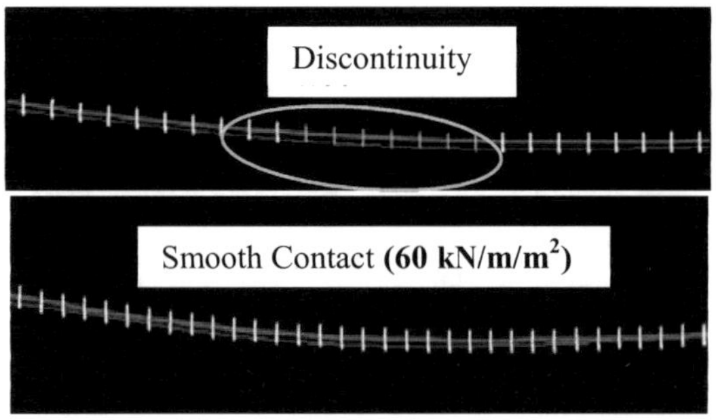

Figure 7. SCR TDZ, Trenched Seabed Profile

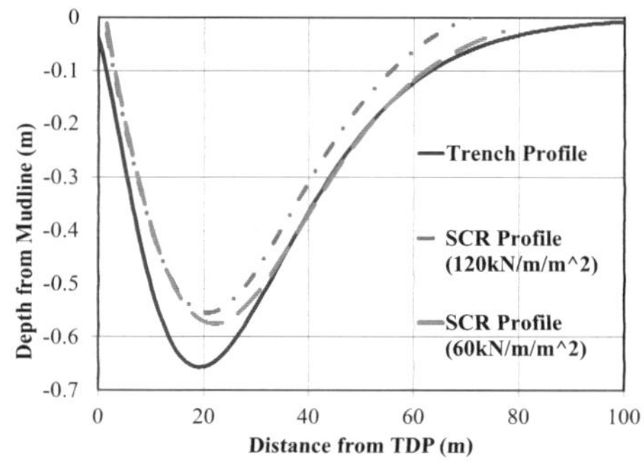

Figure 8. SCR TDZ, Trenched Seabed Profile

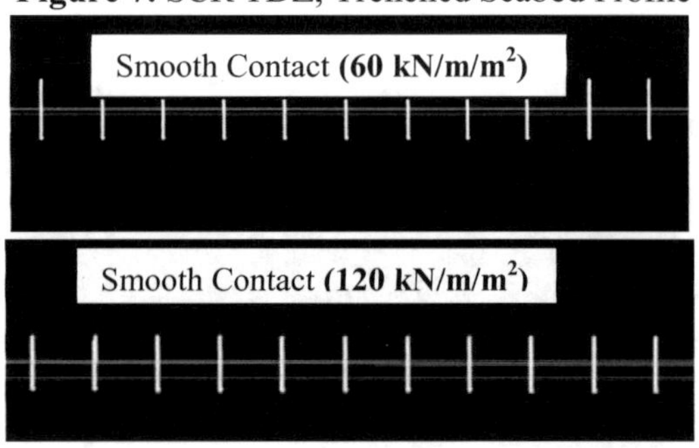

Figure 9. SCR TDZ, Flat Seabed Profile

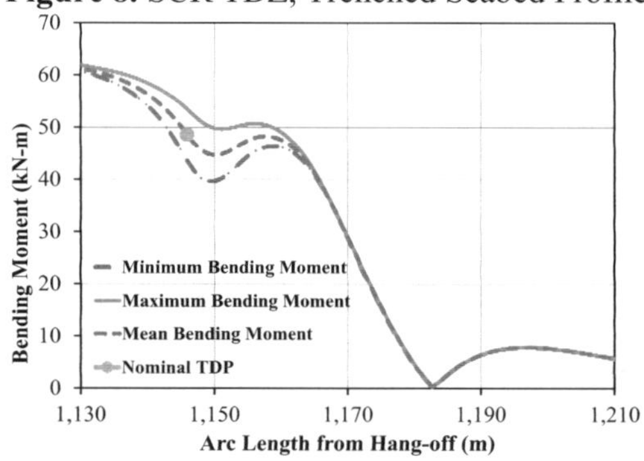

Figure 10. Bending Moment Envelope, Trenched Seabed Profile, 120 $kN/m/m^2$

6. Analysis Results

Results are given in this section considering flat and trenched seabed surfaces for both soil stiffness values. The bending moment envelope for the trenched and flat seabed profiles using soil stiffness of 120 kN/m/m^2 is shown in Figure 10 and Figure 11, respectively. The bending moment envelope for the trenched and flat seabed profiles using soil stiffness of 60 kN/m/m^2 is shown in Figure 12 and Figure 13, respectively. It can be seen that the trenched profile experiences a larger bending moment envelope than the flat seabed profile for both soil stiffness values analyzed. The trenched profile is experiencing a larger bending moment range due to the following: (a) The shape of the SCR in the global model is slightly inconsistent with the analytically derived shape in the TDZ; and (b) Although the SCR shows continuous contact in the TDZ with the smaller soil stiffness, the gradient of seabed resistance is high near the location along the SCR with the lowest fatigue life (see Figure 14).

A stress range comparison between the trenched and flat seabed profiles is given in Table 4. The stress ranges given are at the most damaging location along the SCR, which occur just beyond the nominal TDP. A larger stress range is experienced for the trenched SCR profile when compared to a flat seabed.

Larger stress ranges experienced by the SCR in a trenched seabed profile lead to lower fatigue life, as detailed in Table 5. The fatigue life for an SCR in a trenched profile has decreased by 27% and 31% for the two seabed stiffness values considered when compared to a flat seabed surface, respectively. As expected, the fatigue life reduction increased more when using the larger seabed stiffness.

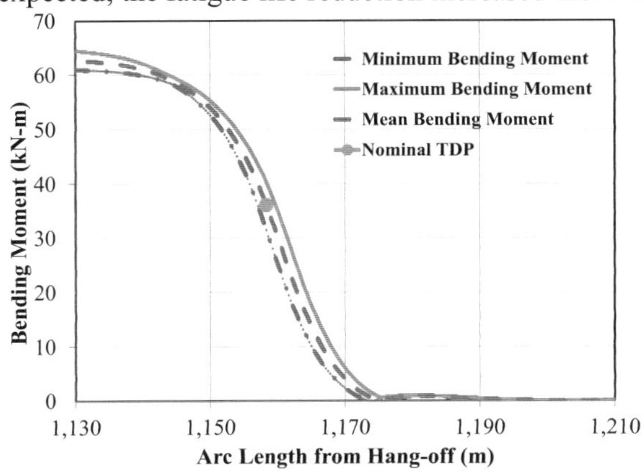

Figure 11. Bending Moment Envelope, Flat Seabed Profile, 120 kN/m/m^2

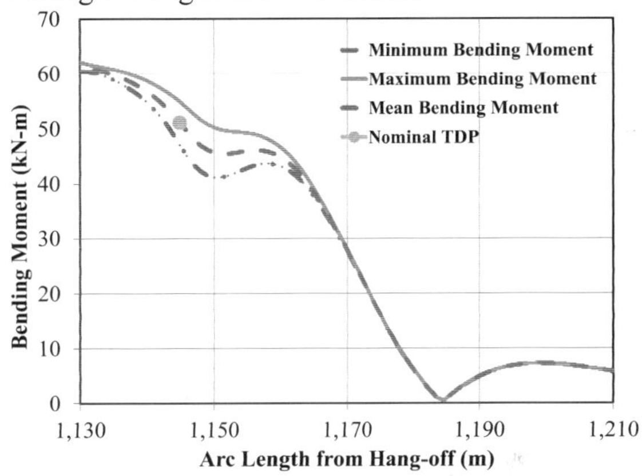

Figure 12. Bending Moment Envelope, Trenched Seabed Profile, 60 kN/m/m^2

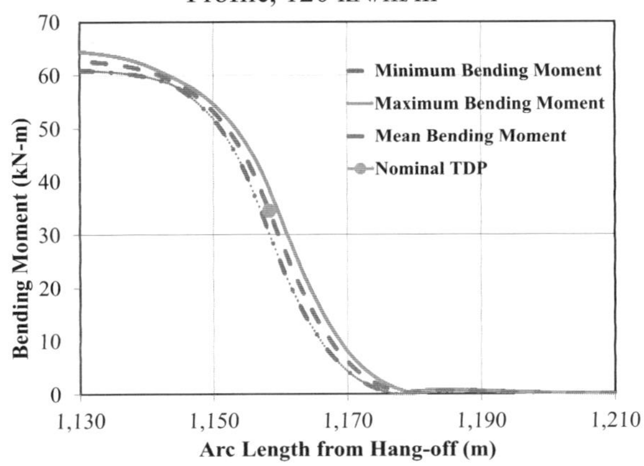

Figure 13. Bending Moment Envelope, Flat Seabed Profile, 60 kN/m/m^2

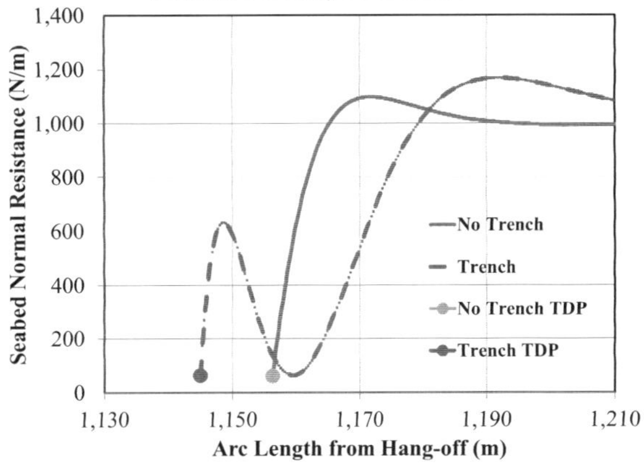

Figure 14. Seabed Normal Resistance, 60 kN/m/m^2

IOP Conf. Series: Earth and Environmental Science **171** (2018) 012020 doi:10.1088/1755-1315/171/1/012020

Table 4. Axial Stress Range Comparison

Seabed Soil Stiffness (kN/m/m2)	Stress Range (MPa)		Difference (MPa)
	Trenched Seabed	Flat Seabed	
60	14.28	12.94	1.34
120	16.57	15.19	1.38

Table 5. Fatigue Life Comparison

Seabed Soil Stiffness (kN/m/m2)	Fatigue Life (Year)		Difference (%)
	Trenched Seabed	Flat Seabed	
60	266	367	-27
120	156	225	-31

7. Parametric Analysis Results

7.1. Refined Trenched Seabed Profile

In this analysis, trenched seabed profile is refined to closely match riser configuration. The SCR's TDZ global coordinates from the analytical trenched seabed profile are used as the new seabed coordinates in the TDP to TSP region. This approach helps reduce discontinuities between the SCR and seabed, and also reduce seabed contact force in the touchdown region.

The soil stiffness considered for this analysis is 120 kN/m/m^2. The refined SCR profile is shown in Figure 15. There is a smooth transition between the SCR's natural catenary and TDP as well as complete contact between the SCR and seabed, as illustrated in Figure 16. Refining the TDZ has reduced the seabed resistance gradient when compared to the analytical trench, as shown in Figure 17.

The bending moment envelope for the refined trench profile is shown in Figure 18. The max bending range has reduced when compared to Figure 10. However, the bending moment range for the refined trench still exceeds the flat seabed configuration when compared to Figure 11. Implementing a refined trench has reduced the stress range when compared to an analytical trench, as detailed in Table 6. However, the stress range is still higher than the flat seabed. The fatigue life improves by approximately 28% when compared to the analytical trench, but falls short of the flat seabed fatigue life by 3.9%, as given in Table 7 and Table 8, respectively.

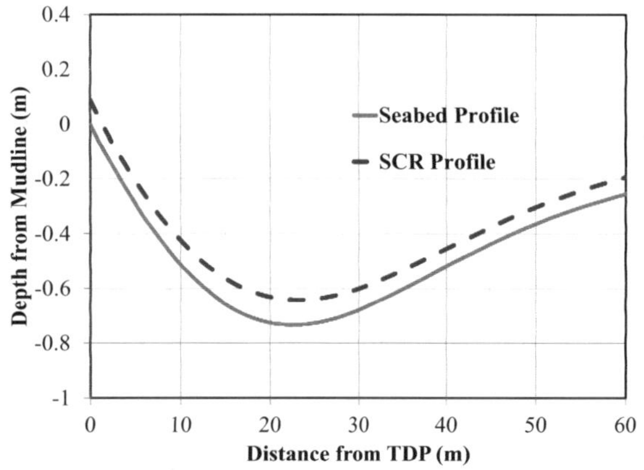

Figure 15. SCR Refined TDZ, 120 kN/m/m^2

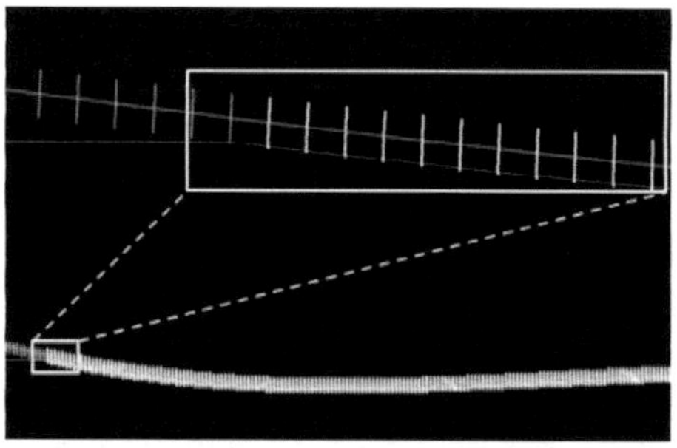

Figure 16. SCR Refined TDZ without Contact Discontinuity, 120 kN/m/m^2

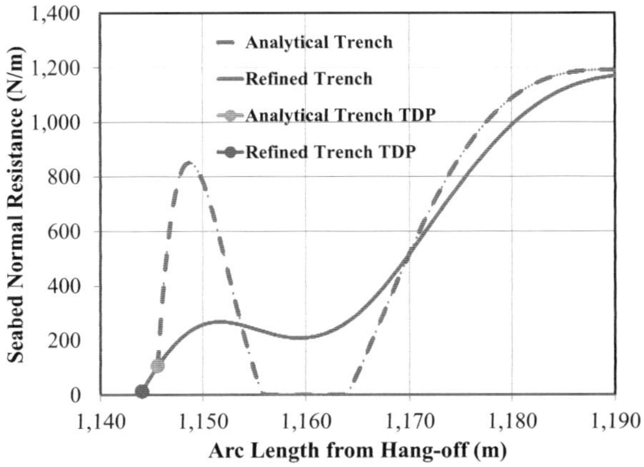

Figure 17. Seabed Normal Resistance Comparison, 120kN/m/m^2

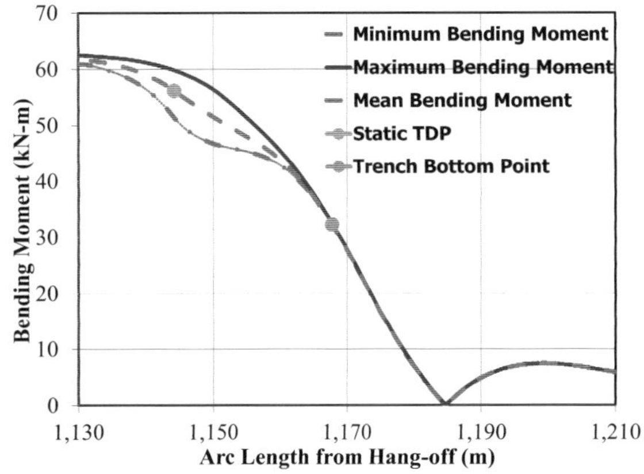

Figure 18. Bending Moment Envelope for Refined Seabed Profile, 120 kN/m/m^2

Table 6. Axial Stress Range Comparison, Refined vs Analytical Trench

Seabed Soil Stiffness (kN/m/m^2)	Stress Range (MPa)		Difference (MPa)
	Refined Trench Seabed Profile	Analytical Trench Seabed Profile	
120	15.31	16.47	-1.16

Table 7. Fatigue Life Comparison, Refined vs Analytical Trench

Seabed Soil Stiffness (kN/m/m^2)	Fatigue Life (Year)		Difference (%)
	Refined Trenched Seabed Profile	Analytical Trench Seabed Profile	
120	218	157	28%

Table 8. Fatigue Life Comparison, Refined vs Flat Seabed

Seabed Soil Stiffness (kN/m/m^2)	Fatigue Life (Year)		Difference (%)
	Refined Trench Seabed Profile	Flat Seabed Profile	
120	218	227	-3.90

7.2. Vessel Offsets with Refined Seabed Profile

Based on the SCR configuration in the refined trench profile, 1% near and far vessel offsets are considered to determine effects on TDZ fatigue. Fatigue lives considering vessel offsets are given in Table 9. A lower fatigue life than nominal is observed for near vessel offset because the SCR is being forced into the trench, hence increasing the curvature in the touchdown region. A higher fatigue life is observed for far vessel offset when compared to a flat seabed. However, this is not representative of a trench in equilibrium conditions because the SCR is lifted from the trench, as shown in Figure 19.

Table 9. Fatigue Life Comparison, Refined vs Flat Seabed, Vessel Offsets

Seabed Type	Fatigue Life (Years)		
	Far Vessel Offset	Nominal Vessel Location	Near Vessel Offset
Refined Trench	290	218	169
Flat	248	227	218

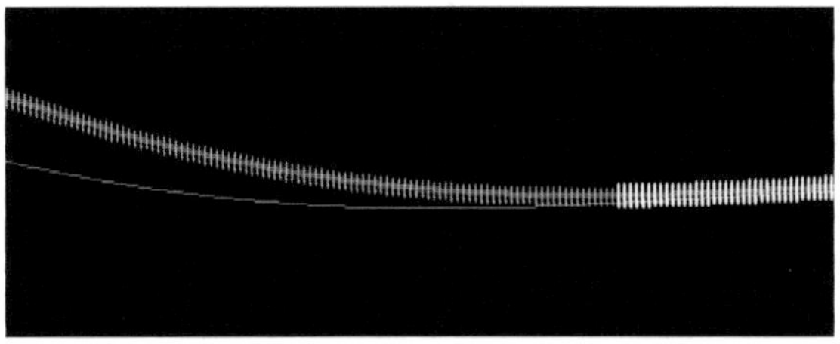

Figure 19. Far Offset TDZ Arrangement, Refined Seabed Profile, 120 kN/m/m^2

8. Conclusions

A linear seabed model that implements an analytical equilibrium trench in an SCR's TDZ derived from non-linear hysteretic model is presented. The stress range, seabed resistance, as well as fatigue life of the SCR in a trenched seabed profile is compared to an SCR on a flat seabed profile. Parametric analyses are also performed by refining the trench profile and applying vessel offsets.

Based on the analysis results presented above, it is found from this study that an SCR's TDZ fatigue life does not improve when implementing an analytically derived trenched seabed. The fatigue life in the touchdown region does not improve when implementing an artificial trench due to the following:

(a) Discontinuity between the SCR and seabed arise which lead to contradictory stresses and large seabed shear gradients; and (b) The gradient of seabed resistance is high near the hot spot of SCR touchdown zone.

It is also observed that vessel offset has significant impact on the fatigue life when trench is modelled. The impact from the vessel offset should be considered as in real life there always be some extent of vessel offset.

Interpreting SCR TDZ soil interaction in analysis models is still very challenging as the modelling of exact profile of an SCR trench in equilibrium conditions is extremely difficult. It is recommended to monitor a full-scale SCR and measure stress response in the TDZ to further understand the trench effect. The data could be used to confirm the results of small scale tests and improve modeling of the TDZ in SCRs. Additionally, further research and field surveys should be conducted to explore the nature and mechanisms of trenching below an SCR.

9. References

[1] Chibueze N., Ossia C. and Okoli J., 2016, On the fatigue of steel caternary risers, Journal of Mechanical Engineering, v62, n12, p751-756.

[2] Clukey E. and Zakeri A., 2017, Recent advances in nonlinear soil models for fatigue evaluation of steel catenary risers, Proceedings of the annual offshore technology conference, v2, p 875-890.

[3] Langner C., 2003, *Fatigue Life Improvement of Steel Catenary Risers due to Self-Trenching at the Touchdown Point*, OTC 15104.

[4] Shiri H., 2014, *Influence of Seabed Trench Formation on Fatigue Performance on Steel Catenary Risers in Touchdown Zone*, Marine Structures 36: 1-20.

[5] Bridge C., 2005, *Effects of Seabed Interaction on Steel Catenary Risers.*

[6] American Petroleum Institute, 2002, *Recommended Practice for Planning, Designing and Constructing Fixed Offshore Platforms – Working Stress Design*, API RP 2A-WSD.
[7] Orcina User's Manual 2010.

Acknowledgements
This study is supported by the National Key R&D Program of China "Research, Development and Project Demonstration of Multipurpose Flexible Pipes for Ultra-Deepwaters (Grant No. 2016YFC0303800)" and the Fundamental Research Funds for Central University (Grant No. 16CX02024A).

Effect of Different Rotational Speed on the Cavitation of Deep- Sea Mining Pump under Multiphase Flow

XU Hai-Liang[1,2] , XU Cong[1] , YANG Fang-qiong[1,2] and CHEN Wei[1]

1 College of Mechanical and Electrical Engineering, Central South University, Chang sha, China
2 State Key Lab of High Performance Complex Manufacturing, Chang sha, China

Abstract. According to the phenomenon of cavitation in the deep-sea mining pump and the instability of the pump's speed, the pump's cavitation characteristic under different speeds is studied. Based on the theory of multiphase flow and cavitation nucleus, the Euler model and Schnerr-Sauer cavitation model are used to simulate the pump's steady state of cavitation under different rotational speed. The pressure distribution and gas volume distribution are analyzed as well as the pump's external characteristics, and it is verified by experimental. The results show that the low pressure and cavitation appears on the inlet of the impeller blade at the pump's first stage. Along the streamline direction, the pressure increases gradually and the gas disappears. With the rotational speed increases from 960r/min to 2000r/min, the area of the low pressure expands rapidly as well as the cavitation, the efficiency of the pump and the *NPSHa* drop to 30% and 1.05m respectively, the *H* adds to 90.89m. The calculation formula of the pump installation height is derived as (21). When the speed is the designed 1450r/min, the pump has the best installation position about 108m under the sea level.

1. Introduction

Cavitation is a phenomenon of mass exchange between the liquid and gas when the pressure reduced, and its occurrence often affects a series of problems such as the normal flow of the fluid in the pump, the change of vibration, noise and external characteristic parameters ,even the entire system would stop working[1]. Now there are many researches about the cavitation in the single-phase flow conditions for different pumps[2-3], but the cavitation in the multiphase flow conditions is in the stage of experimental. Chang Jin-shi got the result that the critical cavitation pressure and the initial cavitation pressure show a linear increase trend with the increase of sediment concentration by experiment[4].Bostjan obtained the variation of the hydrofoil cavitation development with the sediment concentration through the water hole experiment [5]. Madadnia's research show that the destructive effect of cavitation was higher under sandy flow [6]. The experimental study is costly and it has a poor reusability. So the numerical simulation of cavitation under multi-phase flow is of great significance.

As a key part of the deep-sea mining hydraulic lifting equipment, the deep-sea mining pump directly effect the efficiency of the equipment [7]. At present, the research mainly focuses on the structural design of the pump and the flow field analysis of solid-liquid two-phase flow, while there is almost no research on the cavitation characteristics. Yoon C H[8] ,Ding H[9] did the simulation and performance tests about the designed centrifugal pump , it show that solid-liquid slurry pump is difficult to meet the requirements of hydraulic pipeline lifting system. Based on the theory of vertical pipeline hydraulic transportation, Yang Fang-qiong put forward the theoretical formula of the slurry pump's design parameters, and he analyzed the influence of the geometric parameters on the pump performance by using the numerical simulation method[10-11]. The size of nodules which transported by deep-sea mining pump are much larger than sand, and when the cavitation occurs, there is a complex three-phase flow. In most cases, the actual speed of the pump often deviates from the designed. Therefore, Combined the two issues, the cavitation characteristics of the pump under different rotating speeds is explored through simulation, and the pump's best speed is determined, then the installation height of the pump is researched. It has a great significance for the improvement of pump's structural and the operation in the actual project.

2. Numerical Algorithm

2.1. Multiphase Flow

The flow pattern of the solid-liquid flow in the slurry pump belongs to the multi-phase flow. The simulation of the Eulerian phase flow model can be used to get the real flow characteristics of the discrete phase and the continuous flow. The Euler model assumes that the discrete solid phase is the quasi-fluid. The flow is still described by the conservation equation in macro continuum theory . The liquid phase continuity equation of the solid-liquid phase turbulence under the rotating coordinate is:

$$\frac{\partial \rho_{\mathrm{l}}}{\partial t}+\frac{\partial(\rho_{\mathrm{l}} v_{\mathrm{lj}})}{\partial x_{\mathrm{j}}}=0 \tag{1}$$

among them , ρ_{l} is the liquid density; v_{l} is the liquid velocity vector; i, j is the direction of coordinates.

Continuity equation of solid phase is:

$$\frac{\partial \rho_{\mathrm{s}}}{\partial t}+\frac{\partial\left(\rho_{\mathrm{s}} v_{\mathrm{sj}}\right)}{\partial x_{\mathrm{j}}}=\frac{\partial}{\partial x_{\mathrm{j}}}\left(\frac{v_{\mathrm{s}}}{\sigma_{\mathrm{s}}} \frac{\partial \rho_{\mathrm{s}}}{\partial x_{\mathrm{j}}}\right) \tag{2}$$

where the ρ_{s} is the density of solid phase; v_s is the solid velocity vector .

Momentum equation of liquid phase is:

$$\begin{aligned} \frac{\partial(\rho_{\mathrm{l}} u_{\mathrm{i}})}{\partial t}+\frac{\partial(\rho_{\mathrm{l}} u_{\mathrm{i}} u_{\mathrm{j}})}{\partial x_{\mathrm{j}}}=\frac{\partial}{\partial x_{\mathrm{j}}}\left[\mu_{e}\left(\frac{\partial u_{\mathrm{i}}}{\partial x_{\mathrm{j}}}+\frac{\partial u_{\mathrm{j}}}{\partial x_{\mathrm{i}}}\right)\right]- \\ \frac{\partial P}{\partial x_{\mathrm{i}}}+2 \rho_{\mathrm{l}} w r u_{k}+\frac{\rho_{\mathrm{s}}}{\tau_{r s}}\left(u_{\mathrm{si}}-u_{\mathrm{i}}\right) \end{aligned} \tag{3}$$

where w, r, μ_e, u, P and τ_{rs} denote the angular velocity, the radius, the equivalent viscous coefficient, the liquid dynamic viscosity, the equivalent pressure of centrifugal force and the particle relaxation time, respectively.

The momentum equation of solid phase is:

$$\begin{aligned} \frac{\partial\left(\rho_{\mathrm{s}} u_{\mathrm{si}}\right)}{\partial t}+\frac{\partial}{\partial x_{\mathrm{j}}}\left(\rho_{\mathrm{s}} u_{\mathrm{sj}} u_{\mathrm{si}}\right)=\frac{\partial}{\partial x_{\mathrm{j}}}\left[\mu_{\mathrm{s}}\left(\frac{\partial u_{\mathrm{si}}}{\partial x_{\mathrm{j}}}+\frac{\partial u_{\mathrm{sj}}}{\partial x_{\mathrm{i}}}\right)\right]+ \\ \frac{\partial}{\partial x_{\mathrm{j}}}\left[\frac{v_{\mathrm{s}}}{\sigma_{s}}\left(\frac{\partial \rho_{\mathrm{s}}}{\partial x_{\mathrm{j}}} u_{\mathrm{si}}+\frac{\partial \rho_{\mathrm{s}}}{\partial x_{\mathrm{i}}} u_{\mathrm{sj}}\right)\right]-2 \rho_{\mathrm{s}} e_{ilk} w_{\mathrm{l}} u_{\mathrm{sk}} \\ +\frac{1}{2} \frac{\partial}{\partial x_{\mathrm{i}}}\left(\rho_{\mathrm{s}} w^{2} r^{2}\right)+\frac{\rho_{\mathrm{s}}}{\tau_{\tau s}}\left(u_{\mathrm{i}}-u_{\mathrm{si}}\right) \end{aligned} \tag{4}$$

2.2. Cavitation Model

When cavitation occurs, mass exchange between liquid and gas. The main equations include transport equation , mass transfer equation and continuity equation of gas phase[12].

transport equation of gas phase:

$$\frac{\partial}{\partial t}(\alpha\rho_v)+\nabla(\alpha\rho_v\vec{V}_v)=R_E-R_c \tag{5}$$

where v, α, $\vec{V}_v$ denote the gas phase, the volume fraction of the gas phase and the velocity of the gas phase.

R_E and R_c are the mass transfer source, which is related to the growth and collapse of the steam bubble, and describes the mass transfer between the liquid phase and the gas phase, it is modeled based on Rayleigh-Plesset equations for single vapor bubble growth .

mass transfer equation:

$$R\frac{d^2R_B}{dt^2}+\frac{3}{2}\left(\frac{dR_B}{dt}\right)^2=\frac{P_B-P}{\rho_l}-\frac{2\delta}{\rho_l R_B}-4\frac{\mu_l}{\rho_l R_B}\frac{dR_B}{dt} \tag{6}$$

where R_B is the radius of bubble, δ is the surface tension coefficient of liquid, P_B is the the surface pressure of bubble and P is the the far-field pressure of local.

Equation (6) can be simplified as:

$$\frac{dR_B}{dt}=\left(\frac{2(P_v-P)}{3\rho_l}\right)^{1/2} \tag{7}$$

continuity equation:

Liquid phase:

$$\frac{\partial}{\partial}\left[(1-\alpha)\rho_l\right]+\nabla\left[(1-\alpha)\rho_l\vec{V}\right]=-R \tag{8}$$

Gas phase:

$$\frac{\partial}{\partial t}(\alpha\rho_v)+\nabla(\alpha\rho_v\vec{V})=R \tag{9}$$

According to the equation (8) and (9), the continuity equation of mixed phase is got:

$$\frac{\partial\rho_m}{\partial t}+\nabla(\rho_m\vec{V})=0 \tag{10}$$

ρ_m is the density of mixed phase and it is expressed as $\rho_m=\alpha\rho_v+(1-\alpha)\rho_l$

3. Simulation Design of Cavitation

3.1. Geometric Model and Meshing

According to the transportation of big size particles in the pump, we adopt the principle of amplifying flow design to improve the ability of transporting the large particles, the structural is shown in Figure 1, in order to facilitate the installation of pipe, it was designed for the cartridge-type overall structure, the motor,cylinder pump,the inlet flange and the outlet flange were fixed on the same axis through the bolt cylinder. The lifting tube was connected with the inlet flange and the outlet flange. Multi-stage centrifugal impeller and submersible motor connected through the coupling. The main design parameters are as follows: inlet diameter of pump D_0=240mm, diameter of impeller inlet D_l=244 mm, diameter of impeller D_2=401.7 mm, width of impeller outlet b_2=54 mm , the number of impeller blades Z_l=3, the number of guide vanes Z_2=4. Design flow Q=800m^3/h (working flow 420m^3/h), head H=70m, rotating speed n=1450r/min, cavitation allowance $NPSHr$=6m.

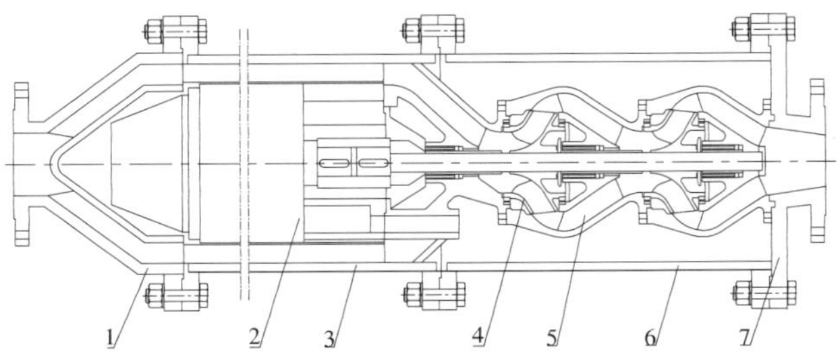

1-inlet flange; 2-submersible motor; 3-motor cylinder; 4-impeller;
5-space guide vane; 6-pump cylinder; 7- outlet flange

Figure 1. Mineral pump structure diagram

The 3D model of deep-sea mining pump was made by Pro/E(Figure 2), and the whole pump is composed of impeller, guide vane, inlet pipe, outlet pipe and shell.

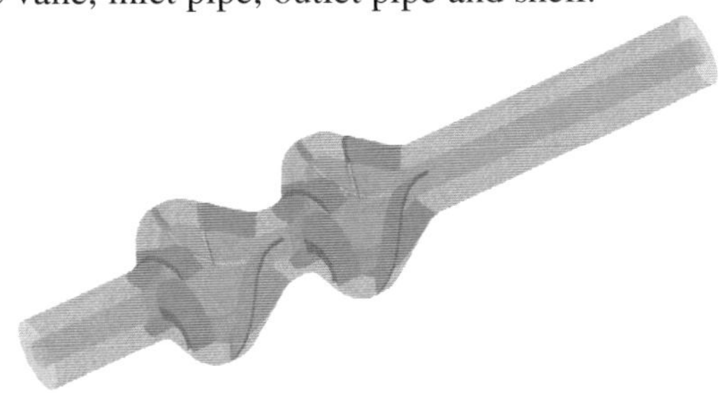

Figure 2. Pump's three-dimensional modeling

ICEM was used to carry out full flow grid. The quality of the mesh has a significant impact on the accuracy of the simulation results. The impeller and guide vane are distorted in the space, so the idea of "bottom-up" topology creation was used to create a topological rectangular block for a part of a single flow channel, then extend the small topology and cut the redundant blocks according to the actual model. At the same time the point, line and surface were associated. Next, the block structure similar to the entity was obtained, and the structure grid was generated by the block. Finally the number of nodes, the number of grids,the node density and the grid spacing near the wall were adjusted. O-type grids are used at the inlet pipe and outlet pipe. The full-channel grids are obtained by assembling the single-channel grids, and the grid quality is 0.6 above (Figure 3).

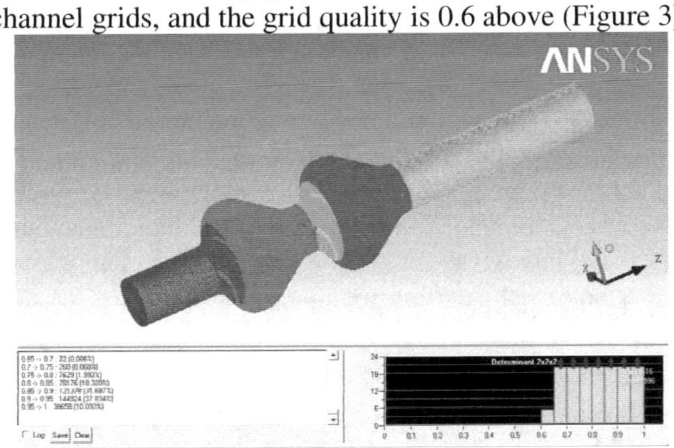

Figure 3. Specific requirements of maps

3.2. Physical Parameters and Boundary Conditions

Deep-sea mining pump delivers slurry containing manganese nodules. The relevant parameters of manganese nodules are based on the technical indicators in the report of the Head of the Tenth Five-year Mining and Testing System Group[13]; the physical properties of seawater are related to their geographical distribution, According to the seawater average salt concentration of 3.5%, refer to the "Chlor-Alkali Industrial Physical and Chemical Constituents Manual" ,the seawater properties were obtained; the nature of the gas phase reference to "Engineering Thermophysical Properties Handbook." The physical properties of each phase are shown in Table 1.

Table 1. Physical parameters

Physical properties	value
Density of seawater(kg.m^{-3})	1025
Viscosity of seawater(Pa.S)	1.673×10^{-3}
Saturated vapor pressure of seawater(Pa)	806.45
Density of air (kg.m^{-3})	0.01927
Viscosity of air(Pa.S)	8.8e-06
Density of Manganese nodules(kg.m^{-3})	2040
Average particle size of Manganese nodules(mm)	20—50
Concentration of Manganese nodules (%)	5—10

Reasonable boundary conditions provide a guarantee for the convergence of the simulation. There were several types of boundary conditions in the deep-sea mining pump: import and export boundary, solid wall boundary and dynamic coupling interface. The inlet boundary took the velocity inlet, the velocity was set perpendicular to the inlet cross-section, and the value was calculated according to equation (11); The outlet boundary was set as free flow because of the outflow condition was unknown ; The impeller wall was set as a rotating wall; and other boundary conditions were set as the stationary wall. Because the impeller is a rotating part, so the interface must be used between the rotating part and the static part.

$$v = \frac{Q}{A} = \frac{4Q}{\pi D_0} \tag{11}$$

where A is the cross-sectional area of inlet.

4. Simulation Results and Analysis

The cavitation and external characteristics of the pump were explored at different rotational speeds under the conditions with the working flow Q= 420 m^3/h, particle size d= 30 mm and solid volume concentration C_v= 8%. The steady-state simulation analysis was carried out with the different rotating speed, because of the article length , there are three conditions to be analyzed, 960 r/min, 1450 r/min and 2000 r/min.

At the speed of 1450 r / min, the static pressure of the pump with the full flow was analyzed. The inlet static pressure and outlet static pressure reach -200000 Pa and 800000 Pa respectively. The pressure at the primary impeller is much lower than the pressure at the secondary impeller, it shows as the figure 4. Then the pump guide vane and impeller are selected to be analyzed, The pressure at the inlet of the first-level impeller blades is the lowest, and it is easy to cavitation. Therefore, this paper analyzes the pressure and gas phase distribution of the first impeller, also calculates the necessary *NPSH* and head *H* of the pump to explore the effect of different rotational speeds on the cavitation characteristics of the pump.

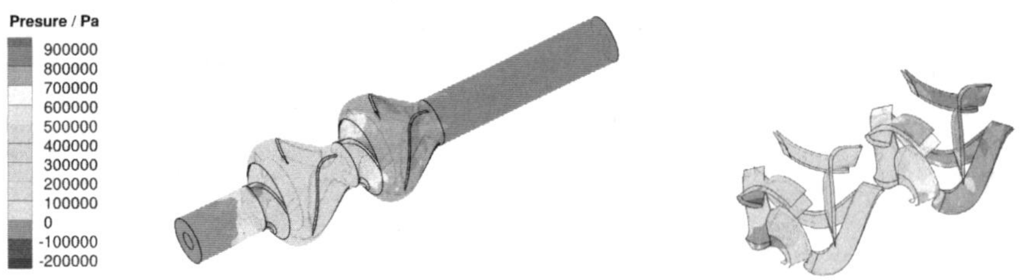

Figure 4. Full-flow static pressure cloud and impeller vane static pressure cloud

4.1. Stress Analysis

It's plotted in figure 5 with the data which was extracted from the first level impeller blade suction surface and the pressure surface of the middle line hydrostatic pressure. It can be seen from the figure that the fluid pressure on the suction side of the impeller is lower than that on the pressure side. Under the three speed the pressure difference is all about 150kpa at the entrance, but it is different at the end of the blade, with the speed increasing the pressure difference increases from 12kpa to 347kpa. The mixed fluid pressure on the pressure side of the impeller blade increases greatly with the increase of the rotational speed. The increase of the fluid pressure on the pressure surface increases from 120kpa at the speed of 960r/min to 460.4kpa at 2000r/min. It can be seen that the greater of the speed, the more effects of the exchange from the kinetic energy to the pressure energy through impeller . When the rotational speed is 960r/min, the pressure of the suction surface of the blade increases with the relative length, and both are higher than 300kpa. From the point of pressure, there is no cavitation under this rotational speed. With the speed increasing to 1450r/min , The minimum pressure of suction surface is -248kpa, it means that the cavitation should occur. At the speed 2000r/min, minimum pressure of the suction is basically the same with 1450r/min, but the trend of pressure rising is slower along the streamline, it means that the blade's suction side has a large area in low pressure , and the distribution of the vacuoles is more extensive.

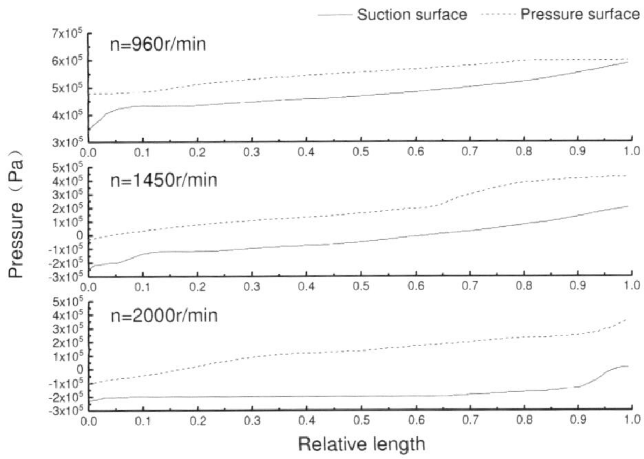

Figure 5. Pressure distribution curves on both sides of blades at different rotational speeds

4.2. Analysis of Gas Volume Fraction

Figure 6 shows the vapor volume fraction of the first stage impeller at different rotational speeds. It can be seen from the figure that cavitation occurs mainly at the inlet of the impeller's back. When the rotational speed is 960 r/min, there is no cavitation occurs, and it's the same as the prediction during the pressure analysis. The kinetic energy is low when the pump at a low speed, and it is difficult to form a low pressure zone which provide the condition for the occurrence of cavitation. During the process of rotating speed from 1450r/min to 2000r/min, bubbles appeared at the inlet of the blade's suction side, and the bubble distribution gradually disappeared along the streamline. When the rotating speed reaches 2000r/min, a large area of bubbles appeared which takes the 3/4 area of the all, and the gas phase volume fraction is mainly about 0.9. The main reasons are as follows: the fluid transported by the pump is a solid-liquid phase flow, and the manganese nodules are not infiltrated and have greater tensile stress than the rupture of the seawater. Under the effect of virtual mass force and high rotating speed , the fluid gets high kinetic energy, and it leads the local pressure dropping down that promotes the occurrence of cavitation; While at a high speed, the particles have a greater inertia. And the velocity difference between liquid and solid phases will be formed because of the different density, it makes the decrease of preasure near the particles. Therefore, the greater rotational speed, the bigger of the velocity difference between the two phases, the more significant the cavitation is.

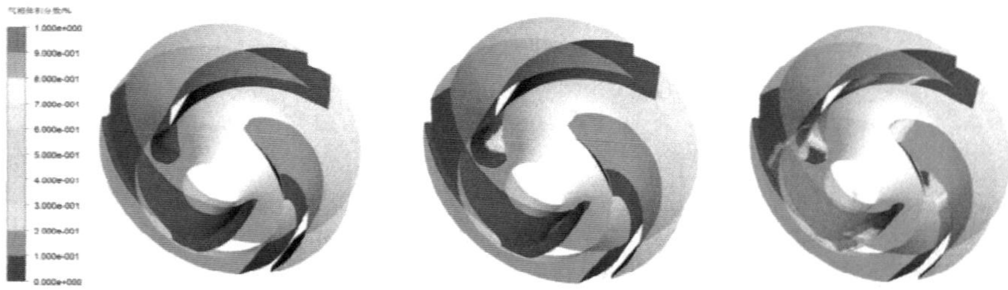

(a). n=960r/min (b). n=1450r/min (c). n=2000r/min

Figure 6. Vapor volume fraction of impeller leaves

4.3. Analysis of Pump Characteristics

1) Lift and efficiency

Head can be calculated by the pump inlet and outlet pressure, the formula is as follows:

$$H = \frac{P_{out} - P_{in}}{\rho g} + \Delta z \tag{12}$$

Δz is the vertical distance of the import and export.

The torque M of the pump's impeller can be obtained in the simulation. And according to the equation (13) , the pump's hydraulic efficiency can be calculated.

$$\eta = \frac{QH\rho g}{M\omega} \times 100\% \tag{13}$$

2) *NPSHa*

The *NPSHa* is pump's device cavitation margin, and it's also called the effective cavitation margin which represents the surplus energy against the pump's cavitation. The greater value of *NPSHa*, less prone to cavitation. And it can be calculated by the following formula:

$$NPSHa = \frac{P_{in} - P_v}{\rho g} \tag{14}$$

According to the formulas, the H and efficiency as well as *NPSHa* were calculated under different rotational speeds, and the external characteristic curve of pump was obtained like figure 7. As can be seen from the figure, at the rotational speed of 960 r/min, the pump's cavitation margin value is up to 55.08m, with sufficient ability to resist the occurrence of cavitation, the efficiency has the highest value of 57.29% too, but the head is only 25.04m; when the rotational speed increases to 2000 r/min, the pressure drops quickly, and the cavitation phenomenon is serious with the *NPSHa* drops to 2.35m rapidly, the loss of pump's energy causes the lower efficiency with a drop of 27% comparing to the low speed. But at a high speed,the fluid has great kinetic energy under the action of the impeller, and it changes into high pressure energy, so the H reaches 90.89m. When the speed is the designed, the H of the pump is 76.64m, which meets the requirement of 70m, the pump cavitation margin value is 5.52m, slightly smaller than the designed, so cavitation will occur, and the pump efficiency drops to 32%. In a summary, the greater speed of the pump, the H will increase sharply, but the cavitation margin decreases rapidly as well as the efficiency.

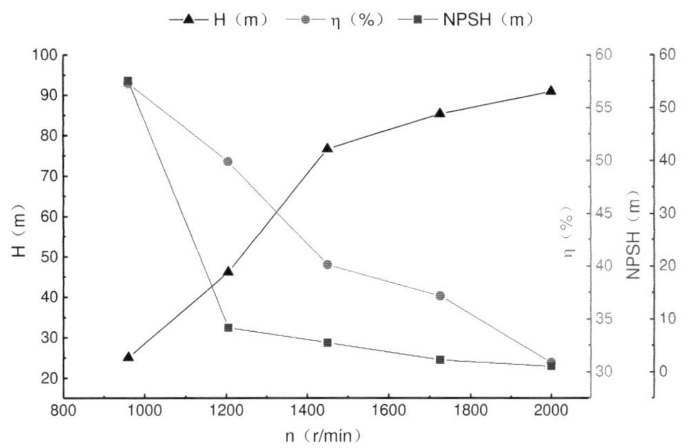

Figure 7. Slurry pump external characteristic curve

4.4. Installation Height of Deep-Sea Mining Pump

In order to avoid the cavitation of deep-sea mining pump, the pump's installation position needs to be discussed. Deep-sea mining pump works below the sea level, and with the increase of depth the pressure increases rapidly, and the lower pump's position, the less prone to cavitation. However, this is not conducive to the pump's installation and maintenance, it also has a high requirements on the pump's material and sealing performance. Therefore, at the premise to meet the delivery system requirements, the higher of the pump's installation, the better. According to the equation (15) of land pump's installation position, it deduced the calculation formula of the installation height of the deep-sea mining pump, and the installation height of the pump under different rotational speeds was discussed [14].

$$H_g = \frac{P_s}{\rho_m g} - \frac{P_v}{\rho_m g} - [NPSH] - \Delta h \quad (15)$$

where H_g and P_s denote the pump installation height and the pressure of pump inlet flange.

$[NPSH]$ is permissible cavitation allowance, which is the cavitation allowance for determining the service conditions (such as installation height) of the pump, and it should be greater than the effective cavitation allowance to ensure that the pump will not cavitate during operation. In general $[NPSH] = NPSHa + k$, k takes (0.3-0.5) m.

Δh represents the hydraulic loss from suction inlet to pump inlet, and it is related with the pressure loss ΔP as: $\Delta h = \frac{\Delta P}{\rho_m g}$

The size of solid particles transported by deep-sea mining pump is about 30mm on average, so Englemann's theory could be used to predict the pressure losses of the system [15]. When the seawater and ore in the pipeline mixed well, the pressure loss in the pipeline which caused by the friction is determined by the following formula:

$$\mathrm{d}P = [(1 - C_v)\rho_l \lambda_l \frac{u_l^2}{2D_i} + C_v \rho_s \lambda_s \frac{u_s^2}{2D_i}]\mathrm{d}X \quad (16)$$

where dX, D_i, C_v, u_l and u_s denote the pipe differential length, the pipe diameter, the ore concentration, the seawater flow rate and the ore particle flow rate.

λ_l 、 λ_s are the drag-line pipe frictional resistance coefficient to seawater and ore particles.

According to the formula above, the loss pressure can be obtained from $X = 0$ to $X(z)$:

$$\Delta P[X(z)] = \int_0^{X(z)} [(1-C_v)\rho_1\lambda_1 \frac{u_1^2}{2D_i} + C_v\rho_s\lambda_s \frac{u_s^2}{2D_i}]\mathrm{d}X \tag{17}$$

In accordance with the Englemann empirical formula, the frictional resistance coefficient has the following relationship:

$$\frac{\lambda_s}{\lambda_1} = 48.9\left(\frac{d_s}{D_i}\right)^{2.1}\left(\frac{u_1}{gD_i}\right)^{-1.6}\left(\frac{m_s^{'}}{m_1^{'}}\right)^{0.7}\left(\frac{\rho_s}{\rho_1}\right)^{2.0} \tag{18}$$

$m_g^{'}$ and $m_1^{'}$ are the weight flow rate of ore particles and seawater respectively, they can be calculated by $m_g^{'} = Q_s\rho_s$ and $m_1^{'} = Q_1\rho_1$.

According to the Nicholas formula:

$$\lambda_1 = \frac{1}{[1.74 + 2\lg(D_i / 2\Delta)]^2} \tag{19}$$

Δ is the roughness of pipe wall, and the value is 0.01 in the deep-sea mining system.

The middle warehouse is the first place to store ore in the deep-sea mining transportation system, and then upgrade, the middle warehouse position is under the sea surface about *L*, the pump inlet position can be approximated as the middle warehouse position. Then the distance from the suction inlet to the pump installation is L-H_g. The pressure loss at this distance is:

$$\Delta P = (L - H_g)[(1-C_v)\rho_1\lambda_1 \frac{u_1^2}{2D_i} + C_v\rho_s\lambda_s \frac{u_s^2}{2D_i}] \tag{20}$$

The deep-sea mining pump installation height can be introduced according to the above formula:

$$H_g = \left(\frac{P_s}{\rho_{m}g} - \frac{P_v}{\rho_{m}g} - [NPSH] - L\times[(1-C_v)\rho_1\lambda_1\frac{v_1^2}{2D_i} + C_v\rho_s\lambda_s\frac{v_s^2}{2D_i}]\right) \times \left(1 - \frac{1}{\rho_{m}g}\left([(1-C_v)\rho_1\lambda_1\frac{v_1^2}{2D_i} + C_v\rho_s\lambda_s\frac{v_s^2}{2D_i}]\right)^{-1}\right)^{-1} \tag{21}$$

According to the formula(21) and the related data obtained from the simulation, the installation altitude of the deep-sea mining pump is calculated at different rotational speeds respectively, and Table 2 is obtained. It can be seen from the table that when the speed is 1450r/min, the installation height of the pump is closest to the sea level about 108m; at low speed conditions, the pump can provide cavitation surplus energy, but the low kinetic energy results a low pressure energy, so the pump needs to be installed deeper to make up for the pressure difference; in high speed conditions, the pump cavitation is serious, and it needs to be installed in the depth to reduce the cavitation damage.

Table 2. Pump installation height

Speed(r/min)	Depth of sea level(m)
960	112
1450	108
2000	119

5. Experiment

Using the simulation experiment system of deep sea mining built by Central South University and Changsha Institute of Mining and Metallurgy, the cavitation performance of the pump was tested. The experimental system consists of five subsystems: a lifting subsystem, a water supply regulator subsystem, an artificial calibration subsystem, a feeding subsystem, a control and measurement subsystem. In the experiment, due to the lack of high-speed cameras and some other equipment, the observation and analysis of the vacuole was not carried out. The E-magE(Electromagnetic Flowmeter),3151SG(Pressure Transmitters) and SINAMICS6150(Torque speed sensor) were chosen to measure the inlet gauge pressure,outlet gauge pressure, torque, shaft power and other parameters. Then according the equations to calculate pump's external characteristics.(figure 8).

(a).deep-sea mining pump (b).riser and return pipe (c).control cabinet

Figure 8. Laboratory equipment

During the experiment, with the increase of speed the vibration and noise of the pump are more serious. And according to the calculate, pump head increases significantly with the increase of speed, though the cavitation margin decreases significantly, from 960 r/min to 1450 r/min the *H* increases 46.6m, the *NPSHa* and efficiency decreases to 3.87m and 38%. When the rotating speed increased to 2000 r/min, the *H*, *NPSHa* and efficiency are 82.3m, 0.93m and 28% respectively. By comparing the simulation with the measurement results (figure 9), we can know that the two results are mostly similar, and the law is basically the same. There are also some different because of the hydraulic loss in the real transportation. It verifies the accuracy of the simulation through the experiment.

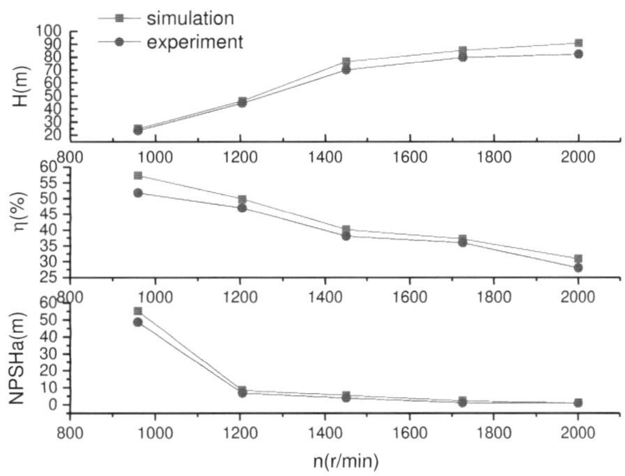

Figure 9. Data comparison between simulation and experiment

6. Conclusion

According to the research, the main conclusions are as below:

1.Cavitation will affect the flow field in the deep-sea mining pump, which should be considered when analyzing the pump. The squeezing between the blades makes the kinetic energy convert to pressure energy gradually, and the pressure of the first-stage impeller blades is lower than the second-stage impeller blades'. The lowest pressure zone is at the inlet of the first-stage impeller blades where is the area cavitation occurs.

2.With the increase of pump speed, the impeller has a great area of low pressure at the first stage and the cavitation becomes serious gradually, it creates a serious impact on the pump's blades at high speed. In general the *NPSHa* and pump efficiency has a negative correlation with speed , *H* has a positive correlation with speed.

3.The formula of the installation height of the deep-sea mining pump is deduced, and the pump has the best installation height when the revolving speed is 1450r/min. Under this speed the *H* and *NPSHa* are good satisfied with the work requirements, that indicates the rationality of the pump speed design.
In this study, numerical simulation and simulation experiments are used to study the cavitation characteristics of deep-sea mining pump, and it makes some progress. In the next step other parameters' effects to the pump's cavitation characteristics will be explored, and it will be further tested in the sea test.

References

[1] Cao Y L, He G and Ming T F 2016 Review on Numerical Simulation of Pump Cavitation *J. Journal of Wuhan University of Technology (Transportation Science and Engineering).* **40**(1):55-59.

[2] Zhang D S, Shi L and Chen J 2016 Experimental analysis on characteristic of cavitation in tip region of axial flow pump impeller *J. Journal of Zhejiang University(Engineering Science).* (08):1585-1592.

[3] Zhang D S, Shi L and Shi W D 2015 Numerical analysis of unsteady tip leakage vortex cavitation cloud and unstable suction-side-perpendicular cavitating vortices in an axial flow pump *J. International Journal of Multiphase Flow.* **06**:244-259.

[4] Chang J S 2010 Cavitation and cavitation erosion of pump and turbine with silt-laden water as working medium J. Journal of Drainage and Irrigation Machinery Engineering. **28**(2):93-97.

[5] Gregorc B, Hribersek M and Predin A 2011 The analysis of the impact of particles on cavitation flow development *J. Journal of Fluids Engineering.* **133**(11): 4—11.

[6] Madadnia J and Owen I 1995 Erosion in conical diffusers in particulate-laden cavitating flows *J. International Journal of Multiphase Flow.* **21**(6): 1253—1257.

[7] Wu B, Chen X M and Tian C 2016 Research Advance on Hydrodynamic Techniques in Deep Sea Mining System *J. Ship Building of China.* **57**(3):204-214.

[8] Yoon C H and Kang J S 2011 Shallow lifting test for the development of deep ocean mineral resources in Korea. *Ninth ISOPE Ocean Mining Symposium. Hawaii: International Society of Offshore and Polar Engineers.* 149-152.

[9] Ding H, Visser F C and Jiang Y 2011 Demonstration and Validation of a 3D CFD Simulation Tool Predicting Pump Performance and Cavitation for Industrial Applications *J. Journal of Fluids Engineering.* **133**(1):011101.

[10] Yang F Q, Chen Q and Zeng Y C 2014 Research on design method of slurry pump for deep-sea mining *J. Journal of Hefei University of Technology (Natural Science).*(12):1413-1418.

[11] Li L, Xu H L and Yang F Q 2015 Three-phase flow of submarine gas hydrate pip transport *J. Journal of Central South University.*

[12] Kazem H, Eslam E and Kasra F H 2015 A comparative study of two cavitation modeling strategies for simulation of inviscid cavitating flows *J. Ocean Engineering.* **16**:257-275.

[13] Tenth Five-year Plan of Mining and Sea Testing System General Team Technical Design of 1000m Offshore Test System for Oceanic Polymetallic Nodules Pilot Mining System. *Beijing: China Ocean Society,* 2004.

[14] Li Q and Li L W 2009 Talking about the Calculation of Pump Installation Height *J.Science and Technology Innovation Herald.***08**:074.

[15] Xu H L 2004 Research On Combined Single Pump and Ore Tank Deep-sea Mining Flexible Pipe Transporting System.*Changsha:Central South University.*

Exploration Application of BIM technology on artificial island engineering

DAI CHAO[1], DONG HAO[1*], ZUO ZHIGANG[1], ZHAO HAILIANG[1], SHANG ZHIHUI[1]

1 No.2618, Xingang No.2 Road, Binhai New District, Tianjin, China.
*E-mail:lanmangyi@mail.dlut.edu.cn

Abstract. With the continuous development of the transportation industry, the project of artificial traffic island is attracting more and more attention. However, the traditional design method has not met requirements for the complex structures of artificial islands. This paper discusses the application of BIM technology in the field of the design for artificial islands. Based on the practical engineering, the paper further proposes how to effectively use the BIM technology in the complex designs of artificial traffic islands, and presents the application effect. Finally, the BIM technology is successfully and effectively applied to the implement of 3D collaborative design, visualization of engineering design, and demonstration of construction progress.
*This paper is Supported by the Fundamental Research Funds for the Central Public Welfare Research Institutes(TKS160301).

1. Introduction

With the continuous development of the transportation industry, great bridges such as the cross-sea bridge is becoming more and more common, and the traffic artificial island is becoming more and more concerned as the important transportation hub of the bridge. The traffic artificial island project generally undertakes the transition joint function of bridge and submarine tunneling, involves many specialized types, many specialized interfaces and complex management, and the traditional design method is difficult to integrate and close the information during the whole design process[1]. As a new design method, BIM technology, based on 3D digital technology, integrate the building life cycle stages of engineering information and include physical information and geometry information, engineering information, manufacturing and assembly of the cost information, etc.; It is support for all participants of the project to invoke, modify and store the project information so as to realize all the phase of the building and the sharing of all the participating engineering information. [2]Using BIM technology to construct artificial island engineering construction will solve many problems existing in the construction of artificial island, and provide data support for operation of artificial island.

The Marine artificial island project generally includes[3] [4]: investigation stage, design stage, construction stage and operation maintenance stage; its construction has a standard complex, complicated construction conditions, construction technology, complexity in construction organization, supporting facilities and engineering scale, high investment cost, etc[5]; at the same time, the key technology lies in the determination of construction standard, location selection, acquisition of basic data, plane form, land domain formation and basic processing, monitoring and control etc.[6]

2. Methods

This article will take a look at the application of BIM technology in the construction of artificial islands and the current application of BIM technology in the practical engineering of the BIM technology, which is to explore the application of BIM technology in the works of artificial islands.

Application point of BIM technology on artificial island engineering:

(1) Application of BIM technology in the exploration phase of artificial island

Based on BIM technology, it can more directly show the terrain geology of the area of the project area by using the software's powerful data processing power, which make the designer could better grasp the three-dimensional geological information and facilitate the design personnel to make reasonable site selection for the project. Because the software has strong data relevance, the modification of data can be timely reflected in the completed 3D model, which saves a lot of manpower due to data modification[7]. At the same time, according to the established 3D model earth volume can be directly calculated by the software, greatly reducing the workload of the designers.

(2) Application of BIM technology in the design stage of artificial island

1) Multi-professional collaborative design applications

There are so many different types of engineering, and there are three major categories of hydraulic, road bridges, and buildings, each of which has more than a dozen separate specialties in each direction. In the construction of the traffic artificial island, each profession often needs to know each other's design intention, so as to carry on the accurate interface design, but the traditional design method, the concept has a great difficulty in the design of reasonable interface. Through BIM technology, a collaborative design platform can be used to establish a project document organization framework and develop a unified workspace and working environment. In the collaborative design platform, each professional can directly reference the established model and promote the orderly design of the project. At the same time, using the real-time update of the collaborative design platform, the modifications of the design can be returned to the associated professional in time, it can save the time of the manual delivery, and avoid problems such as delay in design feedback.

2) Application of BIM design in each specialty

The construction of the traffic artificial island generally takes on the function of island tunneling and bridge tunnel switching. The design of island tunneling and bridge tunneling interface is complicated, and the traditional 2D plane design is difficult to accurately indicate and communicate the design intent of the designer. Based on this complex spatial structure, the 3D space model can be created by using the BIM technology, the "What you see is what you get" is more directly presented in the design intent, which is convenient for the owner to make a prior decision, and reduce the amount of rework designed; based on the strong data correlation of software, the real-time feedback of modification design to the model is realized, and the labor time cost of modifying the design content is reduced. When the model is created, it's a quick way to calculate and produce the materials, and reduce the amount of repetitive labor, it's a lot more accurate than the human calculation. In conclusion, based on the BIM method, it is possible to improve the quality of the drawings by using technology to monitor the quality of the problem, and reduce the "error、leakage、collision and vacancy"[8].

(3) Application of BIM technology in construction stage of artificial island

The construction of artificial island is lack of land support, which makes the construction more difficult, and its construction organization is more complicated. Before the construction of artificial island, BIM technology can be used to simulate the whole construction process by analyzing the construction model, and fully understand the nodes and difficulties of the project construction, so as to make reasonable construction organization plan and guide the construction in a reasonable and orderly manner[9] [10].

(4) Application of BIM technology in operation and maintenance of artificial island

Traffic island as a transport hub plays an important role in the whole traffic routes, guarantee the effective operation is necessary for the whole traffic routes open. As a result, the operation and maintenance phase of possible contingencies for early warning, emergency equipment damage to its important position quickly locate replacement maintenance is very necessary, and the traditional design method has been difficult to meet the requirements of the traffic island in the operating maintenance phase. BIM technology could correlate the complete and accurate correlation between real engineering

IOP Conf. Series: Earth and Environmental Science **171** (2018) 012022 doi:10.1088/1755-1315/171/1/012022

project and data model. Through simulation calculation of data model, it is possible to reasonably arrange the emergency plan, quickly and accurately position the engineering project to age and repair the emergency plan, and improve the operation and maintenance efficiency of the project.

3. Results and Discussion

This is an example of the work of the BIM design process in the construction of the artificial island, and the application of BIM technology in the construction of the artificial island is explained.

(1) Project introduction

The axis of the artificial island is 625m, lateral at its widest point is 456 m, the length of the revetment is 1622m, backfill forms an land area of 13.7 million m3, the structure of the island wall is adopted by inserting steel pipe pile structure + Pavement accropodes scheme.

(2) The establishment of the model

1) Software selection: Bentley digital platform for the series of software

2) Application of BIM technology

a) Set up a project structure and work space

The project uses ProjectWise based on Bentley of digital platform to create the file structure, hydraulic engineering designed by the professional staff will be the model into the appropriate folder for other professional design personnel reference, at the same time, it develops a unified work space and standard, ensure the unified data model.

b) The creation of geological model

The project is modeled on the geology of the project, which is based on the fact that it is based on a Bentley digital platform, and it is a model of the geology of the project, and it is transformed into a digital model, and the design team is able to get an intuitive sense of the region's geological vulnerabilities(Figure 1), and then take the appropriate measures to ensure that the design is in order.

Figure 1. Geological Model

c) Terrain Processing

The project uses the GeoPak which under the Bentley digital platform to process the terrain, and provides the basic conditions for the structure design of the structural profession(Figure 2).

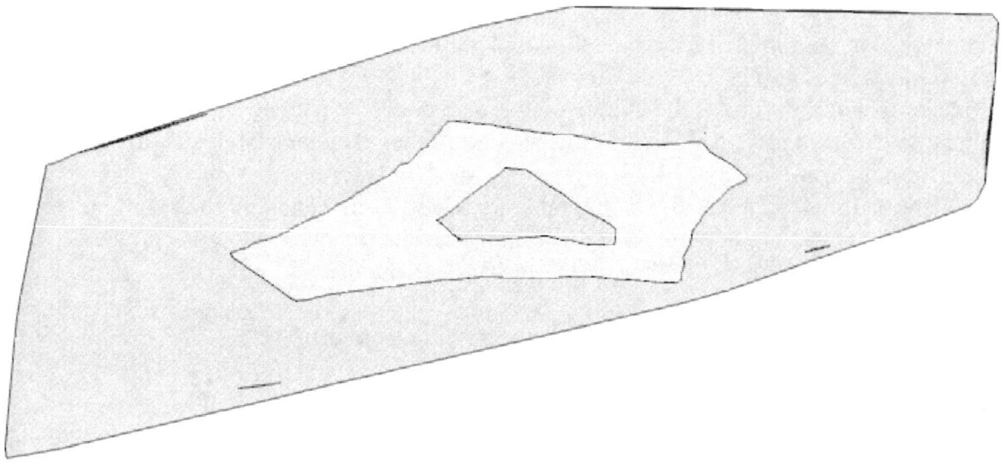

Figure 2. The processing of terrain model

d) The founding of Island structure model
As the main structure of artificial island, the island structure has some technical difficulties in its modeling processing, for example, the intersection between island structure and the terrain after the processing and other structures (such as the structural model of land formation、 the structure model of rescue wharf). At the same time, as the breast wall of the island structure is made of the prefabricated concrete components, the problem of structural joints between breast walls also adds difficulty to model creation. Considering the above difficulties, the project uses Power Civil (referred to as PC) , the software of Bentley digital platform, to buliding the body model of the island, and uses Open Bridge Modeler (OBD) to create a breast wall model(Figure 3). As a result of PC and OBD which are both professional linear modeling tools, the concrete idea of model creation is to draw the cross section of island structure and breast wall structure in the above software and then to form the island structure model by linear pay-off.

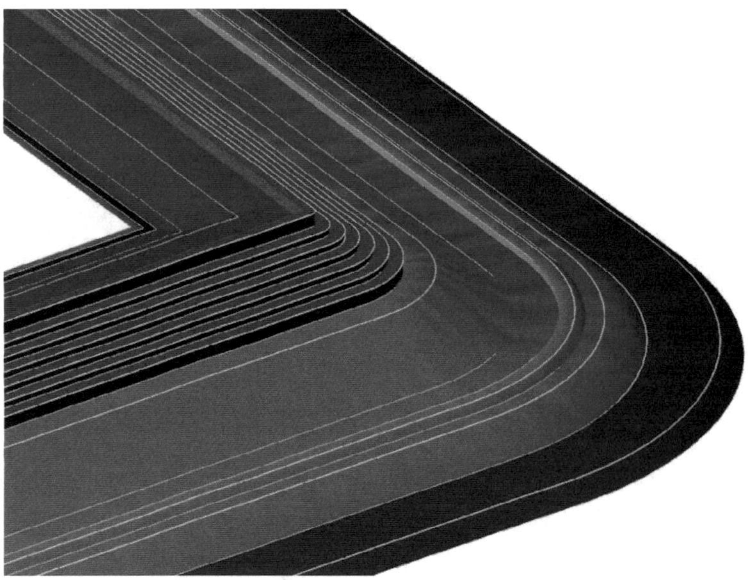

Figure 3. The structure model of island

e) The structural model creating of rescue wharf

The project uses Microstation and AECOsim Building Designer under the software of Bentley digital platform to build model. The special ideas are as follows: the common components such as caissons, bollards, mooring rings and tetrapod block are created into shared units by MS; and the general components such as backfill stones are directly created by the MS into a three-dimensional model; and the attribute information are given to each component by ABD, and finally make up the overall rescue wharf model(Figure 4).

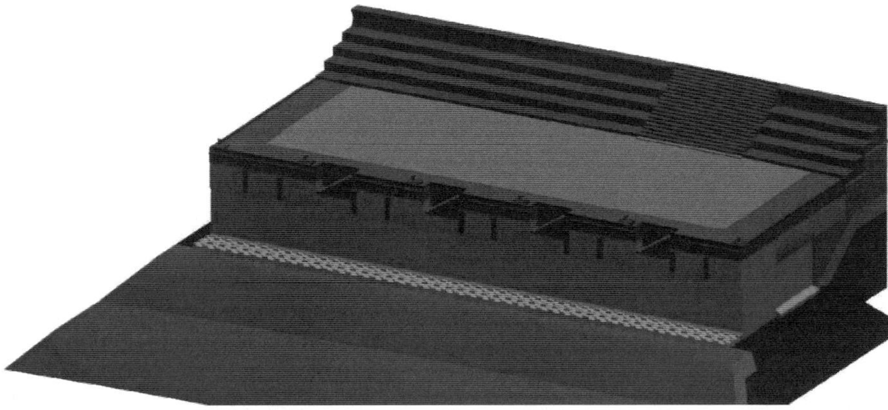

Figure 4. The rescue wharf model

3) The exploration and application of BIM Technology in the project

The project creates overall model by the Bentley digital platform(Figure 5), and it can analyze the created digital three-dimensional model. as a result we can quickly extract the engineering quantity and material list(Figure 6) and cut the model to form the sectional view and carry out the demonstration of construction simulation(Figure 7). Thanks to the BIM, the problems existing in the artificial island construction which because of using the traditional design method are solved, and the artificial island design and construction method are optimized[11]. However, due to the current BIM technology in the field of artificial islands is still in its initial stage and its application is shallow. The application of BIM for engineering design did not carry out a deeper level of application and the application of maintenance phase and operation.

Figure 5. The overall model

IOP Conf. Series: Earth and Environmental Science **171** (2018) 012022 doi:10.1088/1755-1315/171/1/012022

SGMT_Caisson	Prefabricated caisson (quantity)		1 pc	0		0 kg	239897.348	143356.73	7.102
SGMT_Caisson	Prefabricated caisson (quantity)		1 pc	0		0 kg	239892.38	143344.717	7.102
SGMT_Caisson	Prefabricated caisson (quantity)		1 pc	0		0 kg	239902.279	143369.079	7.173
SGMT_Caisson	Prefabricated caisson (quantity)		1 pc	0		0 kg	239889.591	143314.971	7.171
SGMT_Caisson	Prefabricated caisson (quantity)		1 pc	0		0 kg	239887.412	143332.703	7.102
SGMT_Caisson	Prefabricated caisson (quantity)		1 pc	0		0 kg	239911.641	143368.291	7.171
SGMT_Caisson	Prefabricated caisson (quantity)		1 pc	0		0 kg	239882.444	143320.69	7.102
			7 pc	0	0	0 kg			
SGMT_Caisson	Prefabricated caisson (quantity) CX1		1 pc	0		0 kg	239902.279	143369.079	7.173
			1 pc	0	0	0 kg			
SGMT_Caisson	Prefabricated caisson (quantity) CX2		1 pc	0		0 kg	239897.348	143356.73	7.102
SGMT_Caisson	Prefabricated caisson (quantity) CX2		1 pc	0		0 kg	239892.38	143344.717	7.102
SGMT_Caisson	Prefabricated caisson (quantity) CX2		1 pc	0		0 kg	239887.412	143332.703	7.102
SGMT_Caisson	Prefabricated caisson (quantity) CX2		1 pc	0		0 kg	239882.444	143320.69	7.102
			4 pc	0	0	0 kg			
SGMT_Caisson	Prefabricated caisson (quantity) CX3		1 pc	0		0 kg	239889.591	143314.971	7.171
SGMT_Caisson	Prefabricated caisson (quantity) CX3		1 pc	0		0 kg	239911.641	143368.291	7.171
			2 pc	0	0	0 kg			
SGMT_Caisson	Prefabricated caisson (volume)	Above Water Level Fluctuation Area	79.241 m3	0		0 kg	239897.348	143356.73	7.102
SGMT_Caisson	Prefabricated caisson (volume)	Above Water Level Fluctuation Area	79.241 m3	0		0 kg	239892.38	143344.717	7.102
SGMT_Caisson	Prefabricated caisson (volume)	Above Water Level Fluctuation Area	83.999 m3	0		0 kg	239902.279	143369.079	7.173
SGMT_Caisson	Prefabricated caisson (volume)	Above Water Level Fluctuation Area	59.529 m3	0		0 kg	239889.591	143314.971	7.171
SGMT_Caisson	Prefabricated caisson (volume)	Above Water Level Fluctuation Area	79.241 m3	0		0 kg	239887.412	143332.703	7.102
SGMT_Caisson	Prefabricated caisson (volume)	Above Water Level Fluctuation Area	59.529 m3	0		0 kg	239911.641	143368.291	7.171
SGMT_Caisson	Prefabricated caisson (volume)	Above Water Level Fluctuation Area	79.241 m3	0		0 kg	239882.444	143320.69	7.102
			520.021 m3	0	0	0 kg			
SGMT_Caisson	Prefabricated caisson (volume)	Below Water Level Fluctuation Area	107.669 m3	0		0 kg	239889.708	143315.253	1.906
SGMT_Caisson	Prefabricated caisson (volume)	Below Water Level Fluctuation Area	154.127 m3	0		0 kg	239902.579	143368.954	1.873
SGMT_Caisson	Prefabricated caisson (volume)	Below Water Level Fluctuation Area	107.669 m3	0		0 kg	239911.524	143368.008	1.906
SGMT_Caisson	Prefabricated caisson (volume)	Below Water Level Fluctuation Area	154.127 m3	0		0 kg	239897.611	143356.941	1.873
SGMT_Caisson	Prefabricated caisson (volume)	Below Water Level Fluctuation Area	154.127 m3	0		0 kg	239892.643	143344.928	1.873
SGMT_Caisson	Prefabricated caisson (volume)	Below Water Level Fluctuation Area	154.127 m3	0		0 kg	239887.675	143332.915	1.873
SGMT_Caisson	Prefabricated caisson (volume)	Below Water Level Fluctuation Area	154.127 m3	0		0 kg	239882.707	143320.901	1.873
			985.973 m3	0	0	0 kg			

Figure 6. The material list

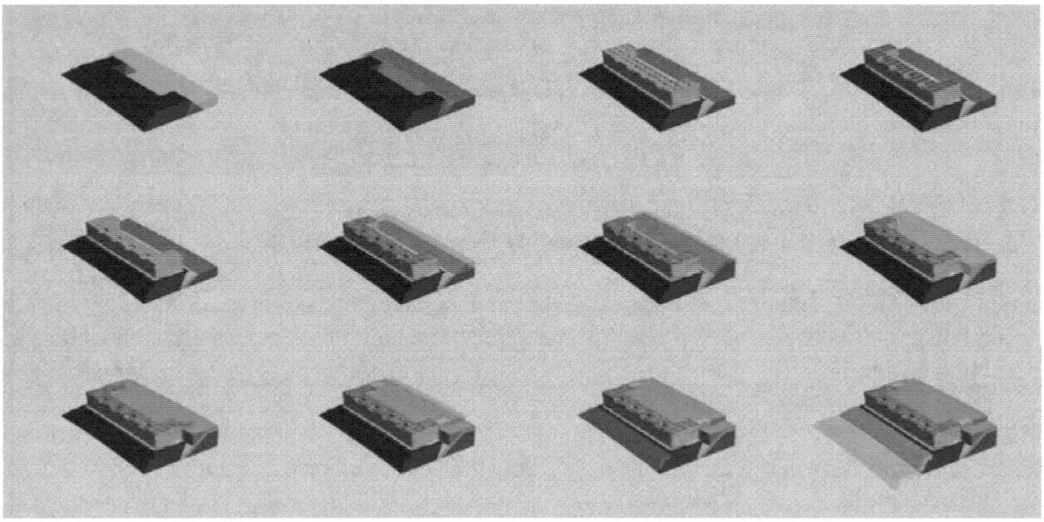

Figure 7. The demonstration of construction process

4. Conclusion

From the current situation and industry trends, BIM technology becomes an important force in engineering construction to achieve a high level of information technology. This paper shows that BIM technology can provide technical support for artificial island engineering in design, construction, operation and maintenance, which make the construction of artificial island more scientific and reasonable, and the operation and maintenance are more convenient and faster. This paper provide some reference for the application of BIM technology in artificial island engineering.

Although the BIM technology is known as a new technological revolution in the construction industry, there is no technical standard for BIM in the field of water transport at present, which has a result in the immature development of BIM technology and the difficulty of comprehensive promotion.

In conclusion, the application of BIM technology in the field of water transport is still in its infancy currently, and the application of artificial island technology is rarely reported. The paper only give a brief analysis on BIM technology about the artificial island engineering application points and the application effect; and it explain the application status of BIM technology in the artificial island field; we can improve the work efficiency of engineering project group by using collaborative design of BIM technology; the 3D digital modeling feature of "what you see is what you get " by using BIM technology can more intuitively present the designer's design idea and at the same time it can avoid the possible mistakes of two-dimensional drawings such as " error、leakage、collision and vacancy ", it also improve the efficiency of the design part of the work; the use of BIM technology on construction progress demonstration function also has a great visual presentation, which provide the construction organization a data support to develop more reasonable construction process. In short, the current application of BIM technology in the design field has a great exaltation on the engineering design、construction efficiency and project quality. However, more deeper application of BIM technology in the artificial island area remains to be explored.

5. References

[1] WANG Jun: Research on the application of BIM concept and BIM software in construction projects, Master's degree thesis of Southwest Jiao Tong University (2011).

[2] Cheng: Applications of BIM Technology in the Rail Transportation Engineering Design, Chinese Journal of Underground Space and Engineering, Vol.10 (2014).

[3] ZHANG Zhi-ming: On key design technology for large offshore man-made island, Port & Waterway Engineering, No.9 Serial No.458(2011).

[4] SUN Cheng-shuang,JIANG Fan, MAN Qing-peng, A Summary on BIM Application in Construction Industry. Journal of Engineering Management (2014)p27.

[5] SAO Xing-hua, Study on Collaborative Design Based BIM, Special Research of Chinese Survey and Design, DOI: 10.3969/j.issn.1006-9607(2015).

[6] YAN Zhi-gang, Development and application of Shanghai Yangtze River bridge BIM management system, Railway Engineering (2016)

[7] GAO Xin-hua,ZHANG Hong-wei,YANG peng-fei, Study On Collaborative Design Based BIM .(2015) p77.

[8] DAILIN Fa-bao, Study on Current Applications and Problems of BIM in Tunnel Engineering, Railway Standard Design (2015) p100.

[9] SUN Shao-nan,ZHANG Hui-jun, Research on the Application of BIM Technology in Water Conservancy Project，Journal of Engineering Management (2015) p77.

[10] SUI Zhen-guo, Ma jin-ming, CHEN Dong, XU Wei. Application Progress on BIM Technology in the Field of Building Construction. Construction Technology(2013) p161.

[11] LIU Bo, LIU Wei, Application Status and Obstacles Research of BIM in Domestic Construction Industry. Construction Economy(2015)p20 .

IOP Conf. Series: Earth and Environmental Science **171** (2018) 012023 doi:10.1088/1755-1315/171/1/012023

Research of Statistical Characteristics of Target Strength of a Single-layer Cylindrical Shell with Random Coating

Hu Bo [1]

1 Engineer, China Ship Developing and Designing Centre, Wuhan, China
E-mail:huboaaaa@163.com

Abstract. That anechoic tiles fall off from the surfaces is a common phenomenon that modern submarines are frequently faced with, which has negative effects on acoustic stealthy performance and increase Target Strength(TS) levels. Based on Monte-Carlo theory, PEM calculations are made by assuming that each anechoic tile fall off randomly and their conditions are tested separately. And the statistical-based values where statistical probability density of the PEM results reaches its culmination are considered as the TS level of a structure that part of its anechoic tiles fall off. The effectiveness of the MC-PEM is validated by the calculation of plate with anechoic tiles, since it achieves good agreement with the results of analytical method. And the relations between TS and the anechoic tiles' exfoliation rate of single-layer cylinder shell are gained by MC-PEM. Finally, log-normal distribution is introduced to describe cumulative failure probability of anechoic tiles. And the dynamic characteristic of TS versus non-dimensional time suggests that it is of significance to overhaul anechoic tiles in the region of 0.3t~0.5t(t represents average lifespan of anechoic tiles) as to keep TS in relative low level.

1. Introduction

Modern submarines are generally equipped with a sound absorbing layer made of rubber and other soft materials, that is, anechoic tile.[1], which will be lost and shedding as submarines carry out mission. Figure 1(a) resembles that for the Virginia-class nuclear submarine of the United States, large-scale exfoliation of anechoic tiles result in returning to repair. Figure 1(b) shows that a typhoon-class submarine shell whose surface anechoic tiles is mostly shedding.

(a)

(b)

Figure 1. Figure of Anechoic Tiles' Exfoliation

On the one hand, exfoliation of anechoic tiles decreases hull fairing[2]., while increases flow resistance and flow noise, and accelerates the corrosion of the hull at the same time. On the other hand, the intensity of the acoustic target will increase, which has a negative effect on the submarines' acoustic stealth performance.

At present, the main use of the high frequency approximation theory in engineering is to obtain the target strength under the state of full rigid surface and full coating, while the problem of the effect of the shedding of the anechoic tiles on the target strength is still lacking. Fan[3] developed the PEM(plate element method) based on the Kirchhoff high frequency approximation theory, which can be then extended to the calculation of the target strength of the immerged double-layer structures[4]. And the accuracy of PEM is consistent with the BASIS algorithm commonly used [5].Considering the anechoic tiles, the equivalent material parameters are back calculated by simulated annealing algorithm combining the acoustic tube test data[6], and the acoustic coefficient is obtained by multilayer medium transfer matrix method[7].

The anechoic tiles' exfoliation can be attributed to the reliability issues of the product. Di YL [8] illustrates several important reliability distributions, such as lognormal, Weibull and double index distributions. Chang Lixin[9] analyzes the influence of the active phased array antenna failure on the performance of the antenna unit obeying the normal distribution. Through the simulation calculation, Niu Cf[10] analyses that when the failure rate of the phased array unit is at 10% level, and the system can still work as usual or downgrade. Zhu Fc[11] analyses the fitting accuracy of several common aviation ammunition storage failure distributions. The mentioned study is of great significance to the estimation of the exfoliation probability and lifespan of the anechoic tile.

Considerable randomness exist in the anechoic tiles' exfoliation. Monte Carlo method, also known as the statistical simulation method, has become the main tool to solve complex statistical models and high-dimensional problems[12], which has been applied in many aspects such as reliability evaluation of generation and transmission system[13], and detecting probability of submarine launched torpedo[14]..

By combining the engineering algorithm predicting target strength of shell with tile, reliability theory and Monte Carlo method, this paper quantitatively analyzes the effect of the anechoic tiles' exfoliation rate on the target strength, and gives the target strength curve in corresponding with the use of time, which has a certain engineering significance for the maintenance work of submarines.

2. Principles

2.1. Monte Carlo - Plate Element Method

Monte Carlo-Plate Element Method（MC-PEM）is a method that combines the plate element method with the Monte Carlo method to analyze the target strength of shell after the random shedding of the anechoic tile. This method is based on the following assumptions.

(1)The probability of falling off of the submarines' anechoic tiles in the same area is consistent.

(2)The thickness, material characteristics and acoustic properties of the anechoic tile in the same area are in accordance with each other, and the exfoliation of any anechoic tiles is independent of each other.

(3)The number of anechoic tiles is enough that each tile has a smaller area relative to the whole surface.

Anechoic tiles exist in two states: normal attachment or detachment. Independently judge the status of each plate, and calculate the target intensity repeatedly a number of times to obtain a series of numerical, mathematical statistics and distribution fitting, the TS values where the maximum probability density occur is obtained as final result.

The above is the core principle of the Monte Carlo algorithm. The PEM is derived from Kirchhoff's physical acoustics method, which is illustrated in Figure 2. The Kirchhoff integral formula for the scattering field of an object on any external surface is:

$$\phi_s(r_2) = \frac{1}{4\pi}\int_s [\phi_s \frac{\partial}{\partial n}(\frac{e^{ikr_2}}{r_2}) - \frac{\partial \phi_s}{\partial n}\frac{e^{ikr_2}}{r_2}]ds \tag{1}$$

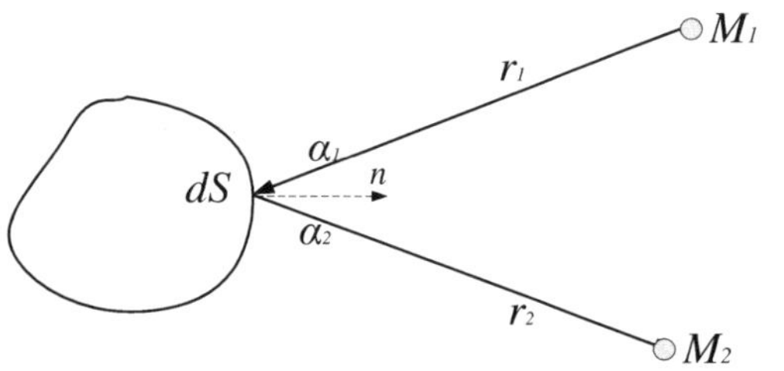

Figure 2. Kirchhoff Approximation Theory

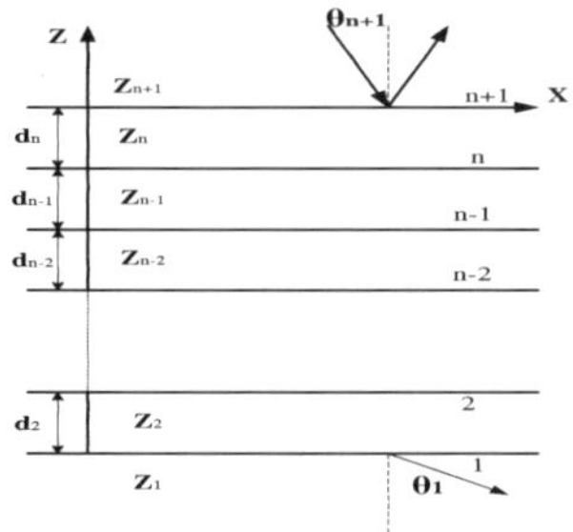

Figure 3. Transfer Matrix of Multi-layers Medium

Φ_i is the incident wave potential, $\Phi_i = (A/r_1)e^{ikr_2}$, Φ_s is the reflected wave potential, $V(\alpha)$ is the sound reflection coefficient, Z_n is the surface acoustic impedance, with the boundary conditions, for the far field transceiver combined situation, the following formula can be derived:

$$\phi_s = \frac{-A}{2\pi}\int_s e^{ik2r}V(\alpha)[\frac{ikr-1}{r^3}\cos\alpha]ds \tag{2}$$

Through the establishment of three-dimensional geometric model, the surface is divided into a number of flat plate. The Gordan formula will converted integral arithmetic into algebraic operation[15], and then we can use the plate element method to calculate the program to obtain the target strength.

With regard to the local acoustic pressure reflection coefficient of the shell, the sound pressure reflection coefficient of the steel shell laying anechoic tile structure is calculated by deducing the transfer-matrix[16] relationship between the stress and displacement of each medium layer and the boundary conditions of the upper and lower interfaces which is illustrated in Figure 3.

2.2. *Reliability Theory*

Cumulative failure probability refers to the probability that a product loses its required function under specified conditions and for a specified period of time, also referred to as the unreliability function F(t).

$$F(t) = \begin{cases} P(T<t) & t \ge 0 \\ 1 & t<0 \end{cases} \tag{3}$$

Anechoic tile can be seen as a product, when it is closely attached to the hull surface, that is effective, when the anechoic tile off, then that failure. Lognormal distribution is an important distribution of reliability analysis, expressed as $LN(\mu,\sigma^2)$,whose density function is

$$f(t|\mu,\sigma^2) = \frac{1}{\sqrt{2\pi}t\sigma}\exp\left\{-\frac{1}{2\sigma^2}(\ln t-\mu)^2\right\} \\ t>0, -\infty<\mu<\infty, \sigma>0 \tag{4}$$

In the reliability analysis, *t* is life expectancy, μ is the mean of natural logarithms of lifetime, and σ is the standard deviation of life logarithms. The cumulative failure probability function is:

$$F(t) = \Phi\left(\frac{\ln t-\mu}{\sigma}\right) \tag{5}$$

$\Phi(x)$ is the distribution function of normal distribution. In this paper, logarithmic normal distribution is used to characterize the shedding life distribution of anechoic tiles.

3. Method Validation

3.1. *Calculation of Anechoic Tiles' Acoustic Performance*

Submarines hulls are made of steel. （ $\rho=7800kg/m^3$, $E=2.1e11Pa$, $\mu=0.3$ ） .The surface of the anechoic tile is porous rubber, the equivalent material parameters of anechoic tile which are shown in Table 1 are obtained by simulated annealing method, and their thicknesses are shown in Figure 4.

Table 1. Equivalent Coefficients of Anechoic Tiles

ρ	v	d
$1140kg/m^3$	0.49	40mm

In addition, reference 5 gives anechoic tiles' Young's modulus, loss factor fitting curves.

$$E = 1.102\times10^{7} + 5447.2f - 3.515\times10^{-2} f^{2} \tag{6}$$

$$\eta = 0.401 + 1.21\times10^{-4} f - 3.29\times10^{-9} f^{2} \tag{7}$$

$$E' = E(1 - i\eta) \tag{8}$$

The relationship between the water-anechoic tile-steel-air acoustical reflection coefficient with the frequency and incident angle which are shown in Figure 5 is obtained by using the multi-layer elastic medium transfer matrix method.

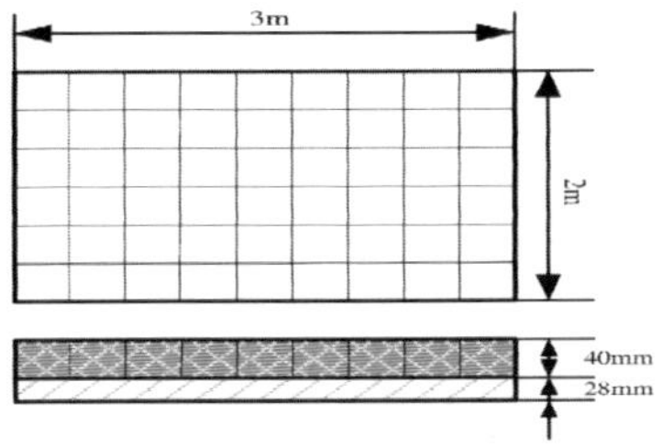

Figure 4. Diameter of a Panel with Anechoic Tiles

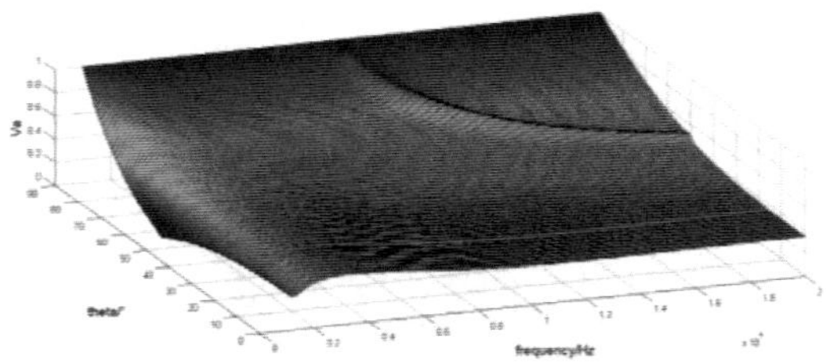

Figure 5. Sound Reflection Coefficients of Water-Rubber-Steel-Air System

3.2. Example for verification

Since the normal direction of the plate is consistent, the target strength of the plate that part of its anechoic tiles have been lost can be calculated by analytical method, and the MC-PEM can be verified. The length of the plate is 3 meters, the width is 2 meters, the thickness of the steel plate is 28mm, the thickness of the anechoic tile is 40mm, the inside is air and the outside is water.

At a certain moment, the probability of failure of a single anechoic tile is assumed to be P, then the number of failure plates which are shown in Figure 6 and Figure 7 obeys the binomial distribution:

$$p(X = k) = \binom{n}{k} p^{k} (1-p)^{n-k} = b(k; n, p) \quad (k = 0, 1, \ldots, n) \tag{9}$$

$$E\xi = np \tag{10}$$

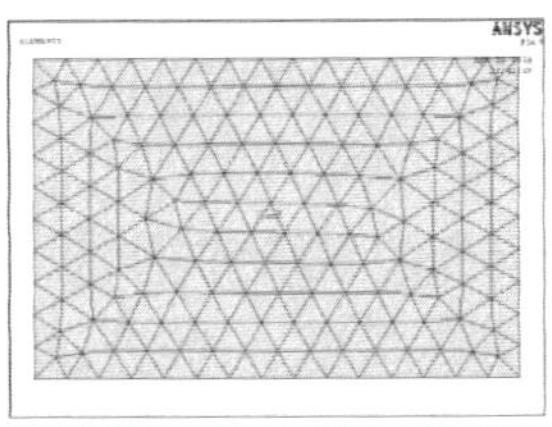

Figure 6. Figure of Panel Meshing

Figure 7. Anechoic Tiles' Falling off (p=0.3)

MC-PEM is a numerical method. When the frequency is 10KHz, the sound wave is perpendicular to the plane of incidence, and the calculation times are 50, 100, 200, 2000, 10000 respectively, the statistical results are shown in Table 2.

Table 2. Statistical Results of MC-PEM Calculation

N	μ /dB	σ /dB

N	μ /dB	σ /dB
50	27.850	0.2262
100	27.866	0.2185
200	27.864	0.2215
2000	27.847	0.2178
10000	27.854	0.2179

As can be seen from the frequency histogram(Figure 8), the discrete TS value has a typical normal distribution, ie $TS \sim N\left(\mu,\sigma^2\right)$. The results were normal fitted to the fitted mean value as part of anechoic tiles fall of via which the target strength are obtained.

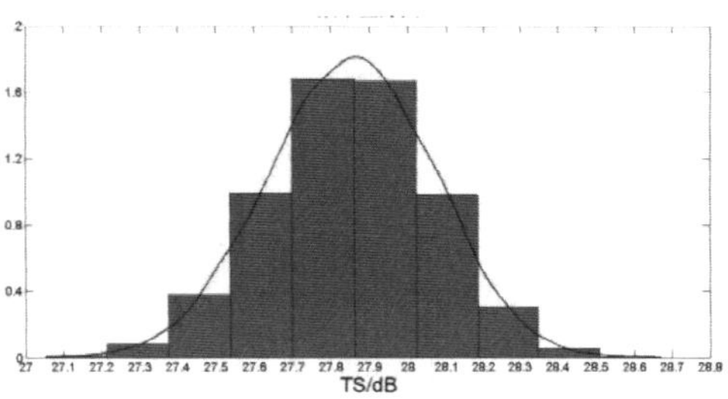

Figure 8. Frequency Histogram (N=10000)

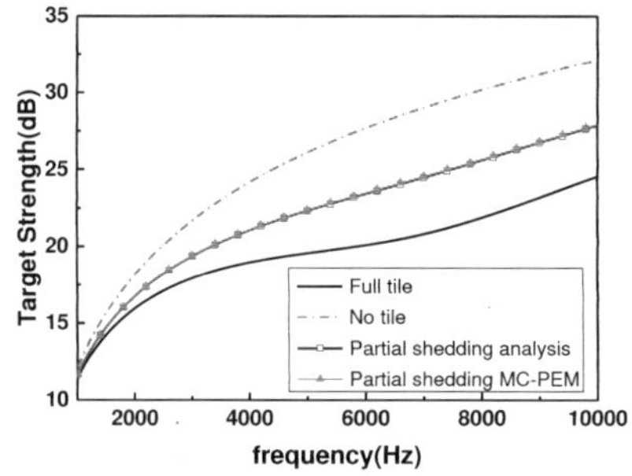

Figure 9. Comparison of the Results of Analytical Method and MC-PEM

For the cladding flat steel plate, the target intensity at the partial shedding can be linearly interpolated between the target intensity value TS_{coat} of the full-blasted tile state and the target intensity value TS_{rigid} of the total reflection state because the incident angles of the acoustic waves are the same.

$$TS = 10 * \log_{10}\left((1-\text{P}) * 10^{\frac{TS_{coat}}{10}} + P * 10^{\frac{TS_{rigid}}{10}} \right) \tag{11}$$

The comparison between MC-PEM analysis and analytic method are illustrated in Figure 9, which shows that the results obtained by the two methods are almost identical. And then the reliability of the Monte Carlo-plate meta-method is established.

4. Example of Single-Layer Cylindrical Shell

A single-layer cylindrical shell can be used as a simplified model of single-hull submarine's main hull. Cylindrical shell echo is generally considered as the main echo source when the submarine is in the transverse direction. Therefore, the single-layer cylindrical shell model is of great significance to the target strength of the submarine. It can be seen from Figure 10 and Figure 11 that the acoustic structure from the outside to the inside of a single-layer cylindrical shell is water-rubber layer-steel-air, and the thickness of the steel plate is 28mm, the thickness of the anechoic tile is 40mm, the length of the cylinder is 40m, the radius is 3.75m.

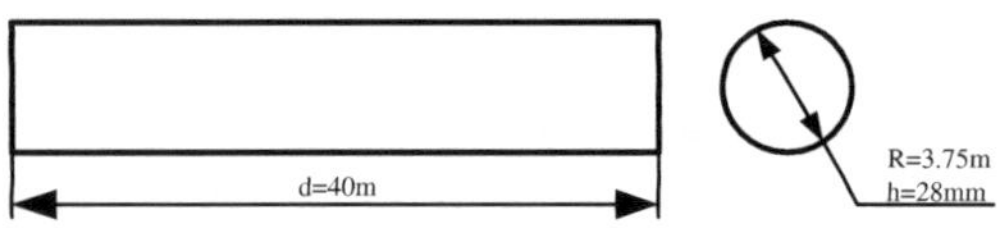

Figure 10. Diameter of Single Layer Cylinder

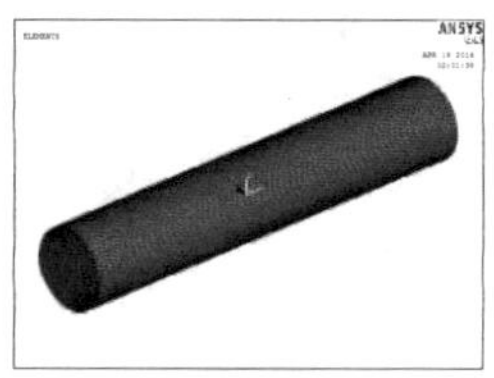

Figure 11. Figure of Single Layer Cylinder's Meshing

IOP Conf. Series: Earth and Environmental Science **171** (2018) 012023 doi:10.1088/1755-1315/171/1/012023

When the frequency is 10kHz and the acoustic wave is incident in the horizontal direction, the relationship between the TS and the exfoliation rate is obtained, and the result is linearly fitted, which are shown in Figure 12 and Figure 13.

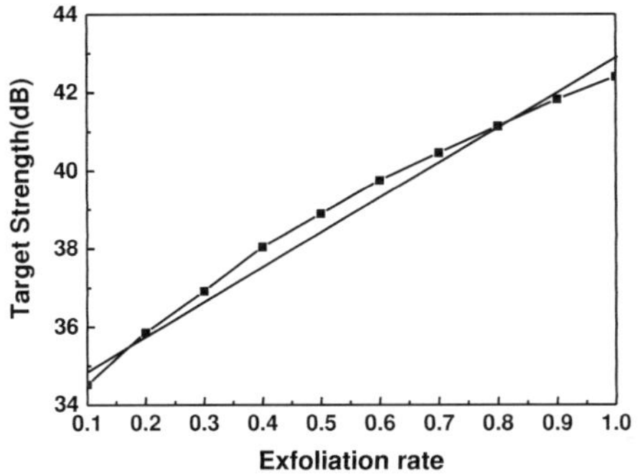

Figure 12. Relationship of Single-layer Cylinder's TS with the Exfoliation Rate

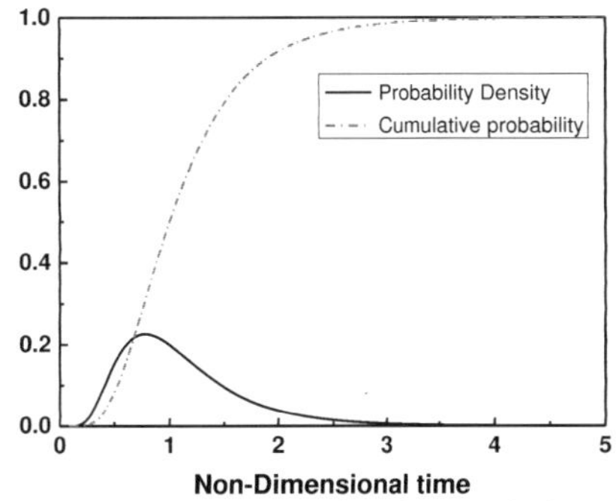

Figure 13. Characteristic of Anechoic Tiles' Failure Distribution

Assuming that the target intensity is TS, the rate of anechoic tile shedding is η and the linear fit is: $TS = 34.25 + 8.606\eta$ .It can be drawn that, as anechoic tile shedding rate increased by 10%, the target intensity increased by about 0.86dB.

5. Time Domain Dynamic Characteristics of TS

Introducing the theory of reliability, the cumulative failure probability function based on logarithm normal distribution is used to describe the failure state of anechoic tile. And then the dynamic characteristics of the single-layer cylindrical shell with time are analysed.

The anechoic tiles' life expectancy is assumed to be μ （$\mu = 4$）,their standard deviation of natural logarithm of life is σ （$\sigma = 0.5$）. The life of anechoic tiles obeys lognormal distribution $LN\left(\ln(4), 0.5^2\right)$. In order to facilitate the analysis, anechoic tile life is dimensionless, defined as the ratio of the use of time and average life expectancy t^*.

$$t^* = t / \mathrm{e}^{\mu} \tag{12}$$

According to the inverse function of the cumulative probability distribution of failure, the dimensionless use time $\hat{t}^*$ corresponding to different shedding rates is obtained.

$$\hat{t}^* = F^{-1}(\eta) / e^{\mu} = e^{\phi^{-1}(\eta)} / e^{\mu} \tag{13}$$

Finally ,the relationship of TS and $\hat{t}^*$ is obtained.

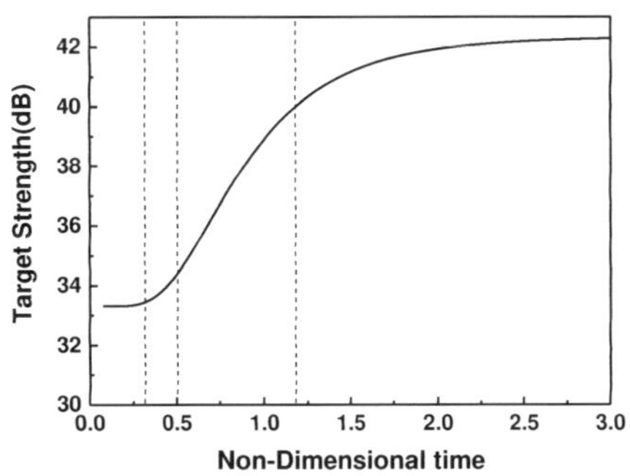

Figure 14. Relationship of TS and Non-dimensioned Time

It can be seen from the figure.14 that, as the anechoic tiles start to fall off, the target intensity curve of the submarine is approximately S-shaped and can be divided into four regions according to the use time.

(1) $\hat{t}^*$ <0.3,initial constant period. As the anechoic tile shedding rate is low, the impact on the TS is not obvious, the target intensity remained almost unchanged.

(2) 0.3< $\hat{t}^*$ <0.5,accelerated increasing period. At this stage, the anechoic tile shedding rate continues to increase, the enhancing rate of TS is also increasing, but the increased value is relatively small.

(3) 0.5< $\hat{t}^*$ <1.2,rapidly rising period. At this stage, the anechoic tiles have the most considerable shedding rate, and the TS of the submarine increases at a constant rate.

(4) 1.2< $\hat{t}^*$ <3,slowly rising period. Muffler tiles have mostly come off, the target intensity slowly increases. When the average life expectancy is more than twice, it increases to the steel boat's target strength level. When the service time exceeds the twice of the average life of anechoic tile, the submarine's TS increases to the high level that lacking anechoic tiles.

Therefore, according to the shedding distribution model of anechoic tiles proposed in this paper, we can draw the following conclusions: In the 0.3 to 0.5 times of the average life of anechoic tiles, detecting and maintaining the anechoic tiles can make the target intensity at a low level.

6. Conclusion

Based on the MC-PEM, this paper describes the mechanism of random laying of anechoic tiles on the target strength of cylindrical shell, and quantitatively analyses the relationship between target intensity and shedding rate. This paper draws the following conclusion:

(1)Linear coefficient of the first-order fitting approximates the sensitivity of substructure's target strength to the shedding of anechoic tiles, that is, the target strength of single-shell cylinder increases by 0.86dB for every 10% change of shedding rate.

(2)By introducing the cumulative failure probability function of anechoic tile based on lognormal distribution, the analysis shows that the target strength of cylindrical shell changes with time in four stages： initial constant period, accelerated increasing period, rapidly rising period, slowly rising period. In the 0.3 to 0.5 times of the average life of anechoic tiles, detecting and maintaining the anechoic tiles can make the target intensity at a low level.

The exfoliating mechanism analysis of anechoic tiles can be carried out in the follow-up work. It is generally believed that the probability of shedding of anechoic tiles is closely related to the distribution of alternating pressure on the surface of the submarine. Based on the CFD simulation technology, it helps to establish a more realistic TS prediction model of submarines coating with anechoic tile.

7. References

[1] Zhang Hao, Fu XY.. A Review of Anechoic Coating Research [J]. *Applied Acoustics*, 2013, 32(4):63-72.

[2] Bao WY. Research on Several Problems in Microwave Non-destructive Testing of Anechoic Tile Quality[D]. *East China Normal University*, 2002.

[3] Fan J, Tang WL, Zhuo LK. Planar elements method for forecasting the echo characteristics from sonar targets [J]. *Journal of Ship Mechanics*, 2012, 16(Z1):171-180.

[4] Hu B, Zhang JP. Effect Analysis for the shape of submarine bow pressure bulkhead upon the acoustic target strength[J]. *Chinese Journal of Ship Research*, 2016, 11(6):22-27.

[5] Zhang JF. Research on Acoustic Scattering Characteristics of Underwater Large-Scale Target[D]. *Harbin Engineering University*, 2013.

[6] Tao M, Zhao Y. Identification method of viscoelastic material dynamic parameters based on acoustic-pipe measurement[J]. *Journal of Vibration and Shock*, 2014(5):97-101.

[7] Zhu T. Research On Low-Frequency Scattering Characteristics Of Underwater Target[D]. *Shanghai Jiao Tong University*, 2008.

[8] Di YL, Zhang ZH, Song Q. Reliability Distribution of Products with Parameters Digression Feature[J]. *Computer & Digital Engineering*, 2016,44(07):1216-1220.

[9] Chang LX, Han GD. Effect of Sub-system Failure on Performance of Active Phased Array Antenna[J]. *Electronic Technology*, 2014, 27(2):71-74.

[10] Niu CF, Wu X, Zhao DH. Influence of Failed Elements on Phased Array Antenna and its Compensation[J]. *Radio-communication technology*, 2013, 39(5):44-46.

[11] Zhu FC, Wang XM, Cui DW. A reliable Storage Life Estimation of Guided Ammunitions Based on Storage And Application Information[J]. *Journal of Projectiles, Rockets, Missiles and Guidance*, 2013, 33(6):197-199.

[12] Shao W. Monte Carlo Methods and Their Applications in Some Statistical Model[D]. *Shandong University*, 2012.
[13] Ren X, Zhao XW, Cai Q. Dynamic Reliability Analysis of Cooling Water System Based on State-Transition Methodology[J]. *Chinese Journal of Ship Research,* 2011, 06(2):81-83.
[14] Chen WW, Min SR, Li MH, Xie HS. Analysis and Modelling of Ship-Borne Anti-Torpedo Torpedo Countermeasures[J]. *Chinese Journal of Ship Research*, 2014, 9(5):110-114.
[15] Wang XN, Fan J. Computation of Target Strength of the steel fish vessels and the influence of water surface [J]. *Journal of Shanghai Jiaotong University*, 2013, 47(10):1515-1519.
[16] Luo Z, Zhu X, Jian LA,. Acoustic Stealth Design of Underwater Sandwich Composite added Structure under Oblique Incidence[J]. *Journal of Vibration and Shock*, 2009, 28(5):49-54.

IOP Conf. Series: Earth and Environmental Science **171** (2018) 012024 doi:10.1088/1755-1315/171/1/012024

The Effect of Increased Temperature and Coral Acclimation of *Sinularia dura*

Z S Juhi and K A Rabbani

Department of Environmental Science, School of Environmental Science and Management, Independent University, Bangladesh
E-mail: rabbani@iub.edu.bd

Abstract. Coral reefs are one of the most vulnerable ecosystems in the world. With evidence of recent and notable mass coral bleaching at the Great Barrier Reef, it is imperative that coral species and their response to increased temperatures be studied in order to identify coral acclimation and the "winners" and "losers" of climate change. *Sinularia dura* is a commons coral species in South Asian and Southeast Asian water. This study has shown that even a 1 to 2 degrees centigrade increase in temperature can have huge effect on the health of *Sinularia Dura* which has implications to the status of coral reefs due to increased sea temperatures caused by climate change.

1. Introduction

Coral reefs are some of the most diverse and vulnerable ecosystems in the world. With ongoing climate change and ocean acidification, coral bleaching events are estimated to effect 90% of corals by 2060 if current emissions continue [1]. Coral bleaching occurs when environmental conditions disrupt the symbiosis, leading the degeneration and/or expulsion of its endosymbiotic algae (Zooxanthellae) from the coral host [2]. With time, as changes in environmental factors pose a challenge to their continued existence, corals can sometimes take major defensive or adaptive actions to ensure their survival [3]. Studies have shown that corals can recover rapidly following a dramatic decline and recoveries of coral cover have been documented at several locations all over the world [4]. Corals withstand heat stress via a number of mechanisms including acquiring thermally resistant symbionts, increasing physiological mechanisms by producing oxidative enzymes or by producing photo-protective compounds [5-9]. There are several biochemical methods of determining coral health and simply counting zooxanthellae density is an established method to monitor the health of corals [10-12]. Though there have been some studies looking at coral health and their relationship to increased temperature, few have been done on common South Asian corals [13].

Corals are a large economical assets for their fishing grounds and tourism in South East Asia and South Asia [14, 15]. It is important to identify species of corals in this region that are susceptible to climate change, and differentiate others to those that are more tolerant to increased temperatures. This study has thus been conducted on the coral species *Sinularia Dura*, also commonly known as cabbage coral. One of the most tolerant species of corals in South East Asia, we have tried to identify the rate of adaptability of this coral to stable sustained increase in water temperature while keeping all other parameters constant. The last part of this paper also attempted to record the recovery of this coral following exposure to elevated temperature to document the recovery this coral. This paper wanted to examine coral acclimation and how a specific hardy species of coral is going to respond to increase in temperature due to climate change.

2. Methodology

2.1. Coral Selection and Collection

All experiments were conducted on cultured soft coral *Sinularia Dura*, commonly known as cabbage leather coral which originates in the warmer water areas of the Indian Ocean. The selection of this specific type of coral has been subjected upon its high tolerance level of water parameters. As well as its higher capability to adapt to sudden changes has made it a highly strong competitor as a coral test subject. This coral was collected from a local aquaria during June 2016 measuring 11.5cm in diameter initially fragged onto a smaller live rock. The aquaria tank maintained a temperature of 25° C.

2.2. Tank Setup

The experiment was conducted at Independent University, Bangladesh during the periods of June to November, 2016. Before the arrival of the coral, a tank, dimensions of 24 in. X 13 in. X 13 in. (15 gallons) was set on cycle with SeaChem Marine Salt in water with a specific gravity of 1.021. The following specifications were acquired to mimic a healthy sea-like aquaria (Table 1).

Table 1. Parameters setup for the tank

Parameter	**Condition in tank**	**Typical ocean values**
Temperature	24.5-28 °C	25-30 °C
Salinity	35 ppt	34-36 ppt
pH	8.1-8.3	8.0-8.5

Temperature changes were recorded with an analogue temperature recorder along with a HACH sensION 156 pH/Conductivity meter for salinity and pH recordings. A liverock was added to the aquaria as an effective bio filter and act as an initial bio-media to form an ideal ecosystem. Additionally, a hang on back power filter with sponge was added for filtration purposes and a small submersible utility pump is added for proper aeration to add flow and circulation for the coral, to form sea like currents.

2.3. Lighting

Three 54W T5 fluorescent lights were added to the coral aquaculture of Red, Blue and White. A light cycle between 10-12 hours per day controlled with a timer was also made to mimic natural photoperiods.

2.4. Zooxanthellae Calculation Procedure

A 2.5 cm sized piece from the coral was cut from its tip and put onto a micro-centrifuge tube using 30μl aquarium water with a micropipette. The Zooxanthellae cell density was calculated using the formula:

Total Cells/ml = Average # cells in each square X dilution factor x 10^4.

A haemocytometer, is used to count the density zooxanthellae cells. It is a slide with grids that can be seen under the microscope for making counting small cells easier and usually used for counting blood samples. 10 μl of the sample was carefully loaded onto the haemocytometer and the slides seen under the microscope. 4 squares from the chamber was counted and recorded with a total of 16 grids in each. The number of zooxanthellae cells was counted in each grid.

3. Results and Discussion

By nature, the coral *Sinularia dura* has a tendency to shed their waxy coating from time to time in order to remove detritus buildup, during which time they will appear to shrink up. This is a natural "cleansing process" that most mushroom corals go through. The polyps of the coral tend to close up in these situations, making the coral seem almost weak, and ash in color. Different water conditions in the wild, or in aquarium tanks will determine how often this coral retracts. Upon visual inspection of 3 months, it was deduced that this coral would shed its coating and retract *once* every month. During that period, the number of Zooxanthellae on average goes through a cycle as well. However, this is a

natural occurrence and does not reflect on the overall health of the coral as they open back up within a few days to a weeks' time. A submersible water heater, was introduced into the aquaria and set at 25° C. During a period of 1-2 days, the temperature was increased by 1 degrees and maintained constant. A sample from the coral was taken to conduct an immediate Zooxanthellae density count using above mentioned procedures. Over the next one month, this gradual increase in temperature was done and recorded. However, at the final increase of 29.5 °C, the heater was taken off after assessing the corals threshold point. The coral was then left to retreat back to its normal temperatures in order to assess any recovery rates.
Sinularia dura Zooxanthellae cell density count

The cell density count period was conducted for a period of 28 days starting from 16th October 2016, until the 13th November, 2016. A total of 10 counts were conducted. The temperature increase in the tanks are also shown in the table below (Table 2). The count decreased with increasing temperature indicating that the health of corals decreased with increasing temperature.

Table 2. Zooxanthellae count of the experiment during the study

Tank temperature	**Zooxanthellae density count**	**Zooxanthellae under microscope**
25° C	12.75 cells/ml	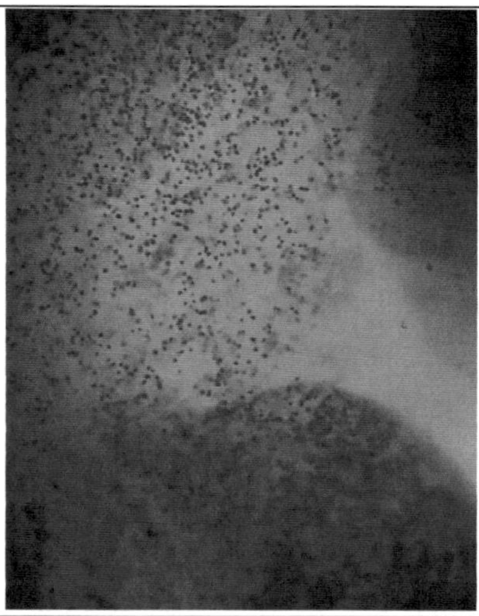
26° C	13.75 cells/ml	
27° C	13.75 cells/ml	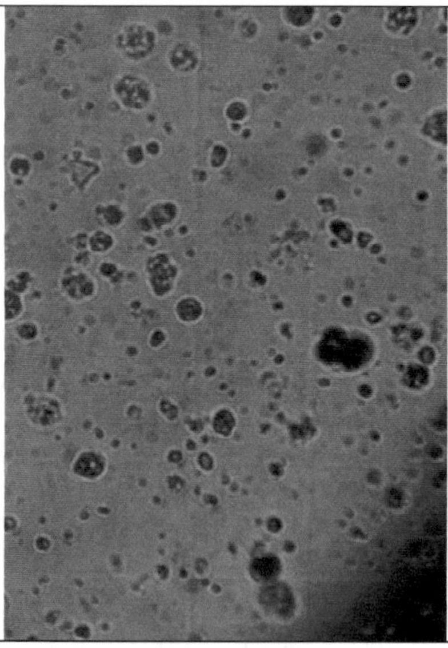

28° C	10 cells/ml	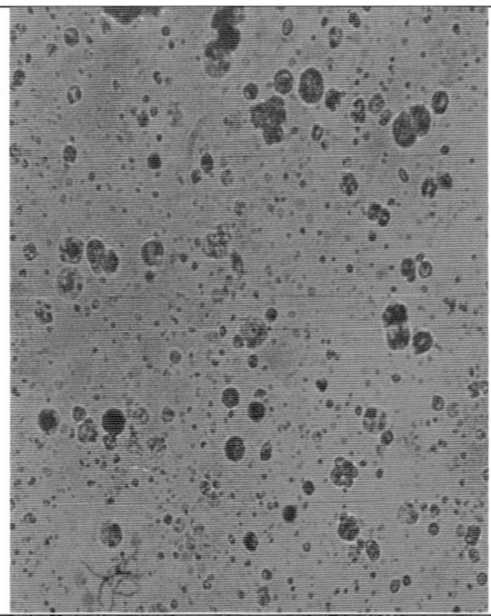
29.5° C	6.5 cells/ml	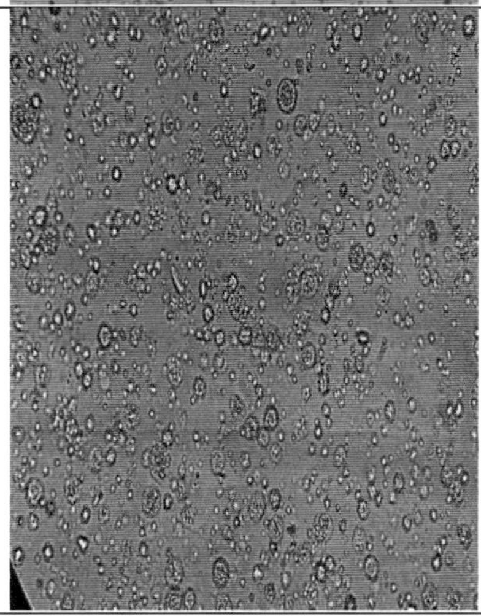

Upon visual and physical examination of the coral to 29.5° C, it was seen that the coral had almost bleached by 80% its patches of its fleshy mass disintegrated. The heater was then removed. By the very next day, the coral had completely disintegrated (Rapid Tissue Necrosis) into the tank with no solid mass left. Given the characteristic of "Hardy" coral, the *Sinularia dura* was objected for a partial recovery since the window of exposure to increased temperature was small.

4. Conclusion

Currently, our ocean average temperatures have reached at 27.6 degrees Celsius. At certain tropical area's the current ocean temperature at coral infused locations have currently reached 33.9°C / 93°F. It is sad that up until the recent mass bleaching of the Great Barrier Reef in Australia, bleaching of coral reefs did not gain mass attention or advertising. Over the last 30-40 years 80% of coral in the Caribbean have been destroyed and 50% in Indonesia and the Pacific. Bleaching associated with the 1982 -1983 El-Nino killed over 95% of coral in the Galapagos Islands and the 1997-1998 El-Nino alone wiped out 16% of all coral on the planet. Globally about 19% of coral is dying out each year (NOAA). The experiment's purpose was mainly to understand a hardy coral species' capability to acclimate/adapt to rapid temperature rises. Although in the end, we may hopefully not be too late. It is quite well known that bleaching signs that do not result in a long-term loss of reef viability. In fact, if proper conditions were to return, coral reefs have a slight chance of recovery. But the process if slow. Yet now, it is not irreversible and we still have a small window of saving left.

5. References

[1] Hooidonk R, Maynard JA, Manzello D, Planes S. Opposite latitudinal gradients in projected ocean acidification and bleaching impacts on coral reefs. Global Change Biology. 2014;20(1):103-12.
[2] Wiedenmann J, D'Angelo C, Smith EG, Hunt AN, Legiret F-E, Postle AD, et al. Nutrient enrichment can increase the susceptibility of reef corals to bleaching. Nature Climate Change. 2013;3(2):160-4.
[3] Comley J, Walker R, Wilson J, Ramsay A, Smith I, Raines P, et al. Malaysia coral reef conservation project: Pulau Redang. Report to the Department of Marine Parks, Malaysia, available at: www dmpm nre gov my/files/malaysia_coral_reef_conservation_project_pulau_redang pdf (accessed 19 October 2015), Google Scholar. 2004.
[4] Halford A, Cheal A, Ryan D, Williams DM. RESILIENCE TO LARGE-SCALE DISTURBANCE IN CORAL AND FISH ASSEMBLAGES ON THE GREAT BARRIER REEF. Ecology. 2004;85(7):1892-905.
[5] Buddemeier RW, Fautin DG. Coral bleaching as an adaptive mechanism. Bioscience. 1993;43(5):320-6.
[6] Rowan R. Coral bleaching: thermal adaptation in reef coral symbionts. Nature. 2004;430(7001):742-.
[7] Coles SL, Brown BE. Coral bleaching—capacity for acclimatization and adaptation. Advances in marine biology. 2003;46:183-223.
[8] Salih A, Larkum A, Cox G, Kühl M, Hoegh-Guldberg O. Fluorescent pigments in corals are photoprotective. Nature. 2000;408(6814):850-3.
[9] Palumbi SR, Barshis DJ, Traylor-Knowles N, Bay RA. Mechanisms of reef coral resistance to future climate change. Science. 2014;344(6186):895-8.
[10] Solayan A. Biomonitoring of Coral Bleaching-A Glimpse on Biomarkers for the Early Detection of Oxidative Damages in Corals. Invertebrates-Experimental Models in Toxicity Screening: InTech; 2016.
[11] Downs C, Fauth JE, Robinson CE, Curry R, Lanzendorf B, Halas JC, et al. Cellular diagnostics and coral health: declining coral health in the Florida Keys. Marine Pollution Bulletin. 2005;51(5):558-69.
[12] Moreira C, Ramos V, Azevedo J, Vasconcelos V. Methods to detect cyanobacteria and their toxins in the environment. Applied microbiology and biotechnology. 2014;98(19):8073-82.
[13] Bolton A, Goodkin N, Hughen K, Ostermann D, Vo S, Phan H. Paired Porites coral Sr/Ca and δ 18 O from the western South China Sea: Proxy calibration of sea surface temperature and precipitation. Palaeogeography, Palaeoclimatology, Palaeoecology. 2014;410:233-43.
[14] Dearden P, Manopawitr P. Climate change–coral reefs and dive tourism in South-east Asia. Disappearing destinations: climate change and future challenges for coastal tourism UK: CAB International. 2011:144-60.
[15] Pauly D. Coastal area management in Southeast Asia: policies, management strategies, and case studies: WorldFish; 1989.

Numerical Simulation of the Oil Spilling Impact on Water Environment

Shanshan Yao [1] , Peng Zhang[2] and Qixiu Pang[1]

1 Coastal & Estuarine Engineering Research Center, Tianjin Research Institute for Water Transport Engineering, Tianjin, China.
2 CCCC Tianjin Dredging Co.Ltd
E-mail: shanshan.hhu@163.com

Abstract. Taking Ganyu Port project (Phase I) as background, based on hydrodynamic and Euler-Lagrangian theory, a 2-D tidal current and oil spill numerical model are established and verified by measured data. According to different meteorological conditions, select representative scenes, the oil film sweep range and diffusion distance are predicted, and the oil spilling impact on surrounding water environment is analyzed. The results show that: the oil sweep range and drift track are closely related to the oil spilling time and wind direction. Under no wind condition, the farthest distance from the oil film edge to the oil spilling point is about 30 km in 72 hours, which will influence the present aquaculture area, but had no effect on the protected area. Under N and WNW adverse wind direction, the farthest distance are respectively 32km and 67km from the oil film edge to the oil spilling point, which will both influence aquaculture area and some individual protected area.

1. Introduction

With the development of offshore oil and the increasing of sea transportation in recent years, oil is a typical marine pollutant, it brings great economic benefits and also increases the hidden dangers of marine oil pollution. Once oil spills occur, the leaking oil will drift and spread with the movement of water flow. It will cause deterioration of marine environment and pollution of water quality, which will endanger marine life and destroy marine ecosystem.

In view of the frequency and severity of oil spill events, since 1960s, foreign scholars have begun to study the numerical simulation of oil spill transportation and diffusion. Reed etc. [1] reviewed the research and development of oil spill model during 20th century, the oil spill model research can be divided into three simulation categories: Fay model, based on Euler's viewpoint to solve the convection-diffusion equation and the "oil particle" model based on Lagrange's viewpoint. After 21th century, some scholars proposed to complement advantages of different simulation methods [2] . At the same time, the oil spill business prediction system is gradually consummated, the relatively high recognition abroad includes the OILMAP system and GNOME system of the United States, and the OSIS system of the United Kingdom, Norway's OILPILL/STAT system and OSCAR commercial software. The oil spill model in China began in early 1980s. In recent years, domestic scholars have carried out numerical simulation of oil spill diffusion for offshore, open sea and pipeline oil spills. Li Yan, Li Huan and Zhang have summarized the progress of numerical simulation of oil spill [3-5]; Lin and Li [6-7] has carried out model research on oil spill diffusion of submarine pipeline, Guo and Yu [8-9] have carried out model research around the near shore and offshore oil film drift diffusion.

In this paper, taking Ganyu Port project (Phase Ⅰ) as background, based on hydrodynamic and Euler-Lagrangian theory, a 2-D tidal current and oil spill numerical model are established and verified by measured data. According to different meteorological conditions, select representative scenes, the oil film sweep range and diffusion distance are predicted, and the oil spilling impact on surrounding water environment is analyzed. Results can provide reference for relevant environmental assessment work.

2. Model Theory

2.1. Hydrodynamic Model

$$\frac{\partial h}{\partial t}+\frac{\partial hu}{\partial x}+\frac{\partial hv}{\partial y}=hS \tag{1}$$

$$\frac{\partial hu}{\partial t}+\frac{\partial hu^2}{\partial x}+\frac{\partial hvu}{\partial y}=f\bar{v}h-gh\frac{\partial \eta}{\partial x}-\frac{gh^2}{2\rho_0}\frac{\partial \rho}{\partial x}+\frac{\tau_{sx}}{\rho_0}-\frac{\tau_{bx}}{\rho_0}+\frac{\partial}{\partial x}\left(hT_{xx}\right)+\frac{\partial}{\partial y}\left(hT_{xy}\right)+hu_sS \tag{2}$$

$$\frac{\partial hv}{\partial t}+\frac{\partial hvu}{\partial x}+\frac{\partial hv^2}{\partial y}=-f\bar{u}h-gh\frac{\partial \eta}{\partial y}-\frac{gh^2}{2\rho_0}\frac{\partial \rho}{\partial y}+\frac{\tau_{sy}}{\rho_0}-\frac{\tau_{by}}{\rho_0}+\frac{\partial}{\partial x}\left(hT_{xy}\right)+\frac{\partial}{\partial y}\left(hT_{yy}\right)+hv_sS \tag{3}$$

Where: $h=\zeta+d$ is the total water depth; ζ is the elevation of water surface; d is water depth; x and y represent the coordinates of the horizontal axis and the longitudinal axis respectively; t is time; g is the acceleration of gravity; u and v are the average velocity of depth along the axis of x and y. f is Coriolis force coefficient; ρ is water density; ρ_0 is the reference density; (τ_{sx}, τ_{sy}) and (τ_{bx}, τ_{by}) are the wind stress and the bottom shear stress along the axis of x and y, respectively. S is source discharge, u_s and v_s are the velocity component along the axis of x and y. T_{ij} is a stress term, including viscous stress, turbulent stress and convection etc., calculated according to the average velocity gradient of water depth.

In this study, nesting calculation of large and small model is adopted. The large model range includes the whole Haizhou Bay, and the small model includes main port area of Rizhao and Lianyungang, China (see figure 1). The open boundary is provided by *ChinaTide* offshore tidal prediction program, which developed by the Ocean University of China [10]. Unstructured triangular mesh are used in the computational domain (see figure 2).

The hydrodynamic numerical model is verified by the hydrographic data measured in 2013. The results show that the tidal level, current velocity and direction are in good agreement with the measured values, which meet the requirements of relevant specification [11]. Therefore, this model can be used in the numerical simulation of oil spill diffusion. In order to save space, the verification is shown in the reference [12].

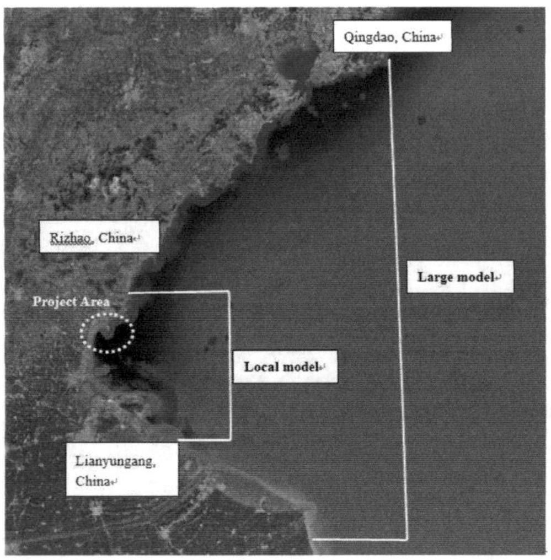

Figure 1. Location of the Project

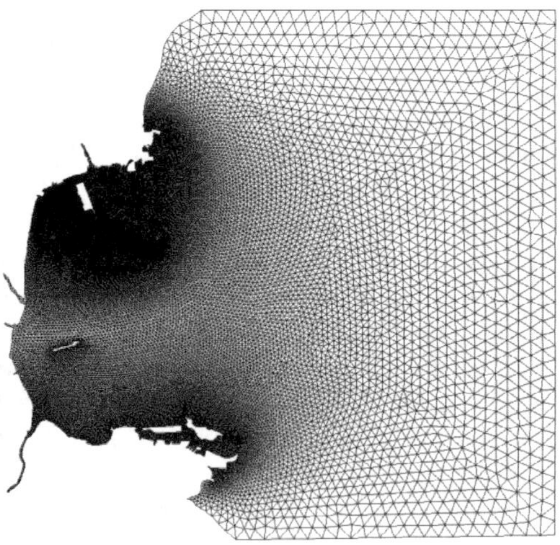

Figure 2. Mesh generation of local model

2.2. Oil Spill Model

MIKE21 SA module is used to predict the oil spill diffusion, which is developed by the Danish Water Environment Research Institute (DHI). Based on the Euler-Lagrangian theory, the variation of the properties of oil particles at each time is calculated in this paper, and the transport process and weathering process are considered in the calculation.

2.2.1. Transport Process

The transport of oil particles includes the processes of expansion, drift, diffusion and so on. These processes are the main reasons for the change of oil particles' position, but the components of oil particles do not change in these processes.

(1) Extended motion
Based on modified Fay Theory, oil film expansion is calculated using Gravity and viscosity Formula.

$$\left[\frac{dA_{oil}}{dt}\right] = K_a \cdot A_{oil}^{\frac{1}{3}} \cdot \left[\frac{V_{oil}}{A_{oil}}\right]^{\frac{4}{3}} \quad (4)$$

Where: A_{oil} is oil film area, A_{oil};$=\pi R_{oil}{}^2$, R_{oil} is oil film radius; K_α is coefficient (0.6 in this research); t is time; V_{oil} is oil film volume; h_s is initial oil film thickness.

$$V_{oil} = R_{oil}^2 \cdot \pi \cdot h_s \quad (5)$$

(2) Drift motion
Drift force of oil particle include water flow and wind drag. The total drift velocity of oil particle is calculated by the following weight formula.

$$U_{tot} = c_w(z) \cdot U_w + U_s \quad (6)$$

Where: U_w is wind speed; U_s is surface current speed; c_w is wind stress coefficient.

2.2.2. Weathering Process

The weathering of oil particles includes evaporation, dissolution and formation of emulsifying substances. During these processes, the composition of oil particles has changed, but its horizontal position has not changed.

(1) Evaporation
Oil film evaporation is affected by oil content, temperature, oil spill area, wind speed, solar radiation and oil film thickness. Assuming that the diffusion inside the oil film is not limited (when the temperature is higher than 0 degrees and the oil film thickness is less than 10 cm), the oil film is completely mixed. The partial pressure of the oil component in the atmosphere is negligible compared with the vapor pressure.
The evaporation rate can be represented by the following formula:

$$N_i^e = k_{ei} \cdot \frac{P_i^{SAT}}{RT} \cdot \frac{M_i}{\rho_i} \cdot X \quad (7)$$

Where: N_e is evaporation rate; k_{ei} is mass transport coefficient; P_{sat} is vapor pressure； R is gas constant; T is temperature; M is molecular weight； ρ is the density of the oil component； X is mole fraction； i represents various oil components。 k_{ei} can be estimated from the following formula：

$$k_{ei} = k \cdot A_{oil}^{0.045} \cdot Sc^{\frac{-2}{3}} \cdot U_w^{0.78} \quad (8)$$

Where: k is evaporation coefficient (0.029 in this research)； Sc_i is the vapor Schmidts number of component i.

(2) Dissolution
Assuming that the hydrocarbon concentration is negligible compared with the solubility, then the solubility of oil in water is expressed as follows:

$$\frac{dV_{oil}}{dt} = K_{si} \cdot C_i^{SAT} \cdot X_{mol_i} \cdot \frac{M_i}{\rho_i} \cdot A_{oil} \quad (9)$$

Where: C_i^{SAT} is solubility of component i, X_{moli} is mole fraction of component i, M_i is molar mass of component i; K_{si} is transfer coefficient of dissolution.

(3) *Emulsification*

Emulsification is the function of a liquid uniformly dispersed in another liquid that is insoluble in small droplets. The movement of oil into water includes diffusion, dissolution and precipitation.
The amount of oil loss from the diffusion of oil film to water body D is:

$$D = D_\alpha \cdot D_b \quad (10)$$

$$D_a = \frac{0.11(1+U_w)^2}{3600} \quad (11)$$

$$D_b = \frac{1}{1+50\mu_{oil} h_s \gamma_{ow}} \quad (12)$$

Where: D_α is the component that enters the water body; D_b is the component that does not return after entering the water body; μ_{oil} is oil viscosity; γ_{ow} is the oil-water interfacial tension.
The return rate of oil droplets is as follows:

$$\frac{dV_{oil}}{dt} = D_\alpha \cdot (1 - D_b) \quad (13)$$

The variation of water content in oil can be expressed by the following equation:

$$\frac{dy_w}{dt} = R_1 - R_2 \quad (14)$$

$$R_1 = K_1 \frac{(1+U_w)^2}{\mu_{oil}} (y_w^{\max} - y_w) \quad (15)$$

$$R_2 = K_2 \frac{1}{As \cdot Wax \cdot \mu_{oil}} y_w \quad (16)$$

Where: y_w is actual moisture content; R_1 和 R_2 are water absorption rate and release rate, respectively; A_s is Asphalt content in oil; W_{ax} is Paraffin content in oil; K_1, K_2 are absorption coefficient and release coefficient.

3. Hydrodynamic Simulation Results

Using the validated numerical model, large tidal current field during spring tide is shown in figure 3, and the analysis is as follows :
In the Haizhou Bay, where Ganyu Port is located, tidal current movement shows reciprocating feature in the direction of NE~SW. Tidal current in Lianyungang and its south sea area is mainly rotating counterclockwise. In the vicinity of Gaogongdao and Xiaoding Harbor, the tidal current presents reciprocating flow along the coast. The velocity distribution shows the trend of north strong and south weak, and the velocity during flood tide is a little larger than ebb tide. The current velocity of most natural sea areas is between 0.4m/s and 0.7m/s, due to the construction of Ganyu Port, the velocity near the buildings is relatively large.

4. Oil Spill Simulation Results

On the basis of hydrodynamic simulation, a 2-D oil spill model in Ganyu Port area was established by using MIKE21 SA module. The oil leakage source is 100 tons and the source is instantaneous. The oil spilling point is located at the front of wood wharf. Oil spilling moment includes high water level(HWL), low water level(LWL), flooding and ebbing (see table 1). According to the surrounding environment sensitive target, a total of 4 protected areas are close to the project, and a large number of aquaculture areas are existed around the project. In this research, N wind and WNW wind are selected as adverse wind to predict the oil spill, and the wind speed is the maximum of 5 class wind, which is about 10.7 m/s. Through analysis, some conclusions are made as follows:
(1) After the oil spill accident, the oil film drifts and diffuses under the movement of current. Under no wind condition (see figure 4 and figure 5), when oil spill occurs at the HWL moment, the oil film

drifts to the open seas under the ebbing flow, thus the influence area is relatively large, when oil spill occurs at the LWL moment, the oil film drifts to the coastal area under the flooding flow, thus the influence area is relatively small. The oil spilling point is located at the front of the wood wharf, the current velocity is generally low, so the oil film sweep area is small in the initial stage of oil spill. The farthest distance is about 30 km from the oil film edge to the oil spilling point in 72h, and the oil spill diffusion have no direct effect on the protected area in 72h. However, due to the existence of a large number of aquaculture in Ganyu Port area, oil spill will inevitably influence the aquaculture areas.
(2) According to the above simulation results, the oil film sweep range is only calculated at HWL moment under adverse wind condition (see figure 6). After 72h of oil spill accident, under adverse wind condition of N, the farthest distance from the oil film edge to the oil spilling point is about 32 km. Under the adverse wind condition of WNW, the farthest distance from the oil film edge to the oil spilling point is about 67km (see table 2). In terms of the impact on adjacent sensitive areas, it can be seen that under the adverse wind condition, oil spill diffusion will have a direct impact on the surrounding protected areas and aquaculture areas. Therefore, in order to protect the marine environment and the surrounding water quality, oil spills should be avoided as far as possible. When the oil spill accident occurs, the emergency response measures should be taken quickly, and combined with the shortest time that the oil spill reach the environmental area, measures such as intercept, clean up, and recover the oil spill should be taken to minimize the serious losses to the sea environment and ecosystem, and to prevent the possible adverse effects of leakage accident on the surrounding water environment.

Table 1. Oil spill prediction condition

Case	Oil spilling point	Wind condition		Oil leakage moment
		Wind direction	Wind speed (m/s)	
1	Wood wharf apron	No wind	-	Flooding
2		No wind	-	Ebbing
3		No wind	-	HWL
4		No wind	-	LWL
5		N	10.7	HWL
6		WNW	10.7	HWL

Table 2. Statistics on oil spill sweeping area and influence distance

Case	Wind conditon		Oil leakage moment	1-72h sweeping area (km^2)	1-72h furthest diffusion distance (km)	Impact on environmental protected area
	Wind direction	Wind speed (m/s)				
1	No wind	—	Flood	390.5	28.2	No effect on protected areas, but influence the aquaculture areas.
2	No wind	—	Ebb	307.9	26.0	The same as above
3	No wind	—	HWL	415.5	29.1	The same as above
4	No wind	—	LWL	102.1	17.2	The same as above
5	N	10.7	HWL	295.5	32.5	Directly influence both protected areas and aquaculture areas.
6	WNW	10.7	HWL	749.0	66.9	The same as above

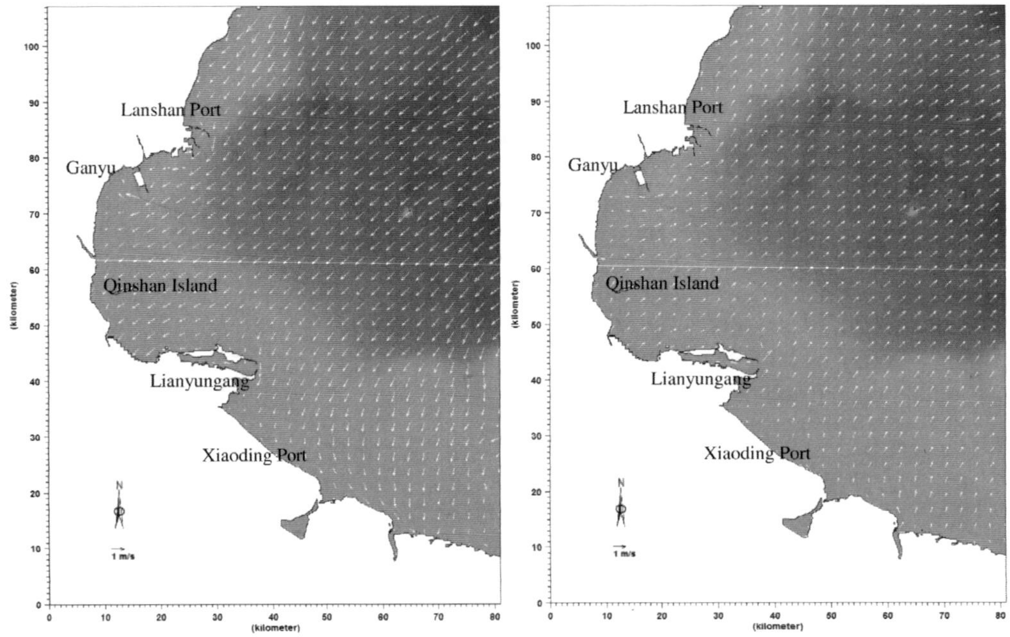

Figure 3. Large tidal current filed (Left: flood; Right: ebb)

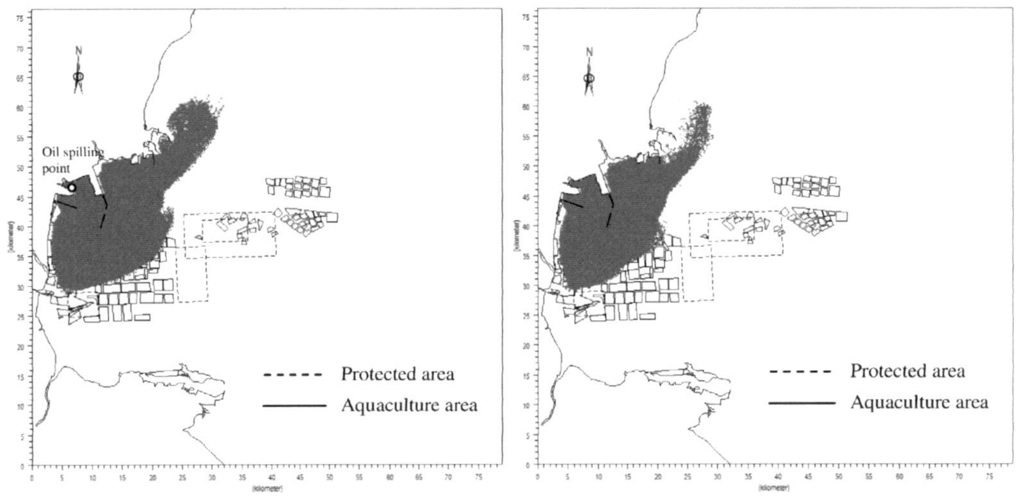

Figure 4. Influence range of oil spill 1-72 hours under no wind condition (Left: flood; Right: ebb)

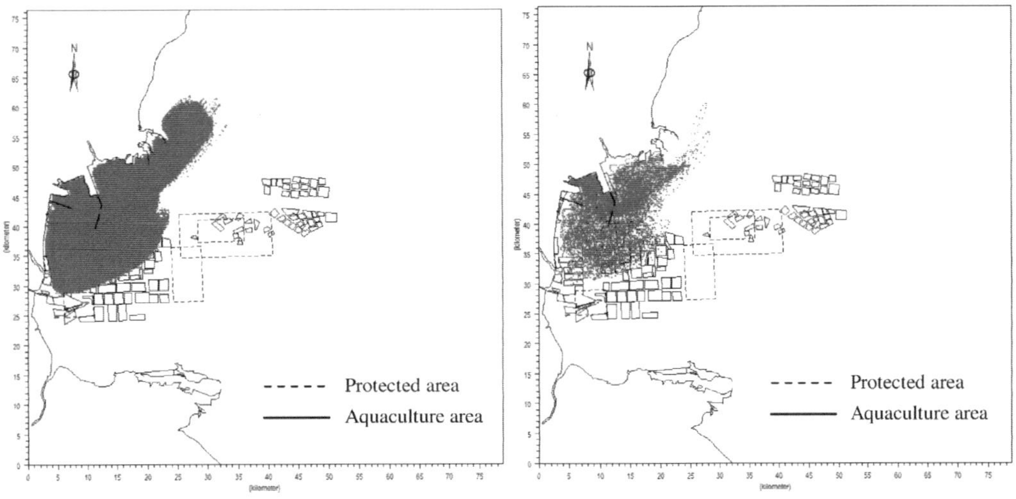

Figure 5. Influence range of oil spill 1-72 hours under no wind condition (Left: HWL; Right: LWL)

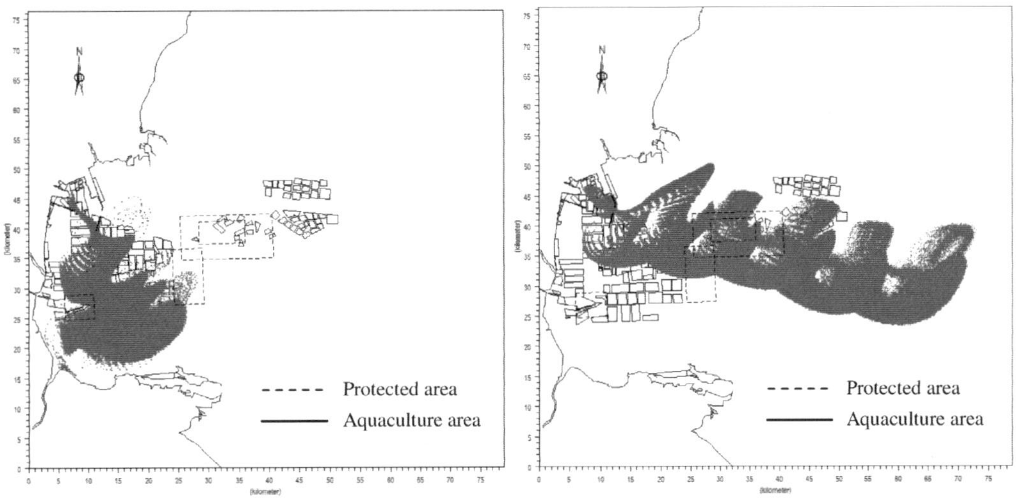

Figure 6. Influence range of oil spill 1-72 hours under adverse wind condition (Left: N; Right:WNW)

5. Conclusions

In this paper, taking Ganyu Port project (Phase Ⅰ) as background, based on hydrodynamic and Euler-Lagrangian theory, a 2-D tidal current and oil spill numerical model are established and verified by measured data. According to different meteorological conditions, select representative scenes, the oil film sweep range and diffusion distance are predicted, and the oil spilling impact on surrounding water environment is analyzed. The results show that:

(1) After the oil spill accident, the oil sweep range and drift track are closely related to the oil spilling time and wind direction. Under no wind condition, The farthest distance is about 30 km from the oil film edge to the oil spilling point after 72h, and the oil spill diffusion have no direct effect on the protected area in 72h. However, due to the existence of a large number of aquaculture in Ganyu Port area, oil spill will inevitably influence the aquaculture areas. Under adverse wind condition of N, the farthest distance from the oil film edge to the oil spilling point is about 32 km. Under the adverse wind condition of WNW, the farthest distance from the oil film edge to the oil spilling point is about 67km.

(2) In order to protect the marine environment and the surrounding water quality, oil spills should be avoided as far as possible. When the oil spill accident occurs, the emergency response measures should be taken quickly, and combined with the shortest time that the oil spill reach the environmental area, measures such as intercept, clean up, and recover the oil spill should be taken to minimize the

serious losses to the sea environment and ecosystem, and to prevent the possible adverse effects of leakage accident on the surrounding water environment.

6. References

[1] Reed M, Johansen Ø, Brandvik P J, et al 1999 *Oil spill modeling towards the close of the 20th century: overview of the state of the art* (England: Spill Science & Technology Bulletin)Vol. 5(1) pp 3-16

[2] Guo W J, Wang Y X 2009 *A numerical oil spill model based on a hybrid method* (England: Marine Pollution Bulletin) Vol.58(5) pp 726-734

[3] LI Y, YANG Y Q, PAN Q Q 2017 *Review on the oil spill numerical forecasting technology* (Beijing: Marine forecasts) 34(5) pp 89-96

[4] LI H, SHAO W Z, LI C 2017 *Research on numerical prediction technologyof oil spill spreading, drift and diffusion*(Beijing: Marine forecasts) Vol.36(5) pp 379-384

[5] ZHAGN A, ZHOU J, LI Y, et al 2017 *Review of numerical model research on oil spill* (Dalian: Marine Environmental Science) Vol.36(2) pp 313-320

[6] LIN M H, HE G X, LI Y S, et al 2017 *A numerical simulation on the spilled oil diffusion of submarine pipelines* (Beijing: Beijing Institute of Petrochemical Technology) Vol.25(4) pp 83-90

[7] LI X J 2012 *Numerical Simulation of the Spread-Diffusion Process of Oil Spilled from Submarine Pipeline* (Qingdao: Ocean University of China)

[8] Guo J 2017 *The Risk Simulation and Environmrent Sensitive area Nearby Effects of Oil Spill in Binhai Offshore* (Shanghai: Shanghai Ocean University)

[9] YU M R, WEI C L, JIANG S C, et al 2017 *A comparison among different oil spill simulation methods, Transactions of oceanology and limnology* (Tianjin: Transactions of oceanology and limnology) Vol. 5 pp 58-62

[10] LI M G, ZHENG J Y 2007 *The application of Chinatide, a tidal prediction software in Chinese waters* (Tianjin: Journal of Waterway and Harbor) Vol.1 pp 65-68

[11] JTS/T 231-2-2010 *Technical specification for tidal current and sediment simulation in coastal and estuary areas*

[12] ZHANG N, YAO S S 2013 *Numerical study on the environmental impact of Ganyu Port Project(Phase I)* (Tianjin: Technical Report of Tianjin Research Institute for Water Transport Engineering, M.O.T., China)

Acknowledgements

This work is financially supported by the National Key Research and Development Program(No. 2016YFC0402605).

IOP Conf. Series: Earth and Environmental Science **171** (2018) 012026 doi:10.1088/1755-1315/171/1/012026

Variation Characteristics of Tropical Cyclones Making Landfall over China between 1951 and 2015 and Their Relationship with ENSO

Chenglin Gu [1,2]; Jiancheng Kang[1*], Guodong Yan [3], Zhiwei Chen [1], Mingxing Niu[1]

1Urban Ecology and Environment Research Center Shanghai Normal University, Shanghai, 200234, China
2Jiamusi University Department of Resources and Environment, Faculty of Science, Jiamusi, 154007, China
3 ShanghaiUniversity of Engineering Science, Shanghai, 200234, China
E-mail: kangjc@126.com

Abstract. Based on tropical cyclone (TC) data provided by the America Joint Typhoon Warning Center, this paper analyzes the variation characteristics of tropical cyclones making landfall over China during the 65-year period of 1951–2015 and their statistical relationship with the El Niño–Southern Oscillation (ENSO). The conclusions are as follows. (1) The landfall frequency during the period of 1951–2015 has strong inter-annual and inter-decadal variability characteristics, and an abrupt change in the landfall frequency occurred in 1988. However, from the long-term variation frequency trend, the landfall frequency decreased slowly from the late 1980s to 2015. (2) The landfall intensity increased rapidly from the late 1950s to the early 1960s, with an abrupt change occurring in 1996. In addition, there is a significant decreasing trend and then an increasing trend from the late 1960s to 2015. (3) There are significant negative correlations between the landfall frequency and the sea surface temperature (SST) in the ENSO-3.4 region. All the sliding correlation coefficients are negative from 1961 to 2002. A strong abrupt change in their correlation occurred in 1970, changing from a weak negative to a strong negative correlation. (4) There are no significant negative correlations between the landfall intensity and the SST in the ENSO-3.4 region, and their sliding correlation coefficient of inter-annual variability shows that there is an alternating appearance of positive and negative correlations. A strong abrupt change occurs in 1999, changing from a weak positive correlation to a strong negative correlation. The significant negative correlation occurred in the 2005. (5) The TC genesis locations move to the southeast in El Niño years and shift to the northeast in La Niña years. (6) TC frequency that influences the Chinese continental coast line in El Niño years is less than that which influences Japan. Meanwhile, in La Niña years, the TC frequency that influences Japan and its surrounding waters, the China Sea, the BohaiSea, and the South China Sea, increases, while it decreases in the TaiwanProvince.

1. Introduction

Tropical cyclones (TCs), called typhoons in Southeast Asia, are a type of extreme weather event that has the potential to cause tremendous losses of human lives and property due to excessive and torrential rainfall, storm tides, landslides, debris flows, flash floods, and strong winds. For example, in 2004, ten TCs caused widespread damage in Japan[1]. Between 1995 and 1999, damages in the

Economic and Social Commission for Asia and the Pacific/World Meteorological Organization (ESCAP/WMO) Typhoon Committee member areas in the western North Pacific were estimated to be approximately \$3620 million (U.S. dollars) with average annual human casualties exceeding a thousand (ESCAP/WMO 2001). Therefore, the changing properties of TC events such as their intensity, frequency, duration, tracks, the effects of global warming on tropical cyclone intensity, and their linkages with large-scale circulation variability are attracting increasing attention from meteorologists, hydrologists, and governments across the globe[2-7].

Many scholars have attempted to test the relationships between TC activity and climatic signals such as the El Niño–Southern Oscillation (ENSO)[8-11]. In addition, some scholars have researched how ENSO, the quasi-biennial oscillation, and the Madden–Julian oscillation affect western North Pacific TC activity. For example, Pudovand Petrichenko[12, 13] found an increase in the intensity of TCs during El Niño years. Since then, many authors have examined the relationship between ENSO and the number of western North Pacific TCs. The results have not been consistent in all cases, presumably due to differences in both the data and the techniques used[14,15]. Li and Wang[16] argue that ENSO affects the large-scale atmospheric circulation, therefore further affecting the activity of TCs in the western North Pacific.

Even though there is no significant linear relationship between the number of TCs and ENSO, a nonlinear relationship between ENSO and the number of TCs has been found in several studies; this is especially evident when the ENSO forcing is strong[17]. Chia and Ropelewski[18] argue that the western North Pacific has a larger number of generated strong tropical storms during El Niño years. However, there are also scholars who think that this trend is not correlated to ENSO.

Currently, most studies have focused on the relationship between TC activity in the western North Pacific and ENSO. Very few studies have tried to examine the relationship between the landfalling TCs' frequencies, intensities, genesis locations, tracks, and observed intensities and ENSO, for which there exist differences.

Based on the TC data provided by the America Joint Typhoon Warning Center, this paper studies the inter-annual and inter-decadal variability trends of TCs making landfall over China during the period of 1951–2015, which shows the statistical relationship between the landfall frequency, the landfall intensity, and ENSO, and further discusses the characteristics of the TC genesis locations, tracks, observed intensities, and space variations. These results could provide a scientific basis for short-term climatic predictions of TCs making landfall over China.

2.Data and Methodology

We used the Joint Typhoon Warning Center TC best-track data covering the period of 1951–2015. These datasets commonly include the locations of the TC centers, the maximum sustained wind speeds at 6-h intervals, the center of the wind speed, and the maximum wind (the 1-min observed center). Wu and Zhao[19] have noted that the Joint Typhoon Warning Center TC best-track data over the western North Pacific is more reliable than other data; and Chan[20] has argued that the Joint Typhoon Warning Center TC intensity data is more reliable. This paper uses TCs that reached at least tropical storm intensity (maximum sustained winds >17.2 m s^{-1} based on the Japan Meteorological Agency/JMA rating) or higher. The main typhoon season in China is from May to November (Fig. 1); therefore, the TC index in this paper is also from May to November.

The SST data are from Had1SST, a dataset provided by the Hadley Center of the UK Met Office (the link for the data is http://hadobs.metoffice.com/hadisst/data/download.html), which contains a unique combination of monthly globally complete fields of SST and sea ice concentration on a 1° latitude–longitude grid from 1870 to the present.

The SST anomaly (SSTA) over the Niño-3.4 region(5°N–5°S, 170–120°W) appears to be a meaningful indicator for ENSO events because the equatorial east–west SST gradient in this area is the largest. In addition, it has been found that the Niño-3.4 SSTA is better correlated with the overall TS activity over the western North Pacific. To further study the relationship between TC activity and the sea surface temperature (SST) during the period of 1951–2015, this paper uses the sea surface temperature variation over the Niño-3.4 region to represent the ENSO variation.

This ensures that the ENSO signal is evident during the peak May–November TC landing season in the western North Pacific. The Niño-3.4 SSTA is therefore used as the reference for classifying the

ENSO events. A year is classified as El Niño (La Niña) if the mean May–December Niño-3.4 SSTA is>0.5°C (<−0.5°C)[14]. The SSTA data were obtained from the National Centers for Environmental Prediction Climate Prediction Center from the reconstruction of sea surface temperature data in the ERSSTv4 extension[21, 22].

This process yields 16 El Niño years (1951, 1953, 1957, 1958, 1963, 1965, 1969, 1972, 1982, 1987, 1991, 1997, 2002, 2004, 2009, and 2015), 15 La Niña years (1954, 1955, 1964, 1970, 1971, 1973, 1974, 1975, 1988, 1998, 1999, 2000, 2007, 2010, and 2011), and 34 ENSO neutral years (1952, 1956, 1959, 1960, 1961, 1962, 1966, 1967, 1968, 1976, 1977, 1978, 1979, 1980, 1981, 1983, 1984, 1985, 1986, 1989, 1990, 1992, 1993, 1994, 1995,1996, 2001, 2003, 2005, 2006, 2008, 2012, 2013, and 2014) (Fig. 2).

The defined El Niño and La Niña years in this study are basically the same as those obtained using Zhao's et al.[23] method of defining El Niño years as the former 25% Niño-3.4 SSTA and La Niña years as the latter 25% Niño-3.4 SSTA. However, other traditional definitions are slightly different.

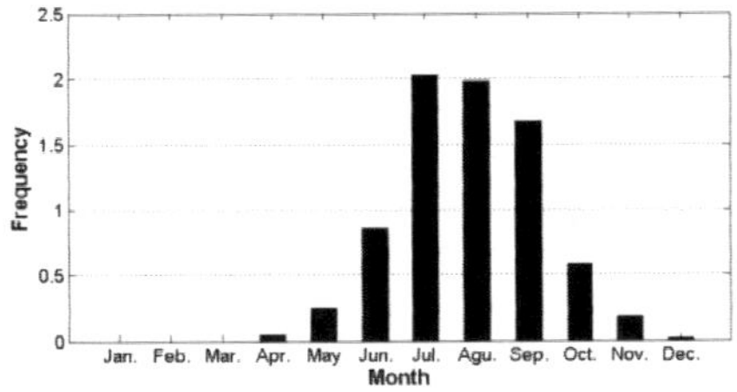

Figure 1. The climatological monthly variation of the frequency of landfalling TCs in China for the period of 1951–2015.

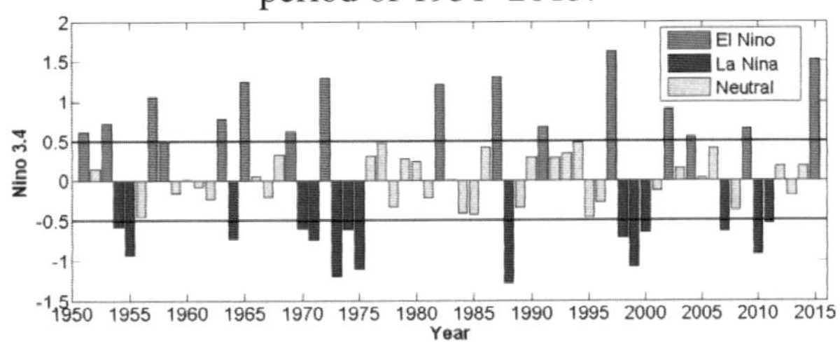

Figure 2. The time series of the ENSO-3.4 SSTA index (°C) for the May–November season in the period of 1951–2015. The horizontal lines show the ENSO-3.4 SSTA index value equal to ±0.5.

3. Annual Variation Characteristics of Landfall Frequency and Landfall Intensity

3.1. Annual Variation Characteristics of Landfall Frequency

Of the TCs (except the tropical depressions) generated over the western North Pacific (including the South China Sea), 498 TCs made landfall over China from 1951 to 2015. Including the islands of Taiwan and Hainan (a TC is only counted once when it makes landfall directly or brushes past the island), the annual landfall rate is 7.7 and the lowest rate was 2, which occurred in 1958, while the highest rate was 17, which occurred in 1974.

Fig. 3a shows that the landfall frequency during the period of 1951–2015 has strong inter-annual and inter-decadal variability characteristics. The period from the early 1950s to the late 1960s was a decreasing period, while those from the mid 1960s to the early 1980s and from the late 1980s to 2015 were increasing periods. This paper uses the general moving t-test to analyze the time series of the landfall frequency. Fig. 3b shows that the statistical value of t is over the significant level test of 0.05 (T0.05=2.101), which indicates that an abrupt change in the landfall frequency occurred in 1988 and then there was an increasing period. However, as a long-term variational trend, the landfall frequency decreased slowly in the inter-annual oscillation from the late 1980s to 2015. Wavelet analysis was carried out on the landing frequency (Fig. 3c and d), landing frequency cycle oscillation for 6 years.

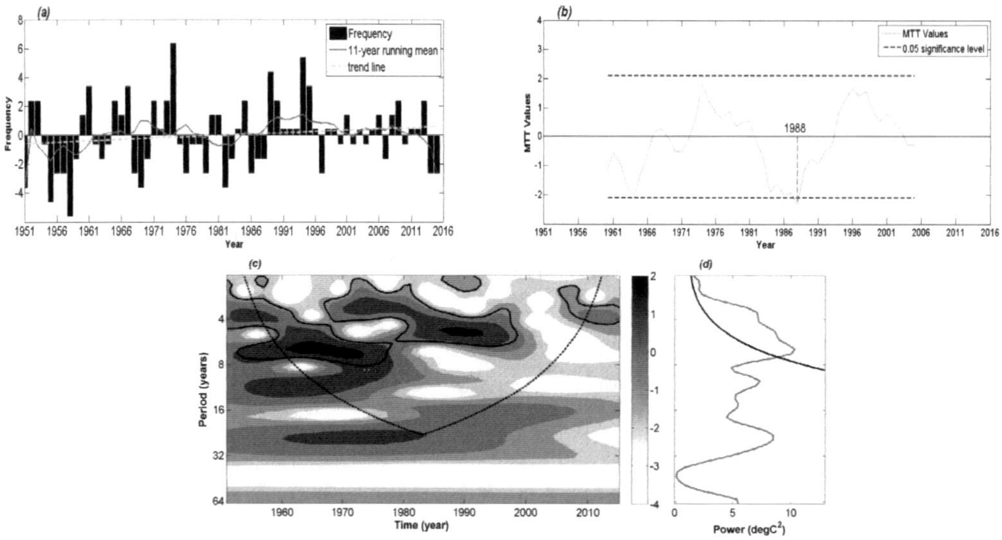

Figure3. (a) the time series of the landfall frequency (the trend line failure to pass 95% confidence tests), (b) the moving t-test of the time series in Figure 3a. (c) wavelet power spectrum, (d) global wavelet spectrum.In panels (b), the black dotted line indicates the critical 0.05 significance level line. In panels (c), the black dotted line indicates boundary effect line. In panels (c) and (d), the black line indicates the critical 0.05 significance level line.

3.2. Annual Variation Characteristics of Landfall Intensity

The landfall intensity of the TCs in a given year is defined as the mean of the intensities at the time when the TC center was closest to the coastal areas on a 6-h basis[24]. From 1951 to 2015, the landfall intensity reached a maximum value of 67.76m/s in 1958, with a minimum value of 24.11 m/s in 1978 and an average value of 33.07m/s. Fig. 4a shows that the period from the late 1950s to the early 1960s was a rapidly increasing period, and there is an obviously decreasing and then increasing trend from the late 1960s to 2015. Using the moving t-test to analyze the time series of the landfall intensity, Fig. 4b shows that the statistic value of t is over the significant level test of 0.01 (T0.01=2.878), which indicates that an abrupt change in the landfall intensity occurred near 1966 and 1987.

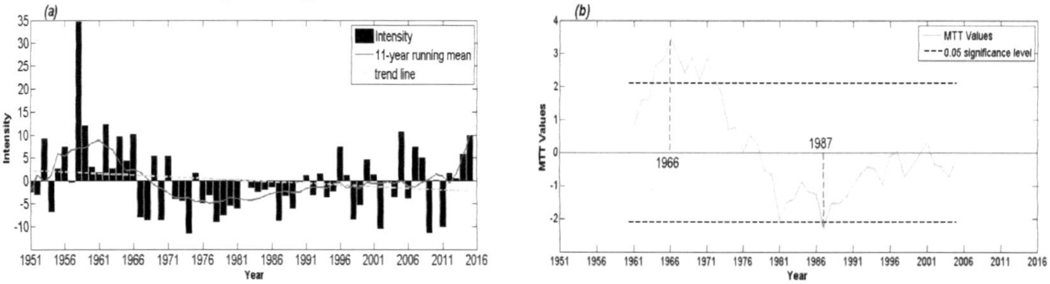

Figure 4.(a) the time series of the landfall intensity (the trend line failure to pass 95% confidence tests), (b) the moving t-test of the time series in Figure 3a. In panels (b), the black dotted line indicates the critical 0.05 significance level line.

4. The Relationship between TC Activity and ENSO

4.1. The Statistical Relationship between the Landfall Frequency, Intensity, and ENSO

Fig. 5a shows the distribution of the correlation coefficient between the landfall frequency and the SST for the Pacific region (January–December). The TC frequency has a negative relationship with the SSTA over the ENSO-3.4 region (green rectangular area), and the correlation coefficient reaches a maximum value of −0.3 (red shading), which indicates that there are less TCs making landfall in El Niño years while the contrary occurs in La Niña years. The result is in agreement with other studies.

This paper uses the sliding correlation method to analyze the long-term variation characteristics of the relationship between the landfall frequency and the SST. Fig. 5b shows the inter-annual variability of the correlation coefficient. All the sliding correlation coefficients are negative from 1961 to 2002. The significant negative correlation occurred in the late 1970s. A moving t-test is made on the sliding correlation coefficient. Fig. 5c shows that the t value surpasses the significant level of 0.05 (T0.05=2.101) in 1970, 1987, and 2000, which means that abrupt changes occurred around 1970, 1987, and 2000. There is a strong abrupt change in 1970, before which the negative correlation was weak and after which there is a strong negative correlation. However, weak positive and negative correlations alternatively occurred in the 2000s.

Fig. 5d shows the distribution of the correlation coefficients between the landfall intensity and the SST for the Pacific (January–December), and there is no significant correlation over the ENSO-3.4 region. When we further study the inter-annual variability of the sliding correlation coefficient of the landfall intensity and the SSTA over the ENSO-3.4 region (Fig. 5e), we see that positive and negative correlations alternatively occur and that periodic characteristics are very obvious. A moving t-test was made on the sliding correlation coefficient. Fig. 5f shows that the t value surpasses the significant level of 0.05 (T0.05=2.101) in 1999, which means that abrupt changes occurred around 1999. A strong abrupt change occurs in 1999, changing from a weak positive correlation to a strong negative correlation. The significant negative correlation occurred in the 2005.

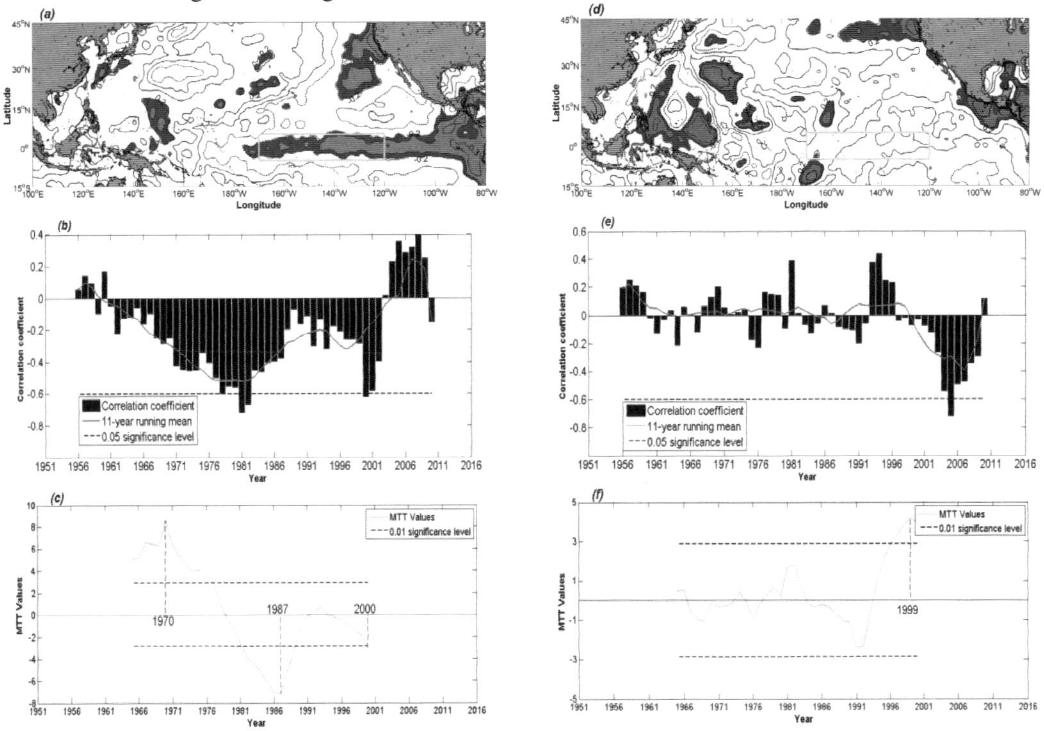

Figure 5. Spatial patterns of the correlation relations between SST and (a) the landfall frequency and (d) the landfall intensity. The time series of sliding correlation coefficients between SSTA in the ENSO-3.4 region and (b) the landfall intensity and (e) the landfall intensity.The moving t-test of the 11 years for the time series (c) in Figure 4b and (f) in Figure 4e. In panels (a) and (d), the light and dark red shading shows differences significant at the 90% and 95% confidence levels, respectively. In panels (a) and (d), the green rectangular box indicates the ENSO-3.4 region. In panels (c) and (f), the black dotted line indicates the critical 0.05 significance level line.

4.2. The Relationship between ENSO and the TC tracks

The TC tracks can be described by the TC passage frequencies. The TC passage frequencies are the number of TCs passing through each 5° ×5° grid area at 6-h intervals, where a TC is only counted once even if it enters the same grid box several times, which could reflect the chances of being influenced by TCs in a certain region.

One major track in Fig. 6a moves from the tropical Pacific to the Philippine Sea and the South China Sea, and another one moves from the tropical Pacific northwestward to Korea and Japan influencing the region around the East Asia coast. The maximum value of the TC frequency occurred over the central region of 118°E 18°N. Fig. 6b of the El Niño versus the neutral years shows that the contour lines are mostly negative over all of the Chinese mainland, which means that the TC frequency is less during El Niño years. This indicates that TCs making landfall over China decrease during El Niño years. Meanwhile, the contour lines are mostly positive in Japan, indicating that TCs making landfall over Japan increase during El Niño years. Fig. 6c contrasts La Niña years with neutral years, and the positive contour lines are located in the area around the Japan Sea, China Sea, Bohai Sea, and South China Sea, which indicates that the TC frequency increases around the Bohai Sea, Jiao Zhou Peninsula, and southern China coastal areas. The contour lines values are negative over the regions of Taiwan and its surrounding areas and the southeastern Chinese coastal area, which indicates that the TC frequency decreases around these regions. Fig. 6d contrasts La Niña years with El Niño years; the distribution shows a pattern decreasing southward and northward and increasing in the middle. The positive values are mostly located over two regions, over the China Sea and the Bohai Sea and over the South China Sea and the Philippines and its surrounding areas, which indicates that the TC frequency increases in the Shan Dong and Jiang Su Provinces of China, while the TC frequency decreases in the southern part of China and over the region of 150°E 7.5° N.

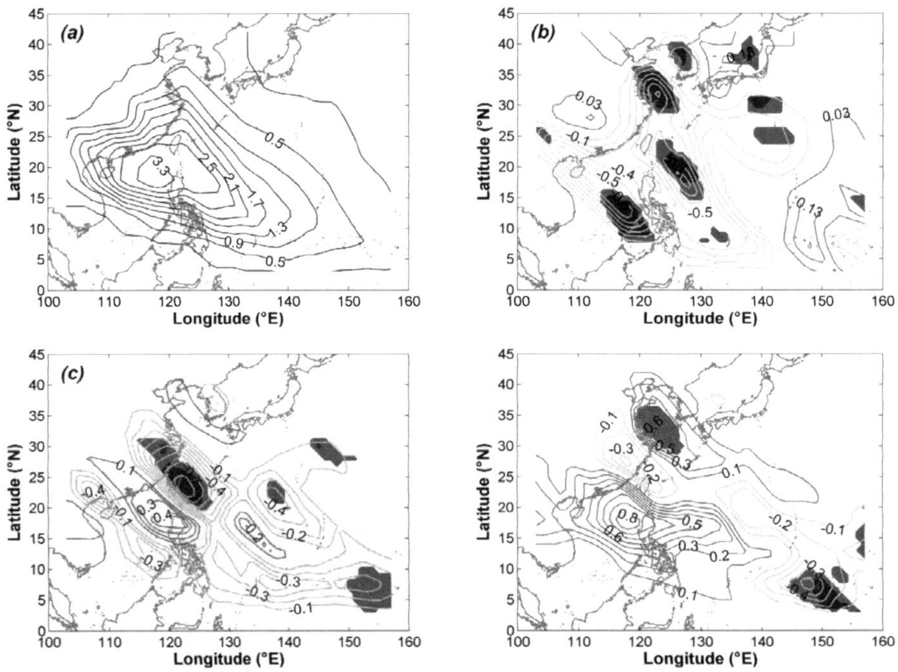

Figure 6. Geographical distribution of the TC passage frequencies (number per year) in each 5°×5°grid area in (a) neutral years, (b) El Niño years minus neutral years, (c) La Niña years minus neutral years, and (d) La Niña years minus El Niño years. The light and dark blue shading indicates areas where the t-test is significant at the 90% and 95% levels, respectively.

5. Conclusions

Based on the TC data provided by the America Joint Typhoon Warning Center, this paper analyzes the trends of the TC frequency, intensity, and inter-annual and inter-decadal variability during the 65-year period of 1951–2015 and then discusses the variation characteristics of the TCs' genesis locations, tracks, and observed intensity and spatial variations. The conclusions are as follows.

(1) The landfall frequency during the period of 1951–2015 has strong inter-annual and inter-decadal variability characteristics, and an abrupt change in the landfall frequency occurred in 1988. However, from the long-term variation frequency trend, the landfall frequency decreased slowly from the late 1980s to 2015.
(2) The landfall intensity increased rapidly from the late 1950s to the early 1960s, with an abrupt change occurring in 1996. In addition, there is a significant decreasing trend and then an increasing trend from the late 1960s to 2015.
(3) There are significant negative correlations between the landfall frequency and the sea surface temperature (SST) in the ENSO-3.4 region. All the sliding correlation coefficients are negative from 1961 to 2002. A strong abrupt change in their correlation occurred in 1970, changing from a weak negative to a strong negative correlation.
(4) There are no significant negative correlations between the landfall intensity and the SST in the ENSO-3.4 region, and their sliding correlation coefficient of inter-annual variability shows that there is an alternating appearance of positive and negative correlations. A strong abrupt change occurs in 1999, changing from a weak positive correlation to a strong negative correlation. The significant negative correlation occurred in the 2005.
(5) TC frequency that influences the Chinese continental coast line in El Niño years is less than that which influences Japan. Meanwhile, in La Niña years, the TC frequency that influences Japan and its surrounding waters, the China Sea, the Bohai Sea, and the South China Sea, increases, while it decreases in the Taiwan Province.

6. Reference

[1] Tu JY, Chou C, Chu PS (2009) The abrupt shift of typhoon activity in the vicinity of Taiwan and its association with Western North Pacific-East Asian climate change. J Clim 22(13): 3617–3628.
[2] Kossin JP, Knapp KR, Vimont DJ, Murnan RJ, Harper B (2007) A globally consistent reanalysis of hurricane variability and trends. Geophys Res Lett 34(4): 344–356.
[3] Wu L, Wang B (2008) What has changed the proposition of intense hurricanes in the last 30 years? J Clim 21(6):1432–1439.
[4] Hong C, Li Y, Li T, Lee M (2011) Impacts of central Pacific and eastern Pacific El Niños on tropical cyclone tracks over the western North Pacific. Geophys Res Lett 38(16):603–13.
[5] Chu HY, Wu RS (2013) Environmental influences on the intensity change of tropical cyclones in the western North Pacific. Acta Meteor Sinica 27(3): 335–343, doi: 10.1007/s-13351-013-0307-4.
[6] Murakami H, Li T, Peng MS (2013) Changes to environmental parameters that control tropical cyclone genesis under global warming. Geophys Res Lett 40(10): 2265-2270.
[7] Zhang L, Zhu H, Sun X (2014) China's tropical cyclone disaster risk source analysis based on the gray density clustering. Nat Hazards 71(2):1053-1065.
[8] Chen G, Tam C-Y (2010) Different impacts of two kinds of Pacific Ocean warming on tropical cyclone frequency over the western North Pacific. Geophys Res Lett 37(1): 70–75L01803. doi:10.1029/2009GL041708.
[9] Kim H-M, Webster PJ, Curry JA (2011) Modulation of North Pacific tropical cyclone activity by three phases of ENSO. J Clim 24(6): 1839–1849.
[10] Zhang W, Graf H, Leung Y, Herzog M (2012) Different El Niño types and tropical cyclone landfall in East Asia. J Clim 25(19): 6510–6523.
[11] Wang C, Wang X (2013) Classifying El Niño Modoki I and II by different impacts on rainfall in the southern China and typhoon tracks. J Clim 26(4): 1322–1338.
[12] Pudov VD, Petrichenko SA (1998) Relationship between the evolution of tropical cyclones in the Northwestern Pacific and El Niño. Oceanology 38(4): 447–452.
[13] Pudov VD, Petrichenko SA (2001) 1997–1998 El Niño and tropical cyclone genesis in the northwestern Pacific. Izv Atmos Oceanic Phys 37(5): 576–583.
[14] Lee HS, Yamashita T, Mishima T (2012) Multi-decadal variations of ENSO, the Pacific Decadal Oscillation and tropical cyclones in the western North Pacific. Progress in Oceanography 105 (2012) : 67–80.

[15] Zhang H, Guan Y (2014) Impacts of Four Types of ENSO Events on Tropical Cyclones Making Landfall over Mainland China Based on Three Best-track Datasets. Adv Atmos Sci 31(1): 154-164.

[16] Li C, Wang, C (2013) Simulated impacts of two types of ENSO events on tropical cyclone activity in the western North Pacific: large-scale atmospheric response. Climate Dyn 42(9): 2727–2743.

[17] Chen TC, Wang SY, Yen MC (2006) Interannual variation in the tropical cyclone activity over the western North Pacific. J Clim 19(21): 5709–5720.

[18] Chia HH, Ropelewski CF (2002) The interannual variability in the genesis location of tropical cyclones in the northwest Pacific. J Clim 15(20): 2934–2944.

[19] Wu L, Zhao HK (2012) Dynamically derived tropical cyclone intensity changes over the Western North Pacific. J Clim 25(1): 89–98.

[20] Chan JCL (2008) Decadal variations of intense typhoon occurrence in the western North Pacific. Proc Royal Soc A 464(2089): 249–272.

[21] Huang B, Banzon VF, Freeman E,Lawrimore JH, Liu W, Peterson TC, Smith TM, Thorne PW, Woodruff SD, Zhang, H (2015) Extended Reconstructed Sea Surface Temperature Version 4 (ERSST.v4). Part I: Upgrades and Intercomparisons. J Clim 28(3): 911–930.

[22] Liu W, Huang B, Thorne PW, Banzon VF, Zhang H, Freeman E, Lawrimore JH, Peterson TC, Smith TM, Woodruff SD (2015) Extended Reconstructed Sea Surface Temperature Version 4 (ERSST.v4): Part II. Parametric and Structural Uncertainty Estimations. J Clim 28(3): 931–951.

[23] Zhao H, Wu L, Zhou W (2010) Assessing the influence of the ENSO on tropical cyclone prevailing tracks in the Western North Pacific. Adv Atmos Sci 27(6): 1361–1371.

[24] Zhang QZ, Zhang W, Lu X, Chen YD (2011) Landfalling tropical cyclones activities in the south China: intensifying or weakening? Int J Climatol 32(12): 1815–1824.

Acknowledgments

This work was supported by the National Natural Science Foundation Research Project (41340045), The Key Disciplines Construction Project of Shanghai Municipal (J50402), and the National Natural Science Foundation Research Project (41301034). The authors are grateful to the editor and anonymous reviewers for their helpful comments and also acknowledge helpful comments by Professor Kang Jian-cheng.

Foundation item. The National Natural Science Foundation Research Project (41340045);The Key Disciplines Construction Project of Shanghai Municipal(J50402); The National Natural Science Foundation Research Project (41301034).

The Variation of Extreme Low Temperature Events in the Northwest Pacific under the Global Warming

Zhiwei Chen [1], Jiancheng Kang [1*], Chenglin Gu[1,2], Yong Xu[1], Ming Tang [1,3] and Kai Lu[1]

1 Department of Geography, Urban Ecology and Environment Research Center, Shanghai Normal University, Shanghai, 200234, China;
2 Department of Resources and Environment, Faculty of Science, Jiamusi University, Jiamusi, 154007, China;
3 Poyang Lake Eco-economy Research Centre, Jiujiang University, Jiangxi, 332000, China
E-mail: Kangjc@126.com

Abstract: Under the background of global warming, the variation of extreme low temperature events in the Northwest Pacific are undergoing significant changes. The extreme low temperature events have obvious monthly, inter-annual and inter-decadal characteristics. Extreme low temperature events were mainly concentrated in January-March. It showed an increasing trend from1982 to 1991, a downward trend from 1992 to 2011 and an upward trend from 2012 onwards. The frequency of extreme low temperature events has significantly reduced. The number of days and the duration of a single event have significantly reduced. The average temperature and the minimum temperature of a single extreme low temperature event show a linear upward trend. The extreme low temperature events have two time scales cycles, with a large cycle of 13-20a and small cycle of 4-10a. Over the past 33 years, extreme low temperature events have changed from small time-scale cycle to large time-scale cycle.

1. Introduction

Changes in the frequency and intensity of extreme weather events are likely to have a huge impact on human society, economy and environment, which has attracted more and more attention from the international community [1-2]. Extreme events are referred to the value of a weather or climatic variable, which is higher than (or below) the threshold near the upper limit (or lower limit) on the range of the observed variable[3].

On November 18, 2011, the Intergovernmental Panel on Climate Change (IPCC) issued a special report --Managing extreme events and disaster risks and promoting climate change adaptation[3]. This report pointed that extreme weather and climate events have changed since 1950. The number of cold days and cold nights is likely to decrease, while the number of warm days and warm nights is likely to increase in the world. In Asia, the rising trend of daily extreme temperature is moderate. Trends in daily extreme temperatures in Africa and South America generally show low to moderate confidence. But globally, the duration or number of warm events, including heat waves, has increased in many regions with adequate information.

What is certain is that on a global scale, it has been observed that the coldest days and nights on most of the land are decreasing while the warmer days and nights are increasing[4]. However, what happened to the extreme temperature events on the oceans has not been observed and studied detailly so far.

2. Purpose
Utilizing the NOAA Optimum Interpolation 1/4 Degree Daily Sea Surface Temperature Analysis data for the period 1982–2014, on the basis of determining the threshold of extreme temperature, the temporal and spatial variation characteristics of extreme low temperature events in the Northwest Pacific during the last 33 years were discussed by a variety of statistical diagnostic methods, which laid the foundation for monitoring and forecasting extreme weather and climate events.

3. Study Areas
In this research, study areas are located between 0° and 30°N along the longitude from 105° to 155°E(Figure 1). The North-western Pacific is a key area that affects China's climate change.

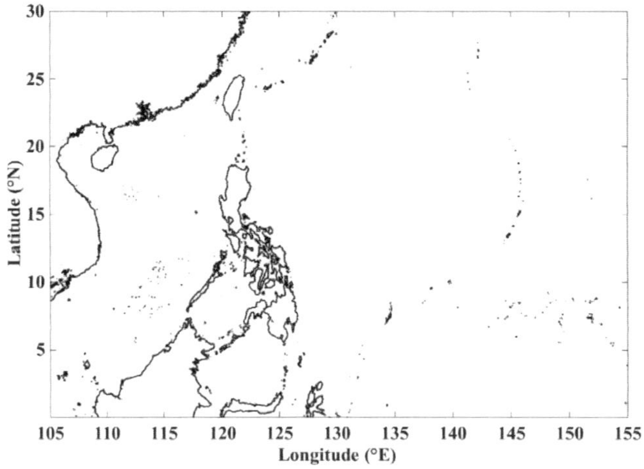

Figure 1. Research area

4. Data and Methods
Sea Surface Temperature (SST) data comes from the NOAA Optimum Interpolation 1/4 Degree Daily SST Analysis Data V2.0. The time range covers 1982.01–2014.12, at a spatial resolution of 0.25°×0.25°, including SST and SST anomalies.

The sea surface temperature anomaly index comes from the ERSSTv4 Climate Prediction Centre Extended Bureau America reconstruction of sea surface temperature data[5-6].

Based on the probability distribution of the meteorological elements, the fourth report of the IPCC defines an event less than or equal to the 10th percentile (greater than or equal to 90th percentile) as an extreme case [7]. Some scholars have pointed out that in the study of climate change, they usually take 5% / 95% or 10% / 90% as the extreme value threshold in order to reflecting the variation trend of extreme events [8-10]. Taking into account the temperature data in this paper over a short time span, we use the 10% / 90% percentile as the threshold for extreme events and use the definition of extreme events proposed in the IPCC Fourth Report.

5. Results and Discussions
The daily average temperature of Northwest Pacific from 1982 to 2014 are arranged in ascending order, with the 10th percentile as the extreme low temperature threshold, which is 25.79℃. Extreme low temperature ranged from 24.86-25.79℃, and the minimum sea surface temperature appeared in March 5, 1986, which is 2.73℃lower than the average daily temperature over many years. Extreme cold events mainly concentrated in the January-March, with the highest occurring in February, accounting for 47.96%, followed by 26.06% and 25.98% respectively in January and March.

In the recent 33 years, the number of extreme cold days in the Northwest Pacific were 1202, corresponding to an average of 36.4 per year. According to the fitted curve, the annual variation of extreme low temperature events was great, with the most number of days in 1993 (81 days) and the least (2 days) in 2007, showing a rapid linear decrease trend with a reduction rate of 27d / 10a and also an obvious stage fluctuation. It showed an increasing trend from 1982 to 1991, a downward trend from

1992 to 2011 and an upward trend from 2012 onwards. The frequency of extreme low temperature events in the western North Pacific has decreased significantly.

During 1982 to 2014, 41 extreme low temperature events occurred in the North-western Pacific, with an average duration of 29.3 days. The longest and shortest durations of single extreme low temperatures were 81 days and 2 days respectively (Figure 2). The duration of a single extreme low temperatures in recent 33 years showed an overall linear decrease trend with a decrease rate of 1.6 days / time. The average temperature and minimum temperature of a single extreme low temperature event also showed a linear upward trend with the average temperature rising rate of 0.0087 ℃ / time and the minimum temperature rising rate of 0.017 ℃ / time.

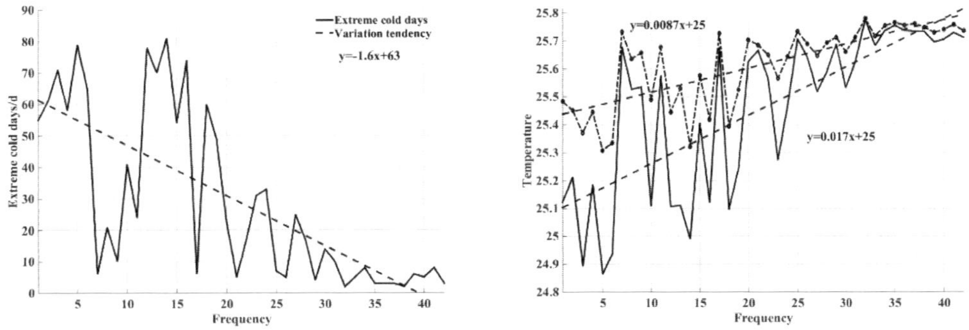

Figure 2. The duration and mean (minimum) temperature of extreme low temperature events

Figure 3 shows the change cycle of extreme low temperature events throughout the year (Fig. 3a) and from January to March (Fig. 3b-3d) during 1982-2014 in the North-western Pacific. It can be seen that in the past 33 years, there were two time-scale cycles of extreme low temperature events. In the large-scale period of 13-20a, there were six stages of multi-little-multi-little-multi-multiple. In 1982-1987, 1993-1997 and 2002-2007, there were relatively more extreme low temperature events. While there were less extreme low temperature events with the main period of 15 years in 1988-1992, 1998-2001 and 2008-2014. Before 2000, there were both medium time-scale cycle of 4-10a and large time-scale cycle, and then medium time-scale cycle had weakened and disappeared. It can be concluded from the above analysis that during the past 33 years, the extremely low temperature events in the North-western Pacific have undergone a transition from small-scale cycles to large-scale cycles.

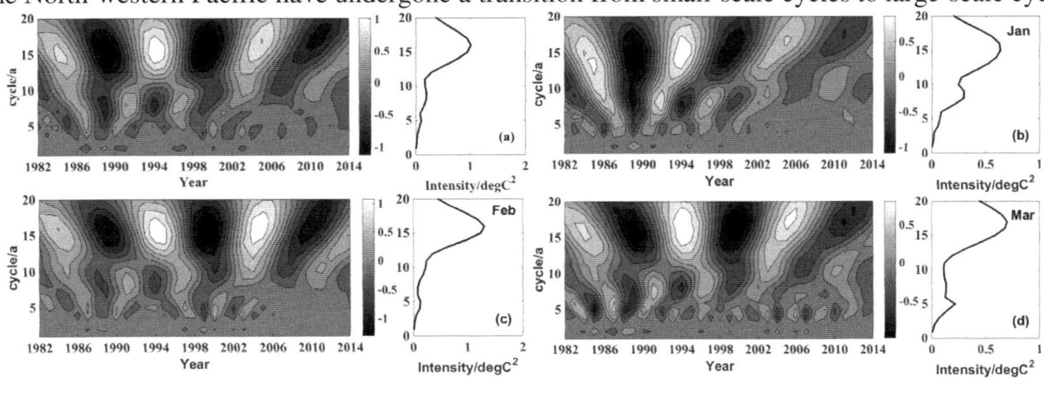

Figure 3. The cycle variation of extreme low temperature events

6. Conclusions

Analyzing and discussing the extreme low temperature events that occurred in the Northwest Pacific from 1982 to 2014, and the following conclusions are as follows:

(1) The extreme low temperature events in Northwest Pacific from 1982 to 2014 have obvious monthly, inter-annual and inter-decadal characteristics. Extreme low temperature events were mainly

concentrated in January-March. It showed an increasing trend from1982 to 1991, a downward trend from 1992 to 2011 and an upward trend from 2012 onwards.
(2) The frequency of extreme low temperature events has significantly reduced. The number of days and the duration of a single event have significantly reduced, and the average temperature and the minimum temperature of a single extreme low temperature event show a linear upward trend.
(3) There are two time-scale cycles of extreme low temperature events, which is the large cycle (13-20a) and the small cycle(4-10a). In the past 33 years, the extremely low temperature events in the North-western Pacific have undergone a transition from small-scale cycles to large-scale cycles.

7. References

[1] IPCC 2001*Climate Change 2001, Contribution of Working Group 1 to the Third Assessment Report of the Intergovernmental Panel on Climate Change*(Cambridge: Cambridge University Press) P155-164
[2] Organization W M. WMO statement on the status of the global climate in 2010 2011 *J. Environmental Policy Collection,* 40(2):144-151
[3] IPCC 2011*Managing the risks of extreme events and disasters to advance climate change adaptation (SREX)*(Cambridge: Cambridge University Press)
[4] Shen Y P, Wang G Y Key findings and assessment results of IPCC WGI Fifth Assessment Report 2013*J. Journal of Glaciology and Geocryology*,35(5):1068-1076.
[5] Huang B, Banzon V F, Freeman E, et al. Extended Reconstructed Sea Surface Temperature Version 4 (ERSST.v4). Part I: Upgrades and Intercomparisons2015*J. Journal of Climate*, 28(3):911-930
[6] Liu W, Huang B, Thorne P W, et al. Extended Reconstructed Sea Surface Temperature Version 4 (ERSST.v4): Part II. Parametric and Structural Uncertainty Estimations 2015 *J. Journal of Climate,*28(3):931-951
[7] Solomon S,Qin Dahe *Climate Change 2007:The Scientific Basis Contribution of Working Group I to the Fourth Assessment Report of the Intergovernmental Panel on Climate Change*(Cambridge: Cambridge University Press)
[8] Hao X, Zhu Z, et al. Characteristics of extreme weather in Xilin river and their relation with regional climate change nearly 50 years 2017 *J. Journal of Arid Land Resources and Environment*,31(7):114-120
[9] Li D H, Zhou L W, Zhou T J Changes of extreme indices over China in response to 1.5°Cglobal warming projected by a regional climate model 2017 *J .Advances in Earth Science*, 32(4) : 446-457
[10] Yan Z, Jones P D, Davies T D, et al. Trends of Extreme Temperatures in Europe and China Based on Daily Observations 2002 *J. Climatic Change*, 53(1-3):355-392

Acknowledgements
The research was supported by the National Natural Science Foundation of China (41340045), the Key Disciplines Construction Project of Shanghai Municipal (J50402).

IOP Conf. Series: Earth and Environmental Science **171** (2018) 012028 doi:10.1088/1755-1315/171/1/012028

The Climatic Characteristics of Surface Salinity in the South China Sea and the Adjacent Northwest Pacific Ocean

Mingxing Niu, Jiancheng Kang*, Zhiwei Chen

Geography Department, Shanghai Normal University, Shanghai 200234, China
*Corresponding kangjc@126.com

Abstract. Based on the high-resolution SODA oceanographic hydrological reanalysis data, analyzing the climatic characteristics of the surface salinity in the South China Sea and the adjacent Northwest Pacific Ocean during the new climatic baseline period from 1981 to 2010 in this paper. The results showed that: Sea surface salinity (SSS) in the research area could be obviously divided into two areas, the South China Sea and the Northwest Pacific Ocean. The SSS in the South China Sea was lower than that in the Northwest Pacific Ocean. The annual amplitude of SSS in the South China Sea varied in different sea areas, but it was smaller and almost same in the Northwest Pacific Ocean area. The inter-monthly variations of SSS were the same at continental shelf side and the sea area near the island in the South China Sea, however differences existed with that in the central of the South China Sea. The inter-monthly variation of SSS in the Northwest Pacific Ocean varied from low latitude to high latitude and the salinity field showed "high-low-high" distribution. According to the climatic characteristics of salinity, the research area could be divided into five climatic zones. Among them, the South China Sea was divided into areas A and B, and the Northwest Pacific Ocean was divided into areas C, D, and E. As water invading from the Northwest Pacific Ocean to the South China Sea area that caused a higher salinity in area B than in area A. The northern equatorial current influenced on the Northwest Pacific Ocean, that caused the salinity in area D lower than in areas C and E.

1. Introduction

Climate change posed an imminent and irreversible threat to human society and the planet (Paris Agreement, 2015). Nowadays climate change has become the most important environmental issue in our society (United Nations Global Environment Outlook 5 (GEO5), 2012).

Research on climate change has become one of the hottest issues of the international community [1, 2]. In order to meet the needs of accelerating climate change research, in 2015, the World Meteorological Organization (WMO) Commission for Climatology formally proposed to adopt the 1981-2010 period as the latest baseline for climate change.

The ocean was an important part of the climate system, which played a crucial role in the energy balance and material balance of the climate system. Marine salinity was one of the most important indicators of climate change and also an important regulator of the climate system, which had a significant impact on local and global climate change. Making researches on the structural features of ocean salinity climate was of great significance for understanding climate and explaining climate change.

The predecessors had analyzed the monthly average salinity change using the SODA dataset from 1972 to 2010 [3] and discussed the variational characteristics of salinity in the sea area and the relationship between salinity and ENSO according to the seasonal, inter-annual and even inter-decadal scales [4, 5]. However, there was a lack of a clear and systematic study of the climatic structure and the monthly variation of the salinity field, especially a lack of in-depth analysis of the climatic structural characteristics of new climate change baseline period.

So, in this paper, the structural features of the salinity field in the South China Sea and the adjacent Northwest Pacific Ocean over the new climatic base years analyzed mainly, and quantitatively analyzed the inter-monthly variation of the salinity structure at the surface layer, which provided a baseline for studying regional climate change and an effective reference for "understanding the past, present and future climate change process."

2. Research Area

In this paper, the South China Sea and the adjacent Northwest Pacific Ocean were chosen as the study area whose latitude and longitude range were 105.25-140.25°E and 0.25-25.25°N respectively, as shown in Figure 1.

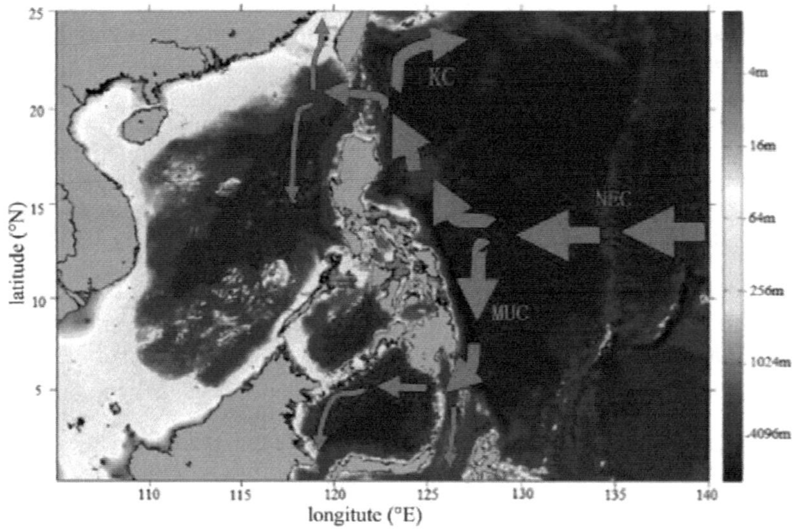

Figure 1. Schematic diagram of the study area (color bars represent seawater depth; red arrows represent the main currents in the area).

3. Data and Methods

3.1. Salinity Data

Simple Ocean Data Assimilation (SODA_2.2.4 monthly average data) quoted in this paper was a reanalysis product and collected by Global Simple Ocean Data Assimilation System which was derived from the University of Maryland and Texas A & M University. The time span was from January 1871 to December 2010 and the horizontal grid was 0.5° × 0.5° and salinity data was vertically divided into 40 layers of unequal spacing. The dataset included seven variables such as temperature, salinity, latitudinal current velocity, Longitude current velocity, zonal sea surface wind stress, longitude surface wind stress and sea surface height.

3.2. Topographic Data

Topographic data were derived from the 1-Minute Gridded Global Relief Data (ETOPO1) released by the National Geophysical Data Center (NGDC) of US in 2009. The grid of the database was 1'x 1', and the depth of the grid was 1 m [6].

3.3. Analytical Methods

- Surface data was extracted from the latest climate baseline period (January 1981- December 2010). In the study area, annual variation of surface salinity and the monthly salinity in climatic distribution chart at surface layer were drawn to quantitatively analyze the variational characteristics of salinity field.

- The annual amplitude of surface salinity was calculated by extracting the maximum and minimum values of surface salinity from January to December in the latest climatic baseline period for each point in the study area. The annual amplitude map could intuitively reflect the variation range of salinity in different sea areas and explore the spatial distribution of salinity in each area in detail.

4. Results Analysis

4.1. The Climatic Distribution of Salinity at Surface

Using the surface salinity information of SODA dataset, it could be drawn the climatic structure chart of surface salinity and surface salinity annual amplitude map for the study area, respectively were shown in Figure 2, with the isohaline space 0.25psu.

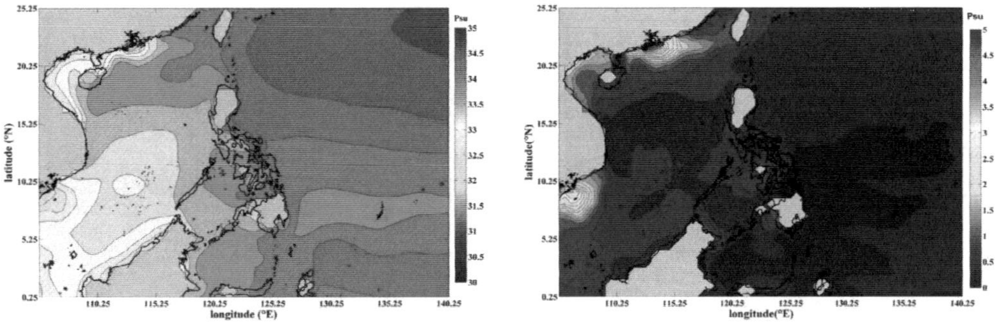

Figure 2. The climatic status map of surface salinity in the study area (left); Yearly variation of surface salinity in the study area (right).

Figure 2 (left) showed the climatic structure of surface salinity in the study area. It could be seen from the figure, salinity ranged from 31.5 to 34psu in the South China Sea. Salinity distributed on the one side of the continental shelf and near the island showed a lower value, with a higher value in the northeast sea area which presented a downward trend from the northeast to the southwest. The SSS values in the central South China Sea gradually decreased from high latitude to low latitude and ranged from 33.2 to 34psu. The isohaline presented a belt distribution feature. Among them, there was a low-salinity nucleus with 33.25psu near the southeast location (112.75°E, 10.25°N). It could be seen that the salinity of the sea areas to a great extend affected by the river runoff was low and the isohalines are dense. In the central sea area, the salinity was relatively high and the isohalines are sparse.

In the Northwest Pacific Ocean, surface isohaline generally presented a belt distribution feature along the latitudinal direction and closed off near the coast of the Philippines. From low latitude to high latitude, the salinity field showed a "high-low-high" distribution characteristics. Low-salinity nuclei occurred near the 7.75°N cross section in the central Northwest Pacific Ocean at about 34psu.Salinity was between 34-34.85psu in the north of the low-salinity nuclei area and 34-34.5psu in the south of the low-salinity nuclei area.

Figure 2 (right) was the surface salinity's annual amplitude in the study area. As seen in the figure, in the South China Sea, salinity's annual amplitude varied in different sea areas. The maximum amplitude appeared in the vicinities of Hong Kong and Macao, reaching as high as 4.66psu, followed by the northern Gulf of Thailand and the Beibu Gulf with a variation of 1.5-3.5psu, all appeared around the estuary of the rivers. The minimum annual amplitude of salinity appeared in the middle of the South China Sea at about 0.5psu.

In the Northwest Pacific Ocean, the annual amplitude of salinity was less than 0.5psu and whose distribution was uniform. In the north area of 15°N, the annual amplitude of salinity was about 0.25psu. In the southwestern part of 15°N, the minimum annual amplitude of salinity was 0.08psu.

In contrast, the sea area with large annual amplitude was distributed on the continental shelf side, especially near the estuaries largely affected by the river runoff, indicating that the annual amplitude of salinity was mainly related to the inflow of the river runoff.

4.2. Monthly Variation Characteristic of the Surface Salinity

There were 3 regions for South China Sea as the north of 20°N, between 10-20°N and south of 10°N, and similarly 3 regions for Northwest Pacific Ocean as the north of 15°N, between 5-15°N and south of 5°N. As shown in Figure 3.

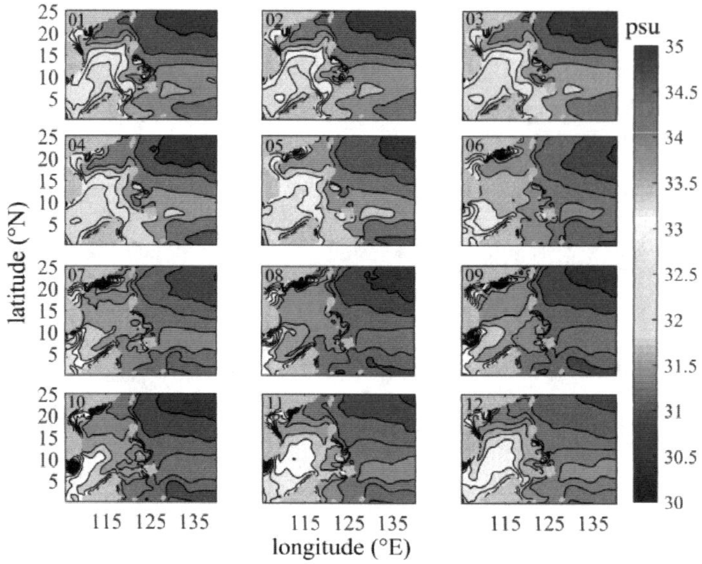

Figure 3. Monthly climatic status map of surface salinity in the study area (color bar indicates salinity range).

- In the north of 20°N for South China Sea, the isohaline had a belt and more dense distribution feature. From January to April, the salinity was constant. The salinity of Beibu Gulf was lower than that of northeastern South China Sea, and the SSS gradually decreased from May to September. The salinity near Hong Kong and Macao sea areas dropped significantly, reaching the lowest in September with 30psu and SSS rose slowly from October.
- Between 10°N and 20°N of South China Sea, SSS decreased from high latitude to low latitude. From January to May, the salinity was constant. From June to August, the SSS increased and distributed in a clumpy structure, which could extend to the Mindoro Strait with around 34psu. From September, the SSS with clumpy structure decreased and gradually presented the belt distribution features again in October.
- In the south of 10°N for South China Sea, salinity was constant from January to May and gradually increased from June to September. However, the salinity in the northern Gulf of Thailand decreased obviously and the low-salinity nuclei continued to expand outward which reached a minimum value of about 30psu in October. From then, the SSS gradually decreased and the salinity in northern Gulf of Thailand increased. It could be concluded that the inter-monthly variation of salinity in the northern Gulf of Thailand was mainly related to the runoff.

In the north of 15°N for Northwest Pacific Ocean, SSS was generally high, above 34.5psu. From January to April, the high-saltwater tongue intruded into the South China Sea from the Northwest Pacific Ocean via the Luzon Strait, reaching a maximum of 111.5°E, which was very similar to the simulation results of the predecessors [7]. In May, high-saltwater tongue disappeared and returned to the east of 121°E near the Northwest Pacific Ocean in September, during which was the involution phase period. In October, high-saltwater tongue reappeared. In November, high-saltwater tongue strongly moved to the west of the South China Sea, reaching about 113°E.

- Between 5°N and 15°N of Northwest Pacific Ocean, the isohaline presented a belt distribution feature along the latitude. From January to May, the salinity was constant and gradually increased from June to October. The salinity began to decrease in November.

- In the south of 5°N for Northwest Pacific Ocean, the isohaline extended westward presenting a belt distribution feature. From January to March, salinity remained unchanged. From April to October salinity was relatively high and then decreased slowly from November.

4.3. The Climatic Zones of Salinity Distribution

According to the climatic states of surface salinity distribution, the climatic state characteristics of monthly surface salinity distribution, and the terrain, ocean currents, and other factors in the research area, the area could be divided into five small climatic regions (Figure 4). Among them, the South China Sea and the Northwest Pacific Ocean was bounded by 121.25°E, the South China Sea was bounded by 16.75°N, and the Northwest Pacific Ocean were bounded by 4.75°N and 14.75°N. They are:
Area A (South of the South China Sea): 105.25-121.25°E, 0.25-16.75°N;
Area B (North of the South China Sea): 105.25-121.25°E, 16.75-25.25°N;
Area C (South of the Northwest Pacific Ocean): 121.25-140.25°E, 0.25-4.75°N;
Area D (Central of the Northwest Pacific Ocean): 121.25-140.25°E, 4.75-14.75°N;
Area E (North of the Northwest Pacific Ocean): 121.25-140.25°E, 14.75-25.25°N.

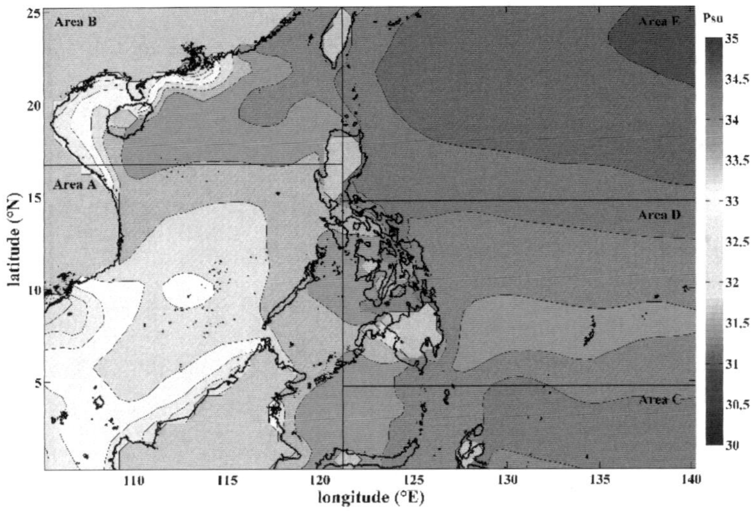

Figure 4. The climatic zones of surface salinity diagram (solid line showed zoning boundary).

From the figure 4 showed that the divisions of areas A and B mainly reflect the influence of the Northwest Pacific Ocean on the South China Sea through the Luzon Strait. Finally, area B was the affected area, and area A was basically not affected by it. The salinity of area B was higher because of the influence of the Northwest Pacific Ocean currents on Area B, and the closer to the Luzon Strait, the higher the salinity; whereas, the salinity of Area A was relatively low due to the influence of precipitation and coastal runoff.

The differences in salinity between the C, D and E areas in the Northwest Pacific Ocean mainly reflected the effects of the Northern Equatorial Current. The salinity of Area D that was the mainstream area of the Northern Equatorial Current was lower than that in areas C and E. The low-salt nucleus appears in the middle of area D and was closed.

5. Conclusion

Using SODA ocean hydrothermal reanalysis data, the climatic state characteristics of surface salinity in the South China Sea and the adjacent Northwest Pacific Ocean were analyzed in this paper. The following conclusions could be drawn:

(1) The salinity field in the study area could be divided into two areas, the South China Sea and the Northwest Pacific Ocean. The SSS in the South China Sea was significantly lower than that of the SSS

in the Northwest Pacific Ocean. The salinity showed a decreasing trend from the northeastern Northwest Pacific Ocean to the southwestern South China Sea. In the South China Sea, annual amplitude of SSS varied in different sea areas. The maximum amplitude appeared at 4.66psu in the area near Hong Kong and Macau on the continental shelf side and the minimum one appeared in the central part of the South China Sea at about 0.5psu. The SSS annual amplitude in the Northwest Pacific Ocean was smaller and even generally consistent, and the minimum value appeared in the southwest at 0.08psu.

(2) In the South China Sea, SSS low value areas were mainly distributed on the continental shelf side, Hong Kong, Maucao, Beibu Gulf and northern Gulf of Thailand, which was between 31.5-33psu with obvious periodic variation. From January to April, salinity increased. From May to October, salinity decreased. From November to December, salinity increased. And isohaline presented belt distribution features. The salinity in central South China Sea from January to May remained basically unchanged with a belt distribution feature. From June to September, salinity increased and showed a clumpy distribution feature. From October to December, salinity decreased and isohaline restored to belt distribution features. The distribution of SSS in the Northwest Pacific Ocean from low latitude to high latitude showed a “high-low-high” trend and isohaline presented belt distribution features. In the south of 4.75°N and in the north of 14.75°N, high salinity area (34.2-35psu) existed. While the area between the north of 4.75°N and the south of 14.75°N was the low salinity areas (33.95-34.4psu).

(3) According to the climatic structure of surface salinity in the study area, the area could be divided into five small climatic zones. In the South China Sea, the salinity characteristics of Kuroshio that originated in the northwestern Pacific Ocean and invaded the South China Sea through the Luzon Strait was different because of the influence of the northwest Pacific. Among them, the influence of the B zone was relatively large, resulted in the salinity in area B was higher than that in area A. In the Northwest Pacific Ocean, the salinity of the D zone that influenced by the Northern Equatorial Current was lower than that of the C zone and the E zone.

6. Reference

[1] Yiyong Luo, Lewis M. Rothstein. Response of the Pacific Ocean Circulation to Climate Change [J]. Atmosphere-Ocean, 2011, 49(3): 235-244.

[2] Xiangying Li, Dahe Qin, Cunde Xiao, etc. Research on recent progress in climate change [J]. Chinese Science Bulletin, 2011, 56(36): 3029-3040.

[3] Haihua Chen, Hongping Li, Linjie He, etc. Analysis on South China Sea surface salinity distribution and long-term trend based on SODA data set[J].Journal of Ocean Science and Technology, 2015, 34(4): 48-52.

[4] Haiyang Li, Qiang Xie, Xiaodong Wang. Analysis on inter-annual variation assimilation data of subsurface salinity in the tropical Pacific Ocean from 1980 to 1999[J]. Acta Oceanologica Sinica, 2006, 28(6): 5-11.

[5] Hua Shen. Analysis of low-frequency variability of mixed layer salinity at the western Luzon Strait [J]. Oceanographic Research, 2014, 32 (1): 11-18.

[6] Amante C, Eakins B W. ETOPO1 1 Arc-Minute Global Relief Model: Procedures, Data Sources and Analysis [J]. Psychologist, 2009, 16(3):20 - 25.

[7] Aimei Wang, Yan Du, Wei Zhuang, etc. Research on the relationship between Seasonal variation of subsurface brine in northern South China Sea and the circulation of the Northwest Pacific Ocean [J]. Journal of Tropical Oceanography, 2014, 33 (6): 1-8.

Acknowledgements

The research was supported by the National Natural Science Foundation of China (41340045), the Scientific Research Development Foundation of Shanghai Normal University (KF201824).

Yanliao Beach Nourishment Methods

Hsing-Yu Wang[1], Hui-Ming Fang[2] and Yun-Chih Chiang[3]

1 Ph. D. Candidates, National Taiwan Ocean University, Keelung, Taiwan
2 Assistant Professor, National Taiwan Ocean University, Keelung, Taiwan
3 Associate Professor, Tzu Chi University, Hualien, Taiwan
E-mail: hmfang@email.ntou.edu.tw

Abstract. Yanliao Beach, part of Taiwan's Northeast Coast National Scenic Area, is a major sightseeing and recreation spot. In recent years, the number of tourists has increased, and maintaining the beach's recreation function has become crucial. In 2007, Typhoon Krosa caused substantial beach erosion at Yanliao Beach; sand dunes collapsed, resulting in a beach scarp and endangering facilities toward the back of the beach. To protect the sand dunes and beach scarp of the post-typhoon topography, based on long-term Yanliao Beach topographic survey information and long-term tidal water level records, this study conducted an integrated artificial beach nourishment method that comprised replenishing the coastal beach and dunes. The planned areas for sand replenishment were those above the mean high water line. This beach nourishment measure protected the dunes from collapsing further and mitigated the risk of beach nourishment polluting the water in front of the beach. Topographic survey results revealed that the large amount of sand deposited in the intertidal zone resulted in the 0 m shoreline advancing toward the seaside. The width of the beach increased, indicating the success of this integrated beach nourishment method.

1. Introduction

Regardless of developing the area or constructing engineering projects under the purpose of coastal disaster prevention, human activities inadvertently influence coastal areas. Changes in coastline occur because of coastal area development involving various coastal constructions and are often attributable to the overdevelopment of coastal areas. Beach erosion, especially in sandy beaches or seaside resorts, has received public attention in Taiwan, notably, that of the receding shoreline in areas of the west and northeast coasts. Natural beach erosion has resulted in heated arguments. Moreover, aggravating greenhouse effects have resulted in extreme weather worldwide and in the increased frequency of abnormal climate. This has led to broad discussions as well as emphasis on coastal restoration and disaster prevention. Both artificial and natural factors contribute to sand erosion in coastal areas; one result is receding shorelines, which sometimes even involve their disappearance. Commonly adopted coastal protection methods include hard and soft engineering methods. Hard engineering methods mainly comprise constructing coastal structures that protect the coast. From the viewpoint of energy conservation, this measure only shifts the focus elsewhere and does not fundamentally solve the problem of coastal protection. When attempting to reestablish a new balance in the nearshore dynamic systems, other coastal protection problems may occur. By contrast, soft engineering methods mainly rely on nourishing the beach by artificially replenishing sand to increase the beach's width, thereby restoring the beach's original function: reducing and eliminating wave energy to protect the coast. Artificial beach nourishment has some strengths; it does not involve hardware construction that protects the coast but destroys the view, and it does not restrict people from visiting the beach or prevent them from engaging in water recreation or sports by the seashore.

Viewpoints on coastal protection no longer stem purely from the perspective of protective engineering. Instead, factors such as the coastal landscape and ecosystem have gradually been included in coastal engineering designs and coastal education thinking. In Taiwan, ideas on coastal protection lean toward employing ecological methods that protect the ecology and landscape. Some relevant studies are as follows. From the perspective of environment sustainability, [1] discussed domestic and foreign coastal protection methods that had the dual functions of disaster prevention and ecology promotion. They examined the design factors to be considered in coastal engineering designs and the importance of their effects. [2] studied the feasibility of employing geotextile tubes in coastal protection planning design. [3] discussed the potential problems of adopting ecological engineering methods in coastal engineering and their solutions. [4] investigated the application of an environmentally friendly coastal protection method combining headland control and beach nourishment for the Nanbin and Baibin coasts in Hualien. [5] demonstrated that a shore-term change of the beach profile was due to significant meteorological events such as typhoon passages, and the long-term evolution of the beach profile was dominated by monsoon waves. To summarize the evolution of coastal protection engineering methods, traditional sea dikes have been replaced by gentle slope-type dikes; conventional detached breakwaters have been replaced by submerged breakwaters or artificial reefs; and straight-type groins have been replaced by T-type or anchor-type groins. These features combined with methods such as artificial beach nourishment constitute the integrated coastal protection engineering methods of today, and the two-dimensional protection methods traditionally applied to coasts have been replaced by three-dimensional protection engineering methods. Fulong Beach in Yanliao, Northeast Taiwan is a major coastal recreation and sightseeing spot visited by numerous tourists every year. In recent years, Taiwanese people's demand for coastal recreation activities has been increasing, and thus, maintaining the leisure and recreation functions of Fulong Beach has gained more attention, especially the post-typhoon beach restoration problem in summer. Therefore, this study set off from the point of view of coastal disaster prevention and coastal preservation, and with the post-typhoon preservation problem as the focus, investigated Yanliao Beach nourishment mechanisms and researched beach nourishment engineering methods as references for future beach restoration projects.

2. Environmental Background Information of the Research Site

2.1. Research Site

This study's research site was the Yanliao Beach Park in the Northeast and Yilan Coast National Scenic Area, adjacent to Aoti and Fulong Beach. It is one of the most famous recreation spots along the Northeast Coast. The topographical features of the Yanliao Coastal region are characterized by the coexistence of sandy shore and reef rocks with bedrock as the seabed foundation. Because of the waves caused by the northeast monsoon and by typhoons in summer, the ocean in this region supplies plenty of sand. Sand drift deposits cover the coastal area, forming a sandy terrain. Because the sand is mainly fine quartz sand from the Shuang River and is of a light yellow color, this area is also called the Gold Coast. Figure 1 shows the research site of this study: in the north is the southern breakwater for the intake of Taiwan Power Company's Lungmen Nuclear Power Plant; in the south is a large reef at the southern end of the beach; in the east, the sea water is 3 meters deep; and in the west, is the junction of the dunes behind the beach windbreak. The research site was approximately 20 hectares in area, as indicated in green in the figure.

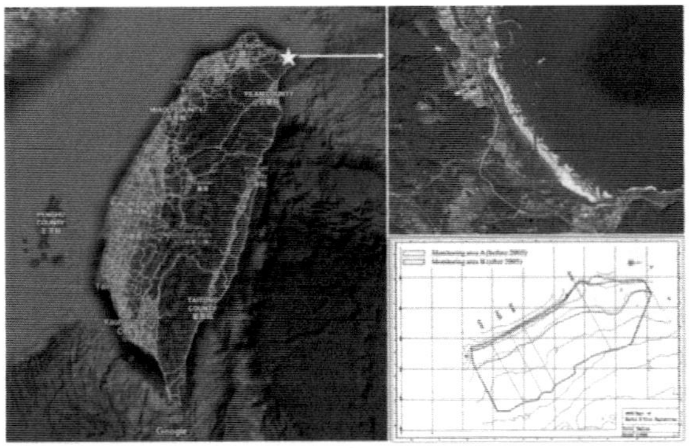

Figure 1. Schematic of relevant locations in the research site

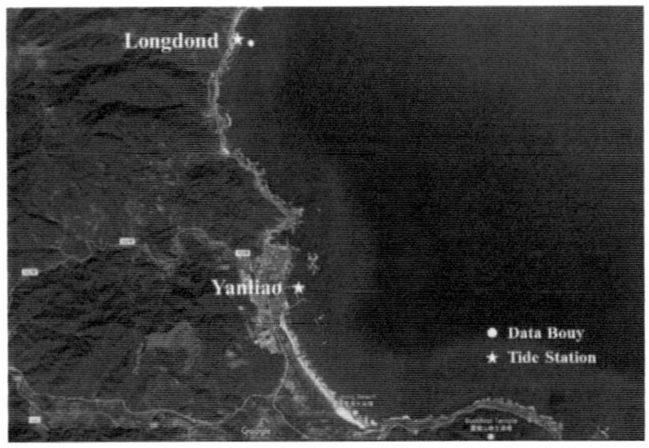

Figure 2. Relevant observation stations near the research site

2.2. Marine Weather Condition

The background information that this study collected for analysis (Fig. 2) was mainly marine weather information from the neighboring Longdong and Yanliao observation stations of the Central Weather Bureau (CWB). Relevant background information is detailed in the following section. Table 1 presents the statistical results from the CWB Longdong Tidal Observation Station and from Yanliao Station. The mean tidal level at the Longdong Tidal Observation Station was -0.01 m, and the mean high water springs was 0.34 m. The mean tidal level of Yanliao Tidal Station was -0.01 m, and the mean high water spring was 0.36 m. Table 2 presents the tidal information from the CWB Longdong Tidal buoy station. During summer (May to July) in the research site from 1998 to 2017, the mean significant wave height was $H_{1/3}$ = 0.63~0.82 m, and the mean significant wave period was $T_{1/3}$ = 5.25~5.75 s. During winter (November to January), the mean significant wave height was $H_{1/3}$ = 1.58~1.80 m, and the mean significant wave period was $T_{1/3}$ = 6.06~6.24 s. Table 3 presents the current observation results from the Yanliao sea area observed by this research team between 2008 and 2012. During spring and summer, the average flood tide velocity was between 0.13 and 0.35 m/s, and the current direction was mainly between north-northwest and north. The average ebb tide velocity was between 0.11 and 0.21 m/s, and the current direction was mainly between west-southwest and north. During autumn and winter, the average flood tide velocity was between 0.16 and 0.42 m/s, and the current direction was mainly between northwest and north. The average ebb tide velocity was between 0.13 and 0.34 m/s, and the current direction was mainly between south and north-northwest. During spring and summer, the stagnation-point flows during flood tide and ebb tide were toward the same direction, between west-southwest and north, whereas during autumn and winter, the flow directions during flood tide and ebb tide were less consistent. Them, according to the CWB database, typhoons affected Yanliao Beach 45 times from 2004 to 2016. Among typhoon occurrences, the most frequent routes were Type 2 and Type 3 with 9 typhoon incidents each, accounting for 40%, followed by Type 1, which occurred seven times (16%). These typhoon invasion routes were referenced from Hsiao et al. (2010). The results indicated that typhoons with Type 2 and Type 3 routes had more noticeable effects on erosion and deposition at Yanliao Beach. Figure 3 illustrates the typhoon routes.

Table 1. Tidal information

	Longdond (2004~2017)	Yanliao (1981~2002)
H.W.O.S.T.	0.34 m	0.36 m
M.H.W.L.	0.27 m	0.29 m
M.W.L.	-0.01 m	-0.01 m
M.L.W.L.	-0.33 m	-0.26 m
L.W.O.S.T.	-0.34 m	-0.34 m

Table 2. Wave information statistics

	1998~2000		2002~2005		2008~2017	
	$H_{1/3}$	$T_{1/3}$	$H_{1/3}$	$T_{1/3}$	$H_{1/3}$	$T_{1/3}$
Summer	0.69 m	5.74 s	0.82 m	5.75 s	0.63 m	5.25 s
Winter	1.80 m	6.24 s	1.60 m	6.11 s	1.58 m	6.06 s

Table 3. Current information statistics

		Avg. Spd.	Main Dir.			Avg. Spd.	Main Dir.
2008/10/10	Flood	0.29	N-NW	2011/07/05	Flood	0.27	NNW
	Ebb	0.13	WNW-NNW		Ebb	0.11	NNE
2009/07/06	Flood	0.35	NNW	2011/11/26	Flood	0.42	NNE
	Ebb	0.11	NNW-SSW		Ebb	0.15	S
2009/12/07	Flood	0.16	NNW-S	2012/09/01	Flood	0.3	NNW
2010/07/15	Flood	0.33	N		Ebb	0.13	NNW
	Ebb	0.21	N	2012/11/21	Flood	0.3	NNW
2010/11/30	Flood	0.31	NNW		Ebb	0.13	NNW
	Ebb	0.25	NNW				

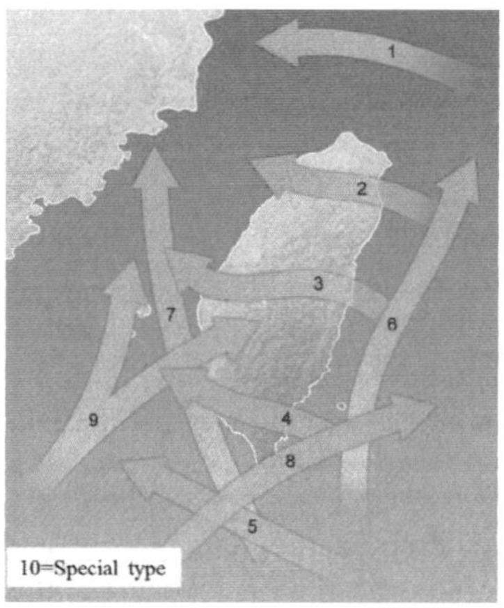

Figure 3. Typhoon route classifications (Central Weather Bureau: Typhoon Database)

3. Topographic Surveying Method

3.1. Beach Topographic Surveying Method

This study conducted a beach topographic survey in the research site that was divided into topographic surveying on land and at sea. A total station laser light theodolite—Leica TC703—was used for land mapping. The coordinates and elevations of the control points were first measured before conducting a beach topographic survey, the results of which are presented in Fig. 4. The mean-section method was employed to calculate the erosion and deposition volumes, with the goal of assessing changes to the beach. For the topographic survey of the sea, remote control vehicles that carried echo sounders were mainly used to map the depth of the topography. This surveying system comprised three parts: a survey vehicle, a total station theodolite, and a navigation integration operation unit. Fig. 5 illustrates the operating principles of the surveying vehicle.

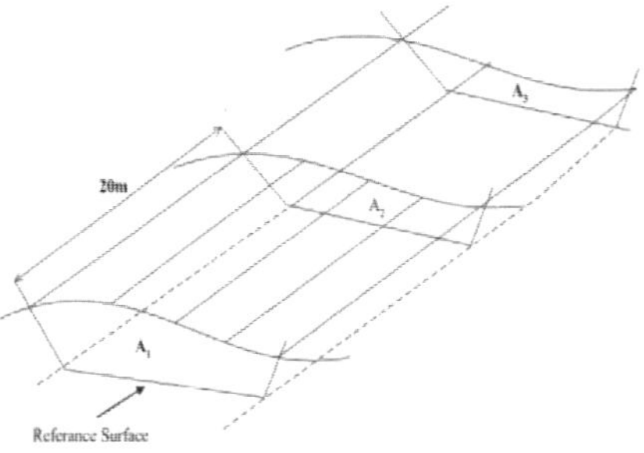

Figure 4. Schematic of earth volume calculation

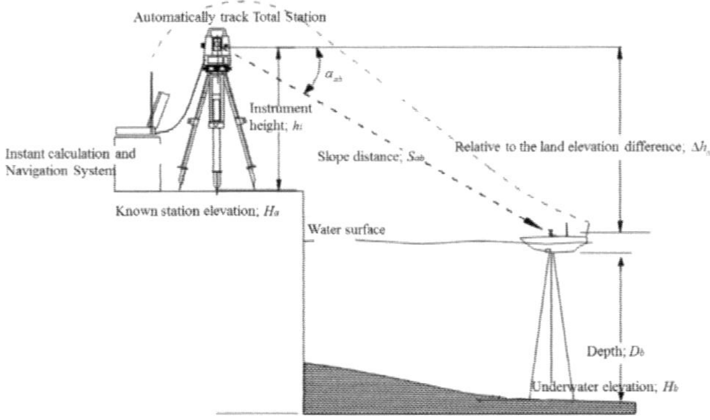

Figure 5. Schematic illustrating the surveying vehicle's operating principles

To record correlated time factors, the surveying vehicle's surveying result data comprised two types: sounding and plane positioning. Subsequently, the two types of data were integrated to become three-dimensional survey point data. The formulas for underwater elevation and position data are as follows:

$$H_b = h_i + H_a + \Delta h_{ab} - D_b - d_p \quad (1)$$

The vehicle's plane position calculation formula is as follows:
Coordinates in the N_b direction:

$$N_b = N_a + \Delta N_{ab};\quad \Delta N_{ab} = \left(S_{ab} \cdot \sin\alpha_{ab}\right) \cdot \cos\phi_{ab} \quad (2)$$

Coordinates in the E_b direction:

$$E_b = E_a + \Delta E_{ab};\quad \Delta E_{ab} = \left(S_{ab} \cdot \sin\alpha_{ab}\right) \cdot \sin\phi_{ab} \quad (3)$$

where H_a is the station elevation, h_i is the height of the total station, H_b is the underwater elevation, $\Delta h_{ab} = S_{ab} \cdot \sin\alpha_{ab}$ is the difference of elevation between the prism on the remote control vehicle and the total station on the ground, α_{ab} is the vertical angle (α_{ab} is the angle of depression; here its negative value is taken), D_b is the depth that the vehicle measured, d_p is the distance between the prism and the transducer, ϕ_{ab} is the azimuth angle (the north is set as 0°), and S_{ab} is the slant distance.

3.2. Discussion on the Reasons for Changes in Topography

From May 2004 to March 2017, the research team surveyed changes in topography. The survey results indicated that in the surveyed coastal area, the areas with relatively severe changes were concentrated near the cross sections in Fig. 6: STA:0+100, STA:0+180, STA:0+240, and the Little Stage STA:0+500. Fig. 7, the monitoring results revealed that as the northeast monsoon weakened, from March, sand was gradually deposited on the Yanliao Beach terrain. By contrast, abnormal marine weather such as typhoons in summer and autumn are short-term variation factors on the monitored beach topography, causing erosion. After typhoons hit and before the northeast monsoon arrived from the north, the beach topography again responded to the summer climate conditions and gradually sand was deposited. In winter, the northeast monsoon is strong. The beach topography in the monitoring area reflected this winter climate condition and was eroded. Later, as the northeast monsoon weakened, beginning the following March, the Yanliao sand beach survey area resumed reacting to the seasonal climate conditions and sand deposition gradually increased. The beach topography indicated the trend of a dynamic balance. This result is supported by the sand erosion and deposition change results in Table 4.

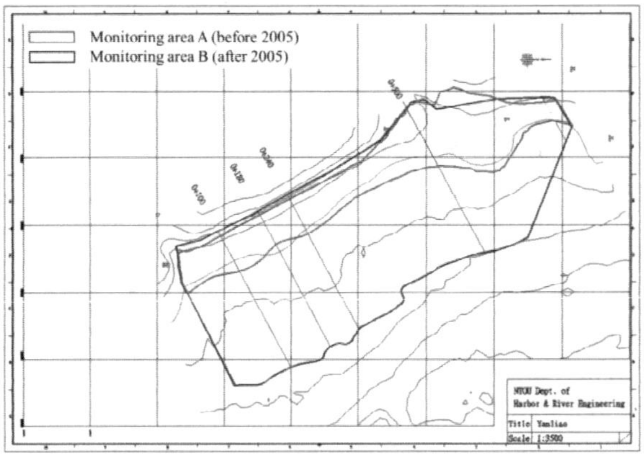

Figure 6. Schematics of the cross sections in the survey area (on site observation)

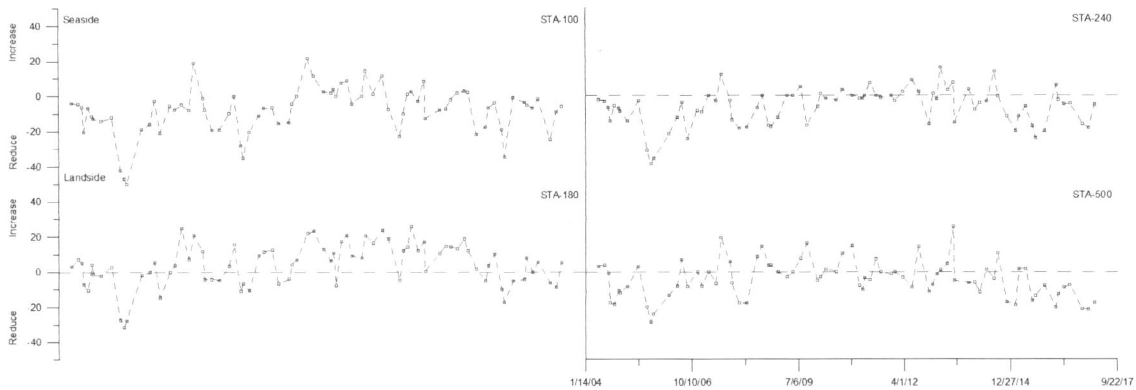

Figure 7. The contour line of each cross section's change over time (on site observation)

Table 4. Statistical information of the current

Year	Season impact	Typhoon impact	Year	Season impact	Typhoon impact
2004	0	-26,096	2011	+12,346	-707
2005[a]	+9,800	-53,582	2012	+4,489	-8,217
2006	+52,981	-4,240	2013	+6,589	-27,077
2007	+9,735	-19,181	2014	+10,491	-7,742
2008[b,c]	+28,357	-15,261	2015	+8,937	-30,715
2009	+20,102	-16,928	2016	+6,875	-30,578
2010	+21,525	-18,613	2017	+31,065	-

[a] 2005 input 25,000 m^3
[b,c] 2008/5/15 input 25,000m^3, 7/15 input 10,000m^3

4. General Discussion

4.1. Coastal Protection after Typhoon Krosa

After Typhoon Krosa hit on October 4, 2007, the width of Yanliao Beach was markedly reduced, and the dunes in the back beach collapsed to form beach scarp. Figures 8 depict Yanliao Beach's topography after Typhoon Krosa. After considering the strengths and weaknesses of hard and soft coastal protection engineering methods, this study determined that artificial beach nourishment would be effective in creating a coast with functions similar to a natural coast in terms of dissipating tidal energy, restoring the beach, and increasing the width of the beach. Moreover, the functions of beach recreation and disaster prevention could be achieved, and balance with the natural ecosystems maintained. To maintain the width of the beach, prevent the back beach from further collapsing and retreating, and satisfy the Yanliao Beach Park's recreational requirements, this beach required fast restoration through adopting soft coastal protection methods.

(a) (b) (c)

Figure 8. Yanliao Beach's topography after Typhoon Krosa, (a) topography near the north side of the reef area (b) topography in the central survey area (c) Topography near the south side of the large reef.

4.2. Planning and Implementing the Beach Nourishment Plan

After organizing and analyzing the aforementioned background information, this study reference the [6] adopted a dune nourishment method (depicted in Fig. 9). This method placed all of the beach nourishment sand along the area of the beach marked for nourishing, forming a continuous dune along the land side of the area. This beach nourishment method was adopted to resist the erosion caused by extreme waves and storm surges, and ensure the safety of buildings behind the coastal beach. Sand from dredging conducted elsewhere and leftover earth from construction sites can be placed near the beach in the form of dunes for maintenance or reconstruction after storms affect the beach. When storms occur, the dunes that serve as blocking facilities will be eroded; however, the dunes also serve to block the waves, thereby reducing costs incurred by assembling workers for emergency repairs. Therefore, some beach nourishment plans use dredged sand to form dunes, thereby reduce the need for emergency repairs when storms erode the coast. When the weather is dry, the dunes may be blown into the wind and cause air pollution, causing negative effects on nearby residences; thus, fences should be erected and vegetation should be planted to enhance the stability of the dunes.

Past topographic survey results revealed that the changes in topography of Yanliao Beach reflect seasonal climate factors and a dynamic balance. However, Typhoon Krosa caused severe erosion and the beach lost a large amount of sand. To prevent the back beach dunes from further collapsing and retreating, artificial beach nourishment through sand replenishment was conducted. According to the topographic change assessment results after Typhoon Krosa, adopting beach dune nourishment could form a continuous dune on the landward side, relying on natural forces to spread out the sand on the beach. However, natural forces would be slow in carrying the sand to form a natural beach, and be unable to immediately form a platform. By contrast, employing the coastal beach nourishment plan could rapidly restore the platform and create a beach for recreation. Therefore, considering the premise of protecting the post-typhoon beach scarp on the back beach as well as maintaining both the beach's recreation function and intertidal water quality, the artificial beach nourishment plan should combine beach dune and coastal beach nourishment. Specifically, this would involve beach nourishment sand being spread across the beach above the mean high water line for beach restoration. Comparing the topographic change observation results between May 2004 and late March 2017, reflected in the forms of the 0 m contour line of the beach cross sections STA:0+100, STA:0+180, STA:0+240, and STA:0+500, revealed that areas above the mean high water line where sand replenishment was conducted for beach nourishment in 2008 were not affected by the tidal water's constant impact. Waves brought by Typhoon Kalmaegi compacted the sand placed above the mean high water line, resulting in an extension of the 0 m shoreline toward the sea. This result satisfied the original beach nourishment plan and proved the effectiveness of this artificial beach nourishment project.

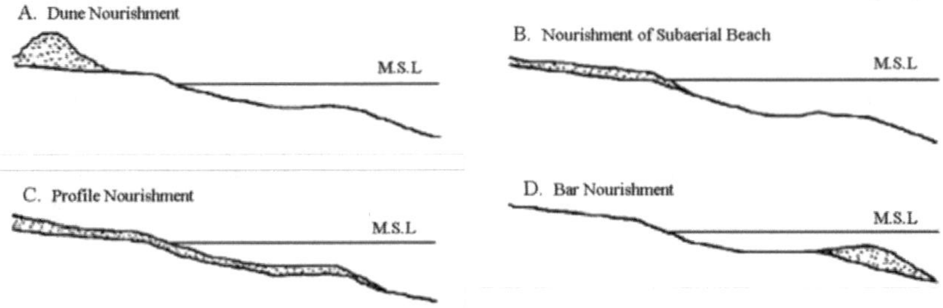

Figure 9. Schematic of artificial beach nourishment (NRC, 1995)

5. Conclusion

This study investigated coastal protection problems after Typhoon Krosa and their solutions. Based on the aforementioned discussion, the following conclusions were reached:

- Past observation results revealed that the survey area's beach topography reflected seasonal marine weather conditions and presented dynamic changes in erosion and position. Artificial beach nourishment through sand replenishment is recommended for maintaining beach width only if the beach has been affected by typhoons that resulted in large-scale changes to the beach topography.
- In 2007, Typhoon Krosa caused severe erosion to the dunes on the back beach of Yanliao Beach. The dunes collapsed and formed beach scarps. Beach nourishment plans for both coastal beach and dunes were adopted; the plans involved replenishing sand in areas above the mean high water line. They proved to be effective in protecting the dunes on the beach and maintaining the beach's width.
- Combining coastal beach and dune nourishment methods could protect beach and dune terrain. In addition, this combination could prevent coastal beach nourishment methods from degrading foreshore water quality. The results may serve as a reference for future beach nourishment plans.

6. References

[1] Yi-Yu Kuo, Hsien-Kuo Chang, Jin-Cheng Liou, 2002, Coastal Eco-Technology, *Coastal New Methods and Ocean Regeneration Conference*, pp. 34-54. (in Chinese)

[2] Chang-Tai Tsai, Tai-Wen Hsu, Jyh-Cherng Ju, Chun-Hung Weng, 2002, Introduction and application of Geotextile Tube, *Coastal New Methods and Ocean Regeneration Conference*, pp. 65-85. (in Chinese)

[3] John Rong-chung Hsu, 2002, Headland control with Nourishment, *Coastal New Methods and Ocean Regeneration Conference*, pp. 86-111. (in Chinese)

[4] Hsia-lan Chiou, 2004, *Application of Headland Control with Nourishment on a Hualien Coast*, National Sun Yat-sen University Master Thesis.

[5] Hui-Ming Fang, Sung-Shan Hsiao, Cheng-Han Tsai, Hung-Pin Tseng, 2008, Investigation on the Results of Topographic Monitor at Yan-Liao Beach, *Proceedings of the 30th Ocean Engineering Conference*, pp. 499-504.

[6] National Research Council, 1995, Beach Nourishment and Protection *National Academy Press*, Washington, DC.

IOP Conf. Series: Earth and Environmental Science **171** (2018) 012030 doi:10.1088/1755-1315/171/1/012030

Influence of Economic Structure Optimization on Water Consumption in Shanxi, China

Wei Li[1,*], Jiahong Liu[1] and Xifeng Wang[2]

1 China Institute of Water Resources and Hydropower Research, Beijing, China
2 Institute of Quantitative & Technical Economy, Chinese Academy of Social Sciences, Beijing, China
E-mail: liwei2769@qq.com

Abstract. Shanxi province of China is threatened by the water shortage crisis. It is an important means to solve the water shortage problem by rationally water utilization planning and controlling. Using the structural decomposition analysis based on input-output table, the driving factors and driving forces of water consumption in Shanxi is researched, especially the effect of economic structure optimization. On the basis of the input-output table of Shanxi in 2002, 2007 and 2012, a comparable price water input-output model was established in combination with the water use data. According to the study, the improvement of water use efficiency has been the main driving force for reducing water consumption. The change in gross final demand has always been the main driving force for increasing water consumption. The adjustment of manufacturing industry structure, the secondary industry structure, the three industrial structure and the final demand structure have different impacts on water consumption, which has been related to the way of development. On the whole, the optimization of economic structure is changed from water reduction effect to increasing effect.

1. Introduction

Water crisis not only affects water utilization of the society, but also affects national security and development. In particular, the development of cities in northern China has been generally influenced by the shortage of water resources. The development and utilization of regional water resources exceed the carrying capacity. Water for ecological is seriously occupied by the cities for production and living, which bring more and more rivers being dry and polluted. Behind the prevailing water crisis, it is essentially that the way of economic development is not compatible with the capacity of regional water resources.

The economics of Shanxi province base on energy industry. And its development has exceeded the carrying capacity of water resources. According to the *water resources bulletin* of the Shanxi Province in 2015, the total amount of water used in the province is About 7.55 million cubic meters. The annual average water resource in the local area is 12.38 million cubic meters, the 61% of which has been used by the development of society.

In order to solve the problem of regional water shortage, Shanxi has constantly adjusted its economic structure. The water used for agriculture and industry has declined continuously, and the living water has continued to rise. However, Shanxi still faces problems such as overloading of groundwater, serious water pollution and so on. This paper focuses on the impact of the economic structure optimization on the water consumption, and the results will provide reference for the regulation and control of water resources utilization in Shanxi province.

2. Methodology

Input-output analysis is a research method based on systematicness and structure. It fully depicts the relationship between departments, technology and demand. And it is widely applied in the fields of resources research and environment economics. For example, **Li Haiying** [1], and **Zhang Youguo** [2] studied the impact of economic development on China's carbon emission intensity. Structural decomposition analysis based on input-output model is an important way to study the interrelations of resources and environment in economic system. Depending on structure decomposition analysis, we can decompose the economic driving forces behind the utilization of water resources. Using then water comparable input-output model of Shanxi province in 1997~2012 and structural decomposition analysis, the amount of water consumption is related to the economic system.

2.1. Comparable Price Input-Output Model

The basic input-output tables used in this study are from the website of Shanxi Municipal Bureau, including 3 tables in 2002, 2007 and 2012. On the basis of the basic forms, the departments with similarity have been merged. The service industry is divided into three sectors: productive service industry, consumer service industry and public service industry. The model includes 26 departments, and the department code and assortment are shown in Table 1.

In the calculation of comparable prices, 1997 is taken as the base period. Using agricultural products producer price index, industrial producer price index, fixed asset investment and retail price index, the value of agriculture, manufacture (except construction), construction industry, service industry is deflated in 2002, 2007 and 2012. The comparable input-output table of 26 sector is obtained.

Table 1. the department of water input-output model

	Department	Assortment		Department	Assortment
1	Agriculture	Agriculture	14	Manufacture of metal smelting and rolling	Manufacture
2	Coal mining	Mining	15	Manufacture of metal products	Manufacture
3	Oil and gas mining	Mining	16	Manufacture of general and special equipment	Manufacture
4	Metal mining and selecting	Mining	17	Manufacture of transport equipment	Manufacture
5	Non-metallic minerals and other mining	Mining	18	Manufacture of electrical machinery and equipment	Manufacture
6	Manufacture of food products and tobacco products	Manufacture	19	Manufacture of communications equipment, computers and other electronic equipment	Manufacture
7	Manufacture of textile	Manufacture	20	Manufacture of instrumentation and office machinery products	Manufacture
8	Manufacture of wearing apparel and leather products	Manufacture	21	Other manufacturing	Manufacture
9	Manufacture of wood products and furniture	Manufacture	22	Electricity, gas, water and steam supply	Manufacture
10	Manufacture of paper products and printing, education, sporting products	Manufacture	23	Construction	Manufacture
11	Manufacture of petroleum, coking products and nuclear fuel products	Manufacture	24	Productive service	Service
12	Manufacture of chemical products	Manufacture	25	Consumer service	Service

13	Manufacture of non-metallic mineral products	Manufacture	26	Public service	Service

2.2. *Water Input-Output Model*

According to the water resources bulletin of Shanxi, the use of water includes agricultural water, industrial water and domestic water. According to **Li Wei**'s study [3], total water consumption in divided into water use of 26 departments. In this way, the water input-output model including 26 departments was obtained. For the economic structure development has influence mainly on industrial water, the domestic water and environmental water is not considered. The comparable water input-output model is shown in Table 2.

Table 2. Water input-output table

	Intermediate input	Finally demand	Total output
Department	A^iX	Y^i	$X = (I - A)^{-1}Y = LY$
Added value	V		
Total input	X^T		
Water use	W_1	W_2	

2.3. *Structure Decomposition Analysis*

The basic formula of input-output model is as following.

$$X = (I - A)^{-1}Y = LY \qquad (1)$$

X is the total output; A is technology coefficient matrix; Y is the final demand (living consumption, government consumption, fixed assets and inventory, export). L indicates Leon Leontief inverse matrix, which reflect department consumption of others.

The value of water consumption of the society could be as following.

$$W = C^iX = C^iLY \qquad (2)$$

W is water consumption of industry; C^i indicates the input intensity of water resources in industry i (industry water consumption coefficient).

In order to study the influence of economic structure optimization, the structure of economic mode is represented by the final demand structure matrix.

$$Y = MNOSG \qquad (3)$$

M is the structure matrix of manufacture industry measured by final demand; N is the structure matrix of the secondary industry measured by final demand; O is the structure matrix of three industrial structure measured by final demand; S to reflect the final demand structure matrix (structure of consumption, fixed assets and inventory, export); G represents net gross final demand.

$$W = C^iLMNOSG \qquad (4)$$

So the changes of W can be expressed as:

$$\Delta W = W^1 - W^0 = C^1L^1M^1N^1O^1S^1G^1 - C^0L^0M^0N^0O^0S^0G^0 \qquad (5)$$

0 represents parameter in basic year, and 1 represents parameter in contract year.

According to the research of **Dietzenbacher E** and **Los B** [4], the decomposition of formula (5) is not unique, and the number of the results is related to the number of factors n, which is n!. The decomposition of formula (5) is 5040. Research of **Fujimagari D** [5] and **Betts J R** [6] indicated that the result of the bipolar decomposition is very close to other results. The bipolar decomposition method is used to calculate the influence to W.

3. Results

3.1. Results Analysis
The results are listed in Table 3, which indicate the influence to water consumption by the factor with the other factors fixed. C represents the technological progress of water use efficiency. The water use was reduced by 2.08 billion cubic meters between 2007 and 2002 because of C, which had illustrated the increasing output of unit water input.
L represents the technological progress based on the intermediate input. From the result, L has decreased the water consumption by 0.84 billion cubic meters between 2012 and 2007, which means that the progress of technology is conducive to saving water.
M is the influence on water consumption by the structure change of manufacturing industry. In 2007-2002, water use decreased 0.09 billion cubic meters coursed by M. In 2012-2007, it increased 2.29 billion cubic meters. The changes of manufacture structure is not conducive to the water resources utilization.
N is the structural change of secondary industry, namely, the structural change of mining industry, manufacture, construction, electricity, gas, water and steam supply and the structural change in the third industries. It is changing towards the direction of reducing water use. O is the structural changes of the three industries. O drive the water consumption increased by 42.72 billion cubic meters in 2012-2007.
S represents the impact of final demand structure changes (structure of consumption, fixed assets and inventory, export) for water consumption, which forced water use increased by 0.95 billion cubic meters in 2007-2002. G is the effect of gross final demand changes. The increase of gross final demand made water consumption increased by 5.83 and 3.84 billion cubic meters.

Table 3. The influence to water consumption in Shanxi (billion cubic meters)

	C	L	M	N	O	S	G	ΔW
2007-2002	-2.07	-0.31	-0.09	0.03	-4.05	0.95	5.83	0.29
2012-2007	-3.17	-0.84	2.29	-23.45	42.72	-20.38	3.84	1.02

3.2. Influence of Economic Structure Optimization
M, N, O, and S represent the economic structure, and the structure optimization is the transformation of development way in Shanxi. The most influential factor in water use is O (the three industrial structure). In 2007-2002, the water consumption decreased by 4.05 billion cubic meters forced by O, and in 2012-2007 years, it increased by 42.72 billion cubic meters.

4. Conclusion
The structural decomposition analysis based on input-output model is applied to study the effect of economic structure optimization on water consumption, taking Shanxi province of China as an example. On the basic input-output table of Shanxi province in 2002, 2007 and 2012, the water consumption of 26 departments is calculated using data from *water resource bulletin* of Shanxi province. The input-output tables has been deflated by production price while 1997 as the base year. The water resource comparable price input-output model is formed.
The results show that the improvement of water use efficiency and technological progress have been an important factor in the reduction of water consumption. And its driving force is increasing. The increase of gross final demand was the main driving force for increasing water consumption, while the driving force had shown a decreasing trend. The influence of M (manufacture industry structure), N (the secondary industry structure), O (the three industrial structure), and S (the internal structure of final demand) to the water consumption various. In general, the optimization of economic structure is changed from water reducing effect to increasing water effect, which indicates that it is not conducive to the ease of water use problem.

5. Reference

[1] Li Haiying 2014 A study on carbon dioxide intensity in China based on comparable price IO-SDA model *Southwestern university of finance and economic*, 4
[2] Zhang Youguo 2010 Economic development pattern change impact on China's carbon intensity *Economic Research Journal*. 4,120-133
[3] Li Wei, Liu Jiahong, Jia Yangwen, et al. 2016 Attribution analysis of economic driving factors of social water cycle evolution *Journal of China institute water resource and hydropower research*, 14.
[4] Dietzenbacher, E. & Los, B. 2000 Structural decomposition analyses with dependent determinants *Economic Systems Research*, 12.
[5] Fujimagari, D. 1989 The Sources of Change in the Canadian Industry Output *Economic Systems Research*, 1.
[6] Betts, J. R., 1989 Two Exact , Non-arbitrary and General Methods of Decomposing Temporal Change *Economic Letter* , 30

Acknowledgments
This work was financially supported by the Chinese National Natural Science Foundation (No. 51279208, No. 51522907) and the Research Fund of the State Key Laboratory of Simulation and Regulation of Water Cycle in River Basin, China Institute of Water Resources and Hydropower Research (No. 2016ZY02, No. iwhr-skl-201301).

Aquifer/Water Sources Drip Factor as Alternative to Measure Climatic Effects on Disponibility of Water Agricultural Irrigation in Semi-arid/Arid Zones

Norzagaray-Campos Mariano[1], Llanes-Cárdenas Omar[1*] Muñoz-Sevilla Patricia[2], Ruiz-Guerrero Rosario[3], Ladrón de Guevara-Torres María[4] and Armenta-Cabrera José[1]**

1 Instituto Politécnico Nacional-CIIDIR-Sinaloa,
2 Instituto Politécnico Nacional-CIIEMAD
3 Instituto Politécnico Nacional-CIITEC-México and
4 Instituto Politécnico Nacional-CIIDIR-Oaxaca.
*E-mail: oma_llanes@hotmail.com
**Masters student from Instituto Politécnico Nacional-CIIDIR-Sinaloa

Abstract. On semi-arid/arid regions (SA/A), new climatic conditions transformed irrigation water on scarce and to obtain efficiency on its use (Eu) is necessary it dosage (D). Under Crops quality and the aim to avoid losses by drainage, evaporation and runoff is necessary implement alternatives front to risks and encountered mechanisms for counteract H2Oi lack and promote Eu and D. Any aquifer to plan, requires immediately know the volumetric exchange (ΔV(x,y,t)) between regional aquifer and sources hydric. Therefore, the objective was determine the spatial variation of drip factor (Fg) between aquifer/water tributaries and take it as indicator of ΔV(x,y,t) behaviour in SA/A coastal agricultural valley of Northwest of Mexico known as " Heart Agricultural of Mexico". Fg results respond at linear expression y = 26.70 x - 44.15, an annual range average of 413-1813 m/year governed by porous media and recharge dynamics closely linked to transitivity anisotropy low to medium (R^2= -0.6) and included on range 20-70 m2/day. In few occasions the river and streams yield water to aquifer (rainy seasons), and on dry season; only the aquifer gives large amounts to all region tributaries. The low sensitivity response at dripping and climatic presence changes constitutes a high risk for H2Oi and sustainable agriculture due to continuous downward trend of watertable levels that encourage an overexploitation responding at lack water and the occurrence of big agricultural economic losses caused by the last intense droughts continually confronted this region. Knowing ΔV(x,y,t) balance helps at control water losses, at sustainable agriculture and contributes to achieving goals of Eu and D; adapted to new agricultural uses forced by climate change.

1. Introduction

Water is an essential natural resource that in the presence of climate change in some regions of the world increasingly turns out to be a scarce resource. In semi-arid/arid regions (*SA/A*), irrigation and drainage projects handle large amounts of *H2Oi* to promote agricultural production. Depending of the water source (surface or groundwater) is used different types of irrigation that depend on the forms of storage, of the transport and distribution systems and of the delivery methods of application in the field [1]. For a long time, surface water has been used for irrigation (mainly that of rivers), and in some countries, for thousands of years; it still constitutes one of the main investments of the public sector [2], but the risk always is latent because does not mean that they cannot depend on the recharge of the local aquifer. On the other hand, the population growth urban and rural populations requires a higher consumption of *H2O* of quality and the current limitations of availability in terms of *H2O* generate peculiar forms of supply and consumption that prevail in the economy and forecast of the entire culture aimed at the economy of *H2O*. In the current society, there are more and more problems related to the economy in the supply networks, distribution, and especially in the habits of irrigation and consumption. Undoubtedly, given the new conditions imposed by climate change, *H2OR* and *H2O* for human consumption has become a scarce commodity. In this sense, climatic alterations that also affect the hydrological cycle, demand in semi-arid/arid regions (*SA/A*) a relevance to achieve the efficiency in its use (*Eu*) and it is a directly function of the manner that occur the dosage activity (*D*) of *H2O*. Under this scenario; is forces society to seek alternatives to avoid losses by drainage, evaporation and runoff by implementing new alternatives in the ways of exercising agricultural risks (drip, spraying, hydroponics, among others). Notwithstanding the activities carried out to adapt to current climate change, in most regions of the world there is still a clear need for further studies that determine factors that can positively or negatively modify the current degree of overexploitation of a regional basin aquifers, and climate change is one of the aspects that can influence the magnitude of these factors, therefore, in many modifications environmental [3]. In fact, by the year 2027 it is required that all hydrological planning must consider the effects of climate change for the establishment of the water resources/demand balance. Regardless of the 95% of irrigation projects in the world the main method of delivery of *H2O* is superficial and only 5% comes from aquifers, in both cases the water is distributed by gravity in the area that will be irrigated, and depending of season year; in complicated areas as *SA/A*, is necessary all efforts to achieve *Eu* and *D*; despite the fact that activities always face complex variables tending to cause the scarcity of *H2Oi*. Considering that in any area to plan when began the activities to reach *Eu* and *D*, is necessary to know immediately the spatio-temporal volumetric exchange ($\Delta V(x,y,t)$) between the regional aquifer (common place of *H2Oi* extraction) and local tributaries; the objective of this study was to determine the drip aquifer/water sources factor (*Fg*) and take it as indicator of the spatial variation $\Delta V(x,y,t)$ that governs the groundwater transport in porous medium (*TIMP*). To conduct at this proposed objective, the databases of previous work in the study area were used. With their respective methodologies the hydraulic conductivity (*Kz*, LT^{-1}), the thickness of the saturated zone (*m*, L), the hydraulic load of the aquifer (*Hs*, L) and the vertical variation (*h*, L) of the sources with possible water carrying potential to the aquifer (in this case, river and streams) were measured. The *Fg* magnitude was controlled under the criteria of Konikow & Bredehoeft established in 1987 in which the influence of water bodies intervenes in the aquifer and the changes in *h* also intervene with *Hs*, *Kz* and *m* of the same aquifer [4]. The foregoing will allow to establish and provide a monitoring mechanism in this water system, and in the face of the new climatic conditions, contributes with one of the activities for reach *Eu* and *D* and avoid overexploitation of the aquifer of this valley which is one of the most important of Mexico in terms of agricultural production.

2. Experiment

Spatial variation of $\Delta V(x,y,t)$ represented for *Fg* under Konikow & Bredehoeft criteria was determinate with expression $F_g(x,t) = \frac{Kz}{m}(Hs - h)$; where *Kz* was the hydraulic conductivity on aquifer (LT^{-1}); *m* thickness of the saturated zone (L) and *Hs* the hydraulic Head of water sources (L) that in this case river and streams were considered as the principal sources with possible potential to carrier water to the aquifer (and vice versa); *h* was the variation of vertical level water (L).
With the aim to calculate the depth of geohydrological basement (*b*), was used the data of previous geoelectric and applied the Vertical Electrical Sounds (*SEV*) principles) [5]. To obtain the variables used on *Fg* expression were considered the next conditions: (1) the spatial variation of *Kz* was the principal parameter to determine *Fg*, (2) the average hydraulic transitivity (*T*) in the aquifer is a product of *Kz* and *m* (*T* = *Kz.m*) and (3) the spatial average *Kz* value could be obtained as

approximation through the *Kx* and *Ky* behaviours. The *T* magnitude was knower using 72 pumping tests that were performed by "*Theis simplified test*" in different wells within the study area. *Hs* was calculated by subtraction of *Nb* and *Nf*, this activity was completed once the topography of the area was estimated, for which the topographic survey closest to the study area was used (18-24 November of 2017). For the average annual spatial variation of *h* in the river and streams, eight piezometers were placed in different places: 6 in the bed of the Sinaloa and/or Petatlán River, 1 on De Ocoroni stream and 1 on De Cabrera stream (Figure 1).

In order to obtain the real average *Fg* magnitude, was considered that a filtration and water transport process varies respect to precipitation, and in alluvial aquifers; it is usually lower during the dry season and higher in the wetter periods, but if it is combined with the absence of waste management plans and land use plans (which apparently do not exist in region); it is difficult to fight water infiltration to aquifer systems and only enters the *Ne* by areas of favourable grains that form a high porosity being this directly promotional to the volumes captured by the aquifer during each season and accord to the dynamics of TIMP aquifer/water sources.

Therefore, this scenery similar on the study area, occur due to fact there are two season environmental, one where include the rainiest periods from July to September [6], and a large period where no there are any rain Under this conditions, exist in the zone annual changes for water collection governed by four underground scenarios

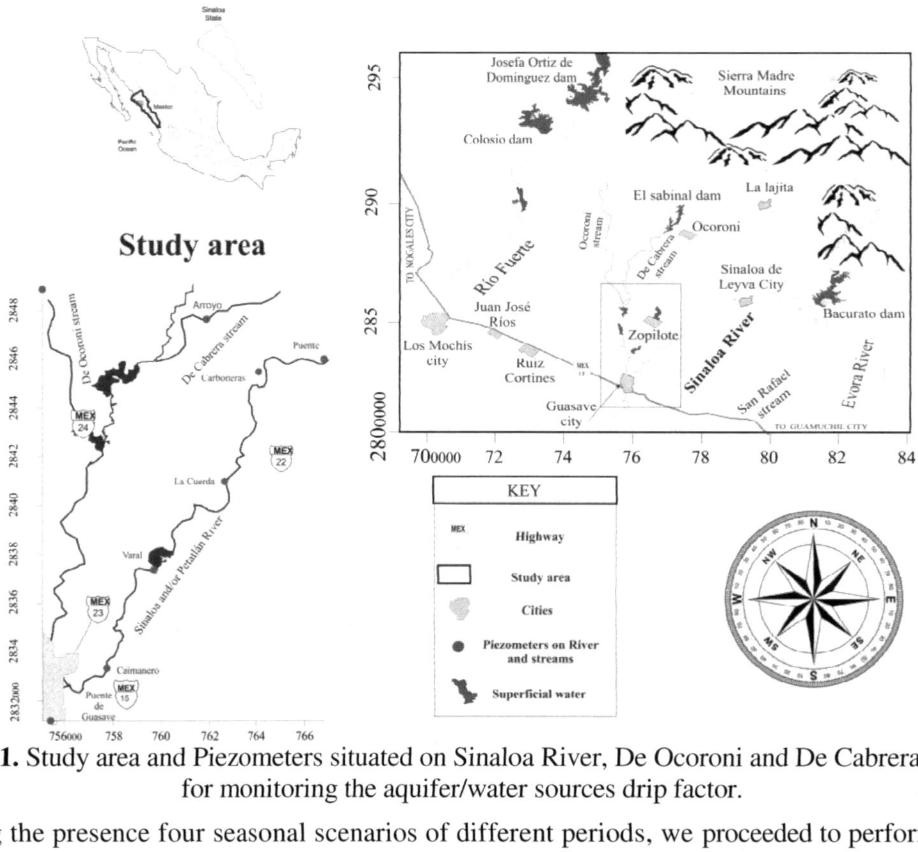

Figure 1. Study area and Piezometers situated on Sinaloa River, De Ocoroni and De Cabrera streams for monitoring the aquifer/water sources drip factor.

Knowing the presence four seasonal scenarios of different periods, we proceeded to perform lectures of h magnitude in periodically form every 8 days for one year (including the rainy and dry season) and so proceed to perform the cabinet work. The places where field measurements were made were positioned using UTM/12 geographic coordinates with a Garmin Olathe GPS of 14 channels. In the SURFER 10.0 program using Kriging method [7], an interpolation of all the measured variables was obtained. The interpolations were stacked on the local urban trace with the aim to obtain the maps that illustrated the annual average spatial variation of each variable that intervened in the determination of *Fg* magnitude. The details of the figures were made in the *Corel DRAW suite X7.0* program and the statistical analysis of the annual average of the measured and calculated results was made using the STATISTICA 7.0 computer program. For the relationship$\left[F_g \; vs. \; \Delta V(x,y,t)\right]$ range variation was established for illustrated the risk zones of the system in the face of the greater to less volumetric ware exchange between aquifer/water sources.

3. Results and discussions

The interpretation of the SEVS defined the presence of a free aquifer with varying depths of 3 to 15 m. Also was present other aquifer confined of irregular geometry located at 40 m depth. The aquifers are divided by a layer of clay. Regardless of the presence of *Fg* that originates in the superficial scenario of aquifer system, and that travels to the water table, it is inferred that exist the presence of an internal drip factor (*Fg'*) between the two aquifers that expands throughout the region; that although it is not the objective of this work if it can be mentioned since it is part of the processes that encompass the superficial *Fg* characterized by the criteria of Konikow & Bredehoeft (1984). The *Fg'* occurrence is function directly and proportional to the genesis of deposition at the beginning formation of the valley, and the SEV sounds; defined the facies properties of both aquifers, however the presence of the clays that divides both aquifers; auspicious the attenuation of the direct current of exploration injected into the porous medium; so that the SEVs opening at *AB/2*=150 m, only reach an average depth of exploration did not go beyond 80 m. However, if is considered the average annual depth of the 8.5 m in *Ne*, the depth of penetration reached by the SEV was useful for the objectives of this work. This was due to the fact that the dynamics of the system aquifer/river /stream was developed in a scenario no more beyond 20 m deep and it includes mostly the free-form aquifer from 3 to 15 m deep. It is important to mention that until now there are no works to determine the drip factor that exists in the Valley, so the results are the first dates that show that there is a volumetric exchange of water between the aquifer of the Guasave Valley and the River Petatlán. Depending on the year season the spatial *Fg* magnitude manifests a variable behaviour, and according at TIMP law; this result is indicator that the scenery is changing and altered, causing that river give water to the aquifer in different seasons of the year (in rainy seasons), and that, in the dry season, will be the aquifer which to give water at the river.

The piezometers of river and streams were positioned near of next populations: Puente de Guasave, Caimanero, Varal, La Cuerda, Carboneras, Puente de Bamoa, Ocoroni and Cabrera. Also on Table 1 is show the variables involved in the determination of *Fg* magnitude are shown in Table 1 The annual average mean of variables of Table 1 used for obtain *Fg* magnitude were: h = 0.42 m, T = 40.6 m^2/day, *Kz* = 1112.12 m/day, *Hs* = 26 m, *m* = 27.37 m, *Ne* = 4.87 m, *Nb* = 32.25 m and *Fg* = 1040 m/year. The results showed a *Fg* that responds in the agricultural zone to linear behaviour of expression $y = 26.70\,x - 44.15$ characterized by annual average value rank of 413 to 1813 m/year and governing by the negative relationships between aquifer/water sources and the recharge dynamics closely linked at anisotropy of a transitivity (R^2 = - 0.6) that goes from low to average within the value range of 20 to 70 m^2/day (Figure 2).

4. Conclusions

Are present the bases for a methodology which periodic can measure *Fg* magnitude and describe it behaviour for on future could help to establish a numerical modelling that allows to understand in a quantitative way the dynamics present in the aquifer/water sources interaction on this region and others basins of world. It also is support as a practical methodology that allows efficient decision making for water resource management when is necessary to get the *Eu* and *D*; principally in regions *SA/A*. The present results could be the initial state of the modelling [8] and it would allow to quantify future impacts that have to do with the exploitation of groundwater mainly on the flow of the rivers and the local effluents hydric [9], as well as inferred the behaviour of $\Delta V(x,y,t)$ that result when is used more the aquifer in comparison with the use of superficial as an underground reservoir of *H2Oi*, and in general to identify the affectation of all anthropic interventions in the exchange of flow between surface currents and groundwater.

Table 1. Annual average mean of variables used for obtain Fg Magnitudes on aquifer/water sources system for an arid/semiarid zone in Sinaloa River Valley situated on Northwest of México.

Number	***Name***	***x***	***y***	***h***	***T***	***Kz***	***Hs***	***m***	***Ne***	***Nb***	***Fg***
		UTM		*L*	LT^{2}	LT^{1}	*L*	*L*	*m*	*m*	L^3T^1
1	Puente de Guasave	755547	2831158	0.45	25	350	17	14	7	21	413.75
2	Caimanero	757729	2833323	0.2	40	840	19	21	4	25	752
3	Varal	759724	2837371	0.2	45	1125	20	25	5	30	891
4	La Cuerda	762639	2840994	0.62	48	912	22	19	6	25	1026.24
5	Carboneras	764100	2845480	0.85	62	1860	30	30	3	33	1807.3
6	Puente de Bamoa	766824	2846044	0.825	35	1120	35	32	5	37	1196.125
7	Arroyo de Ocoroni	761883	2847646	0.175	40	1400	30	35	3	38	1193
8	Arroyo de Cabrera	755078	2848874	0.105	30	1290	35	43	6	49	1046.85

h = water level on vertical variation of hydric effluents (L); T = hydraulic transmisivity (LT-2); Kz = hydraulic conductivity
(LT-1); Hs = piezometric Head (L); m = saturated zone thickness (L); Ne = watertable depth (L); Nb = Level of water holes; Fg = drip factor aquifer/hydric effluents.

Specifically the sprayed *H2Oi* with sprinkles water it droplets on the surface of the earth it simulate the effect of rain; is considered as a new technology, similar than hydroponics, drip irrigation; and others, but relatively, all tools of alternative agricultural irrigation require a large initial investment and more intensive management than surface irrigation, but in face of current climate change exist many problems of water availability and exchange studies must be carried out first to ensure investment, and the most important, is for to give guarantees for sprinkler and drip irrigation that promise a lot potential to optimize the efficiency of water use, and over all, reduce the current problems of *H2Oi*.

Given the risk of the availability of *H2Oi* and given the current climatic conditions, irrigation water has already become a scarce resource, mainly in regions *SA/A*. On this new sceneries is necessary improvements must be made in the *Eu* where achieving it lies in the *D* all with the aim to avoid losses due to drainage, evaporation and runoff.

On the other hand, the lack of quality water attributed to climatic changes also affects the hydrological cycle and oblige to implement new alternatives in the ways of exercising agricultural risks (eg the introduction of focused irrigation systems: drip, sprinkler, hydroponics, among others). Under kind of problems caused by climate change, it is important to mention the existence of different numerical methods to determine recharge and in this case mathematical models for describe the behaviour aquifers could be used as valid tools for their management.

So in the future must be discussed exhaustive which the role that the water´s models can play on management of water resources is necessary.

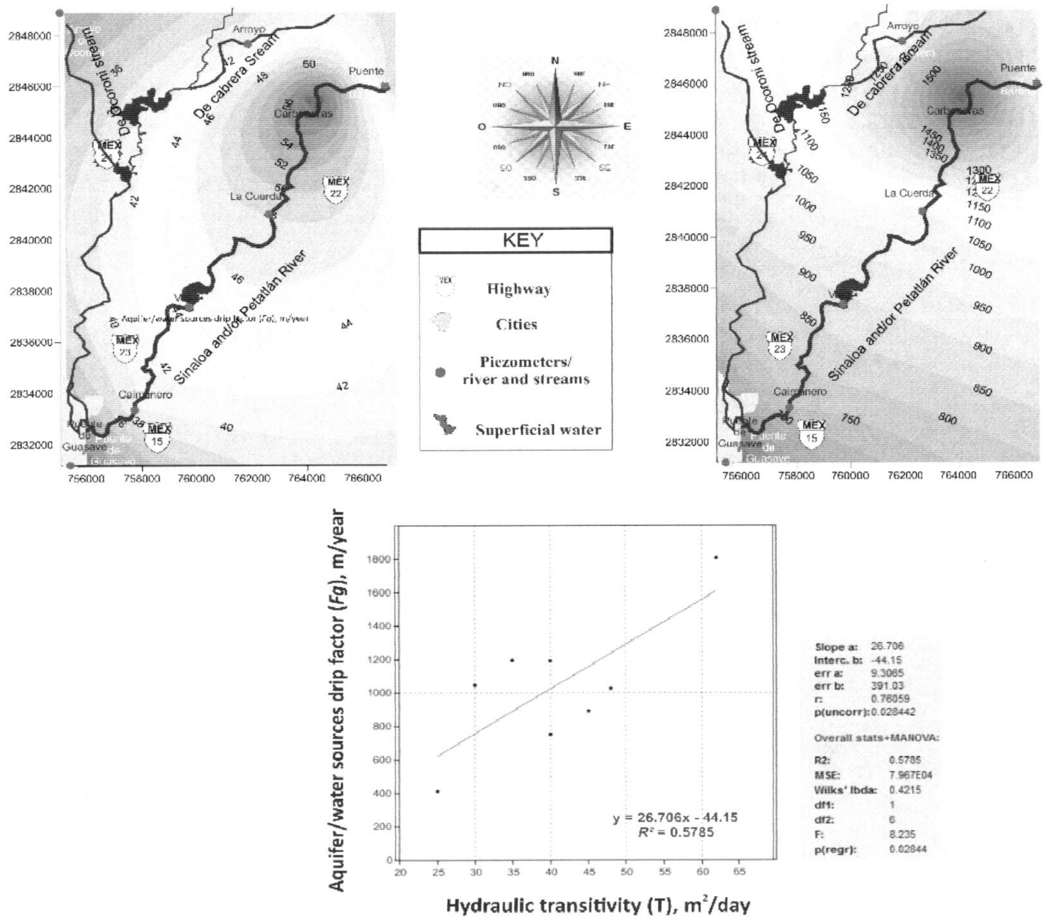

Figure 2. Spatial variation Annual average magnitudes in m2/day for hydraulic transitivity and aquifer/water sources drip factor (m/year) for aquifer River Valley of Northwest of México.

5. References

[1] Medrano, H., Bota J. Cifre J. Flexas J., Ribas-Carbó M. y Gulías J. (2007). Eficiencia en el uso del agua por las plantas. Investigaciones Geográficas, 43 (2007) pp. 63-84 ISSN: 0213-4691

[2] Vandana, S. (2004). Las guerras del agua: contaminación, privatización y negocio. Icaria Editorial, Barcelona España. 159 pp. ISBN 84-7426-698-X

[3] Green, T.R., Taniguchi, M. and Kooi, H. (2007). Potential impacts of climate change and human activity on subsurface water resources. *Vadose Zone Journal*, 6, 531–532 http://www.graphicnetwork.net/wp-ontent/uploads/2014/09/Green_Taniguchi_Kooi_2007.pdf

[4] Konikow and Bredeoeft. (1987). Computer model of two-dimensional solute transport and dispersion in ground-water. Techniques of Water-Resources Investigations of the United States Geological Survey. U.S. Government Printing Office, Washington. Book 7, chap.C2

[5] Loke, M.; Chambers J.; Rucker, D.; Kuras O.; Wilkinson P. (2013): Recent developments in the direct current geoelectrical imaging methods. Journal of Applied Geophysics, 95: 135-156.

[6] Norzagaray, C.; Llanes, C.; Muñoz, S. (2013). Estimación de la dispersividad hidrodinámica usando trazadores. Ed. Publicia, Polonia. ISBN-10: 3639554507, ISBN-13: 978- 3639554502: 100.

[7] Emery, X. (2007). Conditioning simulations of Gaussian random fields by ordinary kriging. *Mathematical Geology*, 39(6): 607-623.
[8] Sahuquillo, A. (2016). Stressed aquifers in the Lower Segura basin and the Vinalopó basin in Easter Spain. Ingeniería del Agua. 20(1): 13-27 pp. file:///C:/Users/mariano/Downloads/4023-15816-1-PB.pdf
[9] Estrada, R.O. (2009). Modelación acoplada de aguas superficiales y subterráneas para la gestión conjunta del recurso hídrico. Tesis Maestría Universidad Nacional de Colombia. 83 pp.

IOP Conf. Series: Earth and Environmental Science **171** (2018) 012032 doi:10.1088/1755-1315/171/1/012032

An Examination of the Validity of MRPS Method for the Detection of Label-free E. Coli and Enterococci in Ships Ballast Water

M M Maw[1], J S Wang[2,*], K Z Yu[1], Y J Wang[2], B W Dai[2], X D Wu[3] and X X Pan[1,*]

1 College of Marine Engineering, Dalian Maritime University, Dalian, 116026, China
2 College of Information and Science Technology, Dalian Maritime University, Dalian, 116026, China
3 Jiangsu Jimbio Technology Co., Ltd, Changzhou, 213022, China
Email: wangjsh@dlmu.edu.cn

Abstract. To test for compliance with the D-2 standard for ballast water management convention of the International Maritime Organization, the bacteriological quality of ship's ballast water is determined to evaluate the risk of invasion of non-indigenous species both onboard ship and in Port. In this article, hydrodynamic flow focusing particles viewed through a microfluidic resistive pulse sensor (MRPS) was used to detect the pathogenic bacteria from ship's ballast water. *Escherichia coli* and *Enterococci* were addressed as bacterial indicators to determine of their characteristic and concentration. The system provided the individual particle-by-particle readout in rather large 20 μm x 10 μm x 8 μm (length x width x height) horizontal rectangular microchannel duct. The average volume flow rate was 4.5 μl/min. However, it returned online and real time results for a complete of 30 μl sample volume during 15 min. The system was low maintenance, high in sensitivity, good accuracy and reliable. Targeted bacteria needed neither labelling nor extracting DNA.

1. Introduction

An estimated over 7,000 marine species are transmitted daily with ship's ballast water to new environments [1]. A few non-indigenous species have already had irreversible impact on human health, ecosystems or economies [2]. Regarding the negative impact of organisms by ship's ballast water discharge into a new environment, the International Maritime Organization (IMO) adopted the "International Convention on the Management of Ships Ballast Water and Sediments" (hereafter it will be used as the convention). In the convention, ships need to manage their ballast water to avoid the uptake or discharge of harmful aquatic organisms and pathogens in ballast water and sediments [3]. Since the convention entered into force on 8 Sept, 2017 [4], all ships need to comply with the D-2 Standard of convention. The convention limited the concentration of indicator microbes for *Escherichia coli* (*E.coli*), *Enterococci* and toxicogenic *vibrio cholera* [5]. The quality of ship's ballast water will be accessed before discharging and it becomes a technical challenge.

The most well-known methods of detection for indicative microorganisms sample from ships ballast water are Adenosine triphosphate (ATP) [6-8], Fluorescein diacetate staining (FDA) [8-10], pulse amplitude-modulation (PAM) fluorometry [8,9,11,12], Flow cytometry [8,9,13,14] and Polymerase chain reaction (PCR) [15,16]. However these conventional methods for monitoring and analysis on bacteria are bulky, labour intensive, time consuming and depend on the land based laboratory. The development methods and portable devices that can rapidly on-line detection and real-time determine the number of seaborne single microorganisms will become necessary.

In the recent era, the advantageous features of microfluidics technologies can characterize the mechanical, electrical and biophysical properties at individual cell level [17-19]. Microfluidic individual cell analysis systems provide for counting, trapping, focusing, separating, sorting, characterization and identification. Many researchers invented and reported microfluidic based analysis tools for single cells by combining the conventional methods of ATP [20, 21], PCR [22, 23], fluorescent in-situ hybridization (FISH) [24,25] and surface enhanced Raman scattering (SERS) [26,27]. Some of their techniques can detect single cell with high accuracy. Nevertheless, the detection can be performed by sophisticated and specialized equipment. Also some techniques need fluorescent activated cell sorting, extraction DNA, and staining bacteria etc. which are done in the laboratory. Among these detection systems, the Coulter counting principle is simple for counting cells [28] in the electrolytic fluid, generally count cells from blood. Debloris and Bean [29] first demonstrated the use of submicron resistive pulse sensor (RPS) which based on Coulter principle for the detection and characterization of nanoscaled objects, including nanoparticles and viruses. The resistive pulse sensing with the demonstration of single stranded DNA detection using a biological pore was introduced by Kasianowicz et al [30]. Due to RPS's sufficient sensitivity simplicity and cost-effectiveness, it has been increasingly applied in bio-sensing. To date, numerous RPS are advanced with synthetic biological pore and microfluidic sensors by many researchers [31-33]. Lab-on-a-chip with resistive pulse sensors have been used for some time and can be applied via a laser or fluorescent light to detect chlorophyll of microalgae from ship's ballast water [34-36].

To extend our experiments in detecting and analysing system which already successful detected of microalgae [37, 38] in the few years ago, now we addressed bacteriological organism detection in this article. Label-free *Escherichia coli* (*E.coli*) and *Enterococcus faecalis* (hereafter named *Enterococci*) were applied as the indicator samples. They are described as indicator microbes in IMO D-2 standard. A combination of microfluidic sample handling chip design and the phenomena of microscale hydrodynamic flow has been used to improve the detecting sensitivity in this paper.

2. Materials and methods

2.1. Sample bacteria preparation

The strains of *Escherichia coli* (*E.coli*) (*ATCC 25922*) and *Enterococcus faecalis* (*ATCC 29212*) were used in our experiments. They were provided by Jie Su et al., Marine Ecology Department, National Marine Environmental Monitoring Centre (Dalian, China). Cultured bacteria were added in the natural marine water from Dalian sea which was taken by sterile flask submerged to 1m below the water surface. The average dimensions of *E.coli* are 0.5 µm ~ 0.78 µm diameter & 1 µm ~ 2 µm length, and *Enterococci* are 0.45 µm ~ 0.65 µm diameter & 1µm ~ 1.5µm length.

2.2. Design of microfluidic chip

A microfluidic chip is the detection platform of the MRPS and the detail design of detecting zone is shown in Figure 1. It was composed of a sample well, a sheath well, a right sensing well, a left sensing well, a main waste well, a complementary waste well, two sheath channels, two detecting channels, a nozzle jet design main microchannel and a detecting zone or sensing gate. The microfluidic chip was mainly made by polydimethylsiloxane (PDMS) (Sylgard 184, Dow Corning, Midland, MI, USA) with the general rectangular size of 40 mm long, 18 mm wide and 3.5 mm thick. The main channel was started from the right quadrants of sample well and it was gradually tapered from 800 µm wide to 400 µm for 6.5 mm long and again suddenly tapered down to 200 µm for 1 mm long to get hydrodynamic focusing particles at detection zone and to avoid the back pressure on the particles in the channel. The width of sheath channels was 200 µm with elliptical shape ways for 13 mm long from the centre of sheath well. The detecting channels width was 100 µm for the first half part and 200 µm for the second half with 45 degrees inclined to the main channel and 6 mm long from centre of sensing well to the main channel. The height of all channels was 8 µm. The channel to complementary waste well was a branch from main channel with 45 degree inclination. All wells were punched the holes with 3.5mm inner diameter on the PDMS. The detection zone was situated at the center between sample well and main waste well. It is a rectangular shape with 20 μm length, 10 μm width and 8 μm height. The fabricated PDMS solid was bounded on a glass substrate which has 50 mm long, 24 mm wide and 1.07 mm thick (CITOGLAS, Suzhou, China) by plasma cleaner.

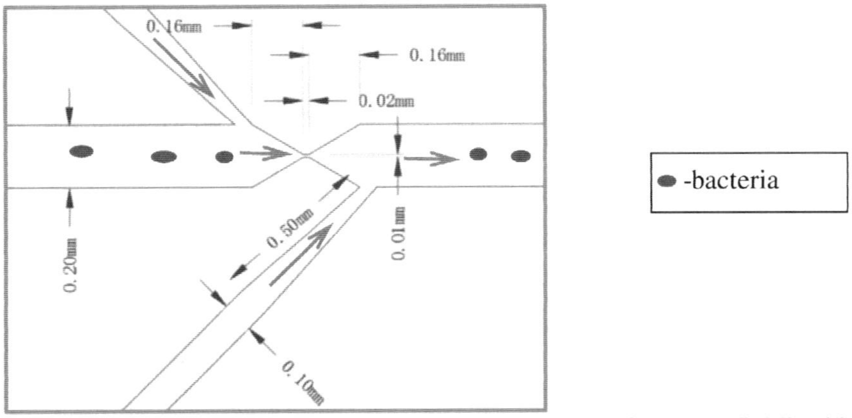

Figure 1. Structural design of detecting zone on a hydrodynamic focusing microfluidic chip.

2.3. Design and working principle of MRPS detection systems

The detecting system was mainly composed of a base unit, power supply unit, signal detection unit, signal processing unit, data acquisition unit and display unit. The principle of MRPS detection system is as shown in Figure 2. The base unit is included a PDMS microfluidic chip and four electrodes. In the power supply unit, the regulated DC power supply (HY1712-2S 0-50V 0-2A, Suzhou, China) was fixed at 24V and the positive charged was connected to the platinum electrode which embedded in the sample well while the negative end was connecting with the silver electrode of the main waste well. In the signal detection unit, the detecting circuit was connected to the electrodes of left and right sensing wells to receive the increase resistance as different potential voltages whenever the bacteria cells or particles were passing at the sensing gate. Data acquisition equipment NI USB-6259 DAQ board (National Instruments, Austin, TX, USA) was used to collect the results output signals from signal processing unit. In the display unit, LabVIEW software (2011 version, National Instruments, Austin, TX, USA) in personal computer received the data from DAQ board and finally displayed the real-time signal pulse results on the screen.

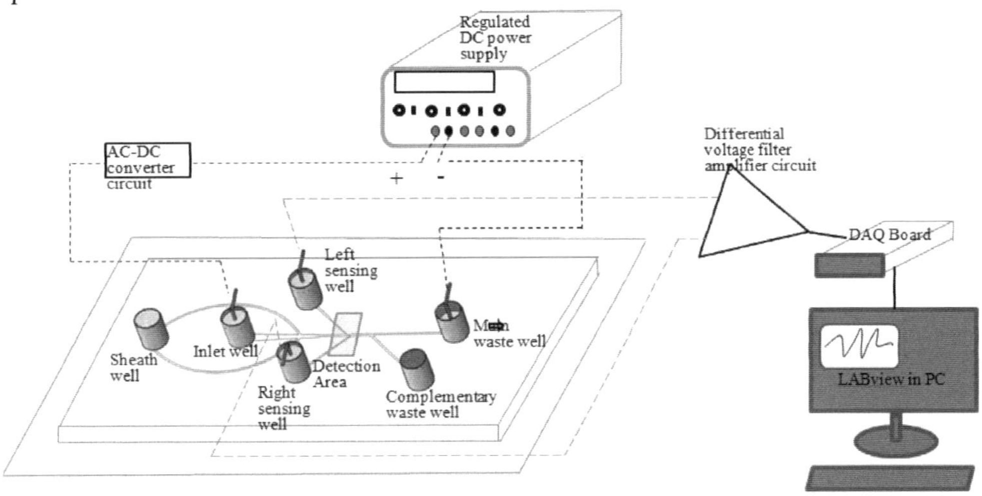

Figure 2. The principle diagram of MRPS detection system.

First of all, the microfluidic chip was placed in the base unit and four electrodes were embedded into the corresponding wells. Secondly the 30 μl volume of sample solution was loaded into the sample well. Then the left and right sensing wells were filled with 30μl phosphorous buffer solution (PBS) while the sheath well was fully added with 35μl of PBS. After that, the main waste well was filled with 10μl of PBS but complementary waste well was intentionally empty. The diffusion of the sample solution in the channels can be seen clearly by naked eyes. At the same time the power was supplied to the whole system as the third step.

When the power supply is applied, the main channel can be equivalent to three resistors connected in series. The main channel section of the sample well to the left end of the detection zone can be equivalent to one resistance (R_1), the detection zone can be equivalent to one resistance (R_2), and the main channel segment of the waste well can be equivalent to one resistance (R_3).

When a cell or particle is aspirated through the detection area, it is replaced in the solution and it blocks the electric field. Then the resulted resistance in the detection area is significantly changed. The total change of resistance at the detection zone is achieved as the change of voltage at two ends detection channels by our system design. The value of total voltage changes can be calculated with the equation (1) [39, 40].

$$\Delta V_{output} = -A\frac{(R_1+R_3)\,\Delta R}{(R_1+R_2+R_3)^2+(R_1+R_2+R_3)\,\Delta R}V \tag{1}$$

where ΔV_{output} is the potential difference between two ends of the detection zone due to increasing resistance by sample, V is the voltage difference between the sample well and the waste well, ΔR is increased resistance by targeted sample when passing through the detection zone, A is the voltage gain by amplifiers

All targeted samples are carried out for detection at the detection zone under real time situation. Data acquisition unit collects the results from the signal processing unit. Finally all detected RPS signal amplitudes are displayed on display unit which is a PC with LabVIEW software. Every test has been repeated 3 times to get the system accuracy.

2.4. Application of Maxwell's approximation theory

According to the Coulter counter theory by Deblois and Bean [29], the resistance (R) of a rectangular detecting channel is,

$$R = \frac{\rho L}{A} \tag{2}$$

where ρ is the resistivity of electrolyte fluid, L is the length of the detecting channel, A is the cross sectional area of rectangular detecting channel

When the detecting channel is filled with a small particle, the resistance will be changed. The expected resistance of a rectangular detecting channel containing a particle is,

$$R_{exp} = \frac{(\rho_{exp})L}{A} \tag{3}$$

By Maxwell's approximation theory for an insulating particle [41, 42], the value of ρ_{exp} becomes:

$$\rho_{exp} \approx \rho\left(1 + \frac{3}{2}v_f\right) \tag{4}$$

where v_f is the ratio of particle volume to detecting channel volume; *d, l* and r are diameter, the length and the radius of particle

Therefore, the increased resistance (ΔR) by particle in the detection channel is achieved by subtracting from expected resistance containing particle to the channel resistance without particle.

$$\Delta R = R_{exp} - R = \left(\frac{3}{2}v_f\right)R \tag{5}$$

According to equations (1) and (5), the amplitude of RPS output in voltage unit can be carried out.

3. Results and discussions

3.1. Effectiveness design results of MRPS system

The detection sensitivity of resistive pulse signal depends on the design and fabrication of the microfluidic chip. It means the chip has a direct impact on the results of the experiments. Its characteristics will result in poor bacteria mobility.

The cross sectional area of sensing zone should be compared with the sample particle size to ensure that the particles can move individually in the microchannel. Both theoretically and practically, the wider area may allow flow particles to overlap. The risk of channel clogging increases with smaller area as well as the back pressure. The physical dimensions of the main channel are much larger than the detecting cross sectional area. This use of hydrodynamic focusing prevents channel blocking.
Moreover hydrodynamic focusing through sheath flows will provide the individual particle to centre position within the flow. Also sheath flows from two channels are kept at uniform velocity and constant pressure by providing one central sheath well. This will align the cells or particles into the narrow stream within a wider channel. The weaknesses of using sheath flow are sample dilution and faster levelling between samples well and waste well. To overcome this issue, a complementary waste well is addressed on our system design with intentionally empty fluid at the beginning of experiment. The flow velocity can be controlled by levelling of all wells simultaneously. Our system allows adding PBS buffer solution while operating the system at the sheath well and two sensing wells. And it also allows taking out the waste solution from the waste well and the complementary waste well without influencing to the experiment results. This way the experiment can be continued detecting sample bacteria until it has finished the sample well. By meeting the fluids from sheath channels and from the sample well at the main channel, an aggregation effect appeared which can prevent the blocking and can avoid the sample bacteria attaching at the wall. PBS solution also helped the bacteria from aggregation in the main channel. Furthermore, embedded sensing electrodes at opposite sides (left and right) provided a homogenous electric field, which is more tolerant to the position of the particle in the main microchannel. The time for the whole experiment depends on sample volume size, sample concentration and sample flow rate. Generally we finished one complete experiment in 15 min for 30 µl sample size with 4.5 µl/min volume flow rate. In addition to, the passage length dimension of sensing gate was determined two times the width to obviate much noise resistance rather than signal peak. This selection got the high signal-to-noise ratio in the detection system.

3.2. Examining the validity of MRPS detection system

To check the validity of the MRPS detecting system, we considered the standard curve for the relationship between sizes of particles, their volumes and amplitudes of RPS output voltage. Five sample of polystyrene particles such as with the diameter of 0.5 µm , 0.6 µm, 0.9 µm, 1 µm and 2 µm were chosen for detection experiments due to the detection allowance percentage of the cross sectional area of the detecting duct. Specific sample particles were spiked separately in PBS solution tube. At first, 0.9 µm particles were pipetted into the sample well and recorded the RPS signal peaks and detail information in excel data sheets with the aid of LabVIEW software. Then 0.5 µm, 0.6 µm, 1µm and 2µm diameter particles were separately loaded into the sample well at different times and recorded the corresponding differential output voltages of all detected sample particles. The experiment results generally give the same height of amplitude of signal pulses for certain spherical shape but there was a range in results sometime due to the case of coincidence or faster speed. So, we calculated the average amplitude value for each sample particles size and based on 15 min duration detection. They are 0.328 V, 0.5364 V, 0.8875 V, 1.554 V and 10 V respectively for 0.5 µm , 0.6 µm, 0.9 µm, 1 µm and 2 µm diameter particles. The volumes of each particle are easily calculated with the mathematical formula of sphere and they are 0.065476 μm^3, 0.113143 μm^3, 0.381857 μm^3, 0.52381 μm^3 and 4.18879 μm^3 respectively. According to Coulter counting base RPS theory, the larger the volume of the particle, the larger the resistance and related differential voltage pulse will be large. The volume of particles is exponentially increasing depending on the size of particle as well as the amplitudes of voltage output which are exponentially increased. The theory was verified using our detection system with a graph demonstrating the relationship between signal pulses amplitude and volume and diameter of particles (see in Figure 3). In short summary, the amplitude of differential voltage output is linearly changed with the volume of particles (see in Figure 4) and this relation is expressed in equation as below.

$$y = 2.3418x + 0.1915 \tag{6}$$

where y = amplitude of increased RPS output voltage (V) , and x = volume of particle (μm^3)

Both of the standard curve graph and linear relationship graph ensure the validity of detection system for real time detecting small individual particles with the maximum system error is lesser than 1%. Therefore, the detection system is reliable to do more experiments.

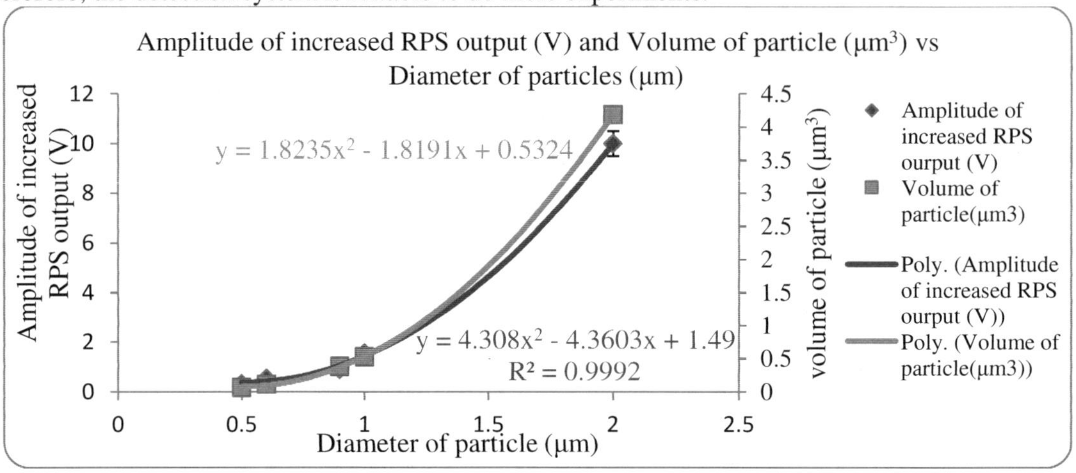

Figure 3. Standard curve for relationship between amplitude of RPS output voltage (V), volume of particles (μm^3) and diameter of particles (μm).

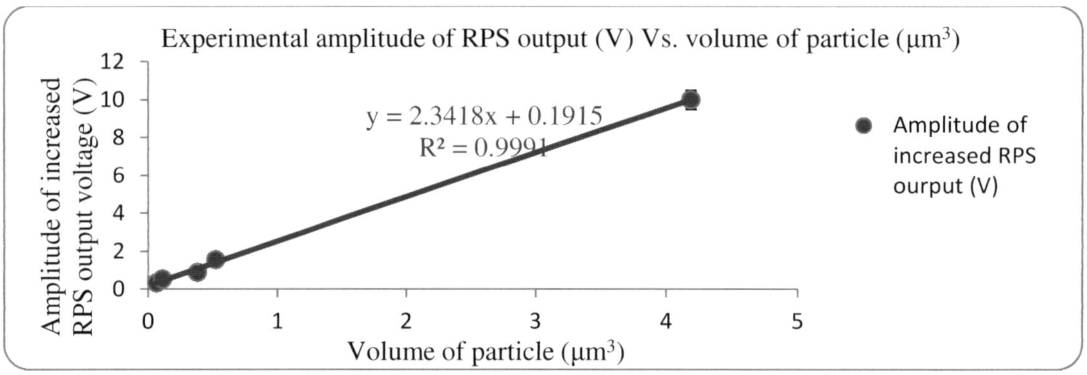

Figure 4. The linear relationship graph between amplitude RPS output voltage (V) and volume of particles.

3.3. Detecting Escherichia Coli and Enterococci

Thus as the system ensured reliability, we started to detect the targeted sample bacteria *E.coli* and Enterococci. Since the *E.coli* bacteria are non-conductible particles, they become resistors in the electrolyte solution and especially significant when they passed in the detection zone. The voltage is increased at the sensing electrodes as well as the resistance increased in the detection zone. According to the development structure of main channel design and especially the sensing gate area, finally the label free *E.coli* were detected and represented RPS output signals which were shown in Figure 5.

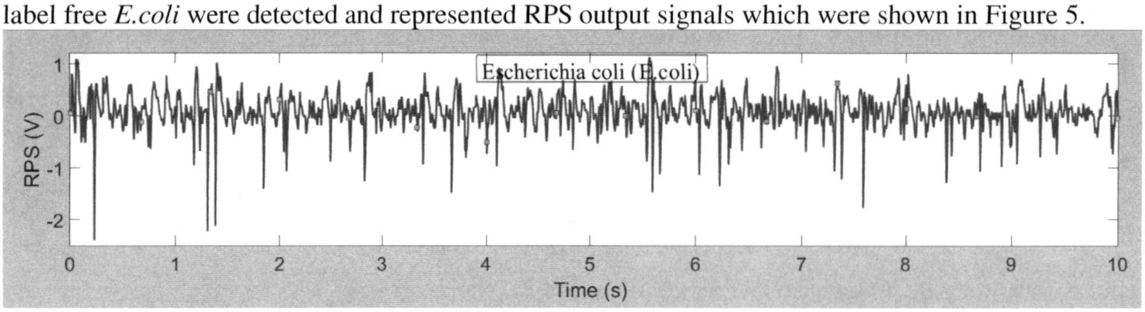

Figure 5. The amplitudes of RPS output voltage of *Escherichia coli* in 10sec.

The peak pulse represents the amplitude of increased output voltage when the bacteria passed into the detecting zone. Similarly label-free *Enterococci* were detected by the same above detection protocols and results output signals were received as Figure 6.

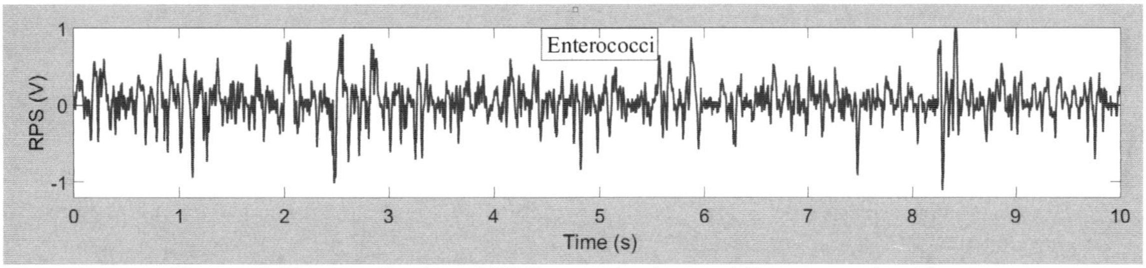

Figure 6. The amplitudes of RPS output voltage of *Enterococci* in 10sec.

3.4. Double checking the results

The results were double checked with the calculation from theoretical point of view. As indicated by equation (5), the change of resistance is directly proportional to the volume of sample while the inversely correlated to the length and the area of the detection zone. The theoretically calculations of ΔV_{output} with equations (1) and (5) by providing the dimension of samples were compared to the experimental result of RPS amplitude as in the following table 1. In addition to, the standard amplitude equation (6) of RPS signal output provides the estimated volume of detected sample and vice-versa.

Table 1. The comparison between theoretical and experimental results of MRPS detection system.

Name of sample	Diameter of sample (d) (μm)	Length of sample(l) (μm)	Volume of sample (v) (μm^3)	Theoretical RPS amplitude ($\lvert \Delta V_{output} \rvert$)(V)	Experimental RPS amplitude ($\lvert \Delta V_{output} \rvert$) (V)	standard curve RPS amplitude ($\lvert \Delta V_{output} \rvert$) (V)
E.coli	0.5	1	0.1963	0.46	0.58	0.65
	0.78	2	0.9557	2.23	2.44	2.42
Entero-cocci	0.45	1	0.159	0.37	0.43	0.56
	0.65	1.5	0.4977	1.167	1.2	1.35

4. Conclusion

Compared with the existing problems of heavy equipment, complicated operation, large energy consumption and long-time using, the microfluidic chip based detection system which is proposed in this paper has a good prospect. The detection system uses the microfluidic chip as the detection platform, so that the entire detection process does not require artificial operation, without the need for complex processing before the test. The entire detection process was automatically completed. Utilizing label free bacteria obviated the need for tagging and thus avoids perturbing the biological system. For practical ballast water analysis, a new method has been developed for simple monitoring of bacteriological quality of water. This method is able to provide rapid results, can be easily performed by an analyst with little or no microbiological training. The reliable are and accurate for enumeration of harmful microorganisms. They do not require standard laboratory equipment. The applicability of a Coulter counter based method can be utilised for ballast water quality monitoring involving the enumeration and identification of microorganisms.

5. Acknowledgments

This research is supported by National Natural Science Foundation of China (51779027，51479020 and 51309042), National Key Research and Development Program of China (2017YFC1404603, 2017YFC1404606), Liaoning Provincial Natural Science Foundation of China (2014025017), the Fundamental Research Funds for the Central Universities (3132017076，3132016325), the Applied

Basic Research Programs of Ministry of Transport of China (2014329225140), and the Basic Research Program of Jiangsu Province (BK20150274).

6. References

[1] Gollasch S, Minchin D and David M 2015 The transfer of harmful aquatic organisms and pathogens with ballast water and their impacts *InGlobal Maritime Transport and Ballast Water Management* 35-58

[2] Nghiem LT, Soliman T, Yeo DC, Tan HT, Evans TA, Mumford JD, Keller RP, Baker RH, Corlett RT and Carrasco LR 2013 Economic and environmental impacts of harmful non-indigenous species in Southeast Asia *PLoS One* **8** e71255

[3] Verna DE and Harris BP 2016 Review of ballast water management policy and associated implications for Alaska *MarPolicy* **70** 13-21

[4] https://maritimecyprus.com/2017/07/11/highlights-of-imo-marine-environmentprotection-committee-mepc-71/ (accessed 7 July, 2017)

[5] David M, Gollasch S and Leppäkoski E 2013 Risk assessment for exemptions from ballast water management–the Baltic Sea case study *Mar pollut bull* **75**, 205-17

[6] Van Slooten C, Wijers T, Buma AG and Peperzak L 2015 Development and testing of a rapid, sensitive ATP assay to detect living organisms in ballast water *J appl phycol* **27** 2299-312

[7] Welschmeyer N and Kuo J 2016 Analysis of Adenosine Triphosphate (ATP) as a rapid, quantitative compliance test for ships' ballast water (Final Report) *Broad Agency Announcement (BAA)* # HSCG32-13-R-R00016

[8] Schillak L 2014 Effective new technologies for the assessment of compliance with the ballast water management convention: final report *BSH Report* Report number: 0014-00007/RO

[9] MacIntyre HL and Cullen JJ 2016 Classification of phytoplankton cells as live or dead using the vital stains fluorescein diacetate and 5-chloromethylflurorescein diacetate *J phycol* **52** 572-89

[10] Akram AC, Noman S, Moniri-Javid R, Gizicki JP, Reed EA, Singh SB, Basu AS, Banno F, Fujimoto M and Ram JL 2015 Development of an automated ballast water treatment verification system utilizing fluorescein diacetate hydrolysis as a measure of treatment efficacy *Water res.* **70** 404-13

[11] Nakata A, Fushida S, Matsuda M and Fukuyo Y 2014 Development of New Method for Estimating Number of Viable Organisms in Ballast Water *T Engineer* **49** 512-17

[12] Gollasch S and David M 2012 On board tests of the organism detection tools BallastCAM, FluidImaging, USA, Hach-PAM-fluorometer, USA, and Walz-Water-PAM-fluorometer. Results and findings *The* Interreg *IVB North Sea Ballast Water Opportunity project*

[13] Olsen RO, Hoffmann F, Hess-Erga OK, Larsen A, Thuestad G and Hoell IA 2016 Ultraviolet radiation as a ballast water treatment strategy: Inactivation of phytoplankton measured with flow cytometry *Mar pollut bull* **103**, 270-275

[14] Bakalar G and Tomas V 2016 Possibility of using flow cytometry in the treated ballast water quality detection *Pomorski zbornik* **51** 43-55

[15] Aridgides LJ, Doblin MA, Berke T, Dobbs FC, Matson DO and Drake LA 2004 Multiplex PCR allows simultaneous detection of pathogens in ships' ballast water *Mar pollut bull* **48** 1096-101

[16] Fykse EM, Nilsen T, Nielsen AD, Tryland I, Delacroix S and Blatny JM 2012 Real-time PCR and NASBA for rapid an sensitive detection of Vibrio cholerae in ballast water *Mar pollut bull* **64** 200-6

[17] Zheng Y, Nguyen J, Wei Y and Sun Y 2013 Recent advances in microfluidic techniques for single-cell biophysical characterization *Lab Chip* **13** 2464- 83

[18] Ma X and Huo YX 2016 The application of microfluidic-based technologies in the cycle of metabolic engineering *Synthetic and Systems Biotechnology* **1** 137-42

[19] Cho YH, Yamamoto T, Sakai Y, Fujii T and Kim B 2006 Development of microfluidic device for electrical/physical characterization of single cell *J Microelectromech S* **15** 287-95

[20] Guan X, Zhang HJ, Bi YN, Zhang L and Hao DL 2010 Rapid detection of pathogens using antibody-coated microbeads with bioluminescence in microfluidic chips *Biomed microdevices* **12** 683-91

[21] Fang C et al 2010 Integrated microfluidic and imaging platform for a kinase activity radioassay to analyze minute patient cancer samples *Cancer res* **70** 8299-308

[22] Sia SK and Whitesides GM 2003 Microfluidic devices fabricated in poly (dimethylsiloxane) for biological studies *Electrophoresis* **24** 3563-76

[23] Park S, Zhang Y, Lin S, Wang TH and Yang S 2011 Advances in microfluidic PCR for point-of-care infectious disease diagnostics *Biotechnol adv* **29** 830-39

[24] Joachimsthal EL, Ivanov V, Tay SL and Tay JH 2004 Bacteriological examination of ballast water in Singapore Harbour by flow cytometry with FISH *Mar pollut bull* **49** 334-43

[25] Meagher RJ and Wu M 2016 Microfluidic Approaches to Fluorescence In Situ Hybridization (FISH) for Detecting RNA Targets in Single Cells *SpringerInternational Publishing.* InMicrofluidic Methods for Molecular Biology 95-112

[26] Walter A, März A, Schumacher W, Rösch P and Popp J 2011 Towards a fast, high specific and reliable discrimination of bacteria on strain level by means of SERS in a microfluidic device *Lab Chip* **11** 1013-21

[27] Knauer M, Ivleva NP, Niessner R and Haisch C 2012 A flow-through microarray cell for the online SERS detection of antibody-captured E. coli bacteria *Anal bioanal chem* **402** 2663-67

[28] Saleh OA 2003 A novel resistive pulse sensor for biological measurements. Doctoral dissertation Princeton University

[29] DeBlois RW and Bean CP 1970 Counting and sizing of submicron particles by the resistive pulse technique *Review of Scientific Instruments* **41** 909-916

[30] Kasianowicz JJ, Brandin E, Branton D and Deamer DW 1996 Characterization of individual polynucleotide molecules using a membrane channel *P Natl A Sci* **93** 13770-13773

[31] Bryan AK, Engler A, Gulati A and Manalis SR 2012 Continuous and longterm volume measurements with a commercial Coulter counter *Plos one* **7** e29866

[32] Blundell EL, Mayne LJ, Billinge ER and Platt M 2015 Emergence of tunable resistive pulse sensing as a biosensor *Anal Methods-UK* **7** 7055-66

[33] Yang L and Yamamoto T 2016 Quantification of virus particles using nanopore-based resistive-pulse sensing techniques *Front microbiol*

[34] Cheung KC, Di Berardino M, Schade- Kampmann G, Hebeisen M, Pierzchalski A, Bocsi J, Mittag A and Tárnok A 2010 Microfluidic impedance-based flow cytometry *Cytom Part A* **77**, 648-66

[35] Wang J, ZhaoJ, Wang Y, Wang W, Gao Y, Xu R and Zhao W 2016 A new microfluidic device for classification of microalgae cells based on simultaneous analysis of chlorophyll fluorescence, side light scattering, resistance pulse sensing *Micromachines* **7** 198

[36] Kozak D, Anderson W, Vogel R and Trau M 2011 Advances in resistive pulse sensors: devices bridging the void between molecular and microscopic detection *Nano Today* **6** 531-45

[37] Wang J, Song Y, Maw MM, Song Y, Pan X, Sun Y and Li D 2015 Detection of size spectrum of microalgae cells in an integrated underwater microfluidic device *J Exp Mar Biol Ecol* **473** 129-37

[38] Maw MM, Wang J, Li F, Jiang J, Song Y and Pan X 2015 Novel electrokinetic microfluidic detector for evaluating effectiveness of microalgae disinfection in ship ballast water *Int j mol sci* **16** 25560-75

[39] Golibersuch DC 1973 Observation of aspherical particle rotation in Poiseuille flow via the resistance pulse technique: I. Application to human erythrocytes *Biophysical journal* **13** 265-80

[40] Maw MM, Pan X, Peng Z, Wang Y, Zhao L, Dai B and Wang J 2018 A changeable lab-on-a-chip detector for marine nonindigenous microorganisms in ship's ballast water. *Micromachines* **9** 20

[41] Russakoff G 1970 A derivation of the macroscopic Maxwell equations. *Am J Phys* **38** 1188-95

[42] Raviart PA and Sonnendrücker E 1995 Approximate models for the Maxwell equations *J comput appl math* **63** 69- 81

IOP Conf. Series: Earth and Environmental Science **171** (2018) 012033 doi:10.1088/1755-1315/171/1/012033

Research on Status Quo of Water Rights Trading in China

WANG Xin[1] and YANG Shizhong[2]

1 Postgraduate at Accounting School, Capital University of Economics and Business, Beijing, China
2 Professor and Ph.D. Tutor at Accounting School, Capital University of Economics and Business, Beijing, China
E-mail: wx19941229@163.com

Abstract. Promoting water rights trading is conducive to optimal allocation of water resources through market mechanisms, and provides an important means to improve water-use efficiency and the benefits from water use. This study explores the background against which water rights trading in China came into being, and sheds a light on the current status of water rights trading in China, which covers topics such as China Water Exchange, the progress of work related to water rights determination and trading in pilot provinces (autonomous regions), and China's efforts to build a water rights system. Furthermore, it identifies a number of problems pertaining to the subjects and objects of water rights trading, and China's efforts in building a network of trading platforms, as well as a water rights system. On that basis, it puts forward countermeasures to address the detected problems, in an effort to facilitate water rights trading.

1. Introduction

The combination of population and economic growth in the recent decades has led to ever-expanding demand for water resources. Nevertheless, water environment problems are increasingly prominent and water shortage has become an issue of global concern. China in particular is faced with a dire shortage of water resources, due in part to low water-use efficiency and prevalent waste of water. As the country continues to develop, the demand for water is likely to be higher, making the situation even worse. To dissolve this crisis, it will not suffice to rely solely on administrative means, and a new trail must be blazed for a return to equilibrium[1].

In China's No. 1 Central Document 2011 and at the Central Water Conservancy Work Conference convened in the same year, it was proposed that a national water rights system be put in place, as part of an effort to leverage market mechanisms for optimized allocation of water resources. In 2012, China's No. 3 Central Document specified that a well-functioning water rights system must be established, and a water market be formed, so that water resources can be reasonably allocated through market mechanisms. In March, 2014, Xi Jinping, General Secretary of CPC Central Committee, put forward "taking systematic approaches to address water issues by placing water conservation as a priority and giving full play to the roles of both the government and the market, for optimized allocation of water resources" as the guiding principle for water conservancy work in the years to come. He noted that a water rights system must be built, according to which water rights are properly determined, and that a water market be created. In 2017, China's No. 1 Central Document stipulated that intensified efforts must be made to foster a water market, in which water-use rights can be determined and trade in water rights can be made. At the 19th CPC National Congress, resources conservation and environmental protection were identified as a fundamental state policy in China's efforts to facilitate ecological reform. As an integral part of the ecological reform, water rights reform is essential in ensuring water conservation and serves as an important means to leverage market mechanisms for optimized allocation of water resources. Therefore, substantial efforts must be made in this regard.

2. Status Quo of Water Rights Trading in China

2.1. Origin and Development

For decades, China has adopted a governance approach featuring the combination of macro-control and specific administrative measures to the exploitation, use, conservation, harnessing and allocation of water resources, with an aim to ensure equal rights to use water[2]. Nevertheless, this approach, without leveraging any market mechanisms, has been plagued by low water-use efficiency. According to the Coase theorem, incorporating property rights into water resources management may create a synergy between administrative means and market mechanisms, thereby conducing to raising management efficiency and optimizing the use of water resources. Such practice was first explored by some foreign localities. One of the most glaring examples is the water bank in California, which is actually a broker scheme. Here is how it works: the water bank kicks in whenever a drought hits. It will start by promptly collecting massive amounts of information concerning water demand, then buy water rights from low-demand subjects through contracts, and at last sell them to those drought-stricken areas and high-demand subjects. This approach helps to both facilitate water conservation and improve water-use efficiency. Against this backdrop, there is an urgent need for China to carry out water rights reform, which shall consist of the following initiatives, aside from total consumption control. First, China needs to determine who are entitled to occupy, use and/or profit from particular water resources. Second, China needs to promote water rights trading, so that optimal allocation of water resources may be achieved according to the forces of supply and demand. Third, China needs to put a limit on the use of water and regulate the water market to ensure adequate supply of water for agricultural and ecological uses and to protect the legal rights and interests of water users.

In promoting water rights trading, China has taken the following measures. First, China has set up pilots. Under the guidance of the Ministry of Water Resources, Zhejiang, Ningxia, Inner Mongolia, Fujian, Gansu, and Xinjiang, among other provincial-level administrative divisions, have started to explore the best practices of water rights trading since 2000. As a result, a number of water rights transactions have been completed. Second, China has set up China Water Exchange (or "CWEX"), a national water rights trading platform. Established in June 2016, the Exchange provides a venue for well-regulated water rights trading activities, where the market plays a decisive role in allocating water resources, and the government complements what the market cannot do. CWEX contributes to improved water-use efficiency and increased benefits from water use, and conduces to sustainable use of water resources and socioeconomic development. Third, China has issued a number of guidance documents to regulate water rights trading, including the *Interim Measures for Managing Water Rights Trading*. CWEX has released the *Rules for Water Rights Trading* and other seven guiding documents. Furthermore, the pilot areas have put extensive efforts to explore the best practices of piloting, water rights determination, and water rights trading.

2.2. Transactions at China Water Exchange

2.2.1. Overview of Transactions. CWEX is an important water rights trading platform. From its inception in June 2016 to March 2018, the Exchange saw 46 completed transactions, involving a total of 1.60 billion tons of water valued at RMB 995.87 million. The highest price in a single transaction was RMB 1.2 and the lowest RMB 0.17 per cubic meter. Participants consisted primarily of government agencies, public institutions, state-owned enterprises and grassroots water-consuming organizations. In sum, despite considerable growth in the number of transactions, the overall transaction size is relatively small.

2.2.2. Types and Forms of Transactions. At CWEX, there are three types of transaction allowed, namely cross-region water rights and water-taking rights transactions, and those between irrigation water users. As shown in Table 1, as of November 20, 2017, the quantity of water involved in cross-region water rights and water-taking rights transactions had totaled 1.18 billion cubic meters, with a value of RMB 744.47 million; and that in transactions between irrigation water users had reached 2.91 million cubic meters, with a value of RMB 652,972.60. In other words, cross-region water rights and water-taking rights transactions take the lion's share. These transactions take two forms: one is transfer through contract, and the other is open-market transaction. As of November 20, 2017, water sold through the had totaled 684.16 million cubic meters, with a value of RMB 445.07 million; and that through the latter had reached 500.31 million cubic meters, with a total value of RMB 300.06 million. As such, both forms are playing an important part in water rights trading at CWEX.

Table 1. Types and forms of transaction.

	Transfer through contract		Open-market transaction		Total	
	Quantity (10,000 m^3)	Amount (RMB 10,000)	Quantity (10,000 m^3)	Amount (RMB 10,000)	Quantity (10,000 m^3)	Amount (RMB 10,000)
Cross-region water rights and water-taking rights transactions	108,656.30	69,387.47	50,000.00	30,000.00	158,656.30	99,387.47
Water rights transactions between irrigation water users	885.31	194.15	31.09	6.22	916.40	200.37
Total	109,541.61	69,581.62	50,031.09	30,006.22	159,572.70	99,587.83

Source: CWEX official website (http://cwex.org.cn/)

2.3. Progress of Water Rights Trading in National Pilot Provinces (Autonomous Regions)

The national pilot provinces (autonomous regions) have made great efforts to promote the related work in accordance with the goals made clear by pilot programs and has made considerable progress and gained work experience especially in areas such as determination of water rights and water rights trading.

2.3.1. Determination of water rights in pilot provinces (autonomous regions). These pilot provinces (autonomous regions) explored the determination of water-taking rights on two levels: regional level and water-user level[3]. On the regional level, water-taking rights and interests were specified mainly through a regional total water consumption quota system and water allocation schemes for specific rivers. For example, Ningxia Hui Autonomous Region allocated regional quotas to lower-level cities and counties, determined cross-region water rights transactions and interests with a step-by-step approach and confirmed nearly 300 water rights certificates to be issued to entities including township governments and water-consuming organizations. Henan Province released a guideline document named *Water Allocation Program for the First Phase of the Mid-route of South-to-North Water Transfer Project in Henan Province* to allocate water consumption quotas to each water areas to serve as the basis for facilitating water rights trading in related cities and counties. Guangdong Province has finished water allocation arrangement for basins of Dongjiang River, Jianjiang River, Hanjiang River and Beijing River. Guangdong also specified total water consumption quotas for the subsequent years until 2030 for all its prefecture-and-above-level cities. On the water-user level, there are two kinds of water users. First, non-agricultural water users who are granted water-taking rights. The water

IOP Conf. Series: Earth and Environmental Science **171** (2018) 012033 doi:10.1088/1755-1315/171/1/012033

administrative authorities that has the rights of review and approval determined water-taking rights by issuing water-taking certificates according to the *Regulation on the Administration of Water-taking Certificates and the Levy of Water Resource Fees*. Jiangxi Province conducted a survey about water-taking rights in Xingan County, Gao'an County and Dongxiang County and organized township governments, reservoir irrigation areas and businesses concerned to fill in survey forms and the outcomes of the survey were publicized and filed. Second, agricultural water users in irrigation areas. The local government or authorized water administrative departments firstly took into account the conditions of water resources, measurement capacity and management needs and then determined water rights by issuing water rights certificates to water user associations or farmers. For instance, in Yidu City, Hubei Province where there are ponds and reservoirs built by the rural collective economic organizations, water rights to the resources concerned were determined to these organizations or farmers through reform in small water conservancy property rights. Gansu Province allocated the irrigation water quotas into specific irrigation ditch and determined the water rights to each water user association.

In general, these pilot regions mainly took four steps to deal with the determination of water rights. First, they determined the total water consumption quota for each administrative area, and allocated the quotas to specific rivers, lakes, reservoirs and groundwater resources. Second, they formulated plans of the water resources allocation among different industries and then based on comprehensive water resources planning and need for future development, specified the share of water distributed to people's daily consumption and major water-consuming sectors including agricultural and industrial sectors. Third, based on the total water consumption quotas, allocation plans for particular industries and allocation standards for individual users, they specified the shares of water for two kinds of users. In terms of non-agricultural water users who were granted water-taking rights, the priority was to regulate the water-taking permission management and strictly verify the amount of water. In terms of agricultural water users in irrigation areas, the priority was to equally distribute water consumption quotas to water user associations or farmers. Fourth, they reviewed the total water consumption quota balance before officially issuing certificates to water users, and after that set up monitoring facilities to ensure that water rights were supervised. In some regions, the outcomes of water rights determination for agricultural water users in irrigation areas were publicized.

2.3.2. Water rights trading in pilot provinces (autonomous regions). At present, there are three models of water rights trading in China, namely, trading between regions, trading between users and purchasing from government authorities repurchased (or withdrawn) water rights[4]. In terms of trading between regions, after total water consumption quotas determination and the distribution of river water were completed, the saved or surplus water can be traded between these administrative regions through water transfer projects. Such intergovernmental transactions are conducted between the local governments or authorized departments. For instance, Xinmi City in Henan Province transacted a total 24 million cubic meters of water with Pingdingshan City in Henan Province through the South-to-North Water Transfer Project with the price of RMB 0.87 per cubic meter and trading period of 3 years. Trading between users consists of transactions between non-agricultural water users, and those between irrigation water users. Non-agricultural water users who gained water-taking rights can sell their saved water. For instance, Zhouta Township of Zhongning County in Ningxia transferred 2.19 million cubic meters of water saved through agricultural water-saving transformation and planting structure adjustment to the county's power plant. In terms of transaction between irrigation water users, these users can sell their saved water rights. For instance, farmers in Shiyang River Basin, Gansu Province carried out water rights transactions. One of the districts in that area, Liangzhou District in Wuwei City completed a total 241 transactions in 2015 with a total volume of 4.68 million cubic meters and a transaction value of about RMB 980,000. In 2015 and 2016, 36 companies (or large agricultural water consumer) in Ningxia were involved in water rights transactions, involving combined irrigated area of 83,000 mu, trading volume of 5.77 million cubic meters, and a total value of RMB 4.22 million. Under the third model, local governments or authorized departments repurchase or withdraw the water saved by users. The water will be first used

IOP Conf. Series: Earth and Environmental Science **171** (2018) 012033 doi:10.1088/1755-1315/171/1/012033

for ensuring people's daily consumption and restoring ecological environment, and the surplus, if any, will be put to the market. In November 2016, the Department of Water Resources of Inner Mongolia Autonomous Region withdrew the spare quota of 20 million cubic meters per year according to *Measures for Disposal of Spare Water Quotas in Inner Mongolia Autonomous Region*, and transferred it to the Inner Mongolia Autonomous Region Water Rights Transfer Center Co., Ltd. The company then listed the rights on the platform of CWEX openly selling them to companies in Ordos City, Wuhai City, and Alxa City. In the end, five companies were chosen as the ultimate transferees in a chronological order with the transaction period of 25 years and down payment of RMB 300 million. By analyzing the statistics from the official websites of departments of water resources in the pilot regions and the files of the National Water Rights Pilot Symposium, we found that the volume of water rights transaction in Inner Mongolia has reached 1117.75 million cubic meters, with a total value of RMB 667.81 million. As shown in Table 2, at present, the seven pilot provinces (autonomous regions), except for Jiangxi Province, have set up water rights trading centers at the provincial or county level. However, only Inner Mongolia has created a website to promote water rights trading – the Inner Mongolia Water Rights Storage and Transfer Center, which presents knowledge of water rights trading and discloses trading information. These specialized trading platforms have furthered the development of water rights trading.

Table 2. Water rights trading centers in pilot provinces (autonomous regions).

National Pilot Provinces	Water Rights Trading Centers
Ningxia	Water Rights Trading Center of Zhongning County
Jiangxi	N/A
Hubei	Water Rights Trading Platform of Jingmen City
Inner Mongolia	Inner Mongolia Water Rights Storage and Transfer Center
Henan	Henan Water Rights Storage and Transfer Center
Gansu	Shule River Basin Water Rights Trading On-line Platform of CWEX
Guangdong	Guangdong Environment Exchange

Source: Official websites of the Department of Water Resources of Ningxia, Jiangxi, Hubei, Inner Mongolia, Henan, Gansu and Guangdong and files of the National Water Rights Pilot Symposium.

3. Problem Analysis

3.1. Insufficient underStanding of Water Rights Trading Subjects

The problem of water rights determination and trading comes partly from local organizations' lack of in-depth understanding of water rights trading, as is the case with county-level departments of water resources in Ningxia water rights trading pilot market. Currently, water resources management departments and all sectors of society have not reached a consensus over the purpose of and the roles of government and market in the establishment of water rights system, with some even consider water rights determination a way to pass responsibilities down to lower levels. Such insufficient understanding hinders the development of water rights determination.

3.2. Difficulty in Determining the Objects of Water Rights Trading

So far, major measures and procedures for water rights determination have taken shape, yet there remain difficulties in recording water rights determination and water resources data as certain types of water use are not easy to meter and monitor. For example, Guangdong pilot market pointed out that as Pearl River Delta adopts tidal irrigation. The measurement of its agricultural water is made mainly through installing meters in the beginning of tidal canals. Such inaccurate agricultural water metering will directly affect water rights determination and the metering and monitoring of trading water volume. Apart from that, how to coordinate water rights determination and the approval of water access rights, and how to determine water rights at different levels of water supply reliability still require further discussion.

3.3. Insufficient and Ill-Structured Water Rights Trading Platform System

The development of water rights trading requires the establishment of specialized trading platforms. Nevertheless, the roles and responsibilities of CWEX and water trading centers of each pilot province or autonomous region are not clearly defined. Nobody knows whether they are to work in parallel or hierarchy. Therefore, currently part of water rights trading in each pilot province (autonomous region) are processed unnecessarily at both CWEX and local trading platforms. Worse still, most of the pilot provinces (autonomous regions) have not established a specialized platform for water rights trading, making it inconvenient for potential traders to access trading information, which, to some extent, impedes the development of China's water rights trading.

4. Policy Suggestions

4.1. Enhancing the Government's Leading Role in Promoting the Water Rights Trading Market

Playing an indispensable part in the development of water rights trading market, the government should further enhance its leading role in guiding local organizations to form an in-depth understanding of water rights determination and trading. The national government should set up a clear accountability system to allow governments at all levels to split responsibilities in the process of water rights determination. Meanwhile, the government should innovate promotion methods and carry out all-round and multi-leveled promotion by making the most of all media including newspaper, television, broadcast and internet, and organize symposiums, seminars and training sessions on water rights determination and trading, educate the public on the importance and necessity of establishing water rights system, increase the engagement of the public, and gradually improve the water rights trading market.

4.2. Filing the Financial Gap through Administrative Incentives and Private Investments

Water rights reform, the metering and monitoring of water resources, the establishment of information platforms, efficient water conservation, and project constructions call for a large number of investment[5]. The determination and registration of water rights and the promotion of water rights trading will be meaningless unless greater efforts are made to enhance the monitoring capacity of water resources by adopting smart monitoring system to enable every water rights owner to meter water use on a household basis. In order to ensure a smooth water rights reform, a reward and punishment system should be established to favor regions in support of the reform in aspects including project and fund arrangements. In addition, private investments should also be encouraged to fill the financial gap in water conservancy by adopting water saving management contract and other cooperation models to achieve win-win results.

4.3. Establishing a Well-Defined System of Water Rights Trading Platforms

Currently, China's water rights trading is in its exploration stage with great potential to be tapped into. In the current water market, there are both intra- and cross-administrative division trades. Therefore, it is suggested that the development of water rights trading platforms be divided into two phases. In the first phase, CWEX and water trading platforms for particular pilot provinces (autonomous regions) work in parallel, with trades within each province or autonomous region made in local trading platforms which should be greatly improved to promote such trades and cross-administrative division trades made in CWEX. In the second phase, as the trade size grows, CWEX as a national water rights trading platform will integrate with all local platforms and serve as the only official platform to provide accessible trading information and complete trading database, in an effort to unleash the vitality of the water market.

5. References

[1] Tian Guiliang, Du Mengjiao, Jiang Yong. Study on water rights trading mechanisms [J]. *Water Resources Protection*, 2016, 32 (05): 29-33+52.
[2] Liu Yunjie, Shi Yubo, Zhang Bin. Study on current status and advancing measures of water rights trading in China [J]. *China Water Resources*, 2016 (21): 1-2+5.
[3] Wang Xiaojuan, Chen Jinmu, Zheng Guonan. Consideration on stimulating water entitlement trade market and recommendations [J]. *China Water Resources*, 2016 (01): 8-11.
[4] Liu Feng, Duan Yan, Ma Yan. Case Study on Cross-region Water Rights Transactions [J]. *Journal of Economics of Water Resources*, 2016, 34 (01): 23-27+83.
[5] Liu Shiqing, Guo Shijun, Lin Rui, Ju Dong. The Characteristics of Water Rights System and the Exploration of Water Rights Reform in China [J]. *Journal of Engineering Studies*, 2016, 8 (01): 12-22.

This article is funded by postgraduate technology innovation project of Capital University of Economics and Business"Research on Accounting Issues of Water Rights Trading—based on the investigation and study of transaction samples in China Water Exchange".

IOP Conf. Series: Earth and Environmental Science **171** (2018) 012034 doi:10.1088/1755-1315/171/1/012034

Developing a Portable Hydroelectric Generator Using Axiomatic Design Method

H Soewardi[1] **and E A Putra**[2]
1,2 Department of Industrial Engineering, Faculty of Industrial Technology, Islamic University of Indonesia, Yogyakarta, Indonesia
Email: [1] hartomo@uii.ac.id, [2] enggangaditya@gmail.com

Abstract. Hydroelectric power plant is the largest renewable energy source in Indonesia. There are several models of hydroelectric power machine applied in Indonesia, one of them is a portable generator. However, there are still some problems in the existing portable design, for example it has a big size and unattractive design; it is unpractical and unsafe to use, that it results in ineffective and inefficient use of the generator as well as results endanger. This reality occurs because the design does not meet user criteria. The purpose of this study was to redesign the portable hydroelectric power generator which can satisfy user requirements. Survey was conducted to identify the customer needs and axiomatic design method was used to determine the design parameter by mapping process from customer attribute and functional requirement. Statistical analysis was conducted to test the hypothesis. The results of this study showed that the proposed design of portable generator can increase user satisfaction as much as 73.67% so it is recognized to be different from the existing design. The design developed is also valid to satisfy user requirement at 5% significance level including the fact that it is durable, waterproof, attractive, lightweight, small in size, and portable.

1. Introduction

Electricity is one of the main points that human requires to help with daily activities [1]. The Indonesian government targets the level of electricity consumption in 2025 reaching 1,500 kWh per capita. Meanwhile, the electricity demand per year is 468,067 GWh. However in the end of 2015, the consumption level of electricity in Indonesia was still about 880 kWh per capita [2]. Thus government is intensively building a new power plant to achieve the target on time. Currently, about 87% of the electricity supply is dominated by the use of fossil fuels especially petroleum and coal [3]. Meanwhile, the unlimited source of renewable energy has not been utilized optimally [4], that is only 17.4% of the 65.764 MW of potential hydroelectric energy is used. Such thing occurs because of high cost and unreachable locations for investment [5]. One of its solutions is by using portable generator [6].

Portable generator is a device to generate an electrical energy by combining an electrical generator and propulsion which is packed simply [7]. The propulsion has a function to trigger the generator by using a high water flow. It is frequently known as portable hydroelectric generator. However there are still some problems in the existing generator. Based on the preliminary study, it is found that 59.2% of the users said that the device is unpractical, 18.8% said that the device has a big size, 12.1% said that the design is unattractive, 9.9% said that this device is unsafe to use. This fact indicates that the portable generator has been ineffective,inefficient, and dangerous to use. Thus this is crucial to improve its design.

Some previous studies had been done. [8] conducted a study about a hydroelectric generator by using low water energy. This research focused on the development of turbine type and generator type, as well as the development cost. In addition, [9] developed a portable hydroelectric generator by using a

IOP Conf. Series: Earth and Environmental Science **171** (2018) 012034 doi:10.1088/1755-1315/171/1/012034

cross flow turbine but they did not discuss the power generated. [10] studied a design of hydroelectric generator by using a spiral shape of hydro coil turbine to produce energy. Portable generator was also developed by [11] by using a polymer electrolyte membrane fuel cell (PEMFC). This prototype could generate energy up to 500 Watt per hour at temperature between 60-80 °C. [12] designed a portable generator based on a chemical reaction from micro direct ethanol fuel cell (Micro-DEFC). The result of this study showed that the chemical reaction produces electricity reaching 0.9 mV andthis chemical reaction depends on temperature. All these studies focus only on the design of portable generator in producing energy regardless of users' requirements in use.
The objective of this study was to redesign the portable hydroelectric generator to have a more effective, more efficient and safer generator by using axiomatic design method.

2. Research method

2.1. Survey
Paper based survey was conducted to identify the customer attribute of portable hydroelectric generator and also to validate the proposed design by distributing a questionnaire.

2.2. Apparatus
Some equipment used in this study, there are :

- SPSS software 20 version was used to conduct statistical analysis.
- Sketch up Pro 2017 software was used to design the virtual prototype.
- Questionnaire, was devided into two stages. The first was to identify the customer needs and the second was to validate the prototype design

2.3. Axiomatic design method
Axiomatic design is a theory that is used to develop a new product that consumers require by mapping into four domains [13], [14]. The four domains used in axiomatic design can be shown in the Figure 1.

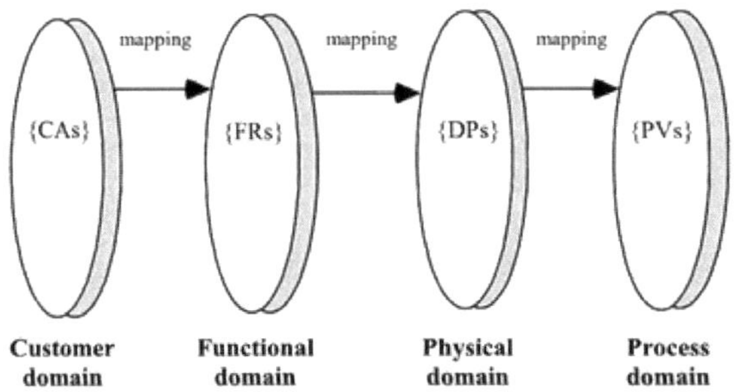

Figure 1. Domain in Axiomatic Design, [14]

Customer domain is a domain that contains customer attribute. Functional domain is a domain that contains all functions to be provided in a product. Meanwhile, physical domain is a domain used to realize the Functional Requirement. In addition, process domain is a domain that explains how to make the product based on physical domain [14]. These domains are connected by using zigzagging process.

2.4. Statistical analysis
There were two kinds of statistical tests used in this study. They were Marginal Homogeneity and Wilcoxon test. The marginal homogeneity test was used to test the null hypothesis (Ho) whether the proposed design was in accordance with the customer requirements [15]. Meanwhile the wilcoxon signed rank test was used to test the null hypothesis (Ho) whether there was a difference between the proposed design and the existing design of the portable hydroelectric generator [16].

IOP Conf. Series: Earth and Environmental Science **171** (2018) 012034 doi:10.1088/1755-1315/171/1/012034

3. Result and discussion

3.1. Result of survey

Paper-based survey found that there are six attributes that is used to redesign the portable hydroelectric generator. They are durable, waterproof, attractive design, lightweight, small size, and easy to carry. According to the Indonesian dictionary, durable defined as the ability of a product to be not easy damaged and has a long life. Meanwhile, waterproof is the addition of synthetic organic polymers into a material to prevent the leakage [17]. While Attractive design defined as a strategy used to increase the value of a product [18]. Lightweight allows the user to lift or carry product easily. Small size allows the user to not require a spacious place to keep or using the product. And easy to carry allows the product being practical to carry before or after use.

3.2. Mapping process of the functional requirement to design parameter.

In the mapping process, customer attribute (CA) was translated into functional requirement (FR) and then the design parameter (DP) determined to realized the Functional Requirement. Mapping process was explained in this following sub-subsections.

3.2.1. Design parameter for satisfy CA1: durable

Figure 2 shows the hierarchical structure of mapping process from FR1: improving the product endurance to DP1: Durable product.

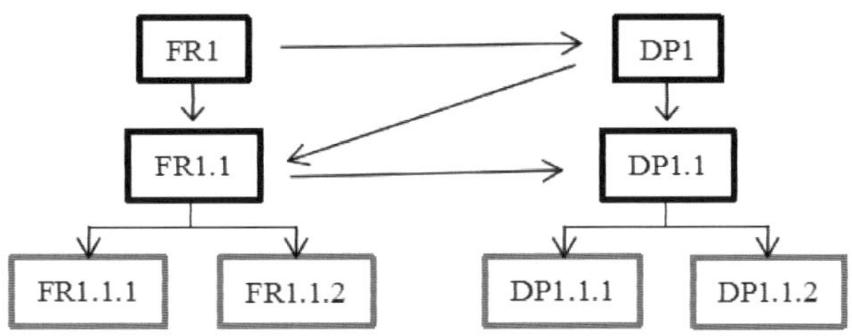

Figure 2. Hierarchical structure of CA1

Based on figure 2, the design parameter of turbine is made from the iron material (DP1.1.2). this material has a function to protect the turbine from the object that can inhibit the turbine spin (FR1.1.2). While the body is of from plastic material (DP1.1.1). This material has a function to protect the body and turbine from cold or hot temperature (FR1.1.1). A sturdy material (DP1.1) is needed to improve the resistance from the environmental changes (FR1.1).

3.2.2. Design parameter for satisfy CA2: waterproof

Figure 3 shows the hierarchical structure of mapping process from FR2: improving the leakage resistance to DP2: waterproof material.

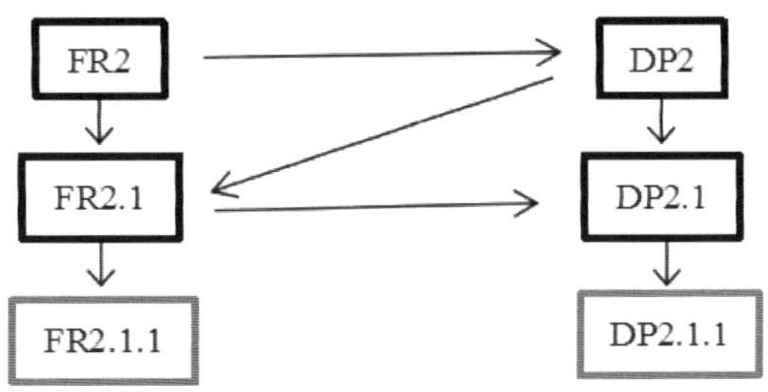

Figure 3. Hierarchical structure of CA2

Based on figure 3, design parameter of the body are made from the abs plastic waterproof material (DP2.1.1). This material has a function of holding water get into the machine (FR2.1.1). Hydroelectric generator requires high fiber material (DP2.1) to improve the resistance on the water flow (FR2.1) and also to improve the leakage resistance (FR2).

3.2.3. Design parameter for satisfy CA3: attractive design

Figure 4 shows the hierarchical structure of mapping process from FR3: Provide the unique in design and features DP3: Innovative design.

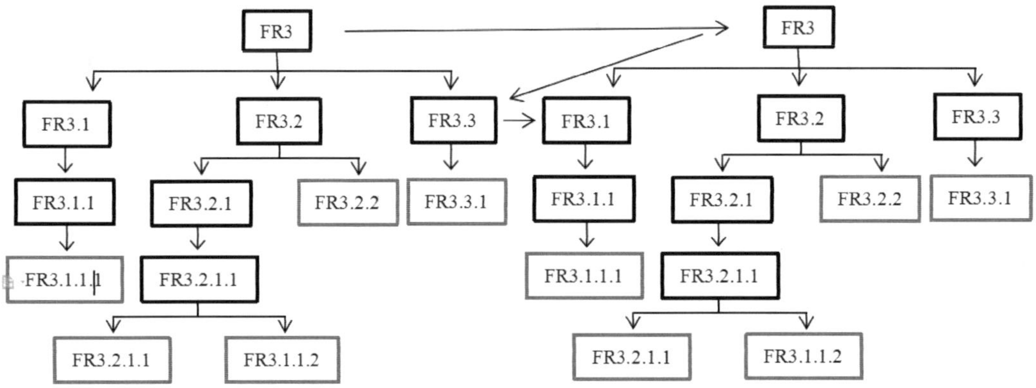

Figure 4. Hierarchical structure of CA3

Based on Figure 4, design parameter of the turbine has 3 blades (DP3.1.1.1) which are used to get an efficient of turbine spin (FR3.1.1.1). This portable hydroelectric generator is designed with a tube shape (DP3.1.1) to maximize the turbine spin (FR3.1.1) and minimize a friction against the water flow (FR3.1). To provided an attractive design color (FR3.4), this product uses a color combination between white (248,243,222) and brown (74,0,12) (DP3.4)

This product used a round power button with a deep basin (DP3.2.1) and white color (DP3.2.1.1.2) to improve the clarity (FR3.2.1.1.2). This power button also used the right dimension (DP3.2.1.1) to minimize a confusion (FR3.2.1.1). The power button has diameter of 2.75 cm (DP3.2.1.1.1) to prevent discomfort able of the fingers (FR3.2.1.1.1) and also used to minimize the miss click intensity (FR3.2.1). Furthermore, this product also used white LED Light (DP3.2.2) to Minimize the energy usage for indicator light (FR3.2.2). The power button and LED Light has a function to provide an easiness in operation (FR3.2).

As the energy storage, this product uses deep cycle battery (DP3.3.1) for an efficient energy storage (FR3.3.1) to maximize the energy storage (FR3.3)

3.2.4. Design parameter for satisfy CA4: lightweight

Figure 5 shows the hierarchical structure of mapping process from FR4: make it easy lift or carry the product to DP3: lightweight design.

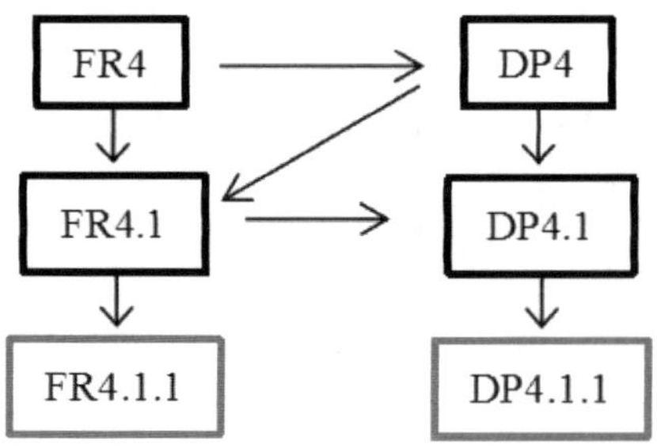

Figure 5. Hierarchical structure of CA4

Based on figure 5, design parameter weight of the product is between 3.5-4 kg (DP4.4.1) to minimize the injury when the product lifted (FR4.4.1). it is important to design a lightweight product (DP4.1) to provide an easiness to lift the product (FR4).

3.2.5. Design parameter for satisfy CA5: small size

Figure 6 shows the hierarchical structure of mapping process from FR5: easily stored to DP3: Adjustable Design.

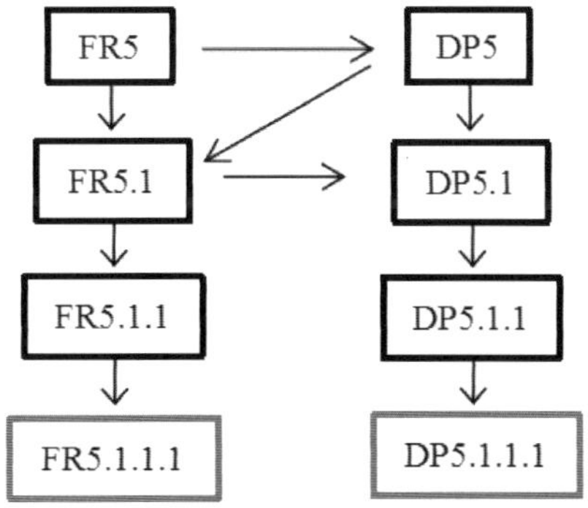

Figure 6. Hierarchical structure of CA5

Based on figure 6, design parameter of the product has 2 levels of form adjustment (DP5.1). They are closed turbine and opened turbine. The closed turbine has length 32 cm (DP5.1.2.1) with the function is to reduce space for saving the product (FR5.1.2.1). and the other form has length 36 cm (DP5.1.1.1) with the function to provide space for the water flow from the turbine section (FR5.1.1.1). The purpose of these 2 levels of form adjustment is to provide the product to be practicality in use (FR5).

3.2.6. Design parameter for satisfy CA6: easy to carry

Figure 7 shows the hierarchical structure of mapping process from FR4: practical to carry before or after use to DP3: ergonomic design.

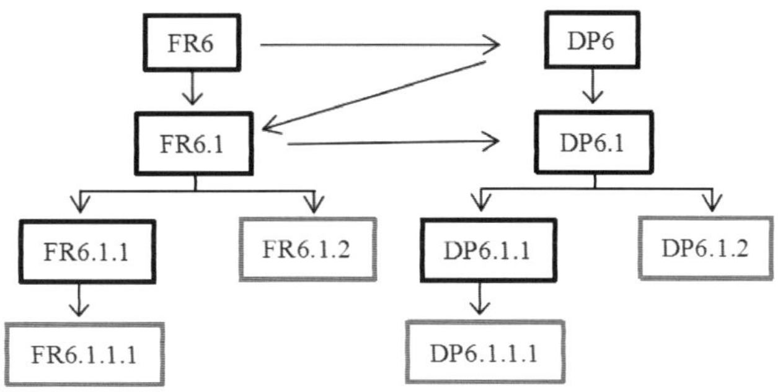

Figure 7. Hierarchical structure of CA6

Based on figure 7, design parameter of the handle are made from the rubber material (DP6.1.1.1). This material has a function minimize the slippery handle when held (FR6.1.1.1) with a dimension of width 7.62 cm (DP6.1.2) and diameter 8.68 cm (DP6.1.1). The handle was designed based on anthropometry data (DP6.1) to provided a comfortable handle (FR6.1) so the handle will be comfortable in use (FR6).

3.3. Virtual design of portable hydroelectric generator

After mapping process, the next step is creating a virtual design of Portable Hydroelectric Generator based on the design parameters that have been mapped in the previous process. The virtual design of Portable Hydroelectric Generator shown on figure 8.

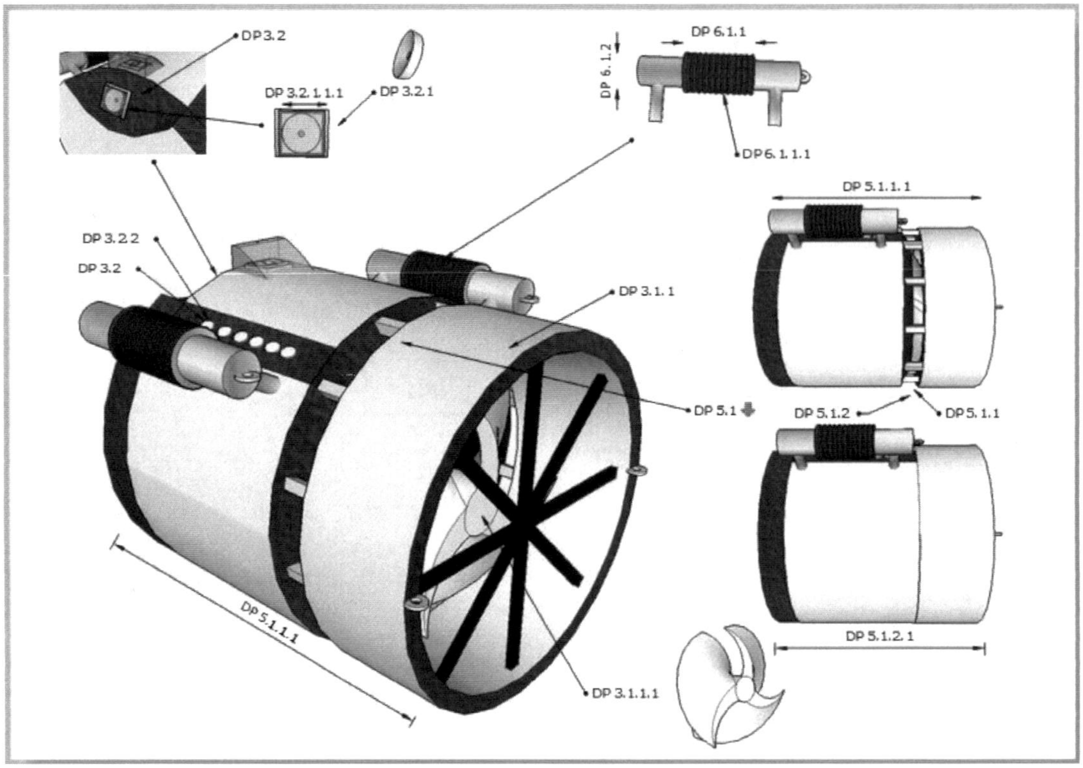

Figure 8. Virtual Design of Portable Hydroelectric Generator

3.4. Result of validation.

There are two kinds of statistical test used in this study. They are Marginal Homogeneity and Wilcoxon test. Table 3.1 present the result of marginal homogeneity test and table 3.2 present the result of Wilcoxon test.

Table 1. Marginal homogenity test result

Attribute	Asymp.sig (2-tailed)
Durable	0.11
Waterproof	0.190
Attractive Design	0.137
Lightweight	0.09
Small Size	0.61
Easy to carry	0.12

Based on the table 3.1, can be seen that the asymp.sig (2-tailed) value of the attributes > 0.05. It means that Ho is accepted so it is possible to conclude that the proposed design of portable hydroelectric generator can meet what the users require.

Table 2. Wilcoxon signed rank test result

Attribute	Asymp.sig (2-tailed)
Durable	0.00
Waterproof	0.03
Attractive Design	0.00

Lightweight	0.01
Small Size	0.01
Easy to carry	0.00

Based on the table 3.2, can be seen that the asymp.sig (2-tailed) value of the attributes < 0.05. It means that Ho is unaccepted so it can concluded that there is a difference between the proposed design and the existing design of portable hydroelectric generator. It means the proposed design is more durable, more waterproof, more attractive, lighter, smaller, and easier to carry such that it can increase satisfaction of users as much as 73,67%.

4. Conclusion

Conclusions of this study are as follow:

There is six attributes that customer required to redesign the portable hydroelectric generator. They are durable, waterproof, attractive design, lightweight, small size, and easy to carry

Design parameter for a durable attribute is an iron material for turbine and plastic material for the body. Then, design parameter for a waterproof attribute is high fiber material that is abs plastic for the body. For attractive design attribute, design parameter is tube form on the body with a color combination between white (248,243,222) and brown (74,0,12), 3 blades on the turbine, a round power button with a deep basin, white LED light, and deep cycle battery as the energy storage. Design parameter for a lightweight attribute is the product weight between 3.5-4 kg. While design parameter for small size attribute has 2 levels of form adjustment. If the turbine used, the length of product 36 cm, but when the turbine is not used, the length 30cm. form adjustment was controlled automatically And design parameter for easy to carry attribute are rubber handle with a dimension of width 7.62 cm and diameter 4.34 cm.

The proposed design of portable hydroelectric generator is valid to satisfy the user requirement and increase satisfaction of users as much as 73,67% which it is more durable, more waterproof, more attractive, lighter, smaller, and easier to carry such that it can increase satisfaction of users as much as 73,67% at 5% of significant level.

5. Reference

[1] Hasanov, F, J., & Mikayilov, J, I. 2017. The Impact of Age Groups of Consumption Residential Electricity in Azerbaijan. *Communist and Post-Communist Studies* 50, 339-351.

[2] Gultom, T, T. 2017. Pemenuhan Sumber Tenaga Listrik di Indonesia. *Jurnal Ilmiah Research Sains* 3(1), 130-138.

[3] BPPT. 2016. *Outlook Energi Indonesia 2016*. Jakarta : Badan Pengkaji dan Penerapan Teknologi.

[4] KESDM. 2015. *Statistik Ketenagalistrikan Tahun 2015*. Jakarta: Kementrian Energi dan Sumber Daya Mineral

[5] BPPT. 2017. *Outlook Energi Indonesia 2016*. Jakarta : Badan Pengkaji dan Penerapan Teknologi.

[6] Shantika, T., & Ridwan, M. 2013. Perancangan Prototipe Pikohidro Portable 200 Watt. *Seminar Nasional XII Rekayasa dan Aplikasi Teknik Mesin di Industri*, 39-48

[7] Balansay, M, L, V., Añonuevo, M, R, Q., Cuenca, R, M., & Garbin, R, V. 2015. Portable hybrid Powered Water Filtration Device. *Asia Pacific Journal of Multidiciplinary Research* 3(3), 136-140.

[8] Zhou, D., & Deng, Z, D. 2017. Ultra-Low-Head Hydroelectric Tecnology : a Review. *Reneable and Sustainable Review* 78, 23-30

[9] Patel, S, U., & Pakale, P, N. 2015. Study on Power Generation Using Cross Flow Water Turbine in Micro Hydro Power Plant. *International Journal of Research in Engineering and Technology* 4(5), 1-4.

[10] Apprilianto, A., Indarto., & Prajitno. 2013. Design of Prototype Hydro Coil Turbine Applied as Micro Hydro Solution. *ASEAN Journal of System Engineering* 1(2), 72-76.

[11] Devrim, Y., Devrim, H., & Eoglu, I. 2015. Development of 500 W PEM Fuel Cell Stack For Portable Power Generators. *International Journal of Hydrogen Energy* 40, 7707-7719.

[12] Saisirirat, P., & Joommanee, B. 2017. Study on the Performance of the Micro Direct Ethanol Fuel Cell (Micro-DEFC) for Applying with the Portable Electronic Devices. *Energy Procedia* 138, 187-192.

[13] Soewardi, H., & Pradana, V. 2016. Developing Features of Water Faucet by Using User Centered Design Approach. *ARPN Journal of Engineering and Applied Sciences* 11(7), 4734-4738.

[14] Lee, D, G., & Suh, N, P. 2006. *Axiomatic Design and Fabrication of Composite Structures: Application in Robots, Machine Tools, and Automobiles*. New York: Oxford University Press.

[15] Yamin, S., & Kurniawan, H. 2009. *SPSS Complete* . Jakarta : Salemba Infotek.

[16] Sheskin, D. 2004. *Handbook of Parametric and Nonparametric Statistical Procedures Third Edition*. Washington : Chapman & Hall/CRC.

[17] Wang, W., Zhao, Y., Liu, H., & Song, S. 2017. Fabrication and Mechanism of Cement-Based Waterpoof Material Using Silicate Tailings From Reverse Flotation. *Powder Tecnology* 315, 422-429.

[18] Ceschin, F., & Gaziulusoy, I. 2016. Evolution of Design for Sustainability: From Product Design to Design for System Innovations and Transitions. *Design Studies* 47, 118-163.

IOP Conf. Series: Earth and Environmental Science **171** (2018) 012035 doi:10.1088/1755-1315/171/1/012035

Microbial Protein Enrichment and Treatment of Winery Residue from Fungi and Yeast by Syntrophic Fermentation

Rui Gao,Qiuying He,Yuanyuan Chen,Yuting Duan, Li Xie[*]

Key Laboratory of Yangtze River Water Environment, Institute of Biofilm Technology, College of Environmental Science and Engineering, Tongji University
E-mail: sally.xieli@tongji.edu.cn

Abstract. Yellow wine lees and rice wastewater are seasonally discharged with high amount of organics. The potential of cultivating *Candida utilis* and *Geotrichum candidum* to produce value- added single cell protein (SCP) and simultaneously bio-transform the wastes was investigated. A series of batch experiments were conducted under sterile condition. Co-culture matrix of *Candida utilis* and *Geotrichum candidum* resulted in the highest biomass and crude protein yield of 4.2g increased biomass/initial biomass and 68.4%, respectively. Response surface analysis was applied to optimize the fermentation process. The optimal conditions for SCP production with 66.3% of crude protein content were fermentation time of 4 days, solid-to-liquid ratio of 6% (w/v), inoculation proportion of 1:1 (ratio of co-cultures (v/v)), inoculum dose of 18% (v/v), and aeration rate of 1.4 volume of air/volume of reactor per minute. The soluble chemical oxygen demand (SCOD) and the total nitrogen removal efficiency were 78% and 55% respectively. Acetic acid was the main species contributing to the residual SCOD. Furthermore, the contents of essential amino acids closely matched commercial standard for fishmeal and soybean meal, providing high possibility of animal food application.

1. Introduction

Certain amount of wine lees and concentrated wastewater are produced during the wine-making process [1]. These residues contain high organic content resulting in a considerable chemical oxygen demand (COD>50,000 mg/L), and low pH values (pH<5) [2]. In addition, the high content of solid waste and wastewater are produced seasonally, mainly during the harvest period. It is similar with the grape wine and Japanese rice wine which are all seasonally generated, leading to the uneven and variation discharge of wine lees and wastewater. Thus the fluctuated discharge limits the application of conventional biological treatment. For example it was found that installation and operation of anaerobic digester for Japanese rice wine lees treatment were very costly [3]. Currently, the winery wastes from sugarcane ethanol, corn ethanol, and grape winery were utilized and bio-transformed for value-added products like single cell protein (SCP) [4, 5].

SCP as a source of protein can be used for animal feed, appearing as an economically and environmentally friendly alternative. Yeast biomass production as SCP using *Candida utilis* species has been extensively studied [6,7]. While syntrophic fermentation with fungi and yeast might be critical for SCP yield improvement [8]. The growth of syntrophic microbial was found to be depended on operational conditions, such as substrate, and aeration rates [4, 5]. The transformation of organics were seldom discussed. More work has to be conducted on it.

Therefore in this study, Chinese yellow wine lees and its rice wastewater were chosen as the typical winery residue and examined. The residues contain high amount of organics, such as amino acids, lactic acids, and protein, which probably are suitable substrates for SCP production. The main objective of this study was to evaluate the production and quantity of microbial biomass in mixed cultures of *Candida utilis* and *Geotrichum candidum*. The fermentation process was optimized, and the pollutants transformation and characterization of the obtained SCP were discussed thereafter.

2. Methods

2.1. Substrate

The yellow wine lees were collected after pressing and the rice wastewater was collected after rice steeping. Both of them were obtained from Tangsong Wine, Shaoxing, Zhejiang Province, China. The yellow wine lees contained 36.3% crude protein, 46.7% total solid and 45.7% volatile solid. Rice wastewater was acidic with SCOD 55800±283 mg/L, total nitrogen 2620±85mg N/L, total phosphorus 3338±3 mg/L, suspended solid 34290±630 mg/L. All collected substrate samples were stored in -4 °C.

2.2. Fungi and cultivation conditions

Freeze-dried cultures of the yeast *Candida utilis* and fungus *Geotrichum candidum* were obtained from the China General Microbiological Culture Collection Center (CGMCC). The cultures were reactivated in Yeast Peptone Dextrose (YPD) liquid medium and grown on YPD solid medium at 28°C for 3 days. To prepare fungal spore suspension, the fungal spores were harvested and kept in YPD liquid medium at 28°C and 150 rpm. After 20h of cultivation, a spore suspension of 10^7 spores/mL can be inoculated into substrate as fungal inoculum.

Yellow wine lees and rice wastewater were mixed together with a solid-to-liquid ratio of 6% (w/v) to prepare the cultivation substrate. The fungal inoculum was inoculated into a 500mL flask containing 300mL substrate to start the fermentation. Fungal cultivation was conducted at 28°C with nutrient supplementation of peptone. Fungal biomass was harvested after 4d of fermentation.

2.3. Analytical method

The daily fermentation samples were analyzed for SCOD, TN, total sugars, crude protein and amino acids. During fungal cultivation, liquid samples were filtered through 0.45μm filters before analysis. SCOD, TN, NH_4^+-N and TP were analyzed using vial kits from HACH (HACH Company, Loveland, CO, USA). The suspended solids analyses were performed as described in Standard Methods [9]. Crude protein was determined by anthrone colorimetric method and semi-micro Kjeldahl method, and expressed as N×6.25, respectively. The dry weights of fungal biomass samples were measured to calculate the fungal biomass yields defined as $g_{biomass\ increase}/g_{initial\ biomass}$.

3. Results and discussion

3.1. Microbial protein yield and protein content of mono and co-culture

The time course of microbial protein production and crude protein content are described in Figure 1 under mono (*Candida utilis*, *Geotrichum candidum* respectively) and co-culture matrix. As shown, lag phase was observed in the first 3 days fermentation, with slow increase of biomass yield and protein content. The biomass yield and protein content achieved the highest amount at the 4^{th}day, and then remained steady after that. Compared to *Geotrichum candidum*, *Candida utilis* presented better performance on microbial biomass production, with the yield of 3.68 g increased biomass/initial biomass and protein content of 60.1%. The biomass yield was further improved to 4.21 g increased biomass/initial biomass and protein content to 68.4% in the co-culture matrix, while the biomass production in the mono cultures (*Candida utilis*, *Geotrichum candidum*) was only 1.07 and 3.68 g increased biomass/initial biomass with protein content of 60.1% and 52.4%, respectively.

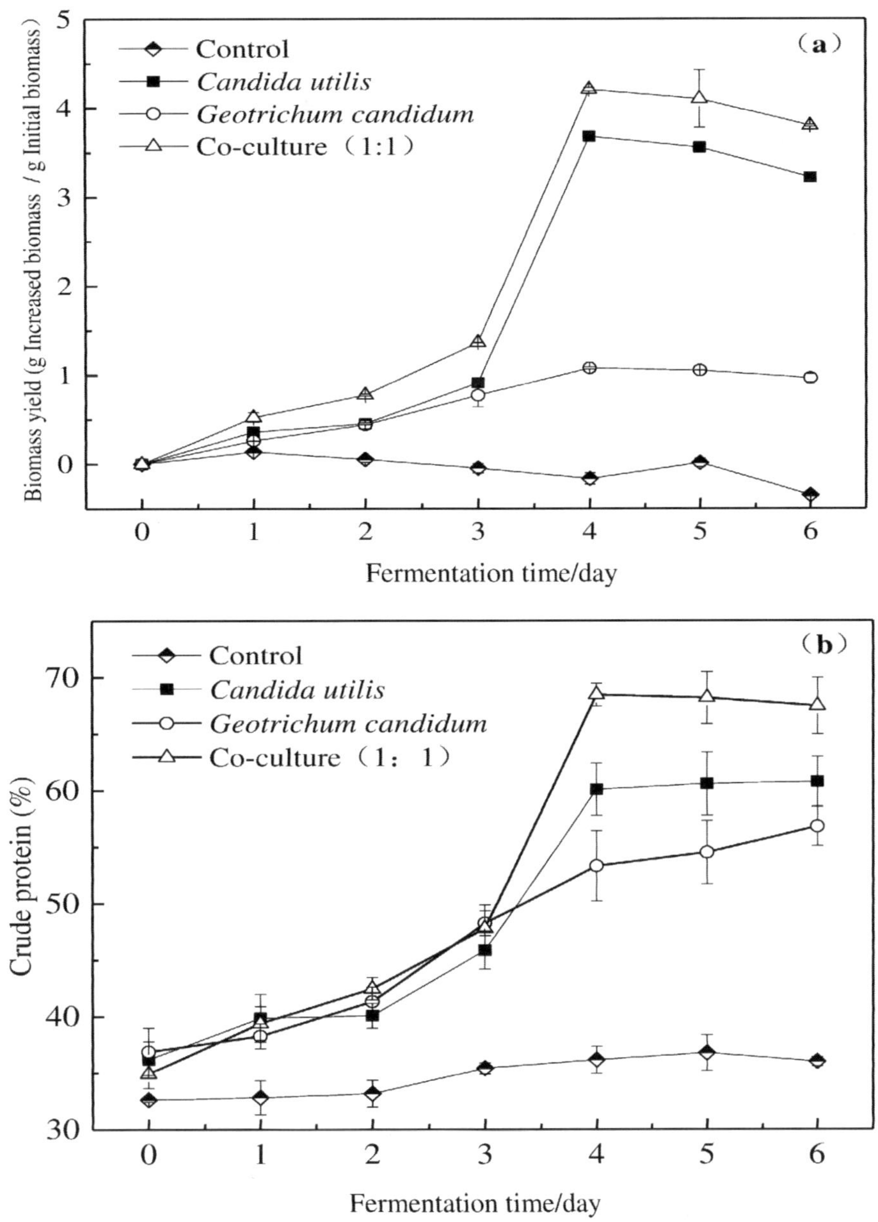

Figure 1. Effects of mono and co-culture on biomass yield (a) and crude protein content (b)

The results indicated that the maximum biomass and protein production was achieved by co-culture of *Candida utilis* and *Geotrichum candidum*. These were in agreement with the study of Yao [10], that co-culture system was beneficial for SCP production. In their study, co-culture of three species, *Candida tropicalis*, *Geotrichum candidum* and *Candida utilis* can produce the highest microbial biomass from molasses wastewater. The fungal assimilation of cellulose provided monosaccharide and disaccharide, which probably contributed to better growth of the yeast. Therefore, co-culture matrix of *Candida utilis* and *Geotrichum candidum* and fermentation time of 4 days were selected for inoculation for the following study.

3.2. Optimization of co-culture ratio, inoculum dose and aeration rate by response surface analysis

The response surface methodology was applied to further optimize the co-culture fermentation process to improve the crude protein content in total fungal biomass. The impact of reaction variables including inoculum dose, co-culture ratio and aeration rate on crude protein content yield are shown in Figure 2 by holding one factor at the zero level and varying another two factors at a time. Figure 2a demonstrated that co-culture ratio and aeration rate were almost independent with respect to the crude protein content and were without interaction effects among each other. It was observed that the crude protein content increased to a maximum of 68.6% with the co-culture ratio of 1:1, aeration rate of 1.48vvm. Figure 2b shows that the decrease of inoculum dose led to the decrease of crude protein content in the range of aeration rate investigated (0.5-2vvm), while it had a greater effect on the crude protein content at lower aeration rate (<1.25vvm). The optimal crude protein content yield was 66.0% when the inoculum dose was 18.65% and aeration rate was 1.42vvm. Similarly from Figure 2c, the decrease of inoculum dose led to the decrease of crude protein content in the range of co-culture ratio investigated (0.3-3), but the effect was more obvious at lower co-culture ratio (<0.3). According to the highest peaks, crude protein content could be as high as 67.9% when the inoculum dose was 16.7% and co-culture ratio was 1:1. It was found that co-culture ratio, inoculum dose and aeration rate as well as interaction between parameters had a positive effect on soluble response.

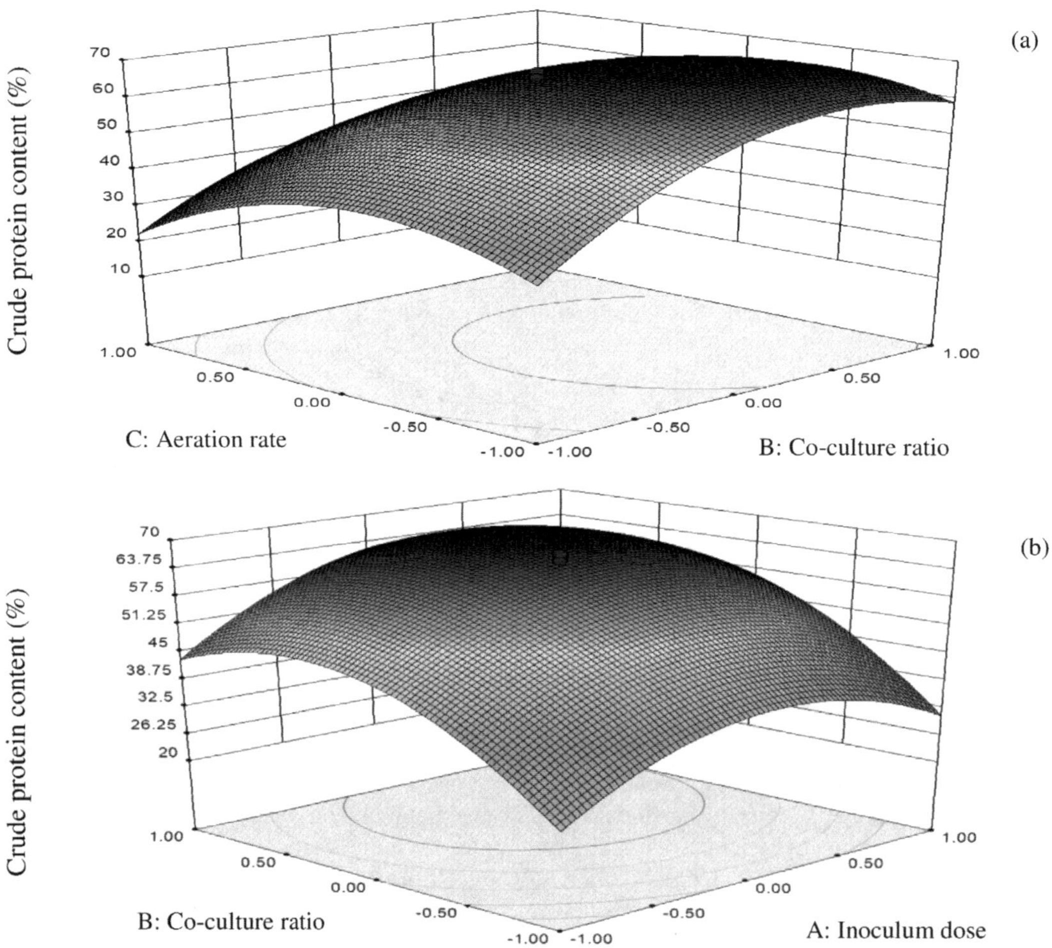

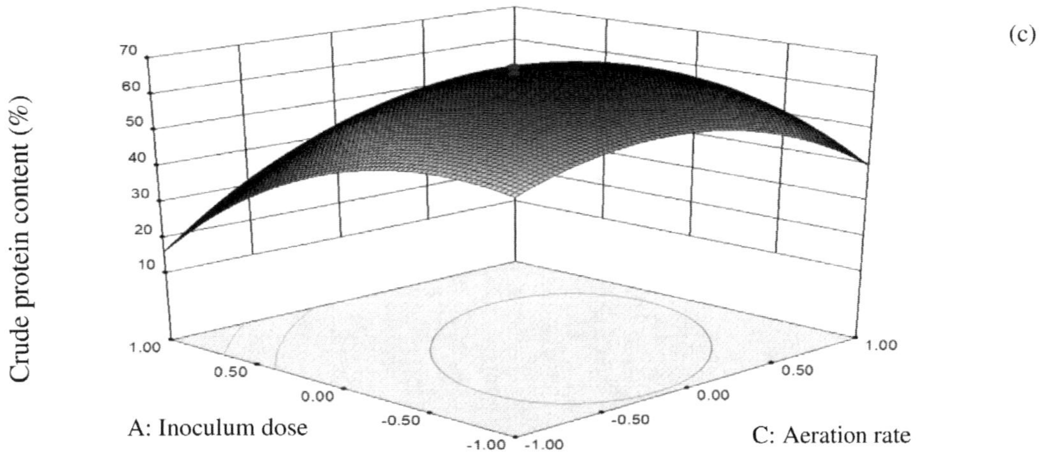

Figure 2. 3D surface plots and contour plots for crude protein content. **a)** Effect of co-culture ratio and aeration rate with constant inoculum dose(%), **b)** Effect of inoculum dose and aeration rate with constant co-culture ratio, **c)** Effect of inoculum dose and co-culture ratio with constant aeration rate(vvm).

The highest peaks could be observed in each response surface plot which means that the maximum SCP production can be obtained inside the experimental range. The maximum crude protein content was estimated by the optimum conditions were: inoculum dose 18.4%, co-culture ratio 1:1 and aeration rate 1.45vvm, corresponding to the highest crude protein content of 68.7%. In order to confirm the effectiveness of optimal condition obtained by RSM, an additional three experiments were conducted. The crude protein content in the experiment was 66.3%, which was close to the predicted result. These results clearly demonstrated the effectiveness of the model.

3.3. Bioransformation of organics and amino acids in SCP

SCOD and TN are the main indicators for organics in wastes. As shown in Figure 3a, SCOD can be effectively removed in both mono- and co-culture matrixes. It was found that the SCOD was reduced from initial 77200 mg/L to 16880 mg/L after 4 days fermentation in co-culture system, with removal efficiency of 78.1%. Most organics was utilized by yeast and fungi to synthesize SCP, and the residual COD was found to be short chain volatile fatty acids, mainly acetate in this study (data not shown).
The performance of TN removal by three culture systems are compared in Figure 3b. The TN removal rate was highest in *Geotrichum candidum* mono-culture, which displayed 67.0% removal efficiency. In co-culture fermentation system, the TN removal efficiency was the lowest, which was different from the COD removal performance. The mechanisms of COD and TN degradation by three culture system may be different, which can be further explored in the future.

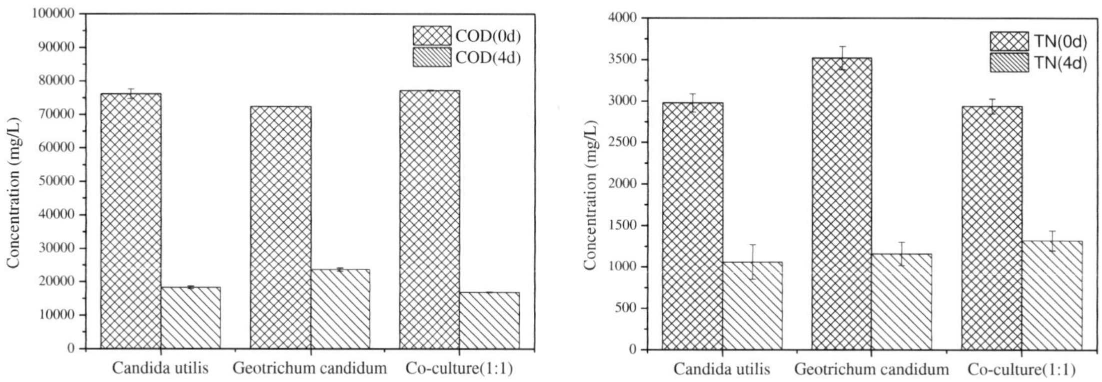

Figure 3. Removal efficiency of COD and TN in the substrate

Fish meal was the most cost-effective animal protein source and as its probable substitute, soybean meal was studied recently [11]. Both contained essential amino acids, among which lysine and methionine are limiting ones for all animals. The composition of amino acids in SCP obtained by fermentation in this study was examined and compared with fish meal and soybean meal. According to the tables of feed composition and nutritive values in China[12], although the compositions of lysine produced in this study(2.20%) were slightly lower than that of fish meal(3.93%)and soybean meal(3.38%), SCP contained more isoleucine (4.12%) and methionine (1.47%), which enabled it to be an efficient nutrition supplement. In conclusion, SCP can be considered as an economic chance for animal food industry by co-feeding with fish meal.

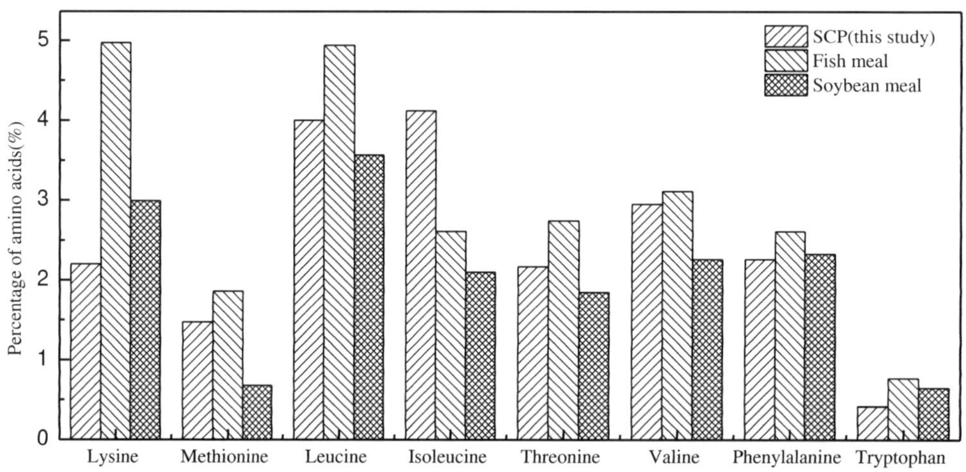

Figure 4. Essential amino acid compositions of this study together with soybean meal and fish meal

4. Conclusions

The feasibility of syntrophic fermentation from winery waste for both SCP production and waste recycling was evaluated. Co-culture matrix of *Candida utilis* and *Geotrichum candidum,* combined with optimizing conditions obtained better results on biomass production and protein content compared to mono-culture. 78.1% SCOD was removed and the residue organics were short chain fatty acids and acetate was the main species. Higher amount of methionine than that of soybean meal was observed in the fungal biomass, which ensures better growth of animals. Furthermore, co-feeding the SCP with commercial animal feedstock can improve the amount of essential amino acid and partly offset the total cost.

IOP Conf. Series: Earth and Environmental Science **171** (2018) 012035 doi:10.1088/1755-1315/171/1/012035

5. References

[1] Da Ros C, Cavinato C, Bolzonella D and Pavan P 2016 Renewable energy from thermophilic anaerobic digestion of winery residue: preliminary evidence from batch and continuous lab-scale trials. *Biomass and Bioenergy* 91 150-9.

[2] España-Gamboa E, Mijangos-Cortes J, Barahona-Perez L, Dominguez-Maldonado J, Hernández-Zarate G and Alzate-Gaviria L 2011 Vinasses: characterization and treatments. *Waste Management & Research* 29(12) 1235-50.

[3] Kanai M, Ferre V, Wakahara S, Yamamoto T and Moro M 2010 A novel combination of methane fermentation and MBR-Kubota Submerged Anaerobic Membrane Bioreactor process. *Desalination* 250(3) 964-7.

[4] Rasmussen M, Khanal S, Pometto A and van Leeuwen J 2014 Water reclamation and value-added animal feed from corn-ethanol stillage by fungal processing. *Bioresource technology* 151 284-90.

[5] Nitayavardhana S, Issarapayup K, Pavasant P and Khanal S 2013 Production of protein-rich fungal biomass in an airlift bioreactor using vinasse as substrate. *Bioresource technology* 133 301-6.

[6] Yunus F, Nadeem M and Rashid F 2015 Single-cell protein production through microbial conversion of lignocellulosic residue (wheat bran) for animal feed. *Journal of the Institute of Brewing* 121(4)553-7.

[7] Zheng S, Yang M and Yang Z 2005 Biomass production of yeast isolate from salad oil manufacturing wastewater. *Bioresource Technology* 96(10) 1183-87.

[8] van Heerden A, van Zyl W, Cruywagen C, Mouton M and Botha A 2011 The lignicolous fungus Coniochaeta pulveracea and its interactions with syntrophic yeasts from the woody phylloplane. *Microbial ecology* 62(3) 609.

[9] Greenberg A 2005 Standard Methods for the Examination of Water and Wastewater. *American Public Works Association/Water Environment Federation* 9-37.

[10] Yao Y, Lin H, Tan L, You S and Zhang L 2014 Research of producing single cell protein with molasses wastewater. *Advanced Materials Research* 838 2785-90.

[11] Wang Y, Yu S, Wang Y, Che J, Zhao L, Bu X and Yang Y 2016 Effect of replacing fish meal with soybean meal on growth, feed utilization and nitrogen and phosphorus excretion of juvenile Pseudobagrus ussuriensis. Aquaculture Research 47(10) 3145-55.

[12] Xiong M, Luo Q, Zhao F and Pang Z 2015 tables of feed composition and nutritive values (26th edition). *China Feed* 21 23-33.

Acknowledgements

The authors gratefully appreciated the National Science Foundation of China (51378373) project and National Undergraduate Training Program for Innovation and Entrepreneurship (201710247045, 2012BAC17B02).

Utilization of Shrimp Shell Waste as Matrix Controller by Using Ionotropic Gelation Method in Slow Release Fertilizer Based on Environmental Conservation

W D P Rengga, M A Mubarok, N S Cahyarini

Department of Chemical Engineering, Faculty of Engineering, Universitas Negeri Semarang, Campus Sekaran Gunungpati Semarang 50229, Indonesia
E-mail : wdpitar@mail.unnes.ac.id

Abstract. The Availability of shrimp waste in Indonesia is huge, especially in shrimp pond areas. However, it has not been utilized optimally and caused environmental pollution. The weight of shrimp waste reaches 30%-40% of its weight. Most of the shrimp waste is dominated with shrimp shells. Shrimp shells containing 60% -70% chitin. It will produce 15-20% yield of chitosan synthesized from chitin. On the other hand, the use of fertilizer continues to increase. Fertilizer is an essential nutrient for plants if lack nutrients to make plants can't grow normally. Preparation of chitosan as matrix controller begins from shrimp shell synthesis which is obtained by deacetylation method with the highest deacetylation degree of 74.5% and the optimum deacetylation is 71.2%. The SEM test which shows the thickening and rougher structure of the fertilizer surface, fertilizer coated by macrospheres has a thickener and coarser surface so as to inhibit the release of elements inside. The shrimp shell waste based coating is able to inhibit the release of NPK contained in the fertilizer. The occurrence of ionic crosslinking shown by the presence of N-H groups and -P=O on material.

1. Introduction

Chitosan is a biopolymer material that consist of β -(1-4) D-glucosamine units obtained by chemical or enzymatic synthesis, by a deacetylation process of the amino acid polysaccharide contained in chitin. Chitin compounds are generally found as structural compounds of arthropods (eg, lobster and shrimp) and are formed from β - (1-4) units of N-acetyl D-glucosamine. Chitosan has biocompatibility and biodegradation in the environment, is non-toxic and has excellent absorption properties when compared to materials made from synthetic plastics [1]. Chitosan has many benefits in the industry, including the food industry, agriculture, water treatment, medical device manufacturing and medicine, synthetic tissue, cosmetics, antifungal and antibacterial products [2], and synthetic membranes [3] [4] [5].

In the fishery sector, the processing industry produces shrimp shell waste. Shrimp shell waste has not been optimally utilized. Waste is only used as a mixture of making shrimp paste and animal feed mixture. Utilization is only able to overcome a small part of the total amount of shrimp waste and has a very small economic value. According to the Badan Pusat Statistik in 2012 noted that shrimp production in Indonesia increased by 5% every year. The Ministry of Marine Affairs and Fisheries also reported that shrimp export volume in 2010-2014 increased from 113 937.0 tonnes to 148 519.4 tonnes, with the foreign exchange value of USD 1 706 784.4 in 2014. However, there are major problems in shrimp processing industry, which produces waste in a very large amount. The waste contains proteins (35-50%), chitin (15-25% of dry weight) and inorganic compounds (calcium carbonate) considered to be the main cause of environmental pollution.

Shown in Figure 1 that describing needs and production of NPK fertilizer in Indonesia from 2007 until 2015. It is necessary to save the use of fertilizer to reduce the increase in the number of fertilizer needs, one way to overcome it is to create innovations in fertilizer, which is making slow release fertilizer. Slow release fertilizer is a fertilizer capable of controlling the speed of release of nutrient elements lost due to dissolve in water, evaporate and the denitrification process against the fertilizer itself. Increasing the effectiveness of the fertilizer by modifying the product into a slow-release fertilizer. Fertilizer of slow modification is made by using chitosan obtained from shrimp waste synthesis as a coating on fertilizer. Thus the release rate of soil nutrients can be adapted to the needs of the plant. In addition to the innovation of low-fertilizer can reduce the amount of fertilizer consumption. The innovation of slow-release fertilizer is made from chitosan, besides can be made using zeolite [6], bentonite and alginate [7], starch-polyvinyl alcohol [8], polymers [9], [10] using kraft lignin pine trees, [11] with acrylamide copolymers, [12] using softwood kraft lignin for coating urea, [13] using sodium alginate and biochar for superabsorbent slow-release fertilizer. Utilization of shrimp shells that are synthesized into chitosan expected to solve the problem of waste shrimp while enhancing the economic value of shrimp shells. In addition, shrimp shells used as a control of soil nutrient release can improve the effectiveness and efficiency of fertilizers. The method used in the manufacture of fertilizer is slow-release is the ionotropic gelation method other than the ionotropic gelation method there are other methods such as matrix hydrophilic, and hydrophobic [14]. [15] encapsulated by starch-based superabsorbent polymer. The basic innovation of this research is that many coatings of fertilizers using polymers are not biodegradable to the detriment of the environment. In addition, the production cost used is also relatively large. The use of ionotropic gelation method is assumed to avoid the possibility of soil contamination.

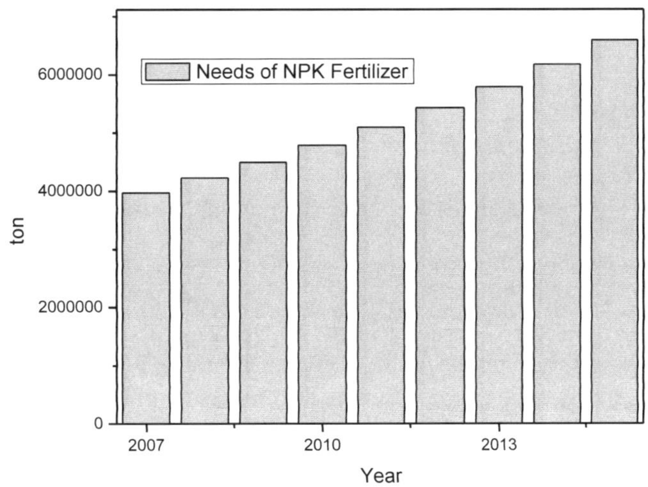

Figure 1. Needs of NPK fertilizer.

2. Materials and method

2.1. Materials

Shrimp shell, HCl 5%, NaOH, pH indicator, acetic acid 1% v/v, potato starch flour, sodium tripolyphosphate 1% wt/wt, NPK fertilizer, aquadest.

2.2. Preparation of slow release fertilizer

Shrimp shells are cleaned and then soaked in 0.5% NaOH solution for 24 hours, wash with water until neutral pH deproteination using 5% NaOH solution for 24 hours, wash water using until neutral pH demineralization with 5% HCl solution for 24 hours , wash again until pH neutral, dry at room temperature then formed chitin, to change chitin into chitosan need to do various stages as follows: soak chitin in 50% NaOH solution, at room temperature for 1 week, done stirring then wash with water until neutral pH dry in room temperature puree. In making Macrospheres by Making chitosan solution of chitosan first 3 grams dissolved in 1% acetic acid as much as 100 mL chitosan solution stirred using magnetic stirrer followed by making potato starch gel, potato starch as much as 8 grams in 100 mL of aquadest, do stirring by using magnetic stirrer and heated until the temperature reaches 76 °C. Preparing STTP solution is done by dissolving STTP as much as 1 gram in 100 mL of aquadest measuring pH to reach pH 8.6. Makes macrospheres coating by mixing chitosan solution with potato starch gel, stirring until homogeneous solution, storing mixture at room temperature for 30 minutes make SRF fertilizer by dipping NPK fertilizer with macrosphere, stirring until homogeneous, mixing macrosphere coated fertilizer with STTP then do stirring 2 hours of draining SRF fertilizer at 85°C for 48 hours up to constant.

The manufacture of matrix controller on NPK fertilizer is made by using chitosan feedstock which is distilled from shrimp waste. The coating of fertilizer with several variables such as chitosan, starch, with the addition of STTP and without the addition of STTP with a certain concentration as follows at Table 1.

Table 1. Variabel used in research.

Chitosan	Potato Starch	Coating (time)			STTP		Fertilizer
1	0	1	2	3	Without STTP	With STTP	NPK
1	1	1	2	3	Without STTP	With STTP	NPK
1	2	1	2	3	Without STTP	With STTP	NPK
1	3	1	2	3	Without STTP	With STTP	NPK

2.3. SEM and EDX characterization

Macrospheres chitosan and chitosan are mounted on aluminum with double tape adhesive and coated with gold. The surface morphology is examined by SEM devices with gun emissions which are operated at 3 kV. The micrograph is taken from an enlargement between 500 X and 15000 X.

2.4. FTIR characterization

The chitosan-starch FTIR spectrum without STTP and chitosan with STTP macrospheres were recorded on FTIR spectrophotometers operating in the range of 4000-400 cm-1 at a resolution of 4 cm-1. The samples were mixed thoroughly with potassium bromide at 1:20 pm (KBr sample) mass ratio. KBr discs are composed by compressing the powder mixture with hydraulic pressure.

3. Result and discussion

3.1. SEM characterization

SEM analysis is used to determine the topography that is the surface characteristics of texture, morphology, and macrospheres composition. Here is a micrograph analysis was taken from an enlargement of 500X to 15000X.

SEM test result known a surface difference between coating fertilizer show in Figure 2 and 3. Coating fertilizer at magnification 500X Figure 2a and 1000X Figure 3a formed shaped granule-shaped granules that indicate the coating on the layer of fertilizer, this layer is able to inhibit the release of nutrients in the fertilizer. While in the sample without coating shown in Figure 2b an 3b is not formed a layer of granule-shaped granules or coating surface so that there is no inhibition of release of elements in it. At magnification 10000 Figure 4b and 15000 Figure 5b can be observed in more detail

IOP Conf. Series: Earth and Environmental Science **171** (2018) 012036 doi:10.1088/1755-1315/171/1/012036

coating fertilizers show a surface with a thicker texture and morphology when compared with a coating without coating.

a) b)

Figure 2. SEM test analysis of fertilizer sample at 500X magnification with coating. a) with coating b) without coating.

a) b)

Figure 3. SEM test analysis of fertilizer sample at 1000X magnification. a) with coating b) without coating.

a) b)

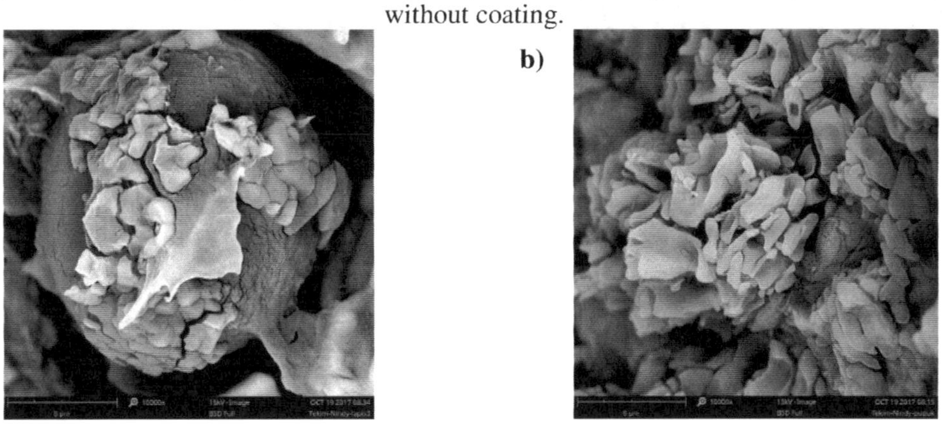

Figure 4. SEM test analysis of fertilizer sample at 10000X magnification. a). with coating b) without coating.

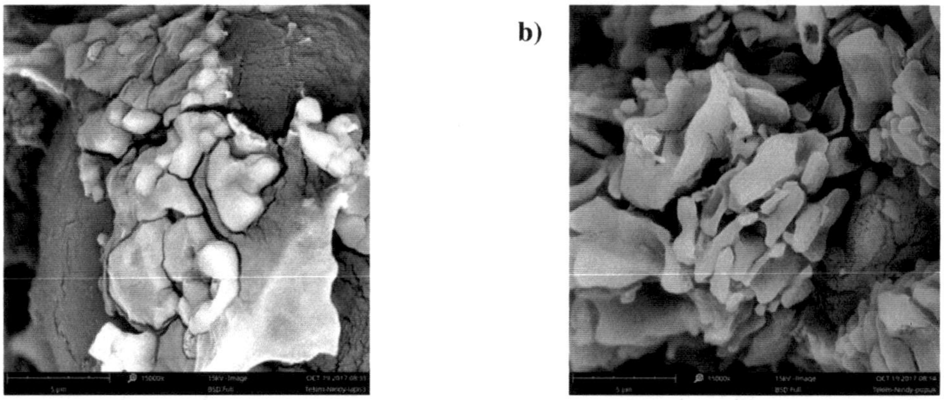

Figure 5. SEM test analysis of fertilizer sample at 15000X magnification. a) with coating b) without coating.

3.2. EDX characterization

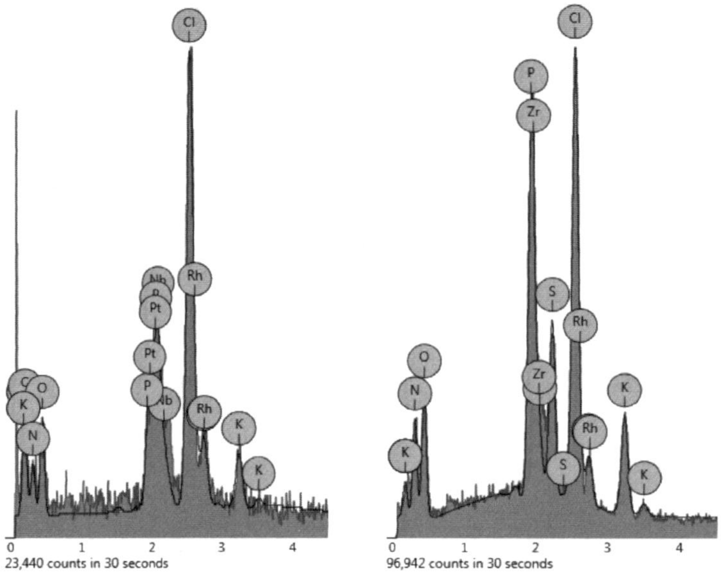

(a) (b)

Figure 6. NPK content analysis on fertilizer a) with coating b) without coating.

Table 2. Result of analysis of fertilizer with coating.

Element Number	Element Symbol	Element	Weight Concentration	Error %
15	P	Phosphorus	1.2	2.6
19	K	Potassium	1.3	0.5
7	N	Nitrogen	6.9	0.5

Table 3. Result of analysis of fertilizer without coating.

Element Number	Element Symbol	Element	Weight Concentration	Error %

15	P	Phosphorus	6.2	0.2
19	K	Potassium	4.2	0.2
7	N	Nitrogen	30.3	0.8

Table 4. Differences of NPK content found on the surface of with coating fertilizer and on the surface of the fertilizer without coating.

Element	Fertilizer with coating	Fertilizer without coating	Ability to inhibit element release
N	6.9	30.3	23.4
P	1.2	6.2	5.17
K	1.3	4.2	3.23

Based on Figure 6 the results of analysis of fertilizer is presented in the following Table 2 and Table 3 show the difference of NPK content found on the surface of with coating fertilizer and on the surface of the fertilizer without coating, presented in the following Table 4. It can be concluded that the content of NPK contained in the surface of with coating fertilizer is smaller when compared with fertilizer without coating so that the release of the elements contained in the fertilizer can be inhibited. Inhibition of release on the fertilizer can be calculated using the ratio value of 3.32 to 5.17 so that the release of elements in the fertilizer can be inhibited ± 3-5 times.

3.3. FTIR characterization

Tests using FTIR tools were performed to gain an understanding of the macrosphere structural changes obtained after the addition of coatings on NPK fertilizers. The spectrum of potato starch FTIR shows a common feature of polysaccharides. In Figure 7 (a) and (b) shows an intense band at 1022 cm-1 and 1736 cm-1 corresponds to the stretch areas C-O and C-H [16]. Shown in Figure 7(a) FTIR test result of the sample with STTP, the pure chitosan spectrum presents the characteristic bands at 3370 cm-1 that are associated with the overlap of the stretching vibrations of the -NH2 and -OH groups [17].

Macrospheres prepared with chitosan only shows the shoulder at 1240 cm-1, which can be given at the stretching vibration of P=O, indicating the presence of phosphate groups as a consequence of ionic crosslinking [18] which a sample is with STTP. A samples using STTP occurs ionic crosslinking. On Figure 7(a) The FTIR spectrum of the matrix obtained from the mixture also revealed the shoulders at 1240 cm-1, indicating the presence of the P-O group and indicating that crosslinking occurs through ionic interactions between the negatively charged group P=O and NH3+ groups in chitosan.

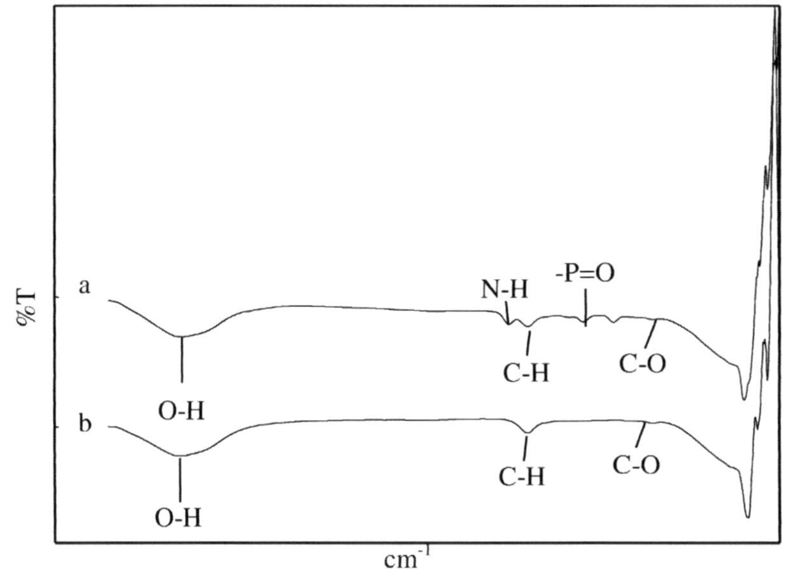

Figure 7. FTIR test result of sample (a) with STTP (b) without STTP.

4. Conclusions

Making chitosan as a matrix controller begins with the synthesis of shrimp shells by deacetylase method. After chitosan was formed making macrospheres by dissolving chitosan and stirring process. The mass of the dry beads is calculated to determine the soluble fraction. Before the test release on fertilizer, the product that has been formed is characterized by various methods such as Scanning Electron Microscope (SEM) test to find out the morphology of the product, there are some other tests to know the characteristics such as FTIR test and EDX after testing is done release test. From the test proved the existence of slow release (slow release) nutrient decomposition on the fertilizer.

5. Acknowledgments

Authors wish to thank PNBP Universitas Negeri Semarang 2017 for financial support through Hibah Peneliti Bagi Mahasiswa.

6. References

[1] Gómez-Ríos D, Barrera-Zapata R and R´ıos-Estepa R 2017 Comparison of process technologies for chitosan production from shrimp shell waste: A Techno-economic approach using Aspen Plus *Food and Bioproducts Processing* **103** 49-57

[2] Peniche C, Argüelles-Monal W and Goycoolea F M 2008 Chitin and Chitosan: Major Sources, Properties and Applications *Monomers, Polymers and Composites from Renewable Resources* 517–34

[3] Mahatmanti F W, Rengga W D P and Kusumastuti E 2017 Adsorption of Pb (II) ion from aqueous solution onto Chitosan/silica/polyrthylene glycol (Ch/Si/P) composites membrane *Materials Science and Engineering* **172**(1)

[4] Mahatmanti F W, Rengga W D P and Kusumastuti E 2017 Chitosan/Silica/Polyethylene Glycol (Ch/Si/P) Composites Membrane as Selective Adsorbent of Rhodamine B from Aqueous Solution *Asian Journal of Chemistry* **19**(2)

[5] Mahatmanti F W, Rengga W D P and Kusumastuti 2018 Dynamic adsorption of mixture of Rhodamine B, Pb (II), Cu (II) and Zn (II) ions on composites chitosan-silica-polyethylene glycol membrane **349**(1)

[6] Rashidzadeh A, Olad A, Salari D and Reyhanitabar A 2014 On the preparation and swelling properties of hydrogel nanocomposite based on Sodium alginate-g-Poly (acrylic acid-co-acrylamide)/Clinoptilolite and its application as slow release fertilizer *Journal of Polymer Research* **21**(2) 1-15

[7] He Y, Wu Z, Tu L, Han Y, Zhang G and Li C 2015 Encapsulation and characterization of slow-release microbial fertilizer from the composites of bentonite and alginate *Applied Clay Science* **109** 68-75

[8] Pon-On W, Charoenphandhu N, Teerapornpuntakit J, Thongbunchoo J, Krishnamra N and Tang I 2014 Mechanical properties, biological activity and protein controlled release by poly(vinyl alcohol)-bioglass/chitosan-collagen composite scaffolds: A bone tissue engineering applications *Material Science and Engineering C* **38** 63-72

[9] Gonzales M E, Diez M C, Cartes P, Monreal C and Navia R 2015 Evaluation of biodegradable polymers as encapsulating agents for the development of a urea controlled-release fertilizer using biochar as support material *Science of the Total Environment* **505** 446-53

[10] Behin J and Sadeghi N 2016 Utilization of waste lignin to prepare controlled-slow release urea *Int J Recycl Org Waste Agricult* **5** 289-99

[11] Guo M, Liu M, Zhan F and Wu L 2005 Preparation and Properties of a Slow-Release Membrane-Encapsulated Urea Fertilizer with Superabsorbent and Moisture Preservation *Ind. Eng. Chem. Res.* **44** 4206-11

[12] Li J, Wang M, She D and Zhao Y 2017 Functionalization of softwood kraft lignin for coating urea as highly efficient nitrogen fertilizer *NWBC* 244-45

[13] Baki M and Abedi-Koupai J 2017 Preparation and characterization of a superabsorbent slow-release fertilizer with sodium alginate and biochar *Applied Polymer Science* **135** 1-8

[14] Azeem B, Kushaari K, Man Z B, Basit and Thanh TH 2014 Review on materials and mthods to produce controlled release coated urea fertilizer *Controlled Release* **181** 11-21

[15] Qiao D, Liu H, Yu L, Bao X, Simon G P, Petinakis E and Chen L 2016 Preparation and characterization of slow-release fertilizer encapsulated by starch-based superabsorbent polymer *Carbohydrate Polymers* **147** 146-54

[16] Capron I, Robert P, Colonna P, Brogly M and Planchot V 2007 Starch in rubbery and glassy states by FTIR spectroscopy *Carbohydrate Polymers* **68**(2) 249-59

[17] Zawadzki, J and Kaczmarek H 2010 Thermal treatment of chitosan in various conditions *Carbohydrate Polymers* **80**(2) 394-400

[18] Pierog M, Gierszewska-Drużyńska M and Ostrowska-Czubenko J 2009 Effect of ionic crosslinking agents on swelling behavior of chitosan hydrogel membranes *Progress on Chemistry and Application of Chitin and its Derivatives. Polish Chitin Society, Łódź*, **75** 82

IOP Conf. Series: Earth and Environmental Science **171** (2018) 012037 doi:10.1088/1755-1315/171/1/012037

Kitchen Waste as a Novel Available Substrate for Prodigiosin Production by *Serratia Marcescense*

Shunxiang Xia[1,2], Enjelia Veony[2] and Qiyin Yang[1]

1 College of Life Science, Nanjing Normal University, Nanjing 210046, People's Republic of China
2 Department of Petroleum Engineering, University of Houston, Houston, TX 77023, USA

Abstract. As a compound with antitumor activities, prodigiosin was produced by *Serratia marcescens* with kitchen waste as a substrate. The initial yield of prodigiosin was 223mg/l and can be further improved with supplementation of nitrogen source and inductor to the medium. A 290% increase of the prodigiosin production was achieved with 1% peptone and 0.2% proline. Physiological fermentation factors such as incubation time (36h), substrate concentration (35g/l), pH of the medium (pH 8.0), and fermentation parameter such as temperature (28°C), agitation speed (150rpm), inoculum level (1%), and liquid volume (20/150ml) all have impact on the prodigiosin production. With optimization, the maximum prodigiosin yield was 890mg/l and kitchen waste was proved to be a novel and economic substrate for prodigison production.

1. Introduction

Although the red pigments produced by a restricted group of *Serratia* genera were discovered at the very beginning of mankind history, further study about the molecular structure and medicinal function of prodigiosin was rudimental until recently. Owing to the rapid advancement in separation science and spectroscopy in the subsequent years, it became clear that "prodigiosin" has a series of close relatives bearing the same pyrrolylpyrromethene ("prodiginine") core with different alkyl substituent [1]. Prodigiosin was also found to be a potent anticancer agent which could induce apoptosis of several in vitro cancer cell lines including hematopoietic cancer cells, colon cancer cells, B-cell and chronic lymphocytic leukemia cells. Many PG--derivatives with lower toxicity like GX15-070 [2] has been used in clinic, and the price of pure prodigiosin was up to $500/mg (Santa Cruz Biotechnology, Inc., U.S.A., 2010). The high price is mainly due to its low fermentation yield and high production cost. Many studies aiming to improve the fermentation yield have been done, including adding amino acids, sodium dodecyl sulfate, lipid into the medium [3] and optimizing fermentation parameter such as pH, dissolved oxygen level, and temperature [4]. But the cost of fermentation medium was still high and little effort to find an inexpensive replaceable medium material was done for prodigiosin production.

Waste recycling represents one of the key elements in many countries to develop a low-carbon economy in the future. Among the total amount of waste, kitchen waste possesses a large part and attracts more attention. Kitchen waste is a kind of organic waste discharged from households, cafeterias and restaurants which was created inevitably every day. With a rapid socio-economic development, the amount of kitchen wastes has been rising in developing countries like China, and is now about 1000t/d in Shanghai, and 1200t/d in Beijing [5]. Recently, widespread public concerns about kitchen waste recycling have risen because of both environmental problems derived from it and

IOP Conf. Series: Earth and Environmental Science **171** (2018) 012037 doi:10.1088/1755-1315/171/1/012037

its potential for utilization. Kitchen garbage has the characteristic of organic component, which make it easier to be deteriorated and hard to handle. On the other hand, the abundant nutrition inside kitchen garbage makes the waste ideal raw materials for producing value added products, such as hydrogen, lactic acid, ethanol, and so on [6]. However more complicated and higher value-added products for example the prodigiosin production from kitchen waste has not been considered yet due to the lack of research on the applicable producing microorganism and fermentation procession.

In this study, the link between prodigiosin production and kitchen waste recycling was established. Kitchen waste could become one available substrate for prodigiosin production by adding only some nitrogen sources. As a result, the utilization of kitchen waste for prodigiosin production could bring down the production cost of prodigiosin and save normal medium material, reducing output of carbon dioxide indirectly, thus it is worth investigating.

2. Methods

2.1. Microorganism and culture conditions

Prodigiosin producing the red wild-type *Serratia marcescens* was isolated in our laboratory from kitchen waste and was maintained on a medium which contains (%) peptone 1.0, and beef extract 0.5, NaCl 0.5, agar 2%, stored at 4°C and sub-culture at one-month intervals.

2.2. Substrate

Kitchen waste material was collected from local kitchen waste handle factory which was then dehydrated and shattered. The waste powder was processed using USA standard sieve set of No.18 and mean particle size of 1.4-1.0mm was obtained.

2.3. Shake-flask culture

Predetermined quantity and concentration of medium was taken into 150 ml Erlenmeyer flasks for culture. In order to optimize the culture medium, different carbon and nitrogen sources were added individually at 1% level to the medium. The medium was adjusted to different pH values by adding NaOH solution to investigate the influence of pH. The medium was mixed thoroughly and autoclaved at 121°C for 20min at 1 kgf/cm^2. Then, the medium in flasks was inoculated with 10-hour grown culture broth under sterile conditions with various inoculation level. Also, in order to study the effect of inductor, predetermined quantity of proline solution was added into the medium through bacterial filter after the medium was autoclaved. Culture condition was optimized by taking flasks under different culture situation including temperature, rotation speed and liquid volume. In the case of incubation time, one-flask contents at every 6 hours were used for extraction and estimation of prodigiosin content. Results reported in this study are the averaged of triplicate findings.

2.4. Prodigiosin extract and content measure

Prodigiosin in the flasks was extracted by acidic methanol (pH 3.0). For each ml of the flask contents, 9 ml acidic methanol was added and shook at 150 rpm for 30 min. Debris was removed by centrifugation at 10,000 x g for 20 min at 4°C, and the concentration of prodigiosin was measured by monitoring the value of absorption at 535 nm with UV–Visible spectrophotometer. The prodigiosin content was calculated using prodigiosin standard curve. The standard prodigiosin used in this research was prepared through silica gel and the prodigiosin was characterized by NMR and MS (date not shown). The purity was determined by HPLC (Agilent HPLC 1100 Series) at 535 nm. Methanol and ultrapure water (8:2 v/v) were used as a mobile phase at a flow rate of 1.0 ml/min through the column (Agilent, Kromasil-C18) at 40°C.

2.5. Measurement of biomass

1 ml of the flask content was added to appropriate water and the absorbance of diluted samples was measured in UV–Visible spectrophotometer at 660 nm. The biomass was calculated using OD660 of broth versus dry cell weight standard curve.

3. Results and discussion

3.1. Evaluation of kitchen waste material for prodigiosin production

The utilization of kitchen waste for prodigiosin production depends on both the microorganism characteristics and the waste composition. *Serratia marcescens* could get a high yield of prodigiosin on medium containing proper carbon and nitrogen sources. Using modified LB medium, Tao obtained maximum prodigiosin production at 583 mg/l [7], and Wei obtained 790 mg/l when the LB was oil [8]. Kitchen waste has a rich carbon source, moderate nitrogen source and low toxic material, which enable it as an ideal substrate for *Serratia marcescens* to grow and produce prodigiosin. When nitrogen source was not added, the prodigiosin production was 223 mg/l and the maximum production 870 mg/l could be obtained when adding peptone 1% and proline 0.1%. Moreover, the cost of fermentation material was brought down sharply as kitchen waste was inexpensive and easy to obtain.

3.2. Role of incubation time

The culture was conducted in rude medium (35 g/l kitchen waste material, pH 8.0) with initial fermentation condition (28 °C, 150 rpm, 1% inoculum level, 20 ml in 150 ml flasks), and the prodigiosin and biomass production curves were shown in figure 1.

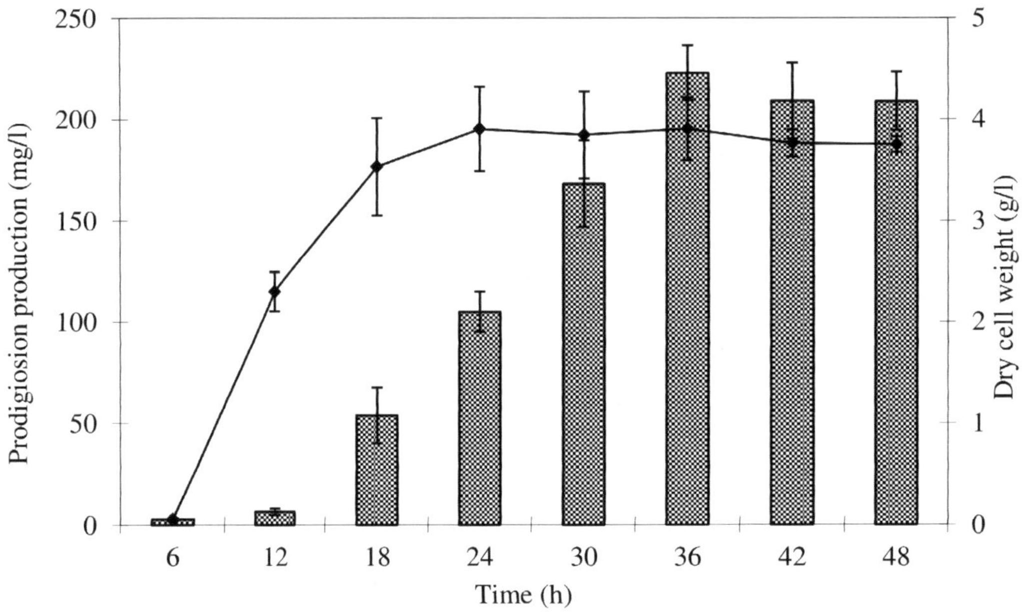

Figure 1. Influence of incubation time on *Serratia marcescens* growth (line) and prodigiosin production (column) with kitchen waste substrate.

Based on figure 1, both prodigiosin production and cell growth increased with incubation time at the beginning of fermentation, and the cell growth had its peak at 24 hour. As one kind of second metabolite [9], prodigiosin production lagged the cell growth and had its peak at 36h and further incubation showed reduction in its production due to the denature of prodigiosin under light. Maximum prodigiosin production was 222.8 mg/l and maximum dry cell weight was 3.9 g/l. This study suggests that prodigiosin production associates with cell growth in nature.

3.3. Optimization of fermentation parameter

Single factor experiments with different fermentation parameters (inoculum level, temperature liquid volume and agitation speed) were conducted to improve the performance of prodigiosin production and the results were shown in figure 2.

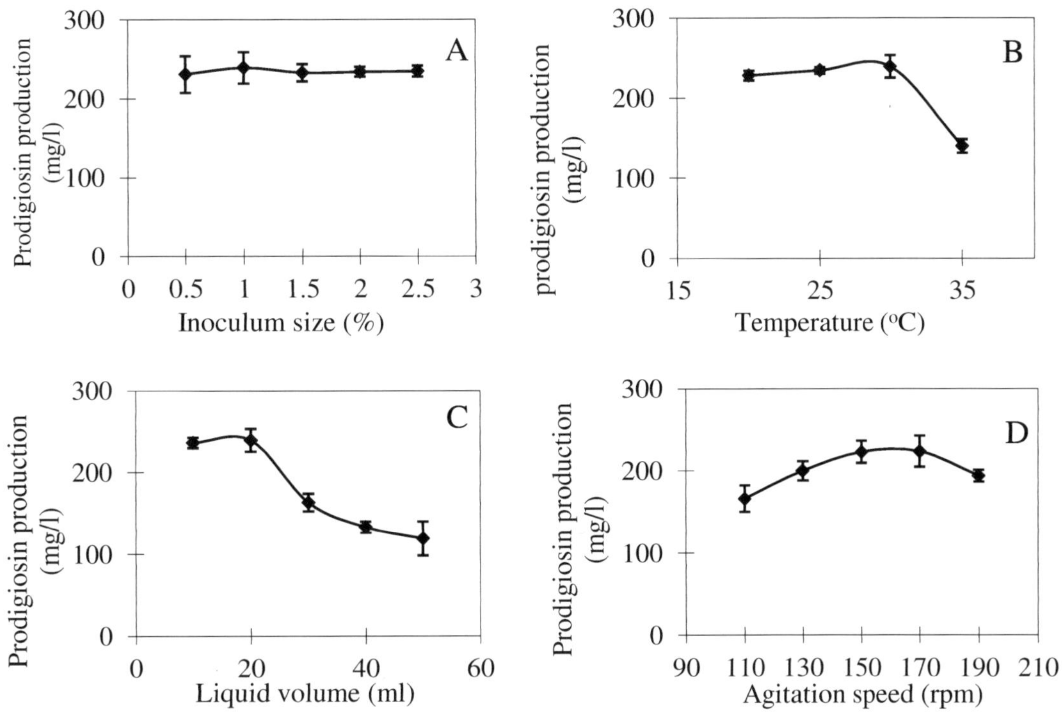

Figure 2. The effect of fermentation parameters (inoculum level (A), temperature (B), and liquid volume (C) and agitation speed (D)) on the yield of prodigiosin.

First of all, as *Serratia marcescens* grew quickly on this medium, inoculum level had little impact on the final prodigiosin yield in this study. Secondly, only a mild fluctuation on prodigiosin production was observed when the temperature was between 20°C and 28°C, but a sharp reduction of yield was occurred with temperature higher than 30°C. Finally, the control of agitation speed and liquid volumes were critical for producing prodigiosin with *Serratia marcescens,* as both factors affected the dissolve oxygen level of the medium. The result was in accord with the report of Bernard Heinemann, who pointed out that the production of prodigiosin required sufficient oxygen supply [10].

3.4. Effect of substrate concentration

A proper concentration of substrate led to successful fermentation. While the too low concentration of substrate cannot provide enough nutrition for the growth of microbiology, an inhibition effect of prodigiosin production was also reported with high concentration of substrate. In order to optimize the substrate concentration, experiments with different substrate concentrations were conducted and shown in figure 3. In present study, maximum prodigiosin production was observed with 35g/l substrate concentration and higher substrate concentration showed reduction in prodigiosin production due to the excess carbon source in the substrate.

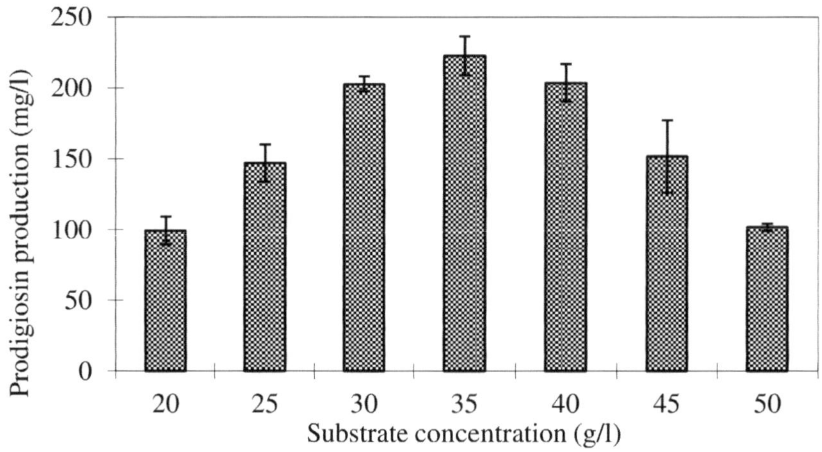

Figure 3. The effect of kitchen waste substrate concentration on prodigiosin production by *Serratia marcescens.*

3.5. Effect of pH

Shake-flask culture experiments were executed with various pH to reveal its influence. Based on figure 4, as long as the initial pH of the media was between 5 and 11, the final pH was in a narrow range, namely 7.8- 8.6, which indicated that *Serratia* genera had a powerful buffering capacity [11]. The maximum prodigiosin was observed with 8.0 initial pH, and little impact on the prodigiosin production was observed when the initial pH was between 7.0 and 9.0.

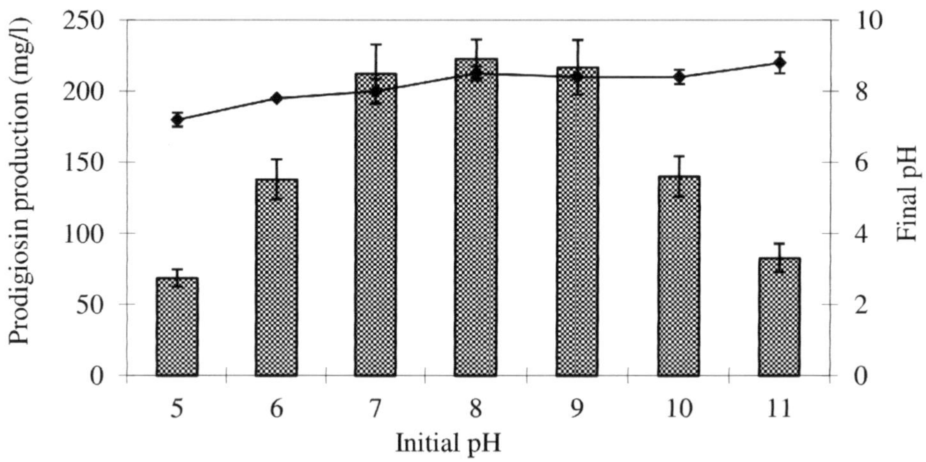

Figure 4. The effect of different pH on prodigiosin production (column) and final medium pH (line) by *serratia marcescens.*

3.6. Effect of different carbon sources

In order to optimize the concentration of carbon source in medium, several carbon sources such as glucose, sucrose, fructose, lactose, glycerol, mannitol, starch, and vegetable oil were added to crude medium at 1% level (figure 5). The study showed that although carbon sources were necessary for the growth of microorganism, supplementing excess carbon sources would affect the prodigiosin production negatively. Especially, with additional monosaccharide such as glucose and fructose supply, small prodigiosin was detected in the medium. Disaccharide and polysaccharides, which were metabolized by microorganism slower than monosaccharide, had less impact on the prodigiosin production. Metabolite derived from carbon source degradation may be the caused for this inhibition effort, which was also suggested by Clements [12].

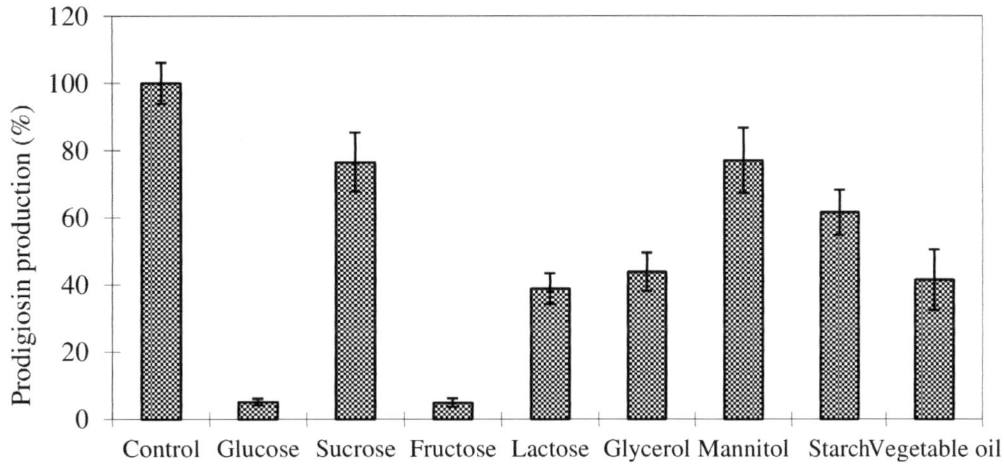

Figure 5. The effect of different carbon sources on prodigiosin production by *serratia marcescens.*

3.7. Effect of different nitrogen sources

Several kinds of amino acid participated in the prodigiosin production as precursor. For example, Lim noticed that proline was the precursor of pyrrolylpyrromethene and alanine also became a part of prodigiosin after decarboxylation. Similar to carbon source, the concentration of nitrogen source in medium was also optimized and shown in figure 6. The results revealed that yield of prodigiosin increased sharply when nitrogen sources were supplemented into the initial medium. Among these sources, peptone showed the best improvement on production. As possessing high level of protease activity,the nitrogen sources in the medium can be utilized and converted to prodigiosin by *serratia marcescens* quickly. To further optimize the peptone concentration, experiments were conducted with various concentration of peptone. The maximum prodigiosin production was 597 mg/l with additional peptone at 1% level.

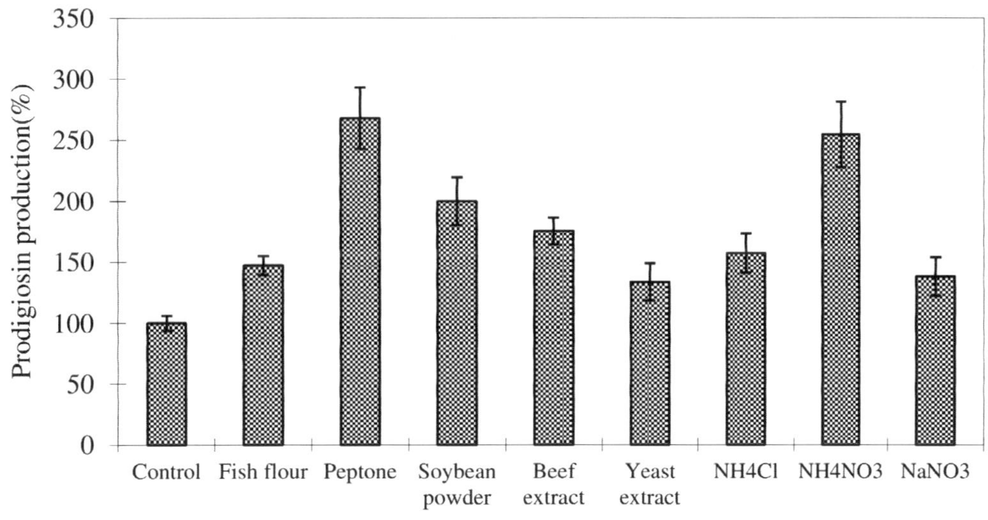

Figure 6. The effect of different nitrogen sources on prodigiosin production by *serratia marcescens.*

3.8. Effect of inductor

Wei reported that the addition of amino acids containing pyrrole-like structures could enhance prodigiosin production. In our present study, proline had the best inducement compared to other amino acids. However, the inducement became not obvious at low concentration of nitrogen source in medium, as the proline was utilized by *Serratia marcescens* as a normal nitrogen instead of an inductor. With additional 1% peptone in crude medium, proline supplementation in the medium at 0.1% level could increase the yield of prodigiosin by 46%, but no further improvement was observed with the concentration was beyond this level. In practice, adding proline may not be economic and necessary due to its high cost. In the end, fermentation was run with optimized fermentation conditions to understand the combinatorial influence of various factors on the prodigiosin production by *Serratia marcescens*. The results were depicted in figure 7. It was evident that at 42 hour, the maximum prodigiosin production was 870mg/l, almost twice more than that with crude medium.

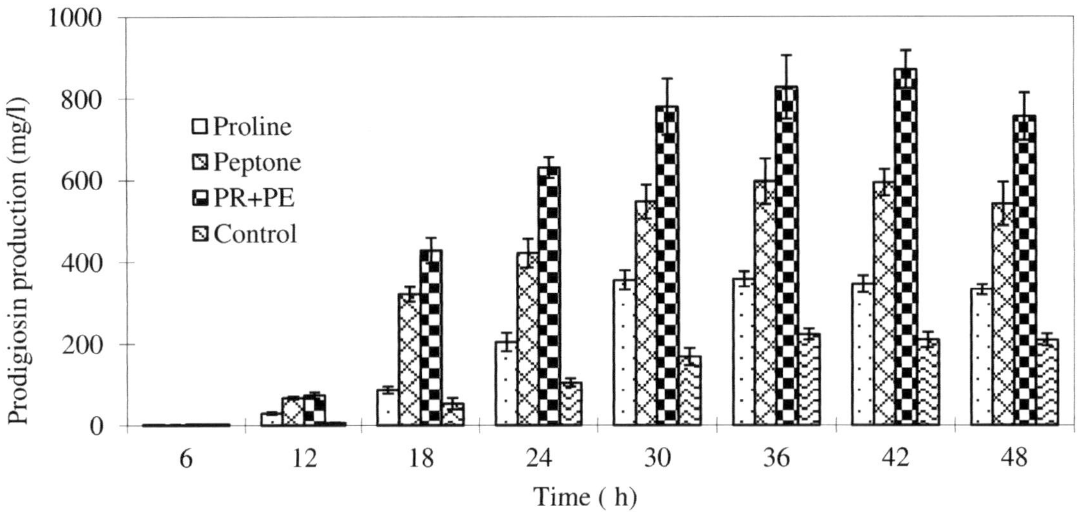

Figure 7. The effect of peptone, proline and their combination on prodigiosin production by *Serratia marcescens* with kitchen waste substrate.

4. Conclusion

Our study first established the link between kitchen waste recycle and prodigiosin production. Prodigiosin production by *Serratia marcescens* was influenced by both the composition of medium and fermentation parameters. Maximum prodigiosin production could be obtained by adding peptone and proline to the kitchen waste medium.

5. References

[1] Alois F 2003 *Angew. Chem. Int. Ed.* 42, 3582
[2] Campas C, Dalmau M, Montaner B, Pons G, Perez R and Gil J 2003 *Leukemia.* 17, 746
[3] Wei Y and Chen W 2005 *J. Biosci. Bioeng.* 99, 616-22
[4] Xu F, Xia S and Yang Q 2011 3rd *International Conference on Chemical, Biological and Environmental Engineering*
[5] Wang Q, Wang X, Wang X, Ma H and Ren N 2005 *J. Environ. Sci. Health A* 40 51–62
[6] Shi Y, Zhao X, Cao P, Hu Y, Zhang L, Jia Y and Lu Z 2009 *Biotechnol Lett.* 31 1327-33
[7] Tao J, Wang X, Shen Y and Wei D 2005 *World J. Microbiol. Biotechnol.* 21 969-72
[8] Wei Y, Yu J and Chen W 2005 *J. Biosci. Bioeng.* 100 466-71
[9] Williams R 1972 *J. Appl. Microbiol.* 25 396-402
[10] Heinemann B, Howard A and Palocz H 1970 *J. Appl. Microbiol.* 19 800-04
[11] Ruis N, Sole M, Francia A and Loren J 1994 *Appl. Environ. Microbiol.* 60 2151-54
[12] Clements S 1976 *Experientia* 31 421-2

IOP Conf. Series: Earth and Environmental Science **171** (2018) 012039 doi:10.1088/1755-1315/171/1/012039

Isolation of Pectin from Carrot Peel as Biosorbent of Pb(II) ion

Budi Hastuti[1*], Saptono Hadi[2], and Fian Totiana[1]

[1]Department of Chemistry Education, Faculty of Teacher Training and Education, Universitas Sebelas Maret, Jl. Ir. Sutami 36A Surakarta, 57126, Indonesia
[2]Department of Pharmacy, Faculty of Mathematics and Natural Sciences, Universitas Sebelas Maret, Jl. Ir. Sutami 36A Surakarta, 57126, Indonesia
E-mail: Budihastuti@staff.uns.ac.id

Abstract. Pectin is a promising bioadsorbents material for heavy metals' removal from aqueous solution. In this study, pectin was isolated from carrot peel by extraction method and evaluated as biosorbent for removal of toxic heavy metals (Pb(II)). The adsorption parameters for removal of Pb(II) were optimized, including pH, adsorbent mass, and contact time. The results shows that pectin can be isolated from carrot peel as was characterized by FT-IR spectra. The optimum condition for Pb(II) were at pH solution 6, Pb(II) with 44% removal capacity. The optimum adsobent mass was 15 mg with 46% removal capacity, whereas the optimum contact time was 120 min with 38% removal capacity.

1. Introduction

Heavy metal contamination is considered to be one of the most ubiquitous environmental issues today. Accumulation of heavy metals in water is of particularly important because it can impact upon human health. In the metallic group, it is known that Arsenic, cadmium, lead, and mercury are listed among the ten chemicals of major public concern. Among them, lead (Pb) is one of the most significant health threats to children and pregnant women. A various adverse effects of Pb has been reported, including: inhibits enzyme activity involved in the formation of hemoglobin (Hb), increase levels of δ-aminolevulinic acid dehydratase (ALAD) and levels of protoporphin in red blood cells, shorten the life of red blood cells, red blood cells and reticulocytes, and increase Fe content in blood plasma.

Treatment processes for heavy metal removal from wastewater has been developed, where adsorption method is among the most effective technique for the removal of heavy metals. Adsorption is a surface phenomenon in which there is absorption or withdrawal of molecules on the surface of the adsorbent. In recent years, the interest in studying modification of pectin has increased. Pectin is an anionic plant cell wall polysaccharide based on a-(1–4) linked D-galacturonic acid [1]. Pectin is widely available in the middle lamella of plant cell walls. Pectin has many active groups. main functional groups contained in the pectin is hydroxyl, carboxyl, amide and methoxy [2]. Currently, pectin was widely used in pharmaceuticals and cosmetics, food industry, and as adsorbent [3,4]. Pectin contains a number of active groups, distributed along the backbone make pectin can be used as a biosorbent [2]. The main functional groups of pectin are hydroxyl, carboxyl, amide, and methoxyl. These functional groups can be used to bind heavy metals, especially hydroxyl groups [5].

Indonesia is a tropical country that has various types of fruits and vegetables which are potential sources of pectin. Isolation and identification of the pectin from pumpkin [6], cocolate peel [7], and durian rind and orange peel [8]. The order of adsorption effectiveness of pectin durian rind toward heavy metal ion are Cu> Pb> Ni> Cd> Zn [8]. Carboxyl groups of pectin has an ability to interact with heavy metal ions to form a complex [5]. The pectin reactivity of heavy metal ions is highly dependent on the degree of esterification. The modification process of durian rind pectin increases binding process of metal ions [8].
In this study, pectin from carrot peel will be isolated and applied as Pb (II) ion adsorbent. Pectin from carrot peel will be tested its absorption capacity, by optimizing the optimum condition in the adsorption process at first. The optimum conditions of pectin carrot peel adsorbent to adsorb Pb (II) ions include of pH optimization, contact time, and optimum mass.

2. Materials And Methodology

2.1. Material
Carrots (*Daucus carota* subsp. sativus) obtained from the local market of Surakarta, Indonesia. The materials used include hydrochloric acid, nitric acid, alcohol (96%), alcohol (70%), Pb(II) were analytical grade and purchased from Sigma-Aldrich.

2.2. Methodology

2.2.1. Isolation of pectin from carrot peel by extraction method
Carrot peel were washed carefully with tap water to remove dirt soil from the surface and dried under shade for 24 h, further dried at 30-40°C until constant weight was obtained. Dried fruit peel was cut into pieces and powdered. The dry powder was sieved using a stainless steel and the obtained powder was stored in the air-tight container until further use. Pectin was extracted from the carrot powder under reflux in a condensation system at 90°C for 60 min (solute/solvent 1:20), using water acidified with hydrochloric acid to pH 2.0. The obtained pectin was then dried and sieved using a stainless steel.

2.2.2. Precipitation and washing of pectin.
The acid pectin extract was precipitated with acidic alcohol solution (1 L of alcohol 96% was added with 2 mL of concentrated HCl) at a ratio 1:1 (v/v) for 24 h. The floating pectin precipitate was then filtered and separated from the filtrate. The pectin precipitate was washed with alcohol up to several times. The pectin was dried to constant weight at 40-50°C for approximately 6, cooled in a dessicator, and hard pectin cake was later grounded and sieved in order to obtain powdered pectin. The pectin yield was calculated on a dry weight basis (initial weight of sample).

2.2.3. Optimum condition of adsorption.
Study of the adsorption process was performed by analyzing several parameters: pH, adsorbent dose, and contact time. It was conducted in laboratory scale with variations in pH of solution (4, 5, 6, 7, and 8), adsorbent dose (5, 10, 15, and 20 mg), and contact time (30, 45, 60, 120, 240, and 360 min). Initial concentration of Pb(II) was 50 ppm by the volume of 10 mL. Analysis of the adsorbed Pb(II) ion was carried out with atomic absorption spectrophotometry (AAS). For those purposes, 20 mg pectin was immersed in 20 mL Pb(II) solution. Concentrations for each solution were 5, 10, 15 and 20 ppm. After being interrupted for 120 min, the contacting filtrate was filtered, and then tested with AAS.

3. Result and Discussion
In this study, pectin-based absorption material from carrot peel was isolated. Pectin has several active groups, which were confirmed by FTIR spectra. From Figure 1, the spectra of carot pectin (A)

shows a broad absorption band at 3404 cm^{-1}, indicates the range of the OH group vibrations. Absorption in the region of 1751 cm^{-1} indicates the presence of an esterified carboxylic group and peak at 1636 cm^{-1} of carboxylic groups. Furthermore, peak at 1636 cm^{-1} and 1064 cm^{-1} indicate the presence of –COOH group and the stretching vibration of the –CO, respectively. The FTIR spectra of carot pectin in general is comparable with those of commercial pectin (B).

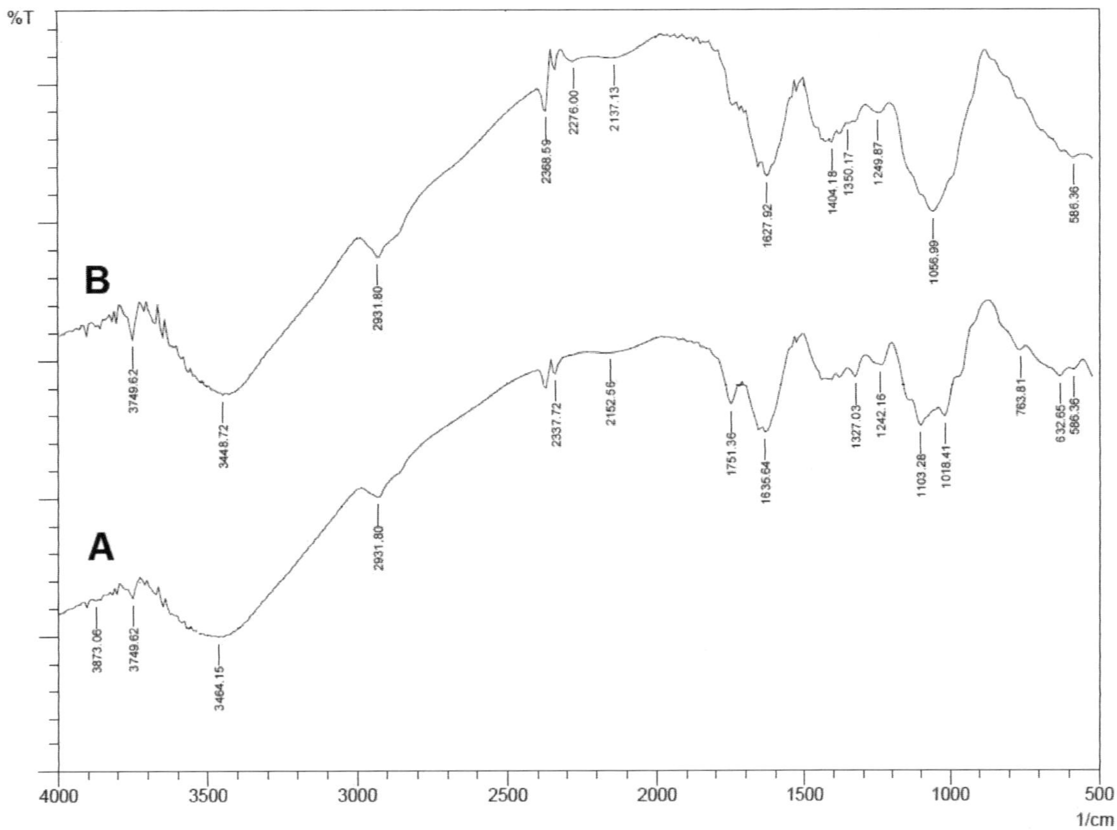

Figure 1. FTIR spectra of A) carrot pectin and B) commercial pectin

3.1. The parameter of optimum condition of carrot peel pectin in adsorption of Pb(II)

3.1.1. Determination of optimum pH

Figure 2 illustrates the influence of pH on the removal of Pb(II). The pH of the solution is one of the important factors that determine the performance of the adsorbent in the adsorption process. If the pH value is too low or too high, the adsorption process cannot be maximized. From Figure 2, within the range tested shows that on pH 4 adsorption percentage of the adsorben is 17 % with adsorption capacity is 2.0 mg/g. Meanwhile, the percentage of adsorption tends to rise on pH 6. The pH 6 is found to be optimum condition, in which the percentage of Pb (II) absorbed is 44 % and the adsorption capacity is 6.0 mg/g. Furthermore at pH 7, the adsorption percentage decreased to 34 % and the adsorption capacity is 4.4 mg/g. At higher pH, the solution has more -OH ions in the solution. The existence of these ions causes Pb(II) ions are hydrolyzed and forms $Pb(OH)_3$. The adsorption process on pH 8 more decreasing with percentage of adsorption is 32% and the adsorption capacity is 4.2 mg/g.

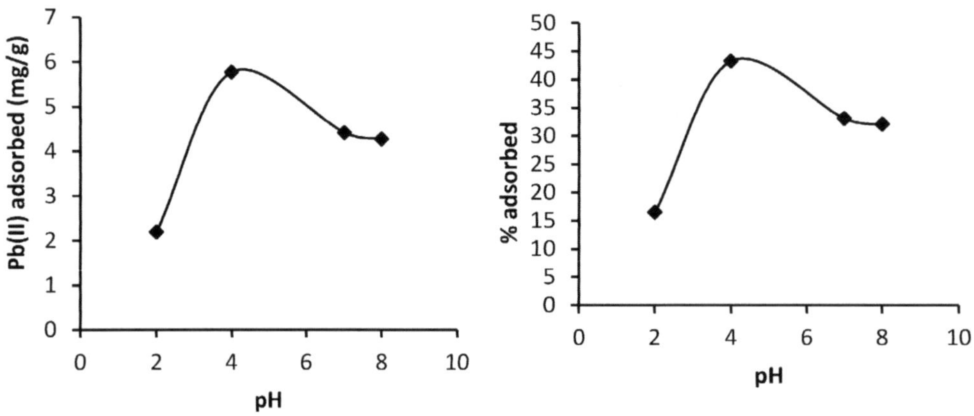

Figure 2. Effect of pH solution on adsorption capacity of Pb(II) on carrot pectin

3.1.2. Determination of optimum adsorbent dose

The amount of adsorbent significantly affected the adsorption rate of Pb (II). In this study 10 mL of Pb (II) ion solution at optimum pH (pH 6) was added with variation of adsorbent mass 5, 10, 15, 20 mg with contact time 120 min. From Figure 3, it can be seen that there is an increase in percentage of absorption due to increasing of adsorbent mass, but after passing the optimum point there is a decreasing in percentage of absorption toward Pb(II). It occurs because more amount of pectin used as an adsorbent, will provide more active sites that bind to Pb (II) ion, so the chances of binding of Pb ions are greater. However, on 20 mg there is a drastic decreasing, it may surface of an adsorbent have been saturated. The optimum mass adsorbent occurred at 15 mg. After the optimum condition is obtained, contacting the optimum condition is 10 mL of Pb(II) solution at pH 6 added with 15 mg carrot pectin and contacted for 120 min show absorption of pectin to Pb(II) is 6.13 mg/g and 46 % adsorbed.

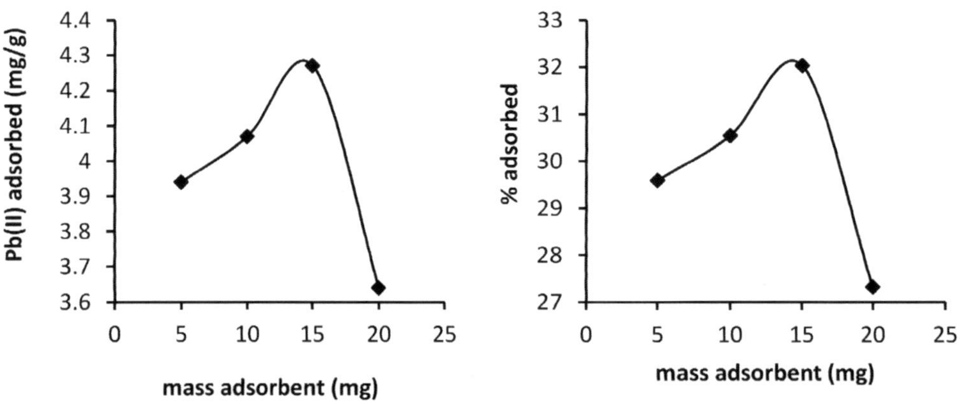

Figure 3. Effect of mass adsorbent on adsorption capacity of Pb(II) on carrot pectin

3.1.3. Determination of optimum contact time

The graph of the relationship between the contact time of Pb(II) ion with adsorption capacity of carrot pectin is show in Figure 4. Contact time is important to determine of the adsorption process, because it allows better mechanism the diffusion process and binding of adsorbate molecules. Variation contact time used was 30, 45,60, 120, 240, and 360 minutes. From Figure 4, it can be seen that within 30-360 min, the largest absorption occurs at 120 minutes. After reaching the optimum point there is a decrease in absorption capacity. This due to 120 of minute carrot peel pectin adsorb Pb(II) perfectly than 30, 60,

IOP Conf. Series: Earth and Environmental Science **171** (2018) 012039 doi:10.1088/1755-1315/171/1/012039

240 and 360 min. The optimum adsorption of Pb(II) by pectin carrot peel occurs on 120 minute with adsorption capacity 4.80 mg/g and 38 % Pb(II) adsorbed.

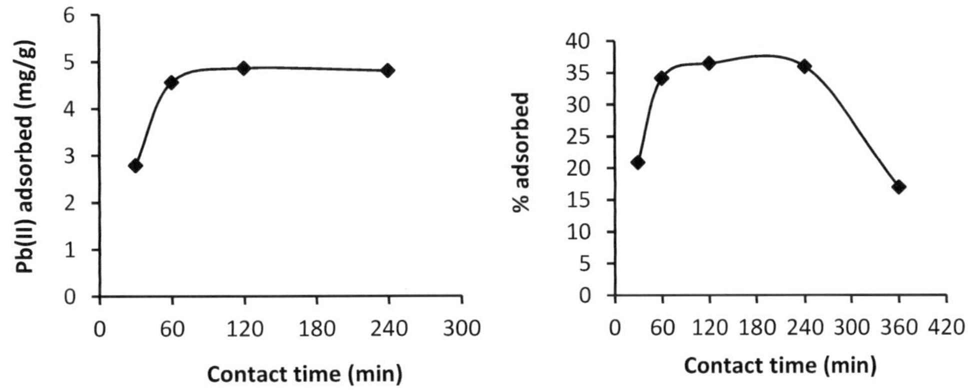

Figure 4. Effect of contact time on adsorption capacity of Pb(II) on carrot pectin

4. Conclusion

It can be concluded that pectin can be isolated from carrot peel proved by FT-IR that found peak at 3404 cm^{-1} of –OH group, 1636 cm^{-1} of -COOH group, and 1064 cm^{-1} prove of the stretching vibration of the -CO. The optimum condition adsorption of pectin carrot peel to Pb(II) metal ion was at pH of solution 6 with contact time 120 min and the adsorbent dose 15 mg. Pectin carrot peel can be used to adsorb Pb(II) in solution.

5. References

[1] Renard, C.M.G.C., Crepeau, M.J. and Thibault, J.F. 1995, *Carbohydrate Research*, 275,155–165

[2] Mata, Y.N., Blázquez M.L., Ballester, A., González, F. and. Mu˜noz, J.A., 2009, *Chemical Engineering Journal.* 150,289–301

[3] Fares, M.M., Tahboub, Y.R., Khatatbeh, S.T. and Abul-Haija, Y.M., 2011, *J Polym Environ* 19, 431–439

[4] Schiewer, S. and Iqbal, M., 2010, *Journal of Hazardous Materials,* 177, 899–907

[5] Kupchik, L.A., Kartel, N.T., Bogdanov, E.S, Bogdanov, O.V., and Kupchik, M.P., 2006, *J,Macromol. Chem. And Polym. Mater,* 79, 457-460

[6] Rosmiati, T., 2000, *Isolasi Dan Identifikasi Pektin Dari Labu Siam (Sechium edule* SW*)*, FMIPA, Univeritas Diponegoro, Semarang, Indonesia

[7] Akhmaluudin dan Kurniawan, A., 2005, *Pembuatan Pektin dari Kulit Coklat dengan Cara Ekstraksi*, FMIPA, Univeritas Diponegoro, Semarang, Indonesia

[8] Wong, W. W., Abbas, F.M., Alkarkhi, and Azhar, M.A., 2009, *Comparing biosorbent ability of modified citrus and durian rind pectin*, School of Industrial Technology, University Sains Malaysia, Penang, Malaysia

IOP Conf. Series: Earth and Environmental Science **171** (2018) 012040 doi:10.1088/1755-1315/171/1/012040

Functionalization of Recycled Diatomite for Green, Stable, and High-performance Phase Change Material (PCM) Composite

Yeng-Fong Shih[1], Chih-Hung Wang*[1], Zong-Gi Tson[1], Yu-Cheng Zhang[1] and Hong-Yuan Lian[2]

1 Department of Applied Chemistry, Chaoyang University of Technology, No. 168, Jifeng E. Rd., Wufeng District, Taichung 41349, Taiwan
2 Department of Advanced Coating, Division of Applied Chemistry, Material and Chemical Research Laboratories, Industrial Technology Research Institute, Taiwan

Abstract. This study reports on the functionalization of recycled diatomite (DT) for preparing green and shape-stabilized phase change material (SSPCM); the DT-based SSPCM can be employed in HDPE composite for high latent heat and good thermal conductivity. After purification, the purified DT (P-DT) adsorbed polyethylene glycol (PEG) by the straight dipping process for producing SSPCM. P-DT showed high surface area of 58 m2g-1 and low organic impurity (<1%); the PEG/P-DT SSPCM exhibited high latent heat of 45 Jg-1 and low leakage (<0.3%). By adding PEG/P-DT SSPCM into the HDPE, the SSPCM/HDPE composite improved both the heat deflection temperature (HDT) and maximum decomposition temperature (Tmax) to 89.18°C and 500.2°C, respectively. In seeking to enhance tensile strength and thermal conductivity, maleated polyethylene (MAPE) and alumina (Al2O3) were studied in the SSPCM/HDPE composite. In the end, the SSPCM/HDPE composite exhibited great heat resistance, mechanical property and thermal conductivity.

1. Introduction

With the high demand for energy along with the depletion of conventional energy sources, development of new energy sources and improved utilization of existing energy sources have become increasingly important.[1] In regard to industrial energy conservation, materials for energy storage and efficiency have attracted the most attention; among these, low-cost phase change material (PCM) has become popular because of its high storage capacity and thermostatic effect during energy storage. PCM can absorb or release heat, change its physical state (e.g. from solid to liquid, and vice versa) and provide latent heat without change in temperature.[2] In order to enhance energy utilization efficiency, PCM has been widely used in air-conditioning, building materials, energy-saving equipment, food preservation and warm supplies, and is regarded as an environmentally friendly energy-saving material.[3-5]

To avoid PCM leakage, PCM is introduced into porous solids to prepare shape-stabilized phase change material (SSPCM) with stable structure. Also, the SSPCM composite can be synthesized by easy and low-cost methods. In Taiwan, the beer breweries discard 4,500 tons of diatomite (DT) every year, and the DT is reused for planting soil or buried in landfill. DT has good adsorbability, stable chemical properties and high porosity; it can adsorb PCM after proper purification pre-treatment. In this study, we purified the recycled DT for adsorbing PCM, and synthesized the SSPCM composite with stable structure, high phase enthalpy, and good thermal conductivity, at low cost.

IOP Conf. Series: Earth and Environmental Science **171** (2018) 012040 doi:10.1088/1755-1315/171/1/012040

2. Experiment

2.1. DT purification

For removing the organics and impurities in the DT pores, the recycled DT was calcined and washed by acetone. The purified DT (P-DT) was oven dried, and sieved by mesh screen.

2.2. *Synthesis of PEG/P-DT SSPCM*

The PEG solution was prepared by PEG600 and PEG1000 in the ratio of 4:1, and the P-DT was then straight dipped in the PEG solution. To observe the structural stability of the PEG/P-DT SSPCM, different PEG/P-DT concentrations (10~50%) were studied in the straight dipping process. The PEG/P-DT mixture was oven dried at 60°C for 30 min, and the dried PEG/P-DT SSPCM with good structure stability was further used in fabricating PCM composites. [6-7]

2.3. *Preparation of SSPCM/HDPE composites*

PEG/P-DT SSPCM and HDPE were oven dried at 60°C for 1 d before composite fabrication. The composite components, as shown in Table 1, were blended at 170°C in a counter-rotating internal mixer for 400 s, and then granulated. In the end, the SSPCM/HDPE composites were prepared by compression molding at 170°C.

Table 1. Formulation of SSPCM/HDPE composites.

Sample	HDPE (g)	SSPCM (g)	MAPE (g)	Al2O3 (g)
HDPE	50	-	-	-
PEC	30	12	-	1.5
MPEC	27.6	12	2.4	1.5

3. Results and Discussion

3.1. Physical properties of diatomite

The structure parameters and thermal properties of DT and P-DT are summarized in Table 2. Before purification, the raw recycled diatomite (DT) had a BET surface area of 32.5950 m2/g and pore volume of 0.1277 cm3/g. After purification, the purified diatomite (P-DT) increased the BET surface area and pore volume to 58.2470 m2/g and 0.1788 cm3/g, respectively. Therefore, the calcination and washing treatments successfully removed the organics and impurities inside the DT, and increased the porous space in the end. From the TGA analysis shown in Figure 1, P-DT retains much higher weight than DT after thermal decomposition, indicating that the P-DT is composed of pure inorganic content. Therefore, P-DT can serve as an ideal adsorbent for PEG in the PCM preparation.

Table 2. Physical properties of purified diatomite.

Sample	Structure parameters		Thermal properties
	BET surface area (m2/g)	Pore volume (cm3/g)	Char Yield at 550oC (%)
DT	32.5950	0.1277	87.09
P-DT	58.2470	0.1788	99.05

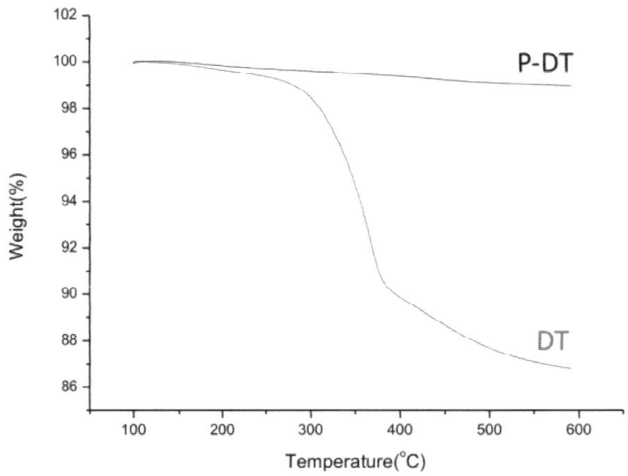

Figure 1. TGA analysis of DT and P-DT.

3.2. Structure stability of PEG/P-DT PCM

As shown in Figure 2, the PEG/P-DT PCM exhibits different structure stability with various PEG contents. The PEG leakage from P-DT was observed after the PEG/P-DT SSPCM was placed in an oven at 60℃ for 30 min. In Figure 2(e), the high PEG content resulted in serious leakage, indicating that the P-DT could not adsorb the excess PEG. At 40% PEG content (Figure 2(d)), only 0.2% of the PEG leaked out. With PEG content lower than 30%, the PEG/P-DT PCM forms a stable structure without PEG leakage.

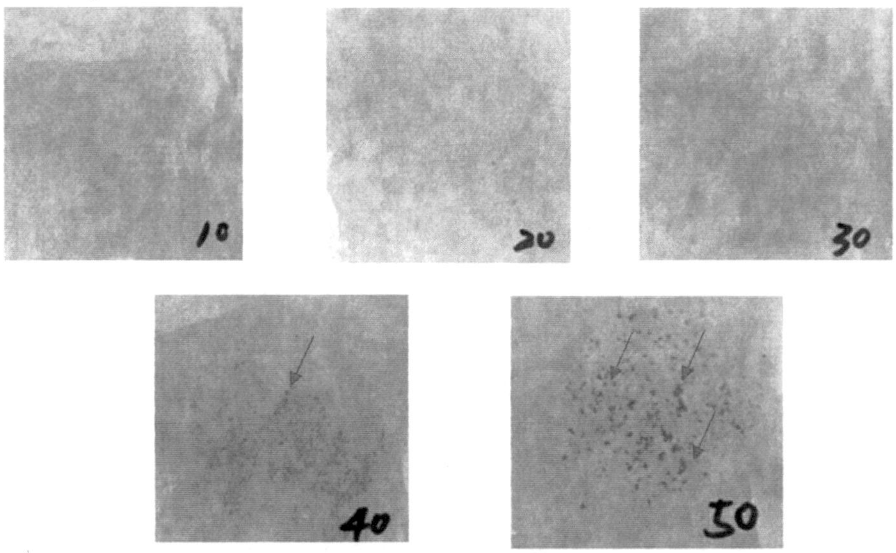

Figure 2. Structure stability of PEG/P-DT SSPCM containing (a) 10% PEG, (b) 20% PEG, (c) 30% PEG, (d) 40% PEG and (e) 50% PEG.

3.3. DSC analysis

The DSC thermograms of PEG and MPEC are exhibited in Figure 3. In Figure 3(a), the PEG consisted of PEG600 and PEG1000 in the ratio of 4:1, and performed phase change from -10℃ to 40℃, where

IOP Conf. Series: Earth and Environmental Science **171** (2018) 012040 doi:10.1088/1755-1315/171/1/012040

the PEG showed latent heat of 104.9 J/g. The HDPE holds Tm of 133.16°C and enthalpy of 140.2 J/g. After introducing the PEG/P-DT SSPCM, the MPEC exhibited latent heat of 6.741 J/g around 24°C.

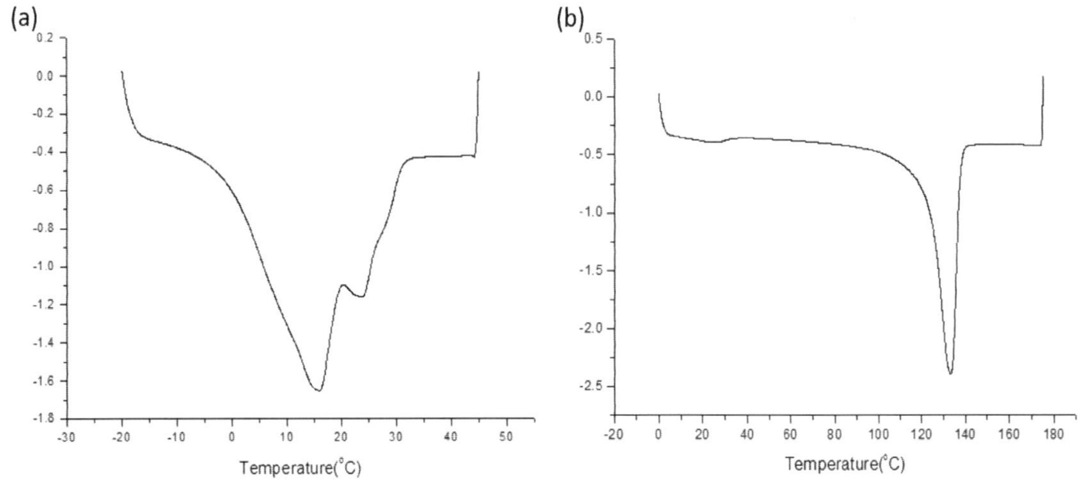

Figure 3. DSC thermograms of (a) PEG and (b) MPEC.

3.4. HDT analysis

The heat deflection temperatures (HDT) of the HDPE and its SSPCM composites are summarized in Table 3. After adding PEG/P-DT SSPCM, the PEC composite elevates the HDT from 75.90°C to 81.77°C because the P-DT porosity can insulate heat and improve thermal resistance. By adding the MAPE for improving the compatibility between the HDPE and PEG/P-DT PCM, the MPEC composite further enhanced the HDT to 89.18°C.

Table 3. Physical properties of HDPE and its SSPCM composites.

Sample	Heat resistance	TGA analysis		Mechanical properties		Thermal conductivity
	HDT (°C)	Td5 (°C)	TMax (°C)	Impact strength(J/m)	Tensile strength(MPa)	k (W/mK)
HDPE	75.90±1.8	432.9	473.0	117.35±2.7	25.25±3.1	0.5196
PEC	81.77±0.4	369.5	403.1 489.1	41.15±1.4	17.75±0.4	-
MPEC	89.18±4.81	386.7	396.2 500.2	34.43±2.4	21.89±0.3	0.6078

3.5. TGA analysis

Thermogravimetric analysis (TGA) and derivative thermogravimetric (DTG) analysis are shown in Figure 4. HDPE has Td5 of 432.9°C. After adding the PEG/P-DT SSPCM, the PEC and MPEC, had Td5 of 369.5°C and 386.7°C, respectively, and the decreasing Td5 was donated by the PEG decomposition at high temperature. From the DTG results, the PEC and MPEC had two TMax peaks.

The first TMax peaks of PEC and MPEC were 403.1℃ and 396.2℃, respectively, and were close to the TMax of the PEG (400.1℃). Accordingly, the first TMax peaks of PEC and MPEC were contributed by the PEG adsorbed inside the P-DT. The second TMax peaks of PEC and MPEC were respectively 489.1℃ and 500.2℃, and were higher than the TMax of the pure HDPE (473.0℃). The elevation of second TMax peaks resulted from the P-DT and MAPE. P-DT is heat-resistant inorganics, and MAPE can improve the compatibility and thermal stability of PCM composite; therefore, P-DT and MAPE enhanced the second TMax of PCM composites.

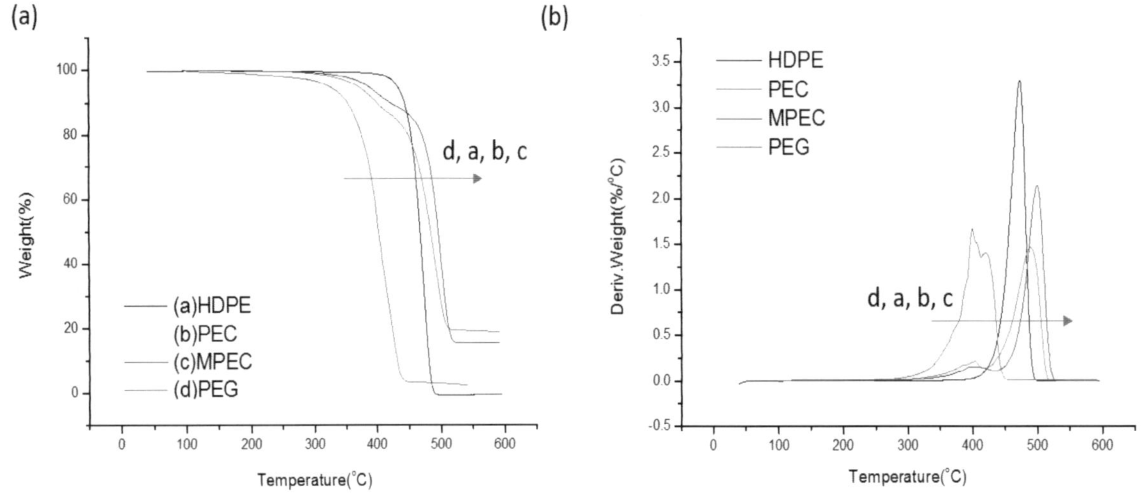

Figure 4. (a) TGA and (b) DTG analysis of HDPE and its SSPCM composites.

3.6. Mechanical properties and thermal conductivity

The impact and tensile strengths are summarized in Table 3. P-DT is a kind of inorganic porous material, so it significantly enhances thermal resistance instead of mechanical properties. HDPE has impact strength of 117.35 J/m and tensile strength of 25.25 MPa. After adding the PEG/P-DT SSPCM, the PEC decreased the impact strength and tensile strength to 41.15 J/m and 17.75 MPa, respectively. By adding MAPE to improve the compatibility and stiffness, MPEC's tensile strength rose to 21.89 MPa.

For quickly conducting heat from environment to SSPCM, Al2O3 is employed in MPEC composite. After adding Al2O3, the MPEC composite elevated the thermal conductivity from 0.5196 W/mK to 0.6078 W/mK.

4. Conclusion

This study successfully purified recycled DT and functionalized PEG/P-DT as SSPCM material. After purification, the P-DT served as an ideal adsorbent for accommodating PEG with great structural stability. By introducing PEG/P-DT SSPCM into HDPE, the SSPCM/HDPE composites improved the thermal properties like HDT and Tmax. Other functional materials were also used in the composites, for example, MAPE for better compatibility and tensile strength, and Al2O3 for higher thermal conductivity.

5. References

[1] Amaral C, Vicente R, Marques P A A P and Barros-Timmons A 2017 Phase changematerials and carbon nanostructures for thermal energy storage: A literature review, *Renewable and Sustainable Energy Reviews.* 79 1212-28.

[2] Biwan X and Zongjin L 2014 Paraffin/diatomite/multi-wall carbon nanotubes composite phase

change material tailor-made for thermal energy storage cement-based composites,*Energy.*72 371-380

[3] Rathod M K and Banerjee J 2013 Thermal stability of phase change materials used in latent heat energy storage systems: A review. *Renewable and Sustainable Energy Reviews.* 18 246-258

[4] Farid M M, Khudhair A M, Razack S A K, Al-Hallaj S 2004 A review on phase change energy storage: materials and applications. *Energy Conversion and Management.* 45 1597-1615

[5] Kousksou T, Bruel P, Jamil A, El Rhafiki T, Zeraouli Y. 2014 Energy storage: Applications and challenges. *Solar Energy Materials and Solar Cells.* 120 59-80

[6] Yeliz K, Orkun E and Ozgur G 2015 Easy and industrially applicable impregnation process for preparation of diatomite-based phase change material nanocomposites for thermal energy storage . *Applied Thermal Engineering*.91 759-766

[7] Xiangyu L, Jay G S and John L W 2014 Fabrication and stability of form-stable diatomite/paraffin phasechange material composites. *Energy and Buildings.*76 284-294

Acknowledgments

The authors thank the Ministry of Science and Technology of Taiwan for its financial support (MOST 106-3114-E-324-001) and the Wuri Brewing Company (Taichung, Taiwan) for supplying the diatomite used in this study.

Eco-friendly Modification for the Cellulose Nanofibers Derived from Pineapple Leaves for High-performance Nanocomposite

Yeng-Fong Shih*[1], Zong-Gi Tsou[1], Chih-Hung Wang and Hong-Yuan Lian[2]

1 Department of Applied Chemistry, Chaoyang University of Technology, No. 168, Jifeng E. Rd., Wufeng District, Taichung 41349, Taiwan
2 Department of Advanced Coating, Division of Applied Chemistry, Material and Chemical Research Laboratories, Industrial Technology Research Institute, Taiwan

Abstract. This study extracted cellulose nanofibers (NFC) from pineapple leaves, and then employed eco-friendly aqueous system to modify NFC with a layer of polystyrene (PS). The successful modification was confirmed by FTIR analysis, after acid-base treatment can effectively reduce the lignin content and better water transport, and the PS layer could lower NFC's surface polarity for improving the interfacial compatibility in PS matrix. By introducing NFC into PS matrix, the PS-modified NFC showed better interfacial adhesion and uniform dispersion, and resulted nanocomposite with higher tensile strength, char yield and transparency. Therefore, the eco-friendly modification could help NFC reinforce nanocomposite for enhancing mechanical property, thermal resistance and transparent appearance.

1. Introduction

Traditional petroleum-based plastic has good plasticity, high strength and durability, and is widely used in commodity and industry. Regarding to the unpredictable petroleum price and supply, it becomes important to develop biomass materials for replacing part of petroleum and creating high-performance composite. Natural fibers are considered as ideal reinforcement candidate due to stiffness, abundant and sustainable sources, and biodegradability.[1]

Among natural fibers,[2] cellulose nanofibers (NFC) has attracted much attention because of light, excellent strength, high surface area, and high aspect ratio. However, cellulose surface is very hydrophilic and hard to be well-dispersed in polymer matrix. Generally, surface polarity of NFC can be adjusted by modification, which is usually based on organic solvent system.[3] For environmental protection and low cost, this study established extraction of NFC from pineapple leaves,[4][5] and aqueous-based modification to functionalize NFC with a layer of polystyrene (PS). The PS-modified NFC was also used to fabricate PS-based nanocomposite with high mechanical property, thermal resistance and transparency.[6]

2. Experiment

2.1. Preparation of NFC (named as OF)

The raw pineapple leaves were chopped, sun-dried, crushed, and further treated with a NaOH solution to remove lignin, hemicellulose and impurities. Subsequently, it was neutralized by HCl solution as shown in **Scheme 1**, washed by water, and then dried for obtaining pineapple leaves fiber (PLF). PLF

was further bleached in the acetate buffer solution containing NaClO2 at pH 4.8 and 60°C for 3 h. In the end, the bleached PLF was purified by water and stored at 4°C before further treatment.
The bleached PLF (0.7 g) were suspended in a sodium phosphate buffer (90 mL, pH 6.8) containing TEMPO, NaClO2, and NaClO. The TEMPO-mediated oxidation was started and kept at 60°C by stirring at 500 rpm for 3.5 h. After washed by water, the TEMPO-mediated oxidation and purification were repeated again. In the end, PLF was nanofibrillated by homogenization at 6000 rpm, and the obtained NFC (named as OF) was stored after freeze-drying.

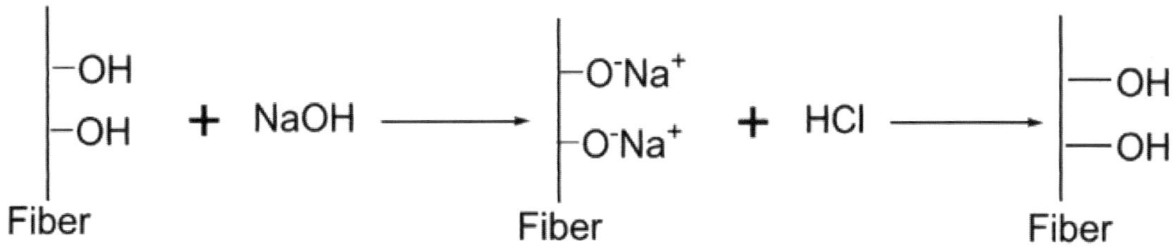

Scheme 1. Chemical pretreatment of PLF.

2.2. *Surface modification*
Dried OF, 0.6g of styrene monomer, 0.018g of initiator (AIBN), and 0.0124g of dispersant (PVA) were mixed in distilled water, and the solution was stirred continuously at 75°C for 6 h to execute suspension polymerization. The PS-modified NFC (named as PF) was obtained after filtration, washing by distilled water, and freeze-drying.

2.3. *Preparation of nanocomposite*
The PS pellets, NFC (PF or OF) were mixed according to the formulation shown in **Table 1**. The NFC (PF or OF) and PS pellets were blended at 160 °C in a counter-rotating internal mixer with a rotation speed of 50 rpm. After being well blended for 5 minutes, the PS nanocomposite was granulated by compression molding at 160°C.

Table 1. Formulation of nanocomposites

Sample	PS (wt%)	PF (wt%)	OF (wt%)
PS	100	-	-
PONFC0.5	99.5	-	0.5
PONFC1.0	99.0	-	1.0
PONFC3.0	97.0	-	3.0
PSNFC0.5	99.5	0.5	-
PSNFC1.0	99.0	1.0	-
PSNFC3.0	97.0	3.0	-

3. Results and Discussion

3.1. FTIR

The FTIR spectra of styrene, OF and PF are shown in **Figure 1**. Styrene monomer exhibits C-H peaks at 695 cm-1 and 906 cm-1, and C=C peaks at 1494 cm-1 and 1630 cm-1. OF and PF have a broad peak at 3400-3600 cm-1, which is contributed by the OH stretch. In OF, the peak donated by lignin at 1735 cm-1 disappears, indicating the acid/base chemical pretreatment and TEMPO-mediated oxidation successfully remove lignin and impurity from PLF. The existence of CH3 and CH2 peak

(2950 cm-1), and C-O-C peak (1059 cm-1) demonstrates the successful carboxylation of OF by TEMPO-mediated oxidation. In PF, the aromatic C-H vibration (696 cm-1 and 900 cm-1) and aromatic C=C (1492 cm-1 and 1601 cm-1) peaks explain the existence of polymerized styrene on NFC surface.

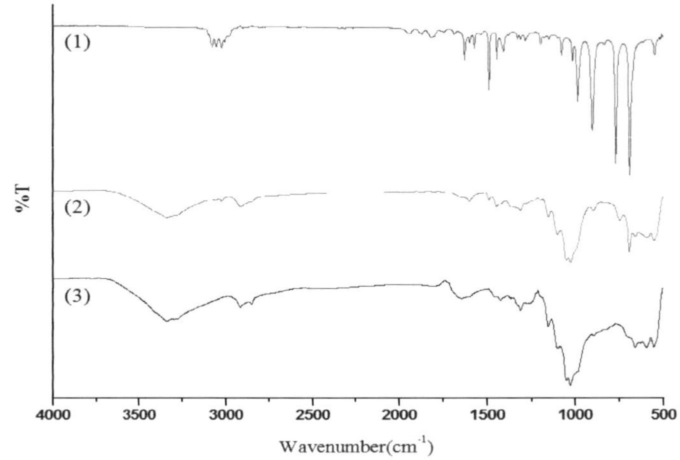

Figure 1. FTIR spectra of (1) styrene, (2) OF and (3) PF.

3.2. Contact angle

Surface polarity of NFC can be represented by surface hydrophilicity, and here the surface hydrophilicity is characterized in terms of contact angle, as shown in **Figure 2**. OF has low contact angle of 12.02°, indicating extremely high hydrophilicity and surface polarity. After PS-modification on PF surface, the contact angle increases to 114.49°. PF surface displays highly hydrophobic and low polarity, and that are helpful to improve compatibility between PF and PS matrix in nanocomposite.

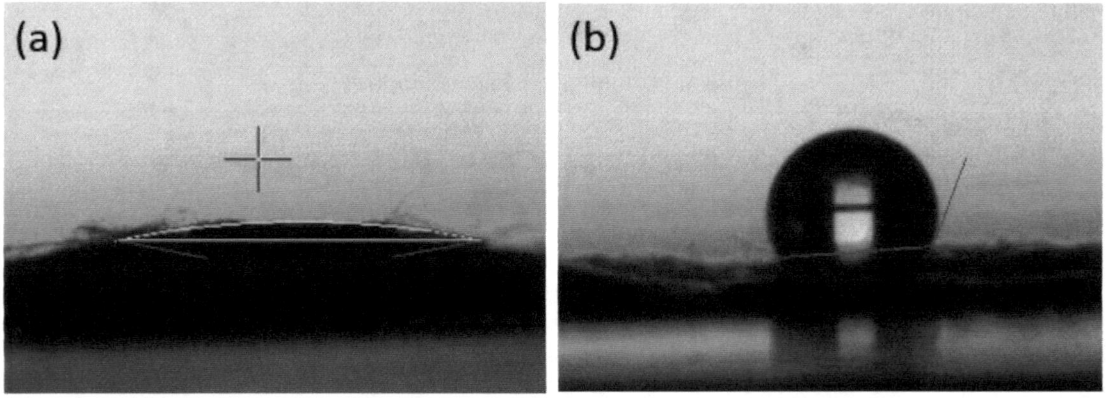

Figure 2. Contact angle of water droplets on the surface of (a) OF and (b) PF.

3.3. Tensile strength

The tensile test of nanocomposites containing 0.5-3% NFC is shown in **Table 2**. Pristine PS performs tensile strength of 23.1 MPa, and all the nanocomposites containing OF or PF can improve tensile strength. At same NFC content, PF reinforces nanocomposite more significantly than OF does. NFC has high tensile strength, aspect ratio and surface area, so it can effectively strengthen nanocomposite. In PONFC nanocomposites, PONFC1.0 performs the highest improvement of 15.6%; in PSNFC nanocomposites, PSNFC exhibits the best improvement of 34.6%. By modifying with a PS layer on the NFC surface, the interfacial compatibility and adhesion between PF and PS matrix are enhanced. As a result, PF can disperse homogeneously in nanocomposite and boost mechanical property.

Table 2. Physical properties of nanocomposites

Sample	Mechanical properties		Heat resistance		TGA analysis	
	Tensile strength (MPa)	Increment (%)	HDT (°C)	Increment (%)	Td5 (°C)	Char yield at 550°C (%)
PS	23.1±0.5	-	90.3±0.60	-	389.3	0.38
PONFC0.5	24.6±0.7	6.5	92.9±0.70	2.88	378.9	0.43
PONFC1.0	26.7±1.1	15.6	92.7±0.70	2.66	373.9	0.74
PONFC3.0	25.2±1.1	9.1	92.6±0.89	2.55	338.7	1.70
PSNFC0.5	26.4±1.7	14.2	92.2±0.21	2.10	384.3	0.46
PSNFC1.0	31.1±2.0	34.6	92.8±0.74	2.77	329.8	2.41
PSNFC3.0	29.4±1.1	27.1	92.6±0.64	2.55	314.6	4.18

3.4. Thermal properties

As shown in **Table 2**, pristine PS has HDT of 90.3℃, and OF and PF can slightly improve nanocomposites' HDT about 2-3%. So, the increasing HDT means that nanocomposites perform better thermal resistance.

The TGA analysis is shown in **Figure 3**, and the decomposition temperatures at 5% weight-loss (Td5) and char yield are summarized in **Table 2**. Pristine PS has Td5 of 389.3℃ and char yield of 0.38%. After introducing OF or PF into nancomposite, the Td5 decreases and char yield gets higher. The Td5 decrement is proportional to the NFC amount. NFC, including OF and PF, has lower decomposition temperatures than PS, and results nanocomposites with lower Td5. In addition, NFC has high specific surface area, and that increases the heated area for early decomposition at lower temperature. Also, OF and PF decompose to form a carbon layer and elevate char yield, that can prevent PS matrix from further decomposition. Basically, more NFC amount generates higher char yield. For example, PSNFC3.0 has char yield of 4.18%.

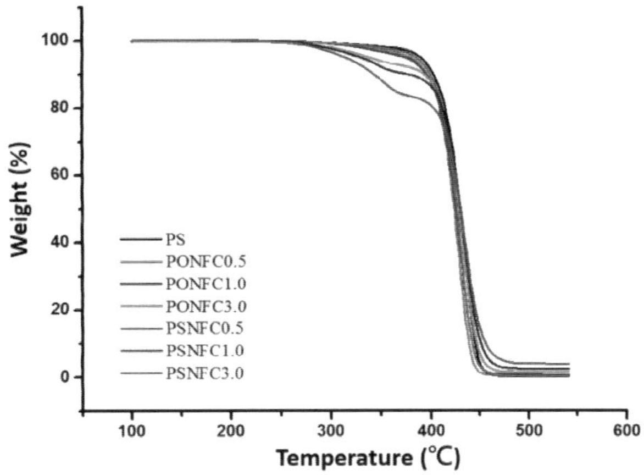

Figure 3. TGA analysis of nanocomposites.

3.5. Transparency

It is observed in **Figure 4** that PSNFC nanocomposites are more transparent than PONFC nanocomposites. OF has high surface polarity as confirmed in contact angle test, and it easily aggregates in PS matrix. As a result, the OF disperses heterogeneously and generates blurred appearance. On the other hand, PF is modified a PS layer on NFC surface, and successfully decreases surface polarity as confirmed in contact angle test. The low surface polarity improves the compatibility between PF and PS matrix, and therefore PF disperses homogeneously in nanocomposite. In consequence, the well-dispersed PF makes the PSNFC nanocomposite very transparent to the naked eye.

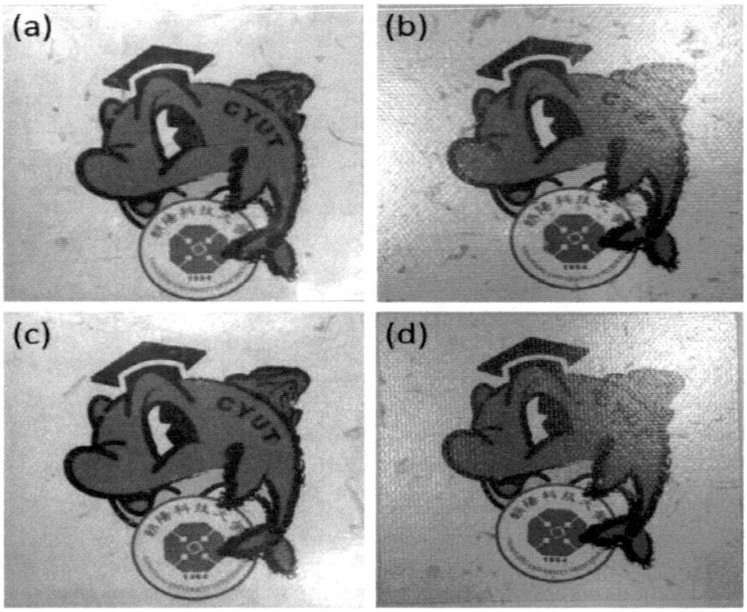

Figure 4. Nanocomposites' appearance and transparency: (a) PONFC0.5, (b) PONFC1.0, (c) PSNFC0.5, and (d) PSNFC1.0

4. Conclusion

This study successfully used eco-friendly synthesis to modify NFC surface with a PS layer, as confirmed by the existence of aromatic functional groups in FTIR analysis. The modified NFC (PF) increased surface hydrophobicity, and the contact angle of water droplet reached 114.49°. The surface modification improved interfacial compatibility and adhesion between NFC and PS matrix, and made PF uniformly dispersed in nanocomposite. As a result, the PF-based nanocomposite performed transparency, high tensile reinforcement (34.6% improvement at 1% PF), and good thermal resistance.

5. References

[1] 2012 *TAPPI Conference on Nanotechnology for Renewable Materials* (Canada: Quebec)

[2] Alila S, Besbes I, Vilar M R, Mutje P and Boufi S 2013 Non-woody plants as raw materials for production of microfibrillated cellulose (MFC): A comparative study. *Industrial Crops and Products.* 41 250-259

[3] Siro I and Plackett D 2010 Microfibrillated cellulose and new nanocomosite materials: a review. *Cellulose.* 17 459-494

[4] Herrick F W, Casebier R L, Hamilton J K and Sandberg K R 1983 Microfibrillated cellulose: morphology and accessibility. *Journal of Applied Polymer Science. Applied Polymer Symposinm.* 37 797-813

[5] Henriksson M, Henriksson G, Berglund L A and Lindstrom T 2007 An environmentally friendly method for enzymeassisted preparation of microfibrillater cellulose(MFC) nanofibers. *European Polymer Journal.* 43 3434-3441

[6] Lijun Q, Jianhui Q, Mingzhu L, Shenglong D, Liang S, Shaoyu L, Guohong Z, Yang Z and Xie F 2011 Mechanical and thermal properties of poly(lactic acid) composites with rice straw fiber modified by poly(butyl acrylate). *Chemical Engineering Journal.* 166 772-778 Herrick F W, Casebier R L, Hamilton J K and Sandberg K R 1983 Microfibrillated cellulose: morphology and accessibility. *Journal of Applied Polymer Science. Applied Polymer Symposium.* 37 797-813

Acknowledgments
The authors thank the Ministry of Science and Technology of Taiwan for its financial support (MOST 105-2221-E-324-024-) and the Chiayi Agricultural Experiment Station of the Taiwan Agricultural Research Institute for supplying the pineapple leaves used in this study.

IOP Conf. Series: Earth and Environmental Science **171** (2018) 012042 doi:10.1088/1755-1315/171/1/012042

An Approach for Analysis of Natural Hazard Impacts on Activated Sludge Wastewater Treatment

P Zlateva[1] and N Dimitrova[2]

1 Institute of Robotics, Bulgarian Academy of Sciences, Sofia, Bulgaria
2 Institute of Mathematics and Informatics, BAS, Sofia, Bulgaria
E-mail: plamzlateva@abv.bg, nelid@bio.bas.bg

Abstract. The activated sludge process is a advansed biological wastewater treatment process for treating sewage or industrial wastewaters by using different aerobic types of microorganisms (biomass). In recent years, the activated sludge processes are shown to be significantly negative affected by the natural hazards (the climate factors) as extreme temperatures, intensive rainfall, etc. An approach for analysis of natural hazard (climate change) impacts on wastewater treatment process with activated sludge is proposed. The approach is based on qualitative analysis of the input-output static characteristics of the activated sludge process, involving uncertainties in the inflow parameters of the aerobic bioreactor. The process model is described by the system from two nonlinear ordinary equations with interval coefficients. The impact levels of one natural hazard on the activated sludge process are defined as a deviation between the nominal and the affected input-output characteristics. The negative effects are described by variation in given intervals of some process coefficients and are calculated as percentages (deviations) from the nominal process coefficients. This analysis of natural hazard impacts on activated sludge wastewater treatment process is necessary in order to operate such system stable and efficiently.

1. Introduction

The wastewater treatment generally involves the sequential carrying out of physical/chemical and biological treatment processes [1, 2]. In particular, the following wastewater treatment processes are caried out: preliminary treatment (physical) to remove large debris and grit, primary treatment (physical) to remove settleable suspended solids, secondary treatment (biological) to remove the remaining particulates and dissolved organic material, chemical precipitation to remove nutrients, tertiary filtration (physical) to remove remaining fine particulates, and chemical or ultraviolet light disinfection [3, 4, 5].

The activated sludge process is a advansed biological wastewater treatment process for treating sewage or industrial wastewaters by using different aerobic types of microorganisms (biomass) [6, 7]. This process can accomplish the oxidizing carbonaceous biological matter, oxidizing nitrogenous matter: mainly ammonium and nitrogen in biological matter, removing nutrients (nitrogen and phosphorus) [8, 9].

In recent years, many studies are established an increase in the negative impact of climate change on the biological wastewater treatment process [10, 11]. These problems due to the climate change are directly related to the natural hazards [3, 12]. The activated sludge processes are shown to be significantly affected by the natural hazards (the climate factors) such as extreme temperatures (positive and negative), intensive rainfall, prolonged periods of deviations from established average daily temperatures and rainfall [13, 14].

IOP Conf. Series: Earth and Environmental Science **171** (2018) 012042 doi:10.1088/1755-1315/171/1/012042

For example in [15] it is presented an investigation on climate change effects on a wastewater treatment system that receive sewage collected in a combined sewer system in Oslo, Norway, during winter operation. It is established an increasing number of days with temperatures below the critical daily air mean temperature in the area, despite the progressively warmer winter temperatures in the last decade. It is demonstrated that these impacts can deteriorate operation of the activated sludge processes through progressively increasing the relative frequencies of very high influent flow rate and of the very low influent sewage temperature. In [14] the effects of temperature variations on aerobic biological wastewater treatment are evaluated with respect to treatment efficiency, solids discharges, sludge physicochemical properties and microbiology.
The analysis of the theoretical results related to the waste water treatment is shown that mathematical models which adequately describe the potential negative impacts of climate factors on the activated sludge processes are still insufficient [6, 8, 9]. Therefore, the issue of developing innovative approaches to assess the risk related to the negative effects of natural hazards on effective wastewater treatment is very important.
The aim of this paper is proposed an approach for analysis of natural hazard impacts (climate change impacts) on wastewater treatment process with activated sludge. The approach is based on stability analysis of a nonlinear model with interval coefficients, describing the activated sludge process in aerobic bioreactor [16]. The impact levels of one natural hazard on the activated sludge process are defined as a deviation between the nominal and the affected input-output characteristics. The negative effects are described by variation in given intervals of some process coefficients and are calculated as percentages (deviations) from the nominal process coefficients. All computations and visualizations are performed in the computer algebra system *Maple*.

2. Problem statement

The activated sludge wastewater treatment process is generally carried out in a system, which main consists of an aeration tank with biomass (bioreactor) and a secondary settler. It is assumed that the hydraulic characteristics of aeration tank are those of a continuously stirred tank bioreactor with cell recycle. Ideal conditions are assumed to prevail in the settler [6, 9]. A schematic diagram of the activated sludge wastewater treatment process with bioreactor and settler is shown on Figure 1.
Because of the system specificity connected with use of the different microorganisms (biomass), the activated sludge processes are characterized with nonlinear dynamics and parameter uncertainty (parameters of the bioreactor).

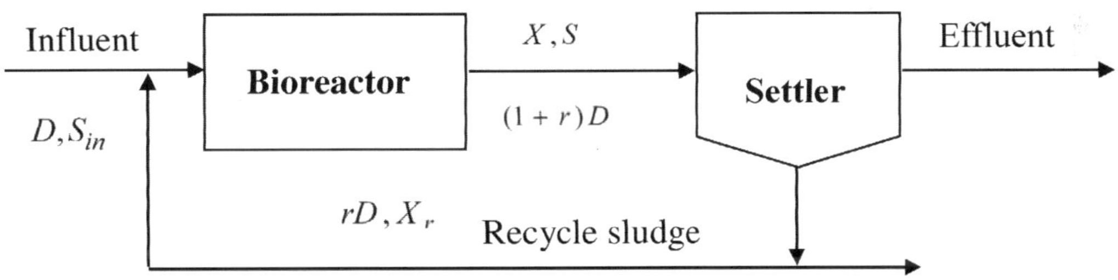

Figure 1. Schematic diagram of the activated sludge wastewater treatment process

The dynamical model of the the activated sludge process in the bioreactor is designed on based of the mass and energy balance equations. In this study, the model is described by the following two nonlinear ordinary differential equations [16]:

$$\frac{dx(t)}{dt} = \mu(s)x(t) + rD(t)x_r(t) - (1+r)D(t)x(t) \tag{1}$$

$$\frac{ds(t)}{dt} = -k\mu(s)x(t) + D(t)(s_{in} - s(t)) \tag{2}$$

where $x(t)$ is biomass concentration (activated sludge) [mg.l^{-1}]; $s(t)$ is substrate concentration (biochemical oxygen demand) [mg.l^{-1}]; D is dilution rate [h^{-1}]; s_{in} is influent substrate concentration [mg.l^{-1}]; x_r is recycle biomass concentration [mg.l^{-1}]; μ is specific growth rate [h^{-1}]; r is sludge recycle ratio; k is yield coefficient.

The specific growth rate $\mu(s)$ is presented by Monod law

$$\mu(s) = \frac{\mu_m.s(t)}{k_s+s(t)},$$

where μ_m and k_s are kinetic process parameters.

It is assumed that the control input is $u = D(t) > 0$ and the output is $y = s(t)$ or $y = x(t)$.

Here is assumed that the main difficulty in controlling the activated sludge wastewater treatment processes comes from the variation of the process parameters and the influent waste load due to negative impacts of natural hazards. These variations induce process state changes that may lead to a reduction of the water treatment efficiency, unless the plant operation is continuously adjusted. Therefore, the steady state analyses should be performed to overcome this problem.

The activated sludge process steady state may critically depend on the process parameters and on the value of the influent waste load. Usually, practical investigations and computer simulations show that the process parameters s_{in}, x_r, r in the model (1)-(2) are unknown but bounded.

It this study, it is used the method for stability analysis of a nonlinear model of wastewater treatment processes, proposed in [16]. It is assumed that instead of numerical values for parameters s_{in}, x_r, r, the parameter intervals $[s_{in}]$, $[x_r]$, $[r]$ are given, respectively.

The main idea of this paper is to study the input–output static characteristics $y_x(u) = x(u)$ and $y_s(u) = s(u)$ involving intervals in the parameters (s_{in}, x_r, r).

Using advanced techniques from recently developed interval analysis [16], it is computed the so called interval static characteristic which presents the set of all functions $y_x(u)$ and $y_s(u)$ for any values of the coefficients in the given intervals ($[s_{in}]$, $[x_r]$, $[r]$).

3. Interval static characteristics

The static model of the process is delivered from (1)-(2) by setting the right-hand side functions equal to zero. Thus it is obtained the nonlinear algebraic system with respect to *s* and *x*:

$$\mu(s)x + rux_r - (1+r)ux = 0$$

$$k\mu(s)x - u(s_{in} - s) = 0$$

or more precisely:

$$\left(\frac{\mu_m.s}{k_s+s} + -(1+r)u\right)x + rux_r = 0 \tag{3}$$

$$k\frac{\mu_m.s}{k_s+s}x - u(s_{in} - s) = 0 \tag{4}$$

By expressing $x = x(s,u)$ from (4) and substituting in (3) it is obtained the following the quadratic equation with respect to s [16]:

$$as^2 + bs - c = 0,$$

where

$$a = (1+r)u - \mu_m, \quad b = \mu_m krx_r - s_{in}((1+r)u - \mu_m) + (1+r)uk_s, \quad c = (1+r)uk_s s_{in}.$$

Its discriminant is given by

$$\Delta(u) = b^2 + 4ac = (\mu_m krx_r - s_{in}((1+r)u - \mu_m) - (1+r)uk_s)^2 + 4\mu_m krx_r(1+r)uk_s$$

Obviously, $\Delta(u) > 0$ for any $u \geq 0$ is valid. The two roots, $s_1(u)$ and $s_2(u)$ are then

$$s_1(u) = \frac{-b+\sqrt{\Delta(u)}}{2a} = \frac{2c}{\sqrt{\Delta(u)}+b}, \qquad s_1(u) = \frac{-b-\sqrt{\Delta(u)}}{2a}$$

Taking into account the biotechnological restriction $0 < s < s_{in}$, the second root $s_2(u)$ is excluded from further consideration. Moreover, $s_1(u)$ satisfies [16]

$$s(u) < s_{in} \qquad \text{and} \qquad \lim_{u\to\infty} s_1(u) = s_{in}.$$

Thus the (biologically reasonable) steady state is $s(u) = s_1(u)$ [16]:

$$s(u) = \frac{2k_s(1+r)us_{in}}{\sqrt{\Delta(u)}+\mu_m krx_r - s_{in}((1+r)u-\mu_m)+(1+r)uk_s} \tag{5}$$

Then (4) implies

$$x(u) = \frac{u(s_{in}-s(u))}{k\mu(s(u))} \tag{6}$$

The functions

$$y_s(u) = s(u) \quad \text{and} \quad y_x(u) = x(u) \tag{7}$$

are called input-output static characteristics of the dynamic activated sludge wastewater treatment process.

It is assumed now that the coefficients s_{in}, x_r, r in the model (3)-(4) are enclosed by the intervals $[s_{in}] = [s_{in}^-, s_{in}^+]$, $[x_r] = [x_r^-, x_r^+]$, $[r] = [r^-, r^+]$, respectively [16].

For arbitrary but fixed $u > 0$ it is considered $y_s(u) = y_s(u;\ s_{in}, x_r, r)$ and $y_x(u) = y_x(u;\ s_{in}, x_r, r)$ from (7) as functions of the variables s_{in}, x_r, r defined on the vector with intervals in the components (interval vector) $[\mathbf{z}] = ([s_{in}], [x_r], [r])$.

As a next step it will be computed the ranges of $y_s(u) = y_s(u;\ s_{in}, x_r, r)$ and $y_x(u) = y_x(u;\ s_{in}, x_r, r)$ on $[\mathbf{z}]$. The main difficulty here is the strong dependence (repeatability) on the uncertain parameters in the expressions for $y_s(u)$ and $y_x(u)$ [17]. To overcome it the monotonicity properties of the functions with respect to s_{in}, x_r, r are used. The gradients of $y_s(u)$ and $y_x(u)$,

$$grad(y_s) = \left(\frac{\partial y_s}{\partial s_{in}}, \frac{\partial y_s}{\partial x_r}, \frac{\partial y_s}{\partial r}\right) \quad \text{and} \quad grad(y_x) = \left(\frac{\partial y_x}{\partial s_{in}}, \frac{\partial y_x}{\partial x_r}, \frac{\partial y_x}{\partial r}\right)$$

have constant signs with respect to each one of the components on $[\mathbf{z}]$, that is

$$\frac{\partial y_s}{\partial s_{in}} > 0, \quad \frac{\partial y_s}{\partial x_r} < 0, \quad \frac{\partial y_s}{\partial r_r} > 0 \qquad \text{and} \qquad \frac{\partial y_x}{\partial s_{in}} > 0, \quad \frac{\partial y_x}{\partial x_r} > 0, \quad \frac{\partial y_{sx}}{\partial r_r} > 0$$

Then the ranges of $y_s(u) = y_s(u;\ s_{in}, x_r, r)$ and $y_x(u) = y_x(u;\ s_{in}, x_r, r)$ on the interval vector $[\mathbf{z}] = ([s_{in}], [x_r], [r])$ are presented [18] by

$$[y_s](u) = [y_s^-(u), y_s^+(u)] = [y_s(u; s_{in}^-, x_r^+, r^+), y_s(u; s_{in}^+, x_r^-, r^-)]$$

$$[y_x](u) = [y_x^-(u), y_x^+(u)] = [y_x(u; s_{in}^-, x_r^-, r^-), y_x(u; s_{in}^+, x_r^+, r^+)]$$

The functions $[y_s](u)$ and $[y_x](u)$ are interval-valued functions of the real variable u. They are called interval input-output static characteristics of the process (1)-(2) with respect to the outputs s and x, respectively. They present the hull of all static characteristics $y_s(u) = y_s(u;\ s_{in}, x_r, r)$ and $y_x(u) = y_x(u;\ s_{in}, x_r, r)$, when the coefficients s_{in}, x_r, r vary in the prescribed intervals.

The interval function $[y_s](u)$ is uniquely defined by its boundary functions $y_s^-(u)$ and $y_s^+(u)$ that is $[y_s](u) = [y_s^-(u), y_s^+(u)]$ with $y_s^-(u) \le y_s^+(u)$ for any $u > 0$. The explicit expressions for $y_s^-(u)$ and $y_s^+(u)$ will be now given. First, it is denoted [16]:

$$\Delta_1(u) = (\mu_m k r^+ x_r^+ - s_{in}^-((1+r^+)u - \mu_m) - (1+r^+)uk_s)^2 + 4\mu_m k r^+ x_r^+ x_r (1+r^+)uk_s$$

$$\Delta_2(u) = (\mu_m k r^- x_r^- - s_{in}^+((1+r^-)u - \mu_m) - (1+r^-)uk_s)^2 + 4\mu_m k r^- x_r^- x_r (1+r^-)uk_s$$

Then the boundary functions are presented by

$$y_s^-(u) = \frac{2k_s(1+r^+)us_{in}^-}{\sqrt{\Delta_1(u)} + \mu_m k r^+ x_r^+ - s_{in}^-\big((1+r^+)u - \mu_m\big) + (1+r^+)uk_s}$$

$$y_s^+(u) = \frac{2k_s(1+r^-)us_{in}^+}{\sqrt{\Delta_1(u)} + \mu_m k r^- x_r^- - s_{in}^+\big((1+r^-)u - \mu_m\big) + (1+r^-)uk_s}$$

The boundary functions $y_x^-(u)$ and $y_x^+(u)$ for $[y_x](u)$ can be computed similarly by substituting, $s_{in} = s_{in}^-$, $x_r = x_r^-$, $r = r^-$ and $s_{in} = s_{in}^+$, $x_r = x_r^+$, $r = r^+$ in $y_x(u)$, respectively.

4. Numerical experiments and discussion

Many simulations are carried out with different nominal values of the coefficients $(\mu_m, k_s, k, s_{in}, x_r, r)$ in the model (1)-(2) in order to analyze natural hazard impacts on activated sludge wastewater treatment processes. All computations and visualizations are performed in the computer algebra system *Maple* [16].

In the computer experiments the nominal values for the process coefficients are considered as centers of the intervals. The radii are given by $\rho_\alpha \alpha$, $\alpha \in \{s_{in}, x_r, r\}$ with $0 < \rho_\alpha < 1$; ρ_α is called deviation from α. Thus

$$[s_{in}] = [s_{in}(1-\rho_{s_{in}}), s_{in}(1+\rho_{s_{in}})], \qquad [x_r] = [x_r(1-\rho_{x_r}), x_r(1-\rho_{x_r})],$$
$$[r] = [r(1-\rho_r), r(1+\rho_r)].$$

Giving different values to ρ_α, intervals $[s_{in}]$, $[x_r]$, $[r]$ with different radii (widths) are obtained [16]. Assume first that only $[s_{in}]$ is uncertain, $s_{in} \in [s_{in}]$ and the other two coefficients x_r and r are exactly known, that is $\rho_{x_r} = \rho_r = 0$ (or equivalently $[x_r] = x_r$ and $[r] = r$). Figure 2 shows the boundary functions of the input-output static characteristics $[y_s]_{s_{in}}(u)$ and $[y_x]_{s_{in}}(u)$ with different deviations, taking into account impact of some natural hazard to the analyzed activate sludge process.

Figure 1 shows that the large increase of u leads to strong decrease of the activated sludge concentration x, so that x may become smaller than a technologically given minimal concentration x_{min}. This may cause instability and inefficiency of the bioreactor. The left picture presents the strong influence of s_{in} on $y_s(u)$ for small values of u, whereas large values for u retain the width $(y_s^+ - y_s^-)$ of $[y_s](u)$ near to the width $(s_{in}^+ - s_{in}^-)$ of $[s_{in}]$.

IOP Conf. Series: Earth and Environmental Science **171** (2018) 012042 doi:10.1088/1755-1315/171/1/012042

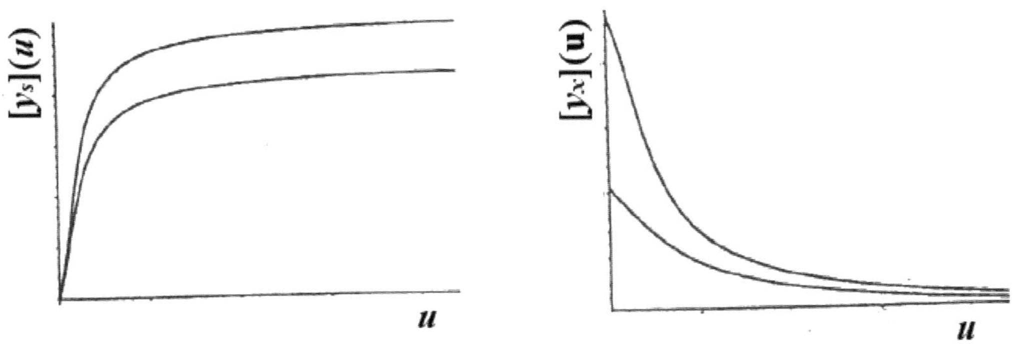

Figure 2. Simulations of the input-output static characteristics $[y_s](u)$ with $\rho_{s_{in}} = 0.1$ and $[y_x](u)$ with $\rho_{s_{in}} = 0.4$ [16].

Let be $x_r \in [x_r]$ and $[s_{in}] = s_{in}$, $[r] = r$. Figure 3 visualizes the interval input-output static characteristics $[y_s]_{x_r}(u)$ and $[y_x]_{x_r}(u)$.

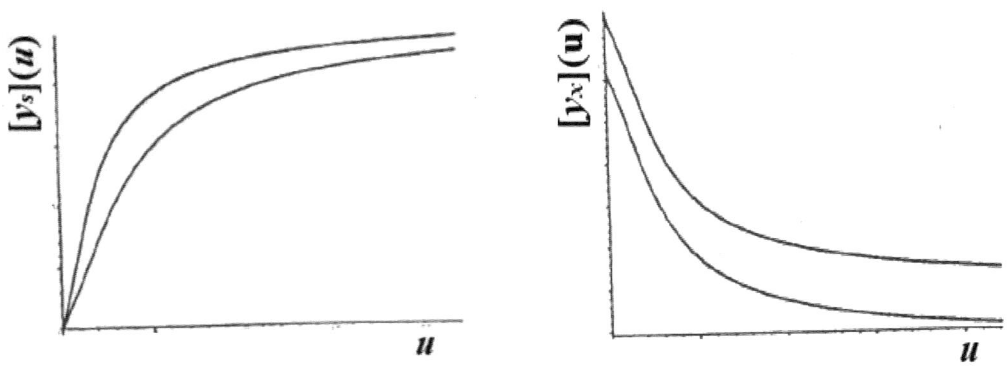

Figure 3. Simulations of the input-output static characteristics $[y_s](u)$ with $\rho_{x_r} = 0.3$ and $[y_x](u)$ with $\rho_{x_r} = 0.007$ [16].

For $r \in [r]$ and $[s_{in}] = s_{in}$, $[x_r] = x_r$ Figure 3 presents the boundary functions of the input-output static characteristics $[y_s]_r(u)$ and $[y_x]_r(u)$.

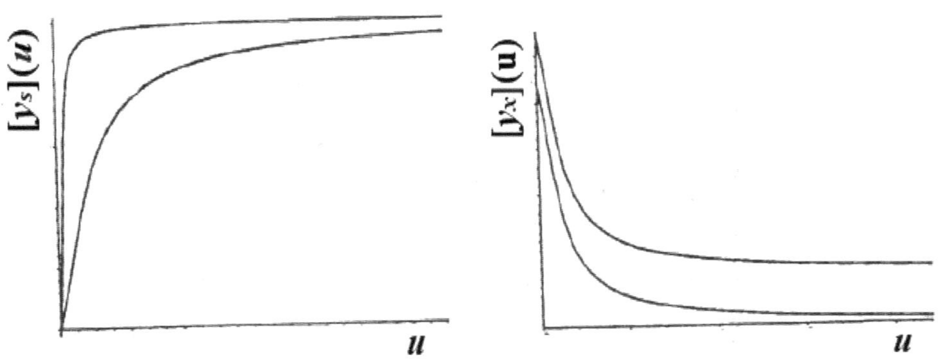

Figure 4. Simulations of the input-output static characteristics $[y_s](u)$ with $\rho_r = 0.9$ and $[y_x](u)$ with $\rho_r = 0.04$ [16].

Figure 3 shows strong dependence of the activated sludge concentration x with respect to recycled biomass concentration x_r. The simulation results of $[y_s](u)$ shows that the deviations in x_r have strong impact on $s(u)$ in the first (exponential) phase, but for large rates u it becomes negligible. Similar effects are observable on Figure 4.
The considered mathematical model of the activated sludge process affected from natural hazard does not require special bounds on the dilution rate u. In practice an operating admissible upper bound $\acute{u}$ for u always exists and depends on the technical characteristic of the bioreactor.

5. Conclusion

An approach for analysis of natural hazard impacts (climate change impacts) on wastewater treatment process with activated sludge is proposed. The approach is based on qualitative analysis of the input-output static characteristics of the activated sludge process, involving uncertainties in the inflow parameters of the aerobic bioreactor. The process model is described by the system from two nonlinear ordinary equations with interval coefficients. The impact levels of one natural hazard on the activated sludge process are defined as a deviation between the nominal and the affected input-output characteristics. The negative effects are described by variation in given intervals of some process coefficients and are calculated as percentages (deviations) from the nominal process coefficients. This analysis of natural hazard impacts on activated sludge wastewater treatment analysis is necessary in order to design and operate such system stable and efficiently.

6. Acknowledgments

The authors wish to thank the Bulgarian National Science Fund for the partial financial support under the Grant № DFNI-I02/15 from 12.12.2014, titled "Information System for Integrated Risk Assessment from Natural Disasters".

7. References

[1] Sperling M 2007 *Basic Principles of Wastewater Treatment*, IWA Publishing, p 200.
[2] Riffat R 2012 *Fundamentals of Wastewater Treatment and Engineering*, Taylor & Francis, p 359
[3] Sokawa H 2016 *Management of Wastewater in Japan*, Japan Sewage Works Association, http://www.jswa.jp
[4] Mihelcic J R and Zimmerman J B 2014 *Environmental Engineering: Fundamentals, Sustainability, Design.* 2nd Edition, John Wiley & Sons, p 704
[5] Mittal A 2011 *Biological Wastewater Treatment, Water Today*, pp 32-44
[6] Olga Sanches (ed.) 2016 *Environmental Engineering and Activated Sludge Processes: Models, Methodologies, and Applications*, Apple Academic Press, p 63.
[7] Brdjanovic D et al (eds.) 2015, *Applications of Activated Sludge Models*, IWA Publishing, p 500.
[8] IWA Task Group on Good Modelling Practice 2012 *Guidelines for Using Activated Sludge Models*, IWA Publishing, p 312.
[9] Meijer S C F and Brdjanovic D 2012 *A practical guide to activated sludge modeling*. UNESCO-IHE Institute for Water Education, p 277.
[10] Tram V T, Ngo H H, Guo W, Zhou J L, Nguyen P D, Listowski A and Wang X C 2014 A mini-review on the impacts of climate change on wastewater reclamation and reuse. *Sci Total Environ* 494-495 pp 9-17.
[11] Tolkou A K and Zouboulis A I 2016 Effect of climate change in WWTPs with a focus on MBR infrastructure, *Desalination and Water Treatment* 57 (5) pp 2344-2354.
[12] Danas K, Kurdi B, Stark M and Mutlaq A 2012 *Climate change effects on waste water* treatment, CEE Jordan Group Presentation. http://courses.washington.edu/cejordan/
[13] McMahan E K 2006 *Impacts of rainfall events on wastewater treatment processes*, University of South Florida, http://scholarcommons.usf.edu/etd/3846
[14] Morgan-Sagastume F and Allen D G 2003 Effects of temperature transient conditions on aerobic biological treatment of wastewater, *Water Res. Sep* 37 (15) pp 3590-601.

[15] Plósz B, Liltved H, Ratnaweera H. 2009 Climate change impacts on activated sludge wastewater treatment: a case study from Norway, *Water Sci Technol.* 60 (2) pp 533-541.
[16] Zlateva P. and Dimitrova N. 2005 Stability analysis of a nonlinear model of wastewater treatment processes, NAA 2004, (Li Z. et al., Eds.), *LNCS* 3401, (Berlin: Springer – Verlag) pp 606-612.
[17] Ratschek H. and Rockne J 1984 *Computer Methods for the Range of Functions*. (New York: Ellis Horwood Publ., Halsted Press).
[18] Dimitrova N. 1996 On a numerical approach for solving a class of nonlinear systems, *Scientific Computing and Validated Numerics*, (Alefeld G. et al., Eds.), (Berlin: Akademie Verlag) pp 147-153.

IOP Conf. Series: Earth and Environmental Science **171** (2018) 012043 doi:10.1088/1755-1315/171/1/012043

Esterification of α-Pinene from Turpentine Oil Using Natural Zeolite Catalyst

N. Wijayati[1], Supartono[1] and E. Kusumastuti[1]

[1]Department of Chemsitry, Universitas Negeri Semarang, Indonesia
Email: nanikanang@gmail.com

Abstract. Turpentine is one of the oils obtained from pine trees with >80% α-pinene composition. Turpentine oil has a high selling value by making α-pinene derivative through an esterification reaction. The esterification reaction was carried out with time variation and reaction temperature using natural zeolite. Characterization of the catalyst using X-Ray Diffraction, IR spectroscopy and Scanning Electron Microscopy. The esterification reaction product was analyzed by IR spectroscopy, and Gas Chromatography - Mass Spectroscopy. The main product of α-pinene esterification reaction was α-terpinyl acetate (21.40%), and the selectivity of 28.87 %, produced at 40°C for 3 hours.

1. Introduction

Most of the pine trees in Indonesia is the kind of Pinus which produces turpentine with a composition of 82% α-pinene and other components.Attempts to get oil terpertin added value is through a distillation process that produces components of turpentine oil.One effort that turpentine oil has a high sales value is by conducting an esterification reaction α-pinene to produce α-terpenil acetate which is used widely in the perfume industry, especially on the soap.However, the presence of the wild-bit terpenil acetate and the need is less economical so as to get it done by isolating it from the plant [1-3].
Natural zeolites are zeolites are widely available on the volcano and mined directly from nature that cost far less than the zeolite synthesis.The main content of natural zeolite mineral morderit, the ratio Si / Al is high so it has high thermal stability [4].
Esterification reaction between α-terpineol and acetic anhydride with a yield of 5.57% and a purity of 67.50% at room temperature.Gainsford et al [1] carry out the conversion of α-pinene into α-terpenil acetate with H-beta zeolite at room temperature in 24 hours resulted in a 29% yield.Liu et al [2] conducted an esterification reaction α-pinene into α-terpenil acetate selectivity α-terpenil highest acetate 27.8% using acidic ionic liquid catalyst is $[HSO_3\text{-}(CH_2)_3\ \text{-Net}_3]\ H_2PO_4$.Li et al [3] conducted a synthesis terpenil acetate using an ionic liquid as a catalyst octadecyl amine ethoxylates, optimum condition is achieved at a ratio of n (α-pinene): n (ionic liquids): n (acetic acid) = 5: 0.3: 20 , where n (α-pinene) = 0.05 mol, carried out at a temperature of 30 ° C with a reaction time of 10 hours to obtainterpenil acetate yield of 35.70%.
This work to study to change the oil component of turpentine in the form of a compound α-pinene to terpenil acetate by the esterification reaction using natural zeolite catalyst.

2. Method

The tools used in the study is a fractional distillation under reduced pressure, IR spectrophotometer, X-ray diffraction, Scanning Electron Microscopy (SEM), and Gas Chromatography-Mass Spectrometer. The material used is turpentine oil, distilled water, natural zeolite, ammonia, pyridine, Na_2SO_4, acetic anhydride, filter paper, dichloromethane, saturated sodium bicarbonate.

IOP Conf. Series: Earth and Environmental Science **171** (2018) 012043 doi:10.1088/1755-1315/171/1/012043

Esterification of α-pinene,
α-Pinene 0.5g were mixed in 10 mL of acetic anhydride, 10 mL H_2O, 10 mL of dichloromethane and then stirred using a mechanical stirrer for 3 hours at a temperature of 40°C with 0.5 g of activated natural zeolite in a closed flask equipped with a thermometer. The reaction products were tested with the GC, FTIR, GC-MS.

3. Result and Discussion

3.1. Characterization of natural zeolite

Natural zeolites are shaped chunks of crushed then pulverized and sieved with a 100 mesh sieve in order to be the same size and the natural zeolite surface area becomes larger. Activation of natural zeolite with acid and salt treatment aims to improve the performance of natural zeolite. Natural zeolites are already activated with HF 1% and then soaking with 6M HCl solution aimed dealuminated natural zeolite can improve the ratio of Si/Al.
The results of the analysis of activated natural zeolite with XRD in Figure 1 shows that there are three highest peaks in the H/ZA is at 2θ = 27,21°; 21,34° and 25,09°. The specific peaks is the peak of morderitie at 2θ = 20,9°; 25,63°; 26°; 26,25°; 27,67°. While the typical peak for morderit located on the highest peak at 2θ = 25,631° and 27,651° (JCPDS No. 700232). Suitability pattern difraktogram H/ZA with morderit can be concluded that the samples H/ZA has type structure mordernit

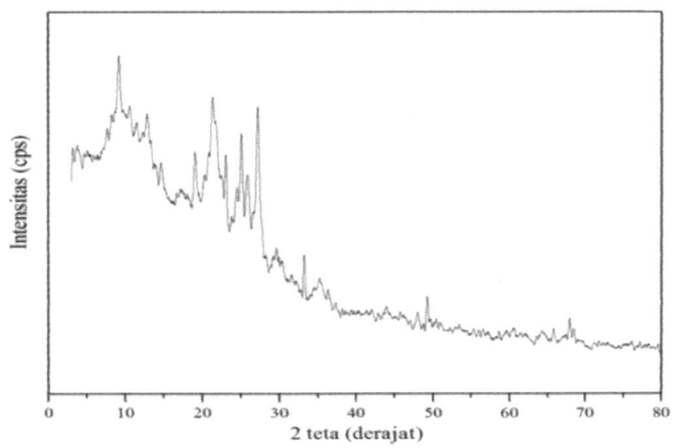

Figure 1. Diffractograms of catalyst H/ZA

Determination of the number of catalyst acid sites H/ZA with ammonia of 0.81 mol / g, whereas with pyridine of 0.79 mol/g. The total number of acid sites with ammonia on the H/ZA catalyst is higher than the number of acid sites with pyridine. This is because ammonia is a stronger base than pyridine. In addition, the size of the ammonia molecule is relatively smaller than the pyridine adsorbed onto the surface in the pores, while the pyridine is only adsorbed on the outer surface (pore mouth).
The results of the scanning using SEM in Figure 2, obtained a morphological picture of H/ZA at 5000x magnification showed that the size of the distribution reached 20 micrometers scale. at 10,000x magnification shows that the distribution size reaches 10 micrometers scale. The molecular size of H/ZA is uneven and irregularly shaped.

IOP Conf. Series: Earth and Environmental Science **171** (2018) 012043 doi:10.1088/1755-1315/171/1/012043

(a) (b)

Figure 2. The morphology of H/ ZA (a) magnification 5.000x (b) magnification 10.000x

3.2. The Esterification from turpentin oil

The main componen of turpentin oil are the monoterpene compounds. The monoterpenes are essential ingredients in fine chemical industry and flavor and parfume industry. The most important constituents of monoterpenes are α- and β-pinene (Figure 3).

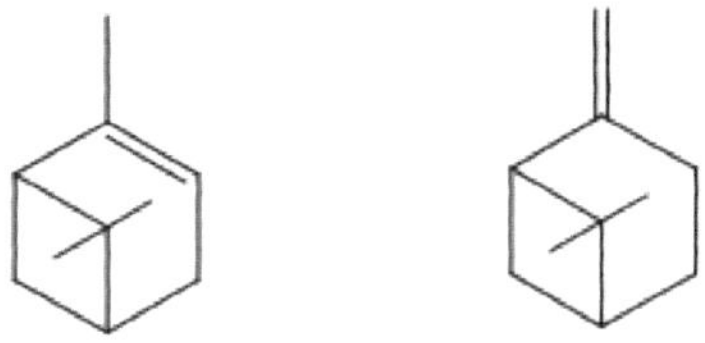

(1) α- pinene (2) β-pinene

Figure 3. (1) α- and (2) β-pinene

The turpentine oil are mobile liquids, usually non-coloured or slighty colored. The boiling point of the turpentine oil between 155-156°C, the specific gravity (20°C) varies between 0.854-0868 g/mL and the refractive index (20°C) between 1.4656. The oils are not soluble in water, but soluble in ethers, dichloromethane and in other oils. Highly pure α- and (2) β-pinene can be obtained by fractional distilation of turpentine oil. The main component of turpentine oil are a-pinene. Indonesia turpentine oil contains about 57-86% α- pinene, 8-12% β-pinene and other groups monoterpenes with a number of minor (Figure 4). The main products obtained from turpenine oil are camphene, limonene, terpinene, terpinolene, α-terpineol, and terpinyl acetate, etc. [5].

The results of the identification of physical compound α-pinene which is a liquid, colored clear and distinctive smell like turpentine. Based on the analysis using FTIR, α-pinene, group C = C at wave number 1650 cm^{-1} showed that α-pinene including alkenes [6-7]. From gas chromatographic yield of α-pinene turpentine oil isolated by 98.10%. The content of turpentine oil is the largest in the compound α-pinene about 82%.

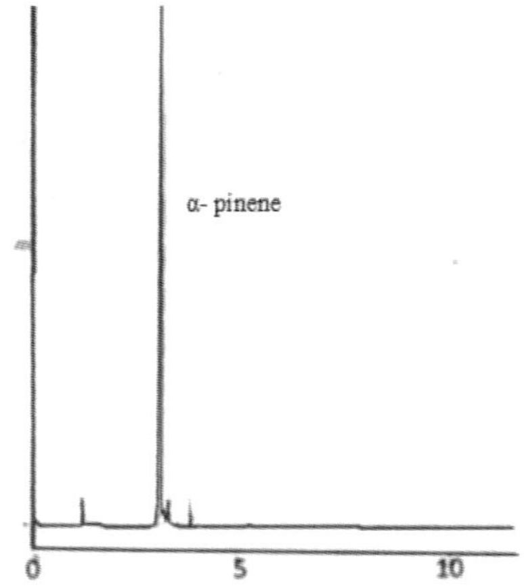

Figure 4. Chromatogram of turpentine oil

Alpha pinene esterification reaction takes place in situ in a reaction that occurs through hydration reactions prior to the addition of distilled water to form α-terpineol then α-terpineol formed will undergo an esterification reaction with acetic anhydride to form α-terpenil acetate. In this case allow for other products in the esterification reaction α-pinene as fencyl acetate, bornil acetate, fencyl alcohol, α-terpineol and borneol.

The α-pinene esterification reaction with variation of reaction time of 2, 3 and 4 hours yields terpinyl acetate with yields of 16.04, 21.40, and 28.77% with the conversion of 51.57, 74.13 and 94.68%.

Table 1. Results of α-pinene esterification reaction time variations

Time (h)	C (%)	Yield (%)						Selectivity
		Fencyl alcohol	Borneol	α-terpineol	Fencyl acetate	Bornil acetate	α-terpinyl acetate	α-terpinyl acetate
2	51.57	1.06	1.44	26.45	2.42		16.04	31,10
3	74.13	1.67	0.44	43.31	3.76	1.02	21.40	28,87
4	94.68	3.16	1.49	55.32	3.62		28.77	30,38

Reaction conditions: 0.5 g of pinene, 10 mL of acetic anhydride, 5 mL H_2O, 10 mL dichloromethane, T = 40ºC

The α-pinene esterification with temperature variations of 30, 40 and 60 ° C with reaction time of 3 hours and the ratio of pinene acetic anhydride compound (1:15) yielded α-terpenyl acetate with 0.18, 21.40, and 14.72% with the conversion of 4.99, 74.13 and 52.44.

The effect of reaction temperature on α-pinene esterification is shown in Table 2. When the temperature was low, the yield of α-terpinyl acetate was low. As the temperature was increased, the yield of α-terpinyl acetate increased accordingly. When the temperature was high, the yield of α-terpinyl acetate tended to decreased. The results showed the yield of α-terpinyl acetat first increased and then decreased with increasing temperature [3].

Table 2. The effect of reaction temperature on α-pinene esterification reaction

Temp (ºC)	C (%)	Yield (%)						Selectivity
		Fencyl alcohol	Borneol	α-terpineol	Fencyl acetate	Bornil acetate	α-terpinyl acetate	α-terpinyl acetate
30	4,99	0,15	0,14	0,74	0,33	0,94	0,18	3,68
40	74,13	1,67	0,44	43,31	3,76	1,02	21,40	28,87
60	52,44	0,35	23,62	1,21	1,65	1,65	14,72	28,08

Reaction conditions: 0.5 g of pinene, 10 mL of acetic anhydride, 5 mL H_2O, 10 mL dichloromethane, t =3 hours

The results of the analysis of esterification of α-pinene using FTIR in Figure 5 that the aliphatic CH absorptions at wave number 2923 cm^{-1}, group C = C at wave number 1617 cm^{-1}, group C = O at wave numbers between 1735-1750 cm^{-1} and the peak of the C-O group at wave numbers between 1050 to 1300 cm^{-1} can be concluded that the compounds produced an ester compound. α-pinene that no group C-O and C = O indicates that the compounds produce ester esterification of alkenes.

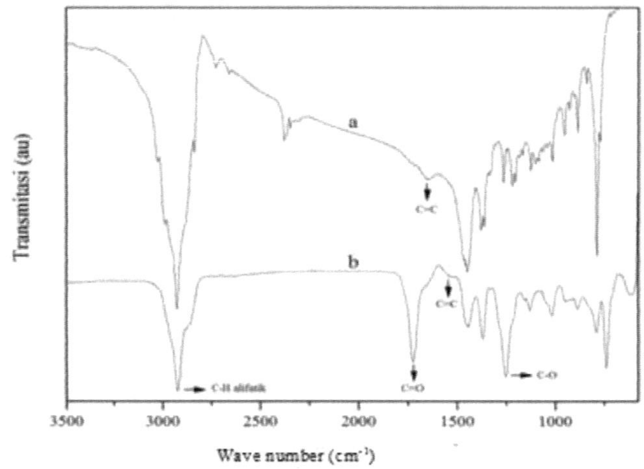

Figure 5. The IR spectrum (a) compound α-pinene (b) the results of esterification results of α-pinene esterification reaction time of 3 hours, a temperature of 40 ° C.

The α-pinene compound has a fairly stretchy structure and the presence of a double-bonded reactive ring within the α-pinene structure makes it easy to participate in cycle-opening reactions and molecular rearrangement reactions. The esterification reaction mechanism of the α-pinene compound is shown in Figure 6.

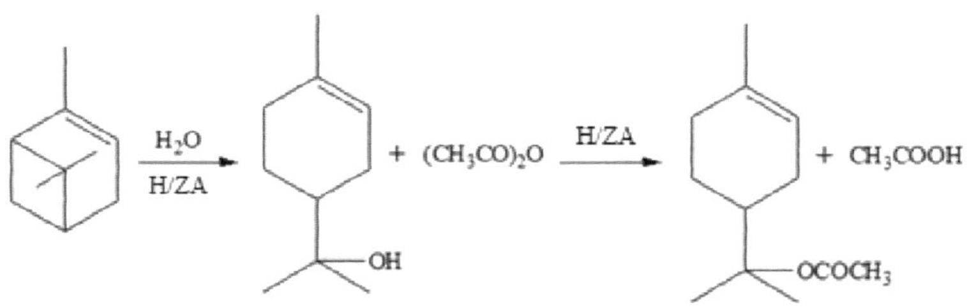

Figure 6. The esterification reaction mechanism of the α-pinene compound

4. Conclution

The α-pinene esterification reactor using natural zeolite catalyst at 40°C for 3 hours with a reactant mole ratio between α-pinene and acetic anhydride (1:15) yielded α-terpenyl acetate of 21.40% and selectivity 28.87 %

5. Acknowledgement

The authors would like to thank Directorate General of Higher Education (DGHE), Department of National Education Republic Indonesia for to support.

6. References

[1] Gainsford, G.J., C.F. Hosie, & R.J. Wetson. 2000. Conversion of α-pinene To Terpinyl Acetate Over H-Beta Zeolites. *Applied Catalysis A: General,* 209: 269–277.

IOP Conf. Series: Earth and Environmental Science **171** (2018) 012043 doi:10.1088/1755-1315/171/1/012043

[2] Liu, S., C. Xie, S. Yu, F. Liu, & K. Ji. 2008. Esterification of α-Pinene and Acetic Acid Using Acidic Ionic Liquids as Catalysts. *Catalysis Communications,* 9: 1634–1638.
[3] Li, L., S. Liu, Y Shi, S. Yu, C. Xie, & C. Qi. 2013. Synthesis of Terpinyl Acetate Using OctadecylamineEthoxylate Ionic Liquids as Catalysts. *Res ChemIntermed 39*:2095–2105.
[4] Wijayati, N., T. Handayani, Supartono. 2017. Isomerization of α-pinene using Zirconiua/Natural Zeolite Catalysts. *Asian Journal of Chemistry*. 29 (8) : 1705-1708.
[5] Wijayati, N., H.D. Pranowo, Jumina, &Triyono. 2013. The Acid Catalyzed Reaction of α-PineneOver Y-Zeolit. *Indo. J. Chem*, 13(1): 59-65.
[6] AvilaM.C.,ComelliN.A.,CastellonE.R.,LopezA.J.,FloresR.C.,PonziE.N., PonziM.I., 2010.Study of Solid Acid Catalysis For The Hydration of α-pinene. *Journal of Molecular Catalysis*, 322(1-2): 106-112.
[7] Avila M.C., Ponzi M.I., danComelli N.A., 2015. Hydration of α-Pinene over Heteropoly Acid H3PW12O40 and H3PMo12O40. *Journal Chromatogr Sep Tech*, 6(7) : 1-6.

IOP Conf. Series: Earth and Environmental Science **171** (2018) 012044 doi:10.1088/1755-1315/171/1/012044

Research on Green Remediation Technology of Oil Spill By Biosorbent

Dan Peng[1,2], Lei Nie[1], Fan Ouyang[1], Liuchun Zheng[3]

1 Department of Transportation and Environment, Shenzhen Institute of Information Technology, Shenzhen, Guangdong 518172, P. R. China
2 The Key Lab of Pollution Control and Ecosystem Restoration in Industry Clusters, Ministry of Education, South China University of Technology, Guangzhou Higher Education Mega Centre, Guangzhou, Guangdong 510006, P. R. China
3 School of Chemistry and Environment, South China Normal University, Guangzhou, Guangdong 510006, P. R. China
E-mail: pengdan987@hotmail.com

Abstract. Oil spillage is a serious problem all over the world. Usage of agricultural by-products for raw materials in biosorbents applications is a promising way to solve this problem. In this paper, a green biotechnological procedure has been developed using *Phanerochaete chrysosporium* to produce an oil-sorbent from corn stalk and corn cob. Without fungal modification, the maximum sorption of oil by raw corn stalk and corn cob was 6.95 g/g and 4.14 g/g, whereas the treatment with *Phanerochaete chrysosporium* increased up to 9.03 g/g and 7.69 g/g, respectively. SEM and XRD were applied to characterization of treated and untreated materials, which showed the changes in the surface morphology and crystallinity of all biosorbents. The agricultural by-products can be utilized as biosorbent for efficient oil removal, and fungal treatment can serve as a mild and green method to increase the sorption capacity.

1. Introduction

Oil is the major source of energy for human life and production, and so many oil spillages are occurred during oil exploration, exploitation, refining, transportation, storage and usage[1]. According to NASA's report, there are more than 700 million gallons of oil end up into the oceans through these ways every year, such as the Arabian Gulf spill during the Gulf War in 1991 and the Gulf of Mexico spill of 2010. Rivers, oceans and coasts are polluted by spilled oil, which endangers human health[2, 3]. The growing number of oil accident is beginning to arise the concern and research of many scholars, so efficient and immediate technologies to remove spilled oil have been developed by various methods[4]. Conventional methods exist on removal of oil, including *in situ* burning, dispersion, skimmers, boomers, sorbents and bioremediation, among those removal techniques, usage of oil sorbents is a promising way because of its effectiveness, economical, and environment-friend[5]. biosorbents derived from agricultural residues are made of cellulose, hemicellulose and lignin, with rough surfaces, pores, and active functional groups, which are in favour of sorption. Cellulose-based raw materials or their modified forms have been explored as biosorbents for removal of spilled oil from oil/water surface by capillary forces and van de Waals forces, such as sawdust, kapok, sisal, cotton and sugarcane bagasse as oil-sorbents. They have the advantages of selective removal of oil over water, biodegradability, relatively low cost, and limited impact on the environment[6].

Corn is one of the major economical crops all over the world, and it produces abundant of byproducts such as corncob, corn leave and corn stalk, so the importance of utilization the abundantly available and inexpensive resource has been realized[7]. The usage of corn stalk as an oil-sorbent in recovery of spilled oils has received attention recently. Considering oil sorption capacity of raw corn stalk is relatively low, some modification methods, mainly physical and chemical, should be taken in order to produce high-efficiency oil-sorbent. Compared to physical and chemical modification, biological modifications, such as fungal cultivation and enzyme treatment, are promising methods that have been used for biosorbent modification, as biological process can avoid use of toxic reagents, and will not cause secondary pollution. White-rot fungi have abundant extracellular enzymes and mycelia which can penetrate into plant cell walls and degrade lignocellulose[8].

In our previous studies, cellulase is utilized as a modifier to improve oil sorption of corn stalk[9]. The conditions of cellulase-modification are mild, and the equipment is simple, and the oil sorption of modified corn stalk is significantly increased about 6 times. This work aims to evaluate the potential of *Phanerochaete chrysosporium* as a modifier to produce a novel biosorbent for oil spill. To our knowledge, this is the first report on the use of white-rot fungal cultivation as a treatment approach to enhance the oil sorption capacity of biomass.

2. Materials and Methods

2.1. *Materials*

Two raw materials, namely corn straw (RCS) and corn cobs (RCC), were collected from Pennsylvania state university (USA). RCS and RCC were washed by distilled water to remove dusts and impurities, and then crushed to particles after air dried. The fraction between 0.25 to 0.85 mm was selected by sieving. The typical white-rot fungus, namely *Phanerochaete chrysosporium* (PC), was taken from Dr. Tien's lab in Pennsylvania state university (USA). The fungus was maintained on potato dextrose agar (PDA) Petri dish at 4 °C and activated before use. Machine oil (density at 20°C=0.87 g/cm, viscosity=0.021 Pa/s) from Shell was used in the experiment.

2.2. *Methods*

2.2.1. Culture condition. The solid medium containing raw materials, millet and wheat bran, with 10% distilled water was autoclaved at 121 °C for 20 min prior to fungus inoculation. *Phanerochaete chrysosporium* was inoculated in Unicorn culture bags, and each bag with 255 g RCS or RCC was inoculated with ten PDA disks (diameter of 2 mm) cut from the margins of actively growing fungal colony. Culture bags were put in an incubator at 37 °C, and culture materials were stirred every week[10].

2.2.2. Material characterization. The morphology and structure of cultivated and uncultivated materials were inspected by SEM (Model S-3700N, HITACHI). The X-ray diffraction properties were measured by a Rigaku D/max-IIIA Diffractometer, and crystallinity index (CrI) was calculated by Segal's equation[11]:

$$CrI = \frac{I_{002} - I_{am}}{I_{002}} \times 100$$

Where I_{002} is the maximum intensity of the 002 peak at 2θ=22.5°, and I_{am} is the lowest intensity of the amorphous area at 2θ=18.7°.

2.2.3. Sorption studies. Based on previous studies, the gravimetric method was used to determine the sorption capacity of oil. All tests were performed at 25°C. To evaluate the oil sorption properties of samples, a 0.2g sample was put on a steel wire mesh, and then placed into a 500 ml beaker containing 150 ml of deionized water with initial oil quality of 20 g. After 60 min, the sample was lifted up and drained for 5 min. The weight of the sample before and after adsorption was measured; the difference between the two weights was oil sorption capacity[9].

2.2.4. Evaluation of capillary rise. The wicking method was used for evaluation the relative wettability of corn stalk and corn cob[12]. Modified and unmodified biomass were packed into a glass column (diameter of 20 mm, height of 200 mm) at room temperature. At the bottom of the column, a glass filter with a pore diameter of 120 mm and a thickness of 2 mm was soldered to keep the sample. The glass tube was immersed in the oil. When the oil surface in the glass tube is consistent with the oil in the beaker, start the timer to continuously measure the height of oil in the tube during the sorption process by ruler[13].

3. Results

3.1. Effect of Modification Time

Effect of inoculation time on the modification of corn stalk and corn cob for oil sorption capacity (OSC) was investigated under the condition of temperature 37 °C during 1 to 4 weeks. As can be observed in Fig. 1, the OSC was increased as the inoculation time rises. After 3 weeks, *Phanerochaete chrysosporium* modified corn stalk (PCCS) and corn cob (PCCC) obtained the highest OSC, which was 9.03, 7.69 g/g. The solid fermentation of corn stalk and corn cob by *Phanerochaete chrysosporium* can improve their oil sorption. Compared with RCS and RCC, the OSC of PCCS and PCCC performed up to 75.3% and 85.7% better, respectively. Corn cob shows a slight advantage in improvement.

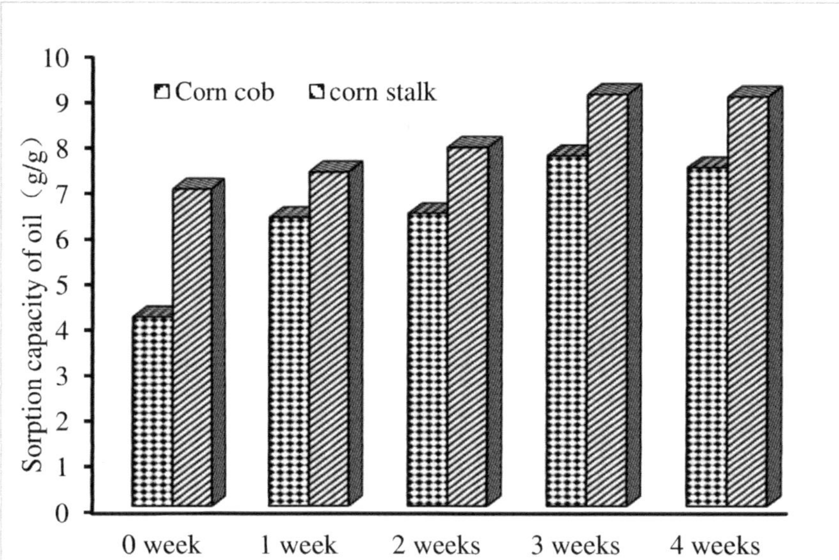

Figure 1. Effect of modification time on the sorption of oil

3.2. Material Characterization

Figure 2 presents the SEM of fungous treated and untreated materials. The surface of RCS is compact and smooth, while the surface of PCCS has formed a lot of grooves. The inner surface of RCS consists of spongy region that compact tightly, after PC treatment, the inner spongy region with a layer structure was expanded, so that PCCS provides more sites for the sorption of oil molecules[14]. As can be seen from Figure 2, the structure of PCCC has a significant change, the inner region showed a porous structure, which also formed a hollow tubular structure, and the compact layer structure was enlarged, which is conducive to accommodate more oil molecules.

Depending on XRD analysis (table 1), the crystallinity of PCCC is the smallest among all materials, and after PC modification, the CrI of PCCS and PCCC decreased by 6.6% and 26.3%, respectively. The decrease of CrI is directly proportional to the increase in the capacity of oil sorption, indicating that the smaller the crystallinity of the biosorbents, the easier the oil molecules to be sorbed. The fungal modification really enlarged the sorption area with spilled oil. Based on all characterization results, the enhanced sorptive capacity of PCCS and PCCC should be mainly attributed to fungal modification.

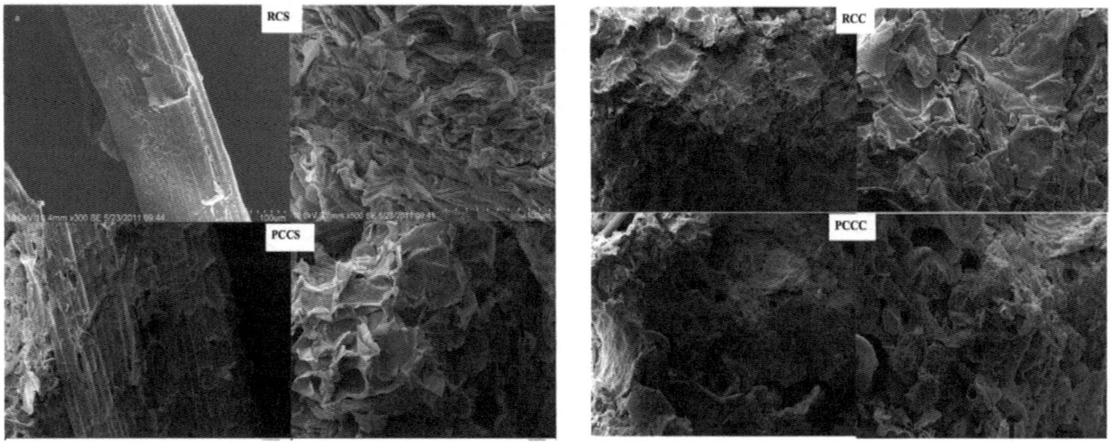

Figure 2. The SEM observation of untreated and treated corn stalk and corn cob

Table 1. Characteristics of untreated and treated corn stalk and corn cob

Biosorbent	CrI (%)
RCS	46.8
PCCS	43.7
RCC	33.5
PCCC	24.7

3.3. Evaluation of Capillary Rise

Capillary rise is one of the main driving forces for oil sorption into biosorbent, which can measure the wettability of biomass. Ribeiro reported that different materials have different capillary forces, and even the different tissues in the same biomass have significant differences in the capillary force[15]. According to this research, Fig. 3 exhibits the relationship between the heights of sorbed oil by capillary rise and the sorption time. The wettability of the materials was proportional to the rate of oil-level rise. It can be seen from fig.3 that the hydrophobicity of PCCC is improved, and the hydrophobicity of PCCS is not significantly changed. The higher capillary rise was observed for PCCC.

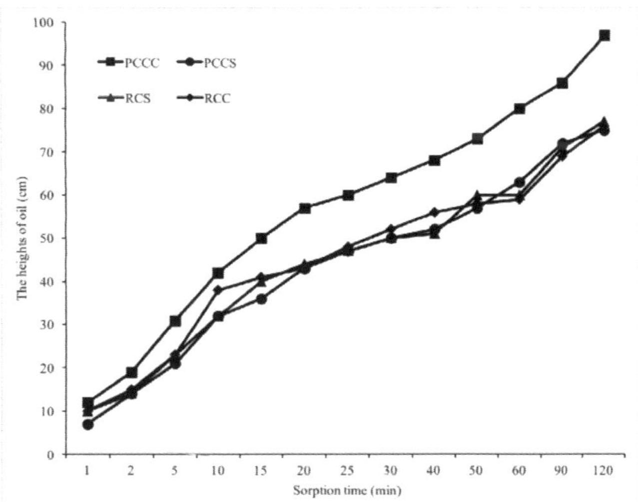

Figure 3. Capillary rises of all biosorbents

4. Conclusions

This study suggests that fungal modification is a green and effective way to produce biosorbent for recovering spilled oil from polluted water. The main mechanism governing the oil sorption is changes in chemical composition, and physical properties in PCCC and PCCS. The maximum sorption of PCCC and PCCS was 7.69 and 9.03 g/g, which increased by 85.7% and 75.3%. The experiment shows that the modification of agricultural wastes by PC has an obvious effect on improvement of oil sorption with no harm to the environment, and also can solve the disposal problem especially for agricultural by-products.

5. References

[1] J.K. Nduka, L.O. Ezenweke, E.T. Ezenwa, Comparison of the mopping ability of chemically modified and unmodified biological wastes on crude oil and its lower fractions, Bioresource Technology, 99 (2008) 7902-7905.
[2] A.A. Al-Majed, A.R. Adebayo, M.E. Hossain, A sustainable approach to controlling oil spills, J Environ Manage, 113C (2012) 213-227.
[3] W. Duan, G. Chen, Q. Ye, Q. Chen, The situation of hazardous chemical accidents in China between 2000 and 2006, J Hazard Mater, 186 (2011) 1489-1494.
[4] G. Hu, J. Li, G. Zeng, Recent development in the treatment of oily sludge from petroleum industry: A review, J Hazard Mater, 261C (2013) 470-490.
[5] S. Sabir, Approach of Cost-Effective Adsorbents for Oil Removal from Oily Water, Critical Reviews in Environmental Science and Technology, 45 (2015) 1916-1945.
[6] H. Liu, B. Geng, Y. Chen, H. Wang, Review on the Aerogel-Type Oil Sorbents Derived from Nanocellulose, ACS Sustainable Chem. Eng., 5 (2017) 49-66.
[7] S. Yang, W. Ding, H. Chen, Enzymatic hydrolysis of corn stalk in a hollow fiber ultrafiltration membrane reactor, Biomass and Bioenergy, 33 (2009) 332-336.
[8] J. Liu, E. Li, X. You, C. Hu, Q. Huang, Adsorption of methylene blue on an agro-waste oiltea shell with and without fungal treatment, Sci Rep, 6 (2016) 38450.
[9] D. Peng, Z. Lan, C. Guo, C. Yang, Z. Dang, Application of cellulase for the modification of corn stalk: Leading to oil sorption, Bioresource technology, 137 (2013) 414-418.
[10] S. Sato, F. Liu, H. Koc, M. Tien, Expression analysis of extracellular proteins from Phanerochaete chrysosporium grown on different liquid and solid substrates, Microbiology, 153 (2007) 3023-3033.
[11] L. Segal, J.J. Creely, A.E. Martin, C.M. Conrad, An Empirical Method for Estimating the Degree of Crystallinity of Native Cellulose Using the X-Ray Diffractometer, Textile Research Journal, 29 (1959) 786-794.
[12] E.V. Lebedeva, A. Fogden, Wettability alteration of kaolinite exposed to crude oil in salt solutions, Colloids and Surfaces A: Physicochemical and Engineering Aspects, 377 (2011) 115-122.
[13] T. Suzuki, E. Takahashi, S. Oishi, M. Endo, M. Inagaki, Evaluation of inter-particle space network of carbon material using capillary rise of liquid, Carbon, 42 (2004) 2771-2773.
[14] S. Kim, M.T. Holtzapple, Effect of structural features on enzyme digestibility of corn stover, Bioresource Technology, 97 (2006) 583-591.
[15] T.H. Ribeiro, R.W. Smith, J. Rubio, Sorption of Oils by the Nonliving Biomass of a Salvinia sp., Environmental Science & Technology, 34 (2000) 5201-5205.

Acknowledgment

This research was financed by the Shenzhen Science & Technology Project (JCYJ20150417094158012; JCYJ20160415114215737; JCYJ20150626102232416; JCYJ201703060-95333329), and the China National Science Fund Program (51508206). The authors are very grateful to professor Ming Tien, department of biochemistry and molecular biology, Pennsylvania state university, for his efforts on the research.

Fabrication of Superhydrophobic-Superoleophilic Cement-Coated Meshes and Their Applications for Oil/Water Separation

Shude Li[1], Sun Jing*[1], Jinlong Song[2] and Ziai Liu[1]

1 School of Mechanical Engineering, Dalian University of Technology, Dalian, China
2 Key Laboratory for Precision and Non-traditional Machining Technology of the Ministry of Education, Dalian University of Technology, Dalian, China
E-mail: sunjing@dlut.edu.cn

Abstract. Extreme wettability materials have shown great ability in dealing with oily industrial wastewater emission and oil spill accidents. Whenas, most of the addressed materials are technically sophisticated or inclined to involve corrosive or toxic chemicals. Herein, we fabricated an easy fabricated, low cost and robust material for oil/water separation. The superhydrophobic-superoleophilic (SOO) cement-coated meshes were made simply by dipping copper mesh in cement paste and surface modification. The micro/nanostructure and low surface energy endowed the mesh with great superhydrophobicity and superoleophilicity in air. The mesh had great oil absorption ability and high separation efficiencies for various oils. They were also admirably recyclable and durable. The meshes could be used to separate oil/water mixture at least for 20 cycles with high separation efficiency, and showed respectable superhydrobocity after being bent for 30 times. This simply fabricated, low cost and robust material will have great potential to be used in industrial wastewater treatment and oil spill in seawater.

1. Introduction

The large amount of oily industrial wastewater emissions and the frequent occurrence of oil spills not only aggravate the situation of water scarcity but also threaten the survival of aquatic organisms[1, 2]. Therefore, how to solve the problem of oil pollution in water effectively has become a hot issue in modern society. Traditional methods including burning, oil containment fences[3], oil containment absorption[4, 5] and biodegradation[6] are generally ineffective facing with serious oil spills, and the separated oil or water are not reusable[7, 8]. Recently, materials with extreme wettablility have attracted increasing attention for oil/water separation due to the different interface effects of oil and water. These super-wetting materials can be obtained by adjusting surface micro/nanostructures and surface energy,[8-10] and can be classified into two main categories, i.e., absorption material and filtration material. Compared with adsorption method[11-15], filtration can be used as a final process of separation because the separated oil or water is almost pure. Extreme wettablility materials for filtration are usually fabricated on three kinds of substance materials, i.e., metal meshes, textiles/fabrics, and polymeric membranes.[16] Because it is mechanically strong and easy to react with other chemicals, metal meshes became the most widely studied substances. The most typical and simple preparation method of extreme wettablility metal mesh is chemical immersion[17-22], for example, Song et al.[18] fabricated a superhydrophobic mesh by immersing stainless mesh in an aqueous solution of 1 M $CuCl_2$ and 1 M HCl, and then modified by stearic acid. The mesh shows superior water repellence and superoleophilicity, thus have high separation efficiency for oil/water mixture. Similarly, Wang et al.[23] fabricated a superhydrophobic and superoleophilic mesh by immersing commercial copper mesh in 4.0 M HNO_3 solution and modified by 1-hexadecanethiol. Although chemical immersion is a simple method and the fabricated meshes have excellent separation effects, it always contains corrosion or toxic chemicals that can cause secondary environmental pollution. In addition, a large amount of other methods were addressed, including chemical vapor deposition[24], spray-coating[25-27], sol-coating[28], electrospinning[29], electrochemical deposition[30, 31] and layer-by-layer assembly methods[31]. Feng et al.[32] fabricated a superhydrophobic and superoleophilic stainless steel mesh by spraying a homogeneous emulsion containing Teflon, adhesive, dispersant, et al. on the surface in 2004. Li et al.[33] fabricated a pH controllable coating by synthesizing polydimethylsiloxane, 2-hydroxyethyl methacrylate and poly (2-(dimethylamino) ethylmethacrylate). However, these methods mentioned above, were almost technically sophisticated, high operation cost, poor corrosion resistance and contain corrosive or toxic chemicals as well. Surely, many researchers have realized the problem, and developed new materials[34, 35], but technically simple and low production cost filtration materials are still solely needed.

Portland cement, a common material of architecture, was found superhydrophilic. Here, we fabricated superhydrophobic-superoleophilic (SOO) cement-coated meshes for oil/water separation simply by dipping copper mesh in the paste of cement, water and silicane, and then by low surface energy modification. The meshes showed great superhydrophobicity and superoleophilicity in air, and had great oil absorption ability. The SOO cement-coated meshes had high separation efficiencies for various floating oils and also were admirably recyclable and durable. This simply fabricated, low cost and robust material has great potential to be used in industrial wastewater treatment and oil spill in seawater.

2. Experimental Section

Please follow these instructions as carefully as possible so all articles within a conference have the same style to the title page. This paragraph follows a section title so it should not be indented.

2.1. Materials

Ordinary portland cement (PO 42.5) was bought from Guangdong Tapai Co., Ltd. (China). Cu meshes were purchased from Shanghai Hengxin Wire Mesh Co., Ltd. (China). Aqueous silicane (DC-30) was bought from Dingsheng New Materials Co., Ltd. (China). Hexane was bought from Sigma-Aldrich (USA). Diesel was purchased from a local gas station. Peanut oil was purchased from Luhua Co. (China) and lubricating oil was bought from CNPC.

2.2. Preparation of the SOO Cement-coated meshes

Cu meshes were firstly cut into circular pieces with diameter of 50 mm, then were washed in detergent, anhydrous ethanol and water, and were dried by absorbent paper. In the meanwhile, a paste composed of aqueous silicane, water and PO 42.5 with mass ratio of 1:1:2 was stirred adequately. Then Cu meshes were immersed in the paste for 30 s, and then pulled out with the velocity of 0.02 m/s. After dried absolutely at room temperature, the mesh was modified by 0.05 M ethanol solution of fluoroalkylsilane for 1 h, and after dried at room temperature, the SOO cement-coated mesh was prepared.

2.3. Oil/Water Separation

The SOO cement-coated mesh was fixed between two acrylic tubes with similar inner diameter of 36 mm with clamping devices. In order to help oils touch the surface of the SOO cement-coated meshes, the device was titled 20 °, and a mixture of 20 mL oils and 5 mL water were used for the separation. The water and oils in the mixtures were dyed differently for better distinguishing. The relevant properties of hexane, diesel, peanut oil and lubricating oil are listed in table 1. After separation, the separated oils were collected to evaluate the separation efficiencies.

Table 1. Properties of oils used in the experiments

Oil	Hexane	Diesel	Peanut oil	Lubricating oil
Density at 25°C [g/cm^3]	0.65	0.84	0.92	0.84
Surface tension at 20°C [mN/m]	17.9	28.3	34.5	40.1
Kinematic viscosity at 40°C [cSt]	0.42	4.33	39.6	74.4

2.4. Characterization
CAs (contact angles) and SAs (sliding angles) of water and oils were measured by an optical contact angle meter (Krüss, DSA100, Germany) using about 5 μL water or oil droplets at room temperature. Surface morphologies of cement meshes were observed by a field emission scanning electron microscope (SEM, JSM-6360LV, Japan). The purity of the separated oils and collected water were examined by Fourier transform infrared spectrum (FTIR, ThermoFisher 6700, America). The separated oils were examined by the FTIR to verify whether it contained water. For the separated water, tetrachloromethane was used to extract the oils in the water, and the absorbance intensity around 2954 cm^{-1}, 2920 cm^{-1} and 2852 cm^{-1} were used for calculating the oil contents. The separation efficiency R was calculated by the following equation,

$$R\ (\%) = (M_a / M_b) \times 100\%$$

where M_a was the mass of the original oil in the oil/water mixture, and M_b was the mass of the separated oil.

3. Results and Discussion
The morphologies of the ordinary Cu meshes and the SOO cement-coated meshes are shown in figure 1(a) to 1(f). The surface of the ordinary Cu mesh was very smooth, but in contrast, the surface of the SOO cement-coated mesh was lumpy. There were lots of triangle-shaped pores and different size of protrusions in the mesh. The longest side of the triangle-shaped pores was around 250 μm, and the surface of the protrusions was covered by lots of leaf-shaped nanostructures. The micro-nanostructures and the low surface modification enabled the SOO cement-coated mesh to have great superhydrophobicity and superoleophilicity. And the triangle-shaped pores could help oil droplets pass through the mesh easily.

Figure 1. Morphologies and chemical composition of the SOO cement-coated meshes: (a)-(c) SEM images of the ordinary Cu meshes, (d)-(f) SEM images of the SOO cement-coated meshes.

The excellent superoleophilicity of the cement-coated mesh is shown in figure 2(a) and 2(b). Droplets of hexane spread rapidly and completely on the SOO cement-coated meshes, and the water contact angle was almost 0°. In contrast, figure 2(c) shows the superhydrophobicity of the cement-coated meshes. The water droplets were almost spherical, and the contact angle was ~ 154°. The SOO cement-coated mesh showed great water repellence when immersed in blue water while the ordinary Cu mesh was dyed blue as shown in figure 2(d). Figure 2(e) and 2(f) shows the touching, detacting and rolling off processes of water droplets on the SOO cement-coated meshes, demonstrating the superhydrophilicity and ultralow water adhesion of the meshes.

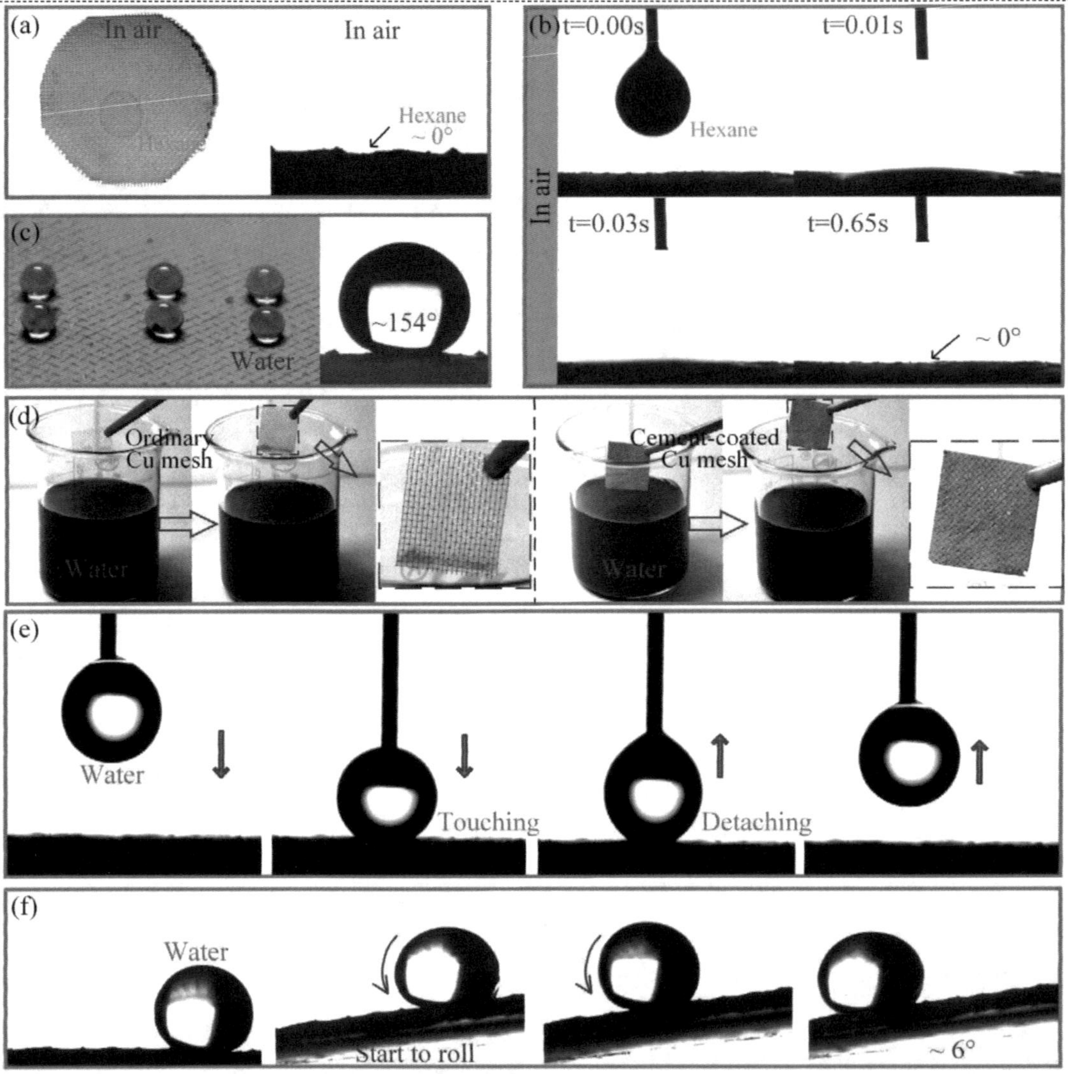

Figure 2. Superoleophilicity and superhydrophobicity of the SOO cement-coated meshes: (a)-(b) photos of hexane droplets on the SOO cement-coated meshes, (c) water droplets on the SOO cement-coated meshes, (d) photos of the ordinary Cu mesh and the SOO cement-coated mesh immersed in blue water before and after, (e)-(f) the touching, leaving and rolling off processes of water droplets on the SOO cement-coated meshes.

Due to the excellent superhydrophobicity and superoleophilicity, the cement-coated meshes showed superior absorption ability of various oils, as shown in figure 3. When the SOO cement-coated meshes were put onto the hexane/water mixture, diesel/water mixture, peanut oil/water mixture and lubricating oil/water mixture, they all absorbed the oils rapidly, and the 20 mm × 20 mm SOO cement-coated mesh could absorb at least 5 mL oil, indicating the excellent absorption ability of the meshes. The SOO cement-coated meshes still had good hydrophobicity after absorbing oils, thus could floating on the water, indicating the excellent water repellence of the meshes.

IOP Conf. Series: Earth and Environmental Science **171** (2018) 012045 doi:10.1088/1755-1315/171/1/012045

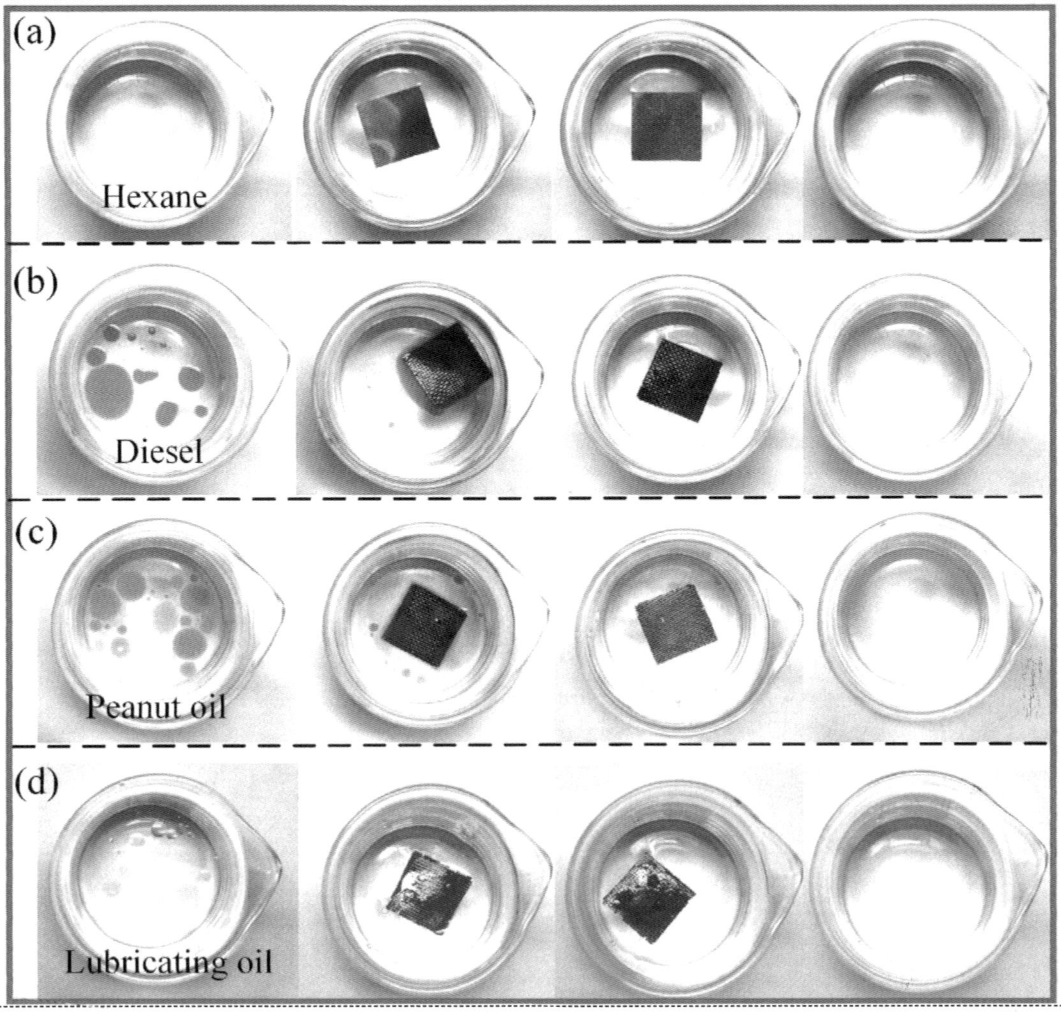

Figure 3. Photos of the absorbing processes of oils using the SOO cement-coated meshes.

The separation processes of oil/water mixtures are shown in figure 4(a) and 4(b). The separation method was an oil-removal process, thus the separation method was suitable for heavy oil/water mixture. However, this separation method could also be used for floating oil/water mixture. The device was tilted about 20 ° to help the oil touch the surface of the SOO cement-coated mesh, thus the SOO cement-coated meshes could separate floating oil/water mixture without water-bridge[36]. Hexane, diesel, peanut oil and lubricating oil were used as floating oils. The mixture of floating oil and water was poured onto the SOO cement-coated mesh. The floating oil touched the surface of the mesh firstly, and passed quickly. Although the mesh was wetted by the oil, it showed good water repellence. The following water could not pass through the mesh, and occupied some area of the mesh. Thus, the oil came lastly passed through the remaining area. The separation efficiencies of the oil/water mixtures were all above 90% and even reached up to 96% for hexane/water mixture as shown in figure 4(c). Moreover, it should be noted that the separation efficiencies we gotten are slightly smaller than their true values because some oil adhered to the acrylic tubes and some oil was absorbed by the SOO cement-coated mesh. The FTIR spectra of the separated oils are shown in figure 4(d). All the absorption bands were corresponded to the characteristic vibrations of alkanes and ester groups. The band corresponded to hydroxy was not observed, indicating the high purity of the separated oil. Moreover, the purity of the collected water was measured by FTIR as well, and the oil contents in the collected water were shown in figure 4(e). The oils contented in the separated water were 12 ppm, 37 ppm, 54 ppm and 64 ppm.

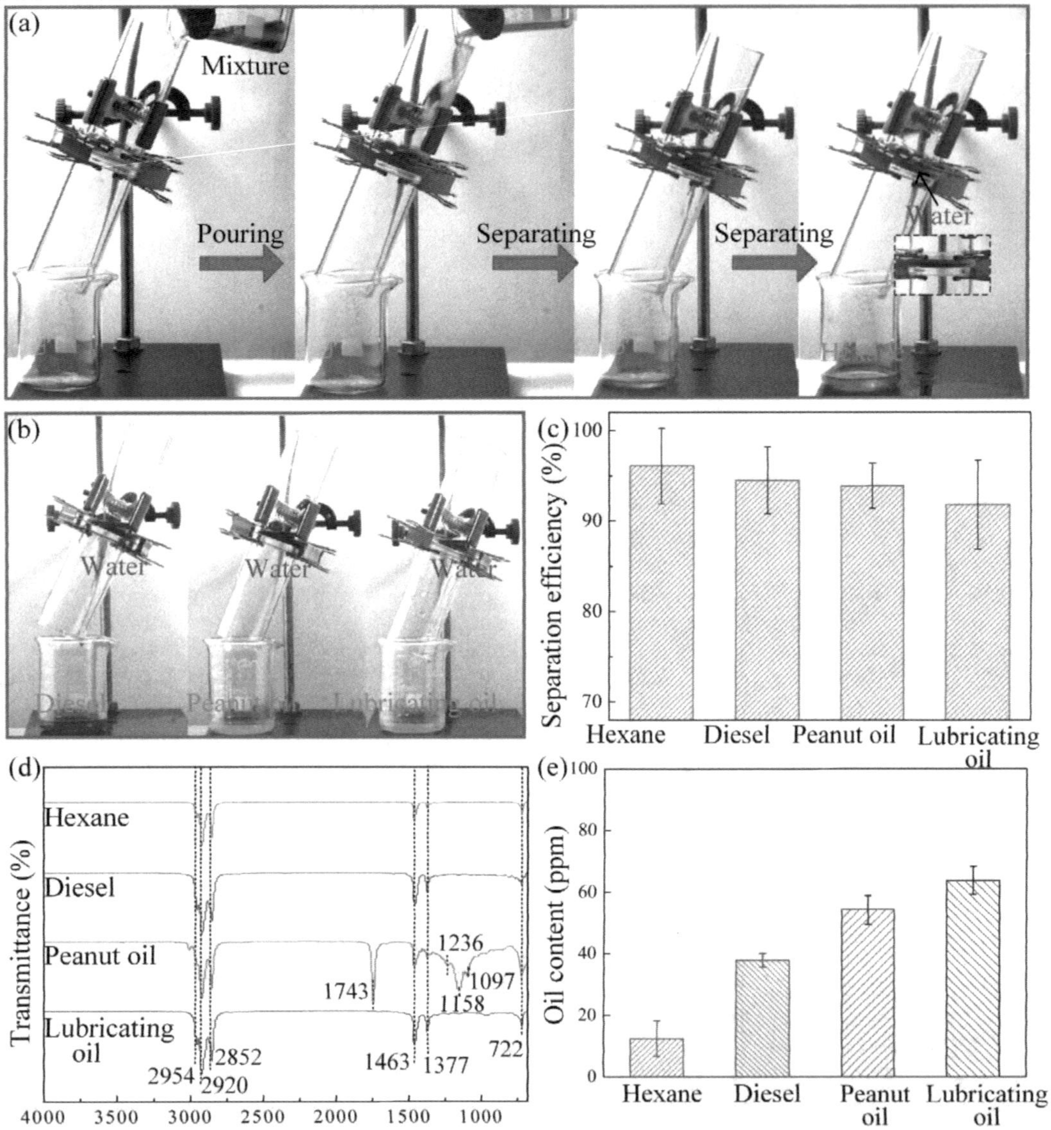

Figure 4. Separation of oil/water mixtures using the SOO cement-coated meshes: (a)-(b) the separation process of hexane/water mixture, diesel/water mixture, peanut oil/water mixture and lubricating oil/water mixture, (c) the separation efficiencies of the SOO cement-coated meshes for different oils, (d) FTIR spectra of the separated oils, (e) the oil contents in the collected water.

The SOO cement-coated meshes had not only high separation efficiency for oil/water mixture but also superior durability. As shown in figure 5(a), the SOO cement-coated mesh could be used to separate hexane/water mixture at least for 20 cycles with high separation efficiencies. Figure 5(b) shows the SOO cement-coated meshes have great stability, the CAs of the water droplets on the mesh were all above 150° and even after stored for 120 h. The process of bending the SOO cement-coated mesh and water droplets on the bent SOO cement-coated mesh were shown in figure 5(c). The water droplets on the mesh that was bent for 30 times were all almost spherical, indicating the respectable durability of the SOO cement-coated meshes.

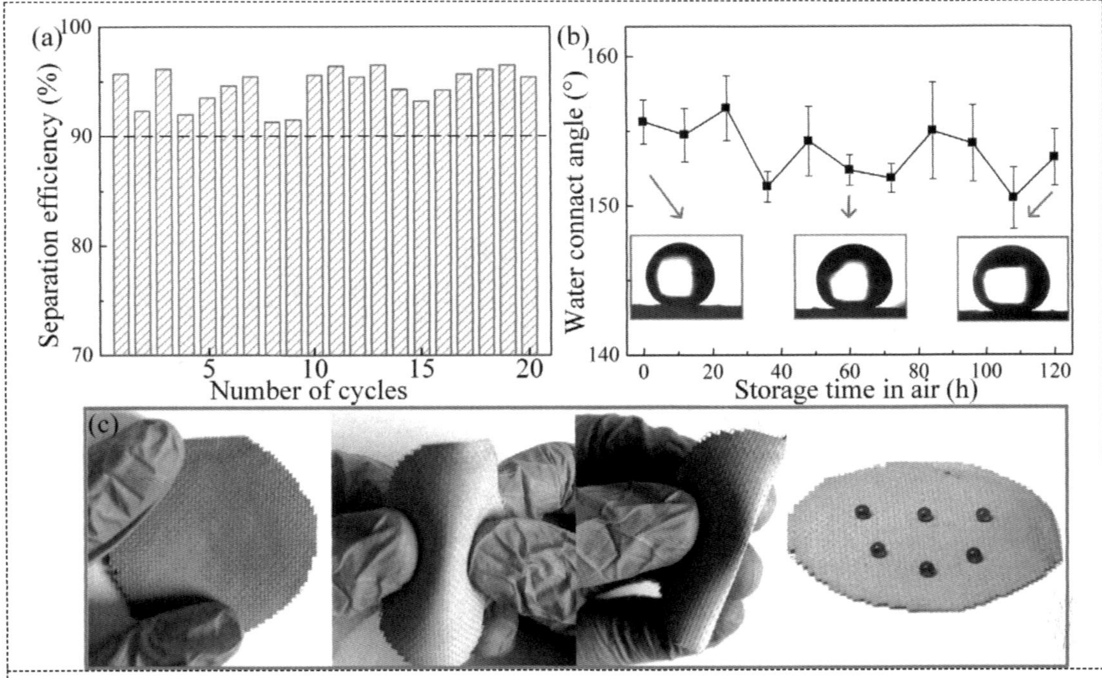

Figure 5. The durability of the SOO cement-coated meshes: (a) the variation of the separation efficiencies of the SOO cement-coated meshes with the hexane/water separation cycles,(b) the variation of contact angles of water droplets on the SOO cement-coated meshes with storage time in air, (c) the process of bending the SOO cement-coated mesh and water droplets on the bent SOO cement-coated mesh.

4. Conclusion

In this work, we simply fabricated a superhydrophobic-superoleophilic (SOO) cement-coated mesh for oil/water separation by dipping copper mesh in the paste of cement and surface modification. The micro/nanostructure and low surface energy endowed the SOO cement-coated meshes with excellent superhydrophobicity and superoleophilicity in air. The meshes had great oil absorption ability and high separation efficiency for various oils, and showed admirable recyclability and durability. The SOO cement-coated meshes could be used to separate oil/water mixture at least for 20 times with high separation efficiency above 90%, and had great superhydrophobicity after being bent for 30 times. The mesh also had unexceptionable stability. The CAs of water droplets on the meshes were all above 150° after stored for 120 h. This easy fabricated, low cost and robust method will have great potential to be used in industrial wastewater treatment and oil spill.

5. Reference

[1] Dubansky B, Whitehead A, Miller J T, Rice C D and Galvez F 2013 *Environ. Sci. Technol.* 47 5074-5082

[2] Lee C H, Tiwari B, Zhang D and Yap Y K 2017 *Environ. Sci-Nano.* 4 514-525

[3] Cunningham C J, Ivshina I B, Lozinsky V I, Kuyukina M S and Philp J C 2004 *Int. Biodeter. Biodegr.* 54 167-174

[4] Kim T H, Jang D J, Yang K U, Na S C and Kim D A 2009 *China Ocean Eng.* 23 657-668

[5] Adebajo M O, Frost R L, Kloprogge J T, Carmody O and Kokot S 2003 *J. Porous Mat.* 10 159-170

[6] Radetic M M, Jocic D M, Jovancic P M, Petrovic Z L and Thomas H F 2003 *Environ. Sci. Technol.* 37 1008-1012

[7] Chen F, Song J, Liu Z, Liu J, Zheng H, Huang S, Sun J, Xu W and Liu X 2016 *Acs. Sustain. Chem. Eng.* 4 6828-6837
[8] Deng X, Mammen L, Butt H and Vollmer D 2012 *Science* 335 67-70
[9] Bhushan B and Jung Y C 2011 *Prog. Mater. Sci.* 56 1-108
[10] Padaki M, Surya Murali R, Abdullah M S, Misdan N, Moslehyani A, Kassim M A, Hilal N and Ismail A F 2015 *Desalination* 357 197-207
[11] Liu Z, Wang X, Gao M, Song J, Huang L, Liu J, Liu X, Xu W and Zhao D 2016 *Rsc. Adv*. 6 88001-88009
[12] Gao X, Zhou J, Du R, Xie Z, Deng ., Liu R, Liu Z and Zhang J 2016 *Adv. Mater.* 28 168-173
[13] Hu H, Zhao Z, Gogotsi Y and Qiu J 2014 *Environ. Sci. Tech. Let.* 1 214-220
[14] Bao X, Cui J, Sun H, Liang W, Zhu Z, An J, Yang B, La P and Li A 2014 *Appl. Surf. Sci.* 303 473-480.
[15] Oribayo O, Feng X, Rempel G L and Pan Q 2017 *Chem. Eng. J.* 323 191-202
[16] Ma Q, Cheng H, Fane A G, Wang R and Zhang H 2016 *Small* 12 2186-2202
[17] Lu Y, Sathasivam S, Song J, Chen F, Xu W, Carmalt C J and Parkin I P 2014 *J .Mater. Chem.* A 2 11628
[18] Song J, Huang S, Lu Y, Bu X, Mates J E, Ghosh A, Ganguly R, Carmalt C J, Parkin I P, Xu W, and Megaridis C M 2014 *Acs. Appl. Mater. Inter.* 6 19858-19865
[19] Wang Q, Zhang B, Qu M, Zhang J and He D 2008 *Appl. Surf. Sci.* 254 2009-2012
[20] Qu M, Zhang B, Song S, Chen L, Zhang J and Cao X 2007 *Adv. Funct. Mater.* 17 593-596
[21] Li L, Breedveld V and Hess D W 2012 *Acs. Appl. Mater. Inter.* 4 4549-4556
[22] Wang S, Feng L and Jiang L 2006 *Adv. Mater.* 18 767-770
[23] Wang C, Yao T, Wu J, Ma C, Fan Z, Wang Z, Cheng Y, Lin Q and Yang B 2009 *Acs. Appl. Mater. Inter.* 1 2613-2617
[24] Lee C H., Johnson N, Drelich J and Yap Y K 2011 *Carbon* 49 669-676.
[25] Yang J, Tang Y, Xu J, Chen B, Tang H and Li C 2015 *Surf. Coat. Tech.* 272 285-290
[26] Chen Q, De Leon A and Advincula R C 2015 *Acs. Appl. Mater. Inter.* 7 18566-18573.
[27] Wang F, Lei S, Li C, Ou J, Xue M and Li W 2014 *Ind. Eng. Chem. Res.* 53 7141-7148
[28] Lakshmi R V, Bharathidasan T and Basu B J 2011 *Appl. Surf. Sci.* 257 10421-10426
[29] Lee M W, An S, Latthe S S, Lee C, Hong S and Yoon S S 2013 *Acs. Appl. Mater. Inter.* 5 10597-10604
[30] Xiao C, Si L, Liu Y, Guan G, Wu D, Wang Z and Hao X 2016 *J. Mater. Chem.* A 4 8080-8090
[31] Zhang L, Zhong Y, Cha D and Wang P 2013 *Sci. Rep-Uk.* 3 1991
[32] Feng L, Zhang Z, Mai Z, Ma Y, Liu B, Jiang L and Zhu D 2004 *Angew. Chem. Int. Edit.* 43 2012-2014
[33] Li J J, Zhou Y N and Luo Z H 2017 *Chem. Eng. J.* 322 693-701
[34] Yu Z W, Yun F F, Gong Z Y, Yao Q, Dou S X, Liu K S, Jiang L and Wang X L 2017 *J. Mater. Chem. A.* 5 10821-10826
[35] Yong J, Chen F, Yang Q, Bian H, Du G, Shan C, Huo J, Fang Y and Hou X, 2016 *Adv. Mater. Interfaces* 3 1500650-1500676
[36] Liu Z, Wang X, Song J L, Roy T, Li S D, Liu J Y, Huang L, Liu X, Jin Z J and Zhao D Y 2017 *Micro Nano Lett.* 12 744-748

Acknowledgments
This work was financially supported by National Natural Science Foundation of China and the Fundamental Research Funds for the Central Universities.

IOP Conf. Series: Earth and Environmental Science **171** (2018) 012046 doi:10.1088/1755-1315/171/1/012046

The Influence of the Storage Strategy on the Complexity of the Container Selection Procedure

A Kuznetsov[1] **A Kirichenko**[2]**and O Izotov**[3]

1 Professor, Admiral Makarov State University of Maritime and Inner Shipping, St.Petersburg, Russia
2 Professor, Admiral Makarov State University of Maritime and Inner Shipping, St.Petersburg, Russia
3 Assistants professor, Admiral Makarov State University of Maritime and Inner Shipping, St. Petersburg, Russia
E-mail: thunder1950@yandex.ru

Abstract. The paper deals with the practically and theoretically important problem of the container selection from the stacks of the container yard. This problem to a great extent determines the operational efficiency of the modern specialized maritime and dry container terminals.The severe ecological pressure exposed on the terminals ruled out the extensive way of the terminal development by the territory expanding. The permanent growth of the container flows leaves the only dimension to expand, to go high. In the same time, the increase in the operational height of container stack causes another big problem, that of container selectivity. With the growth of the number of tiers in the stack, the number of non-productive moves is also bigger, thus deteriorating the operation efficiency of the container yard. The results of the study shows that this problem could be fought by the changing the operational strategy of the box allocation in the container blocks of container yard. The benefits resulting from the changing of the strategy from the traditional horizontal filling of the empty space cell by cell and tier after tier to the vertical allocation of the boxes in sequential columns and rows are displayed on the concrete samples. This problem cannot be solved by the analytical mathematical techniques and should be treated with computer simulation.

1. Introduction

Containerization today is the dominating technology of the industrial goods' transportation, which leaves for the break bulk transportation ever narrowing niche [1,2]. The development of the international trade causes permanent growth of cargo volumes handled in container ports and terminals, while the correspondent growth of vessel sizes aggravates the demands for the operation efficiency in peak surges of shipload storage. The ecological pressure put on container terminals usually excludes any extensive development, since the port territories and water areas approach the boundariesoutlined by the regional master plans [3,4].

Under these circumstances, the focus of this pressure falls on the most area-demanding element of every terminals, i.e. its container yard [5]. With the territory allocated for this site limited, the growth of the operational demands for container yards (playing the role ofa buffer element in the technological chain of container handling system)is inevitably translated into the demand for the increase of the one-time storage capacity. In its turn, this factor becomes a driver for the increase of the stacking density, measured in number of containers stored on area unit. Today, the advanced terminals display the values of this indicator around 1000 TEU per hectare, which assumes the utilization of highly productive and expensive technological equipment[6,7].

At the same time, the standard geometrical dimensions and very strict safety rules restrict the blueprint area of the stack foundation, leaving for the growth only one direction – vertical, demanding for the equipment with a bigger stacking height.

Solving the problem of increasing the stacking density, this approach creates a new problem, even more important from commercial and financial point of view. This is the problem of the container selectivity.

The bigger stacking heightis, the bigger also will bethe probability that a container claimed by the client (called "hot box") would be blocked by other containers atop of it ("cold boxes"). In order to provide the access to the hot box, the cold boxes should be shifted to empty positions of the stack, which increases the labor input of operations and the servicing time. Whatever productive the equipment would be, these non-productive moves could ruin the economic foundation of container handling at the terminal.

A common stack of the container yard usually stores several cargo parties. Newly arrived "cold boxes" usually land over the previously arrived "hot boxes".The latter ones gradually "sink" to the bottom of the stack. When there comes a request for a hot box, it is necessary to "dig" it out of the stack [8,9]. It is not possible usually to predict a moment when a container will be ordered from the stack, so the operational procedure practically is described by pure combinatory schema

In order to fight the damnation of combinatorics, the container operators have developed several empirical strategies, which usually deal not with a single container, but concern general principles of the stack structure's formation.These empirical strategies inevitably form a foundation for the next level of operational strategies aimed to reduce the non-productive expenses for box handling in the stack.

The goal of this study is to verify details of the basic combinatory schema which form the operational tactics and strategy of container handling, to analyze them critically and offer some recommendation for their alternations. The ultimate goal is to develop more effective general strategy of container yard management.

2. Methods and materials

2.1. General description of the problem

The standard marine shipping container has right parallelepiped shape and strict geometric dimensions. The possibility to stack them one atop another provides the formation of the stacks which keeps the conformity of the cargo unit itself, i.e. the container. In other words, in any container handling system practiced by the terminal, the container stack forms a parallelepiped or a block of containers built by several tiers and rows. The cross section of containers in the stack (placed side to side and top to bottom) is a rectangular which is called a bay or section. This is illustrated by Fig. 1.

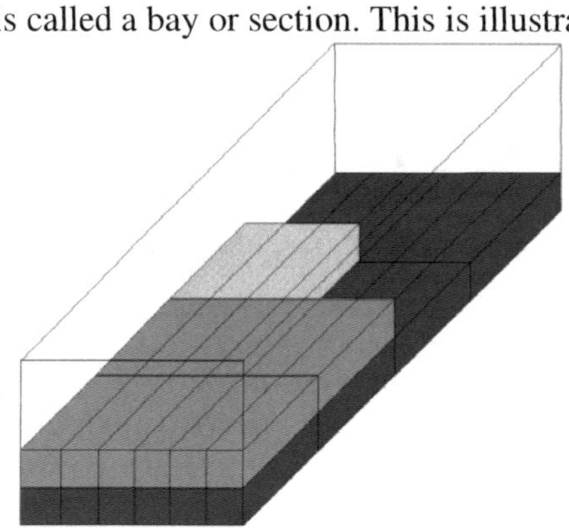

Figure1. General configuration of the container block

IOP Conf. Series: Earth and Environmental Science **171** (2018) 012046 doi:10.1088/1755-1315/171/1/012046

The group of the most advanced and productive equipment providing the maximal stacking density is gantry cranes, rail mounted (RMG) and rubber tired (RMG). The construction of these cranes includes a massive still portal travelling over the container stack (block). In the span of the portal the trolley moves crosswise along the upper beams of the structure, equipped with the special box-catchment device, the spreader. The spreader put on a container can automatically catch it, lift and move to any location in the section of the stack or alongside of it. The transportation between different bays usually takes longer time since it involves travelling of the whole crane structure and precise positioning in a new location. Accordingly, practical reasons encourage avoiding these movements and performing all operations in one location over the same bay.

Accordingly, the problem of container selection as the first approximation (rather sufficient for practical purposes) can be regarded as two-dimensional task of operations in one rectangular cross-section of a block.

2.2. Formal description of the problem

The referenced cargo flow$Q$with a given cargo dwell timeT_{dw}at the terminal allows to evaluate the size of one-time storage capacity of the container yard $\mathrm{E} = \mathrm{Q} \cdot \frac{T_{dw}}{365}$.

This volume, taking into account some external factors, is divided into a certain number of blocks $E = \sum_{k=1}^{K} E_k$. Every block has its width (number of rows$w$), length (number of bays$l$) and maximal height (number of tiers$H$).

The operational features of the vessels handling at the terminal enables us to evaluate an average volume of the cargo party, thus determiningthe average share of this party allocated for a block. Knowing the length of the block, it is possible to determine the average volume of the party Vallocated for one bay of the block.

The external commercial environment of the terminal operation determines not only the cargo dwell time T_{dw}, but also the mean interval between party arrivalsT_{int}. The ration $\frac{T_{dw}}{T_{int}}$ defines the amount of parties $n$with the average volume V to be stored in one section of the block. The total amount of the stored containers $n \cdot V$ divided by the area of the block $w \cdot l$(measured in terminal ground slots) gives the average stacking height $\mathrm{h} = \frac{n \cdot V}{w \cdot l} = \frac{T_{xp} \cdot V}{T_{uнm} \cdot w \cdot l} < H$.

The task is to evaluate the dynamic behavior of this variable against the static average value, the number and the total transportation distance of all moves needed to operate the block under given conditions.

2.3. Necessity and goals of the simulation

The problem of the random selection of elements from a set traditionally belongs to a mathematical branch called combinatorial analysis or combinatorics. The difficulty of this problem is in a very high sensitiveness of the solution to the initial state. Moreover, every move performed for selection of a current element strongly affects the relevant branch space of the problem solution tree.

Additionally, this problems is not a pure mathematical one, since the allocation of arriving parties could follow different rules, as well as the relocation of blocking elements at every step does.

Eventually, the solution of this problem requires toidentify not only dependences of the number and distance of the moves on the numeric values – number of parties, average height, but also from the strategy of the initial allocation and shifting of the boxes at every stepof the handling procedure.

Traditionally, the strategy of the arriving containers allocation assumes keeping the minimal surface elevation ("flat relief"). Under this strategy every arriving container is placed in the lowest free position (slot) in the block. If there are several equal positions, the nearest one to the already formed massive in the tier is selected. The resulting strategy bears the character of epithelial growing of stratum after stratum, as Fig. 1 shows.

The similar strategy is used also for the removal of the blocking "cold boxes": the upper one is shifted to the lowest position in the section; if there are several alternatives positions, the closest is selected. The same procedure is repeated for every next blocking container.

Fig. 2 shows the example of the epithelial (horizontal) allocation procedure for initial allocation and operational shifts of blocking containers.

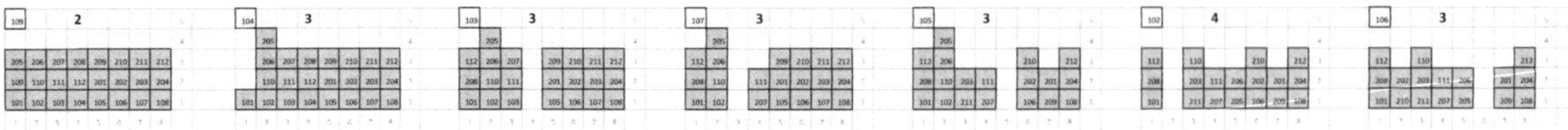

Figure2. Horizontal allocation and operation strategy

The figure displays the current structures of the block, the box selected at the step and number of moves requires to dig it out.
At the same time, it is possible to offer another strategy that seems to be paradoxical from the first glance. This strategy assumes a dense allocation of the arriving containers in a compact block of the maximal height, thus providing the same epithelial allocation procedure, but vertically oriented (Fig. 3).

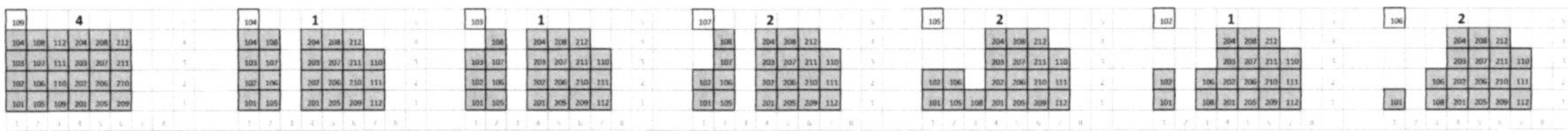

Figure3. Vertical allocation and operation strategy

As one can see from this figure, the amount of moves at every step of the procedure drops significantly in respect to the horizontal strategy.
The preliminary simulation experiments, similar to the ones shown by Fig. 2 and 3,with different block dimensions and numbers of parties, gave the reduction of total moves per selected container in the range of 20-30%, which is quite a significant value.
At the same time it also has been proved that the methodological reasons discussed above exclude any other tools for the study except for the simulation: the analytical toolkit of mathematics really turns out to be powerless for this kind of problems.

3. Results

The results given by the simple models of basic combinatoric dependencies enable to conclude that the primal mechanisms responsible for the operational complexity seem to be studied insufficiently. The reduction of the number of moves for 20-30 %is too significant value to deprive these basic mechanisms of additional fundamental studies.
Simultaneously, in order to conduct the studies of this kind it is necessary to develop the software tools to make massive and different experiments, explore a wide variety of managing strategies. The development of hidden trends and lows also assumes that this software complex include an equally powerful methodological apparatus for statistical data acquisition and interpretation.
The discussed results permit only to define a general direction for this study, to state the hypothesis and evaluate the perceptiveness.

4. Discussion

Even if the full-scaled experiments would show the expected convergence of the results, this will be only the first step in the indicated direction of the study. The transfer from the sections to the blocks, block groups and the whole yard levels could not preserve these promising results.
Many practical restrictions (e.g. aimed to avoid occasional “towers” of containers in the stack, demanding to form “slopes” at the periphery of the blocks, taking into account client ownerships etc.) could also blur the “pure theoretical” effect observed in the modelling experiments.
On the other hand, the degree of these pessimistic assumptions could be assessed only by the further studies and introduction of more adequate models, which in its turn would affect greatly the operational technology of container terminal planning and control.

5. Conclusions

1. The mathematical problem of combinatorial complexity of operations over the simple section of container block is studied.
2. It is proved that the suggested solution depends not only on the geometrical characteristics of the studied segment, but also on the sequence of the previous operations and the strategy of shifting of blocking containers.
3. It is shown that this problem cannot be solved by the analytical mathematical techniques.
4. The goal and tasks for the simulation procedure are discussed.
5. The simulation results have revealed a significant reduction in number of non-productive moves with abandoning traditional strategy of stack planning.
6. The results of the study are critically analyzed to state the goal of the further study.

6. References

[1] Kuzneczov A. Bazovaya model` logisticheskix potokov cherez kontejnerny`j terminal / A. L. Kuzneczov, E. Yu. Kozlova // *E`kspluataciya morskogo transporta*. — 2008. — № 2. — 18–20.

[2] Kuzneczov A. Klassifikaciya i funkcional`noe modelirovanie e`shelonirovanny`x kontejnerny`x terminalov / A. L. Kuzneczov, A. V. Kirichenko, A. A. Davy`denko // *Vestnik Gosudarstvennogo universiteta morskogo i rechnogo flota imeni admirala S. O. Makarova*. — 2015. — № 6 (34). — S. 7–16.

[3] Kuzneczov A. Raschet vmestimosti portovogo sklada s uchetom neravnomernosti raboty` smezhnogo transporta / A. L. Kuzneczov, V. A. Pogodin, Ya. B. Spasskij // *E`kspluataciya morskogo transporta*. — 2010. — № 4. — S. 3–19.

[4] *Port development. A handbook for planners in developing countries*. — Second edition. — New York: UNCTAD, 1985. — 228 p.

[5] Basov E. Sravnitel`naya xarakteristika doxodnosti po operaciyam perevalki i xraneniya kontejnerov na terminale / E. A. Basov // *Transportnoe delo Rossii*. — 2013. — № 6. — S. 197–198.

[6] Waters D. *Logistics. An Introduction to Supply Chain Management*. 2003. PALGRAVE MACMILLAN. Houndmills, Basingstoke, Hampshire RG21 6XS and 175 Fifth Avenue, New York, N.Y. 10010.

[7] Kuzneczov A. Optimizaciya vy`soty` skladirovaniya kontejnerov / A. L. Kuzneczov // *Rechnoj transport (XXI vek)*. — 2008. — №2. — S. 91–93.

[8] Kuzneczov A. Kolichestvo peremeshhenij kontejnerov kak funkciya ot vy`soty` skladirovaniya / A. L. Kuzneczov, V. A. Pogodin // *Kontejnerny`j biznes*. — 2008. — № 7 (17). — S. 26–29.

[9] Kuznetsov A. Do boxes really have a “digging” effect? / *Cargo systems* / A. Kuznetsov. — 2008. — Pp. 42–43.

IOP Conf. Series: Earth and Environmental Science **171** (2018) 012047 doi:10.1088/1755-1315/171/1/012047

Study on the Shear Performance of Different Gradation Types of Asphalt Mixture Based on Triaxle Test

Xiao-yan Wang[1], Kai Gong[2] and Shi-jie Ma[1]

1 Key Laboratory of Highway Maintain Technology Ministry of Communication (Jinan), Shandong Jinan, China
2 Qilu Transportation Development Group Binzhou Branch, Shandong Binzhou 256600, China
E-mail: sonorra@163.com

Abstract. In order to evaluate the shear performance of asphalt mixtures with different gradation types, two kinds of substances, glycerol and asphalt, were used as the media of mixture for triaxle test. The test results show that the shear performance of the mixture is greatly influenced by the gradation of the mixture. It also shows that the more the gradation structure is embedded and squeezed, the greater the internal friction angle, the stronger the shear resistance, and by improving the performance of the cementing material can also improve the shear resistance of the mixture.

1. Introduction
As one of the main deteriorations in highway asphalt pavement, rutting jeopardizes driving easement and traffic safety. Rutting is such a kind of shear flow deformation born of shortage of shear strength resulting in failing to stick up to the shear stress caused by traffic load. However, till now the researches on shear strength of asphalt mixture have not been that consummated to be brought into the design specification. Regarding the severity of rutting distress, the shear properties of asphalt mixtures are studied through triaxle shear test. The shear strength index and criteria are discussed in this dissertation [1]-[2] .

2. Gradation
In order to evaluate and compare the shear performance of asphalt mixture of the different grading types of mixture, this paper chooses the dense gradation(AC-13), broken gradation （SMA-13） [3]and EME-14according to the gradation of asphalt mixture design guidelines for the design of French[4]-[8]. Two different aggregates of basalt and limestone are selected for the type of SMA-13 gradation. Two structures of continuous and discontinuous gradation are selected for EME-14 asphalt mixture, which are recorded as EME-14I and EME-14II. In this paper, five typical gradation structures are selected. The gradation is shown in Table 2.1.

Table 2.1. Mixture synthesis gradation

Gradation	Passing rate （%）										
	19	16	13.2	9.5	4.75	2.36	1.18	0.6	0.3	0.15	0.075
AC-13	100	96	89.4	77.0	46.3	29.8	21.1	16.2	11.7	8.7	5.5
SMA-13	100	100	88.6	67.5	26.5	19.7	17.2	15.8	14.7	14.0	10.8

SMA-13 basalt	100	100	93.9	66.3	26.3	19.5	17.2	15.6	14.4	13.9	10.2
EME-14I	100	100	90.8	74.7	51.4	32.8	24.3	19.8	15.9	13.6	9.1
EME-14II	100	100	84.3	58.5	50.6	32.7	24.2	19.7	15.9	13.6	9.0

3. Triaxle Shear Test [9]

In this paper, the triaxle shear test is used to test the internal friction angle value, and it is more intuitively analyzed and compared the influence of the structure of the gradation on the performance of the mixture. The three axis compression test is carried out by the advanced universal material test system (UTM-100). The test parts are made by vibration molding. Test process and test parts are shown in Figure 3.1. The diameter of the molded specimen is 100mm and the height is 200mm. The loading rate of the test is 1.27mm/min and the confining pressure is 0kPa, 138kPa and 276kPa respectively. The shear test is carried out under these three different confining pressures.

Figure 3.1 Forming and testing before and after test of three axis test

3.1. Determination of the Dosage of Glycerol

In order to compare and analyze the influence of the aggregate structure on the shear resistance of the mixture, the uniform thickness of the asphalt film is 9 μ m. Glycerol instead of asphalt is used in this experiment. The purpose is to reduce the influence of cohesive force C on the shear strength of mixture, so as to highlight the effect of aggregate internal friction angle on shear performance. According to the difference of the structure and composition of gradation, different amounts of glycerol were converted. The conversion formula is as follows.

$$SA = \Sigma(P_i \times FA_i) \quad (1)$$

$$P_{be} = \frac{r_b \times SA \times DA}{10} \quad (2)$$

Where: SA : Specific surface area of aggregate, m2/kg;
Pi: Percentage pass percentage of various particle sizes, %;
FAi : Surface area coefficient of aggregates corresponding to various particle sizes;
DA: Thickness of bitumen film, 9μm;
rb: Relative density of bitumen (25 /25 C), dimensionless;
Pbe: Amount of effective bitumen, %.

From the above formula, the amount of different glycerol in each gradation can be calculated. The result is shown in Table 3.1.

Table 3.1. The amount of glycerol in different gradation structures

Gradation	AC-13	SMA-13	SMA-13 (basalt)	EME-14I	EME-14II
Effective glycerol dosage (%)	4.8	6.8	6.6	6.7	6.7

3.2. Triaxle Shear Test of Glycerol Mixture

Due to the particularity of glycerol, the experiment was carried out at normal temperature. The results of the test are shown in table 3.2 as shown below.

Table 3.2. The data of three axis test results for different gradation of glycerol mixture

Gradation	Confining Pressure (Kpa)	σ_1 (Kpa)	σ_3 (Kpa)	Shear Index		
				C (Kpa)	φ (°)	R^2
AC-13	0	367	0	59.82	33.31	0.990
	138	464.62	138			
	276	724.38	276			
SMA-13	0	417.25	0	56.94	38.52	0.993
	138	641.1	138			
	276	875.5	276			
SMA-13 (basalt)	0	442.75	0	58.84	39.48	0.991
	138	671.55	138			
	276	893.55	276			
EME-14I	0	429.7	0	60.45	37.62	0.990
	138	648.3	138			
	276	796.55	276			
EME-14II	0	469.2	0	61.15	38.29	0.992
	138	629.0	138			
	276	1003.3	276			

From table 3.2, it can be seen that in the three gradation types, the internal friction angle of SMA-13 mixture is the largest, indicating that the grading structure of this type of mixture is interlocked skeleton. For the SMA-13, the internal friction angle of the basalt SMA-13 is the largest, which is 2% more than the limestone mixture. From the results we can say that the aggregate lithology, have certain effects on the frictional resistance of mixture. Because basalt has better angularity and wear resistance than limestone, the internal friction angle of aggregate in its mixture has greater internal friction angle, and its shear performance is better.

For the AC-13 type continuous dense gradation of internal friction angle of the minimum is 33.31%, 18% less than the SMA mixture. This shows that the AC-13 mixture in the inner frictional resistance is relatively small, easy to slip between the aggregate.

In addition, we can see from the table that the cohesive force C is almost the same. It is proved that it is scientific and feasible to study the influence of the friction angle of the aggregate on the strength of the mixture by using glycerol instead of asphalt. The cohesiveness produced by glycerol is very small, about 60kpa. The effect of glycerol is to reduce the effect of cohesive force on the shear resistance of the mixture, so this test of the shear strength of the mixture mainly depends on the internal friction force between the aggregate. At the same time, we see from the experimental data that cohesion C has a certain regularity. Under the same glycerol film thickness, the cohesion of EME mixture is the largest, indicating that cohesive force C has the greatest influence on the shear strength of EME mixtures, and indicating that the shear strength of EME mixtures is mainly from cohesion C. By comparing the C and the phi values of the EME continuous and discontinuous gradations, it is found that the internal friction angle of the EME discontinuous gradation is relatively large. Although the EME mixture gradation structure does not have a large degree of discontinuous SMA mixture, we may conclude that in the case of increasing C value, by optimizing the grading structure of EME mixture and make it has better shear performance.

3.3. Triaxle Shear Test of Asphalt Mixture

In this experiment, three different gradations of AC-13, SMA-13 and EME-14II were selected. Three kinds of gradation were made respectively by 70 ordinary asphalt, SBS modified asphalt and 30 hard bitumen. Triaxle shear test was done under optimum asphalt content. The results of the test are shown in Table3.3.

Table 3.3. The data of three axis test results for different gradation of asphalt mixture

Gradation	NO.	Confining Pressure (Kpa)	σ1 (Kpa)	σ3 (Kpa)	Shear Index		
					C (Kpa)	φ (°)	R^2
AC-13	1	0	630	0	162.7	32.3	0.991
		138	1100.6	138			
		276	1662.3	276			
	2	0	732	0			
		138	1159.8	138			
		276	1686.2	276			
SMA-13	1	0	654	0	171.4	38.7	0.990
		138	1264	138			
		276	1578	276			
	2	0	628.8	0			
		138	1358	138			
		276	1654	276			
EME-14	1	0	634	0	189.7	35.2	0.993
		138	1258.4	138			
		276	1767	276			
	2	0	651.8	0			
		138	1225	138			
		276	1674.8	276			

From Table 3.3, we can see the size of the cohesion C and the internal friction angle are the two main factors affecting the shear properties of the asphalt mixtures. The cohesion C of the EME-14 asphalt mixture is the largest. It is 16.5% and 10.6% larger than the AC-13 and SMA-13 asphalt mixtures respectively. From the analysis data, it can be clearly concluded that the use of low grade 30 asphalt mixture has greater contribution to asphalt mixture strength than that of matrix asphalt 70 or SBS modified asphalt. Low grade asphalt can effectively improve cohesion C value and improve the strength and shear resistance of asphalt mixture. At the same time, it can be seen from the table that the internal friction angle of SMA-13 asphalt mixture is maximum, which is 19.8% and 10% higher respectively than that of AC-13 and EME-14 asphalt mixtures respectively. It shows that the embedded and squeezed structure of the broken gradation of SMA-13 plays a major role. Therefore, we can conclude that the more the grading structure is embedded, the greater the internal friction angle of the aggregate is, the stronger the strength of the asphalt mixture is, and the stronger the shearing resistance is.

4. Hamburg Test

The hamburger test [10] was carried out for the three types of mixture to verify the high temperature resistance performance of the asphalt mixture. The results of the test are shown in Table 4.1.

Table 4.1. The results of the Hamburg test

Mixture Type	Temperature	No.	20000 Rutting Depth /mm
AC-13	60°C	1	5.59
	60°C	2	5.49
SMA-13	60°C	1	4.64
	60°C	2	4.52
EME-14	60°C	1	4.37
	60°C	2	4.41

The experimental results show that the high temperature rutting performance of EME-14 asphalt mixture is better because of the low grade asphalt 30 improving cohesion C and the formation of the skeleton structure. It promotes the strength of asphalt mixture when the cohesive force C and internal friction angle increase simultaneously, and the EME-14 asphalt mixture has a good performance of high temperature resistance to rutting.

5. Conclusion

1. It is scientific and feasible to study the influence of the friction angle of the aggregate on the strength of the mixture as the medium of glycerol instead of the asphalt.
2. There are two ways to improve the strength of asphalt mixture, one is to enhance cohesion C by using low grade asphalt to improve the performance of asphalt binder, the other is to increase the internal friction angle, to enhance the embedment of asphalt mixture grading structure for optimizing the grading structure.
3. Through the study on the gradation structure of the high modulus EME-14 continuous and discontinuous mixture, it is found that the shear performance of the discontinuous graded mixture is better than the continuous gradation. So we can take the optimization of continuous gradation, which is the broken gradation between 6.3mm and 4mm of the sieve to enhance the internal friction angle. At the same time, the relative bituminous binder is used to make the asphalt mixture have high shear strength, and he advantages of water permeable, easy compaction and so on.

6. Reference

[1] Bi Yufeng,Sun Lijun. Research on Test Method of Asphalt Mixture's Shearing Properties [J] Journal of Tongji University：Natural Science Edition,2005,33(8):1036-1040
[2] Peng Yong, Sun Lijun.Shi, Yongjiu. Influence Factor of Shear Resistance of Asphalt Mixture[J]，Journal of Southeast University：Natural Science Edition,2007,37(2):331-333
[3] Trade Standard of the People's Republic of China, Technical Specification for Construction of Highway Asphalt Pavement (JTJ F40-2004), Beijing: China Communications Press, 2004.
[4] Laboratoire Central des Ponts et Chaussées. LPC Bituminous Mixtures Design Guide [M], November 2007.
[5] AFNOR. Couches d'assises: enrobés à module élevé (EME) (In French), NF P98-140, 1999.
[6] Translated and edited by Mafeng, Fuzhen. The application of hard grades of bitumen and high modulus asphalt mixture in France [J]. Chinese-foreign highway, 2008, 28(6): 221-223.
[7] He Xinyuan, Yan Jianhe. Technological application of high modulus asphalt mixture Traffic world, 2009(13), 186-189.
[8] LCPC Bituminous Mixtures Dseign Guide, CEN Bituminous Mixture Specification for hot mixture asphalt.
[9] Trade Standard of the People's Republic of China, Testing Regulations of Asphalt and Asphalt Mixture for Highway Engineering (JTJ 052-2000), Beijing: China Communications Press, 2000.
[10] AASHTO T 324-04, Hamburg Wheel-Track Testing of Compacted Hot Mix Asphalt (HMA).

A Study on Mixing Proportion of Roller-Compacted Concrete Pavement Made of EAF Slag Aggregate and Fly Ash by Using Taguchi Method

My Ngoc-Tra Lam[1,2,*], Saravut Jaritngam[1], and Duc-Hien Le[3]

1 Department of Civil Engineering, Faculty of Engineering, Prince of Songkla University, Hat Yai, Songkhla, Thailand.
2 Faculty of Civil Engineering, Ho Chi Minh City Open University, 97 Vo Van Tan, District 3, Ho Chi Minh city, Vietnam.
3 Sustainable Developments in Civil Engineering Research Group, Faculty of Civil Engineering, Ton Duc Thang University, Ho Chi Minh City, Vietnam.
* Corresponding author, email: ngoctramys@gmail.com

Abstract. The aim of this study is to investigate the effect of factors in mixing proportion on the dry density of Roller-compacted concrete pavement made of EAF slag aggregate and fly ash by using Taguchi method. In this study, EAF slag with size in range of 4.75 – 19 mm was used as a substitute for natural coarse aggregate with three percentages (i.e. 0%, 50% and 100%). Cement was partially replaced by fly ash at three content levels (i.e. 0%, 20% and 40%). Four factors are considered to study including the percentage of EAF slag aggregate, the percentage of binder (cement + fly ash), the fly ash ratio, and the moisture content. And, three levels of four factors were proposed to study the influence of them on the dry density of RCCP. Thus, nine mixtures conforming the orthogonal array L_9 of Taguchi method were prepared to determine the dry density of RCCP by Proctor test. The results were assessed by the analysis of variance (ANOVA). The results of this research indicated that increasing of the percentage of EAF slag as aggregate replacement led to rise the dry density, whereas increasing of the fly ash ratio as cement substitution resulted in the decrease of dry density. The percentage of binder is range of 10-14% and moisture content in range of 6-8% affected slightly on the dry density. When the moisture content exceeded 8% leading to the reduction of the dry density.

1. Introduction

Electric Arc Furnace (EAF) slag is a by-product of steel production, which is generated during the manufacture of crude steel by the Electric Arc Furnace process. World crude steel production reached 1,621 million tons for the year 2015 leading to the world output of EAF slag was estimated in the range of 170 -250 million tons [1], [2]. The traditional stockpile method of EAF slag has created challenging environmental issues such as a lack of waste storage areas and pollution of soil and underground water. Nowadays, many researches on EAF slag as aggregate replacement in conventional concrete, high-performance concrete, and alkali-activated concrete are found in literature. And, scholars have demonstrated that EAF slag can replace natural aggregates in many kinds of concrete [3]–[8]. For example, the replacement of natural aggregates by EAF slag in conventional concrete led to increase compressive strength and splitting tensile strength [3], [4], [6], [7]. Similarly, the increase of mechanical properties in high-performance concrete made of EAF slag aggregate was also observed [5]. Furthermore, alkali-activated concrete containing steel slag as coarse aggregate attained sufficient strength for structure, whereas the values of durability reduced slightly [8]. But very few reports conducted on the using of EAF slag as aggregate in Roller-compacted concrete. Roller-compacted concrete is a stiff and no-slump concrete, which is usually used for pavement (called Roller-compacted concrete pavement) due to its fast construction and cost efficiency. Because aggregate in Roller-compacted concrete pavement (RCCP) mixture consists of 75% volume, the substitution aggregate in RCCP by EAF slag is very important to limit the natural aggregates consumption. In addition, fly ash as cement alternative in RCCP has created several benefits in concrete industry [9]–[11]. Mardani-Aghabaglou et al. studied the mechanical and durability properties of RCCP using 20%, 40%, and 60% of fly ash as a partial cement substitution. It is reported that there was a slight decrease in the strength and durability values as fly ash content increased at early ages. However, at later ages, the development of strength and durability attained satisfactory level for pavement application [9]. The similar results are found in Yerramala and Babu research [10]. It was observed that RCCP mixtures with moderate amount of cement ranging from 150 to 190 kg/m^3 and fly ash percentage ranging from 60% to 70% performed low values of the permeability, absorption, sorption and chloride diffusivity.

Based on the previous research, the incorporation of EAF slag aggregate and fly ash in Roller-compacted concrete pavement (RCCP) is proposed an effective solution to reduce the environmental impacts and develop a sustainable infrastructure. In this study, EAF slag was used as a substitute for natural coarse aggregate with three percentages (i.e. 0%, 50% and 100%). And, cement was partially replaced by fly ash at three content levels (i.e. 0%, 20% and 40%). RCCP is compacted by vibratory rollers to attain a target density and homogeneous surface pavement [12]. RCCP mixture should be dry enough to maintain the stability of vibratory roller and wet enough to distribute the paste. Hence, RCCP mixing proportion has to be calculated carefully. Soil compaction method is the popular approach to determine the optimum moisture content and maximum dry density of RCCP mixture. Typically, mixtures compacted with maximum dry density would provide the highest performance, especially strength. Therefore, an assessment the influence of materials (i.e. aggregate, binder, and water content) on dry density of mixture is important to enhance strength of RCCP.

Taguchi method developed by Genichi Taguchi is an approach provides an efficient way to optimize designs and keep the experimental cost at the minimum level. A set of orthogonal arrays (OA) to design experiments was designed in Taguchi method [13]. The using of OA provides the optimum working conditions of the affected factors. In addition, Taguchi method allows finding out the effective parameters of target value. Taguchi method is suitable for a wide range of applications. In recent years, Taguchi method has been interested by some scholars in civil engineering. For instance, Joshaghani et al. [14] was used Taguchi method to optimize the performance of density, strength, porosity, and permeability on pervious concrete pavement. Furthermore, Ozbay et al. [15] determined the optimum levels of parameters in mix proportion of high strength self- compacting concrete by Taguchi method. Additionally, Tanyildizi et al. [16] conducted the thirty-two experiments conforming the orthogonal array L_{32} of Taguchi method to analyze the effects of the polymerization type, the percentage of silica fume, and heating degree on the concrete strengthened with polymer after exposure to high temperature.

The aim of this study is investigation the effect of factors of mixing proportion (i.e. EAF slag aggregate, cement, fly ash, and water content) on dry density of RCCP by Taguchi method.

2. Method

2.1. Materials

Ordinary Portland Cement (OPC) type I conforming to ASTM C150 was used in this study [17]. Class F fly ash conforming to ASTM C618 [18] replaced cement. Table 1 shows the chemical composition and physical characteristics of OPC and fly ash.

Natural aggregates used in this study with gradation as shown in Table 2 is crushed stone. EAF slag aggregate with size in range of 4.75-19 mm (Figure 1) was used to replace natural coarse aggregate in mixture. The properties of natural aggregates and EAF slag aggregate conform to ASTM C33 [19] for concrete making (Table 3).

Table 1. Chemical composition and physical properties of OPC and fly ash

	OPC, Type I	Fly ash, Class F
Chemical composition (%)		
Silica (SiO_2)	20.7	52.3
Alumina (Al_2O_3)	4.5	24.9
Ferric oxide (Fe_2O_3)	3.3	14.1
Calcium oxide (CaO)	63.0	-
Magnesium oxide (MgO)	1.8	-
Sodium oxide (Na_2O)	0.10	0.67
Potassium oxide (K_2O)	0.74	-
Sulphuric anhydride (SO_3)	2.3	0.47
Loss on ignition (LOI)	2.8	0.15
Physical characteristics		
Fineness (Blaine) (m^2/kg)	347	289
Specific gravity	3.14	2.40
Initial setting time (min)	110	-
Final setting time (min)	170	-
Particle composition		
Retaining on 45 μm sieve (%)	-	7.92
Compressive strength (N/mm^2)		
1 day	14.6	-
3 days	26.2	-
7 days	33.0	-
28 days	43.0	-

Table 2. The suggested limits and the designing gradation of RCCP used in this study

Sieve size	Passing mass (%)		
	The lower limit	The upper limit	The designing gradation
19 mm	95	100	100
12.5 mm	70	95	83
9.5 mm	60	85	73
4.75 mm (No. 4)	40	60	54
2.36 mm (No. 8)	30	50	41
1.18 mm (No. 16)	20	40	30
600 μm (No. 30)	15	30	22
300 μm (No. 50)	10	25	15
150 μm (No. 100)	2	16	10
< 75 μm (No. 200)	0	8	7

Table 3. Physical properties of the crushed stone and EAF slag aggregate

Properties	Crushed stone		EAF slag
	Fine aggregate (4.75 - 0 mm)	Coarse aggregate (19 - 4.75 mm)	Coarse aggregate (19 - 4.75 mm)
Apparent specific gravity	2.71	2.72	3.40
Density (OD) (kg/m^3)	2609	2680	3085
Density (SSD) (kg/m^3)	2644	2691	3176
Bulk density (kg/m^3)	1493	1466	1686
Water absorption (%)	1.36	0.44	2.93
Los Angeles abrasion (%)	NA	13.98	19.37

Note: NA means not available

Figure 1. EAF slag aggregate used in this study

2.2. *Testing Approach*

Four factors of mixture affected on dry density of RCCP in this study are considered:

- The percentage of EAF slag aggregate replaced natural coarse aggregate in the mixture.
- The percentage of binder (cement + fly ash) is a ratio of the weight of binder and total the weight of binder plus the weight of aggregates in the mixture.
- The moisture content of mixture is a ratio of the weight of water and total the weight of binder plus the weight of aggregates in the mixture.
- The fly ash ratio replaced cement in the mixture.

Typically, the weight of binder was in range of 10-14% of total weight of binder and oven-dried aggregates in the mixture. And, the moisture content of the mixture was in range of 6-10% [12]. Therefore, three levels of four factors were proposed to investigate the effect of them on mix proportions as shown in Table 4.

Table 4. Design factors and their levels

No.	Factors	ID.	Level 1	Level 2	Level 3
1	EAF slag aggregate	E	0%	50%	100%
2	Binder	B	10%	12%	14%
3	Moisture content	M	6%	8%	10%
4	Fly ash	F	0%	20%	40%

For four factors and three levels, the proper orthogonal array L_9 of Taguchi method was used to design experiments of this study with the standard L_9 table [13] as listed in Table 5.

Table 5. The standard orthogonal array L_9 (4 factors, 3 levels)

Experiment no.	E	B	M	F
1	Level 1	Level 1	Level 1	Level 1
2	Level 1	Level 2	Level 2	Level 2
3	Level 1	Level 3	Level 3	Level 3
4	Level 2	Level 1	Level 2	Level 3
5	Level 2	Level 2	Level 3	Level 1
6	Level 2	Level 3	Level 1	Level 2
7	Level 3	Level 1	Level 3	Level 2
8	Level 3	Level 2	Level 1	Level 3
9	Level 3	Level 3	Level 2	Level 1

Table 6 shows nine experiments were proposed conforming to the standard L_9 table. From Table 6, the mixing proportion of each mixture presented in Table 7 was calculated to conduct Proctor test in accordance with ASTM D1557 [20].

Table 6. Nine experiments used in this study

Experiment no.	ID. mixture	E	B	M	F
1	C1	0%	10%	6%	0%
2	C2	0%	12%	8%	20%
3	C3	0%	14%	10%	40%
4	C4	50%	10%	8%	40%
5	C5	50%	12%	10%	0%
6	C6	50%	14%	6%	20%
7	C7	100%	10%	10%	20%
8	C8	100%	12%	6%	40%
9	C9	100%	14%	8%	0%

Table 7. RCCP mixing proportions (kg/m^3)

ID. mixture	Coarse aggregate		Fine aggregate	Binder		Water content
	EAF slag	Crushed stone		Cement	Fly ash	
C1	0	842	1163	222.8	0.0	133.7
C2	0	842	1163	219.2	54.8	182.3
C3	0	842	1163	195.8	130.6	233.1
C4	424	424	1172	134.7	89.8	179.6
C5	424	424	1172	275.4	0.0	229.5
C6	424	424	1172	263.1	65.8	140.9
C7	867	0	1197	183.5	45.9	229.3
C8	867	0	1197	169.2	112.8	140.8
C9	867	0	1197	336.0	0.0	192.0

According to ASTM D1557, the mixture is placed five layers into the 6-in (152.4 mm) mold and compacted 56 blows of a rammer per layer (Figure 2). Each mixture is conducted with three trials to determine the average of dry density. Then, the analysis of variance (ANOVA) was applied to assess the contribution of each factor in the mixture.

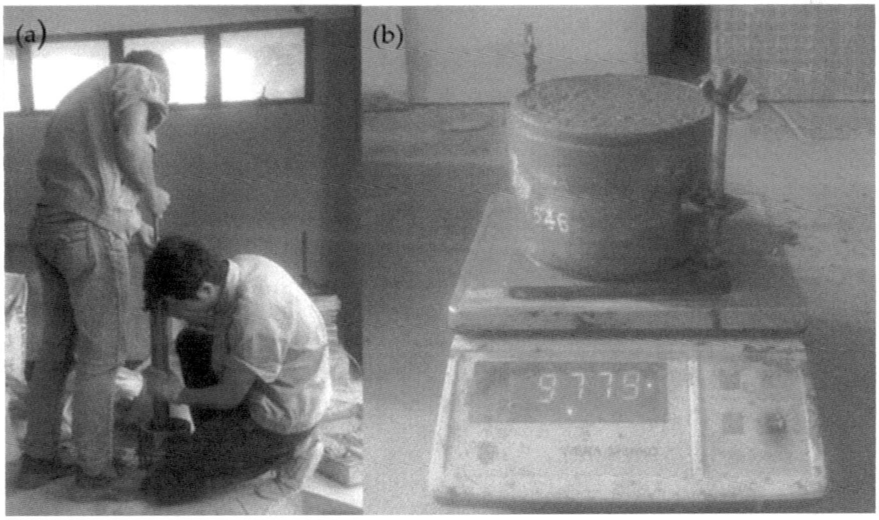

Figure 2. Proctor test, (a) compacting the mixture in mold, (b) determining the density of mixture after compacting.

3. Results and Discussion

Table 8 shows the dry density of nine mix proportions determined by Proctor test. This result was analysed by Minatab software to assess the effect of four factors on the dry density of mixture. Figure 3 presents the result of the analysis of variance (ANOVA).

Table 8. The dry density results of RCCP (kg/m^3)

ID. mixture	Trial 1	Trial 2	Trail 3	Average
C1	2235	2292	2277	2268
C2	2220	2241	2250	2237
C3	2107	2113	2149	2123
C4	2233	2258	2262	2251
C5	2255	2234	2294	2261
C6	2301	2297	2308	2302
C7	2244	2255	2272	2257
C8	2280	2290	2276	2282
C9	2381	2359	2349	2363

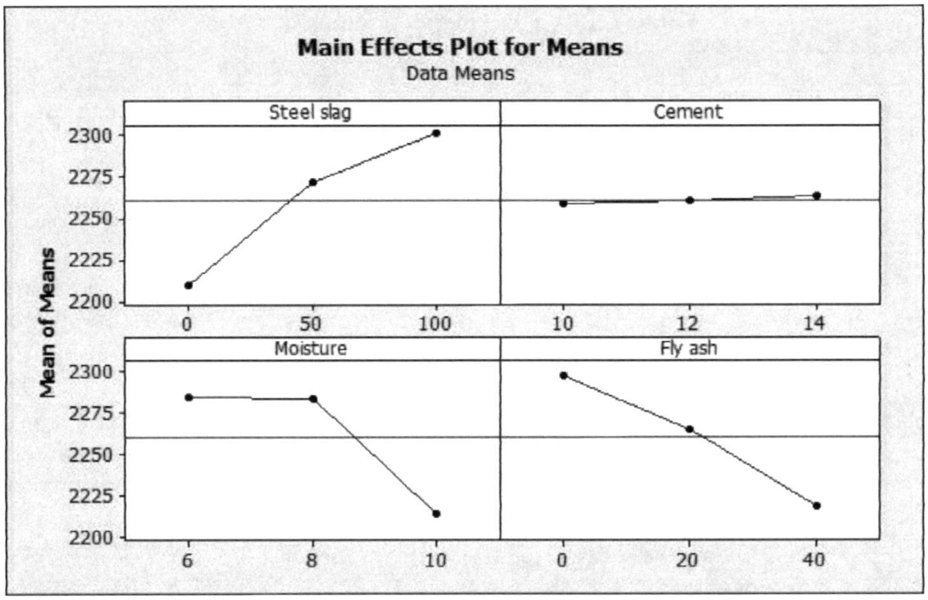

Figure 3. Analysis of Variance of the results

As can be seen in Figure 3, increasing the EAF slag as aggregate replacement in mixture led to rise the dry density. This behaviour resulted from the high-density feature of EAF slag aggregate. Whereas increasing the fly ash as cement substitution in mixture resulted in declining the dry density, because specific gravity of fly ash is lower than that of cement.

Besides, the percentage of binder in mixture hardly affected on the dry density of RCCP. Therefore, when design the mixing proportion, the mass of binder was determined by the strength requirement of RCCP.

On the other hand, the moisture content in range of 6-8% affected slightly on the dry density. And, the moisture content exceeded 8% leading the reduction of the dry density.

4. Concluding Remarks

Based on the experimental work, some of highlighted remarks could be drawn:

(1) The higher percentage of EAF slag coarse aggregate led to the higher dry density of the mixture.

(2) Increasing the fly ash content as cement substitution in the mixture led to decrease the dry density of the mixture.

(3) The moisture content in range of 6-8% affected slightly on the dry density of the mixture. And, there was a dramatic reduction of the dry density when the moisture content exceeded 8%.

(4) The percentage of binder in range of 10-14% hardly affected on the dry density of the RCCP mixture.

5. References

[1] World Steel Association, "World Steel in Figures 2016," *World Steel Assoc.*, pp. 3–30, 2016.
[2] U.S. Geological Survey, "Mineral Commodity Summaries 2016: Iron and Steel Slag," no. 703, pp. 88–89, 2016.
[3] J. M. Manso, J. A. Polanco, M. Losañez, and J. J. González, "Durability of concrete made with EAF slag as aggregate," *Cem. Concr. Compos.*, vol. 28, no. 6, pp. 528–534, 2006.
[4] G. Adegoloye, A. L. Beaucour, S. Ortola, and A. Noumowé, "Concretes made of EAF slag and AOD slag aggregates from stainless steel process: Mechanical properties and durability," *Constr. Build. Mater.*, vol. 76, pp. 313–321, 2015.
[5] F. Faleschini, M. Alejandro Fernandez-Ruiz, M. A. Zanini, K. Brunelli, C. Pellegrino, and E. Hernandez-Montes, "High performance concrete with electric arc furnace slag as aggregate: Mechanical and durability properties," *Constr. Build. Mater.*, vol. 101, pp. 113–121, 2015.
[6] S. I. Abu-Eishah, A. S. El-Dieb, and M. S. Bedir, "Performance of concrete mixtures made with electric arc furnace (EAF) steel slag aggregate produced in the Arabian Gulf region," *Constr. Build. Mater.*, vol. 34, pp. 249–256, 2012.
[7] S. Monosi, M. L. Ruello, and D. Sani, "Electric arc furnace slag as natural aggregate replacement in concrete production," *Cem. Concr. Compos.*, vol. 66, pp. 66–72, 2016.
[8] N. Palankar, A. U. Ravi Shankar, and B. M. Mithun, "Durability studies on eco-friendly concrete mixes incorporating steel slag as coarse aggregates," *J. Clean. Prod.*, vol. 129, pp. 437–448, 2016.
[9] A. Mardani-Aghabaglou and K. Ramyar, "Mechanical properties of high-volume fly ash roller compacted concrete designed by maximum density method," *Constr. Build. Mater.*, vol. 38, no. 1, pp. 356–364, 2013.
[10] A. Yerramala and K. Ganesh Babu, "Transport properties of high volume fly ash roller compacted concrete," *Cem. Concr. Compos.*, vol. 33, no. 10, pp. 1057–1062, 2011.
[11] S. K. Rao, P. Sravana, and T. C. Rao, "Investigating the effect of M-sand on abrasion resistance of Fly Ash Roller Compacted Concrete (FRCC)," *Constr. Build. Mater.*, vol. 118, pp. 352–363, 2016.
[12] D. Harrington, F. Abdo, W. Adaska, and C. Hazaree, "Guide for Roller-Compacted Concrete Pavements," *Inst. Transp. Iowa state univerisity*, no. August, p. 104, 2010.
[13] R. Roy, "A-Primer-on-the-Taguchi-Method.pdf." Society of Manufacturing Engineers, 1990.
[14] A. Joshaghani, A. A. Ramezanianpour, O. Ataei, and A. Golroo, "Optimizing pervious concrete pavement mixture design by using the Taguchi method," *Constr. Build. Mater.*, vol. 101, pp. 317–325, 2015.
[15] E. Ozbay, A. Oztas, A. Baykasoglu, and H. Ozbebek, "Investigating mix proportions of high strength self compacting concrete by using Taguchi method," *Constr. Build. Mater.*, vol. 23, no. 2, pp. 694–702, 2009.
[16] H. Tanyildizi and M. Şahin, "Application of Taguchi method for optimization of concrete strengthened with polymer after high temperature," *Constr. Build. Mater.*, vol. 79, pp. 97–103, 2015.
[17] ASTM C150-07, Standard Specification for Portland Cement.
[18] ASTM C 618-05, Standard Specification for Coal Fly Ash and Raw or Calcined Natural Pozzolan for Use in Concrete.
[19] ASTM C33/C33M-08, Standard Specification for Concrete aggregates.
[20] ASTM D1557-12, Standard Test Methods for Laboratory Compaction Characteristics of Soil Using Modified Effort (56,000 ft-lbf/ft^3 (2,700 kN-m/m^3).

Acknowledgments

This work was supported by the Higher Education Research Promotion and the Thailand's Education Hub for Southern Region of ASEAN Countries Project Office of the Higher Education Commission

(Contract No. TEH-AC 032/2015) and PSU.GS. Financial Support for Thesis scholarship. Besides, we gratefully acknowledge the travel funding to attend conference of Faculty of Engineering, Prince of Songkla University, Hat Yai, Songkhla, Thailand.

Institute of Physics
Dirac House, Temple Back
Bristol BS1 6BE UK

CURRAN ASSOCIATES INC.
proceedings.com

ISSN: 1755-1307
ISBN 978-1-5108-6781-9
9 781510 867819